Achieving Sustainability in Ukraine through Military Brownfields Redevelopment

NATO Science for Peace and Security Series

This Series presents the results of scientific activities supported through the framework of the NATO Science for Peace and Security (SPS) Programme.

The NATO SPS Programme enhances security-related civil science and technology to address emerging security challenges and their impacts on international security. It connects scientists, experts and officials from Alliance and Partner nations to work together to address common challenges. The SPS Programme provides funding and expert advice for security-relevant activities in the form of Multi-Year Projects (MYP), Advanced Research Workshops (ARW), Advanced Training Courses (ATC), and Advanced Study Institutes (ASI). The NATO SPS Series collects the results of practical activities and meetings, including:

Multi-Year Projects (MYP): Grants to collaborate on multi-year R&D and capacity building projects that result in new civil science advancements with practical application in the security and defence fields.

Advanced Research Workshops: Advanced-level workshops that provide a platform for experts and scientists to share their experience and knowledge of security-related topics in order to promote follow-on activities like Multi-Year Projects.

Advanced Training Courses: Designed to enable specialists in NATO countries to share their security-related expertise in one of the SPS Key Priority areas. An ATC is not intended to be lecture-driven, but to be intensive and interactive in nature.

Advanced Study Institutes: High-level tutorial courses that communicate the latest developments in subjects relevant to NATO to an advanced-level audience.

The observations and recommendations made at the meetings, as well as the contents of the volumes in the Series reflect the views of participants and contributors only, and do not necessarily reflect NATO views or policy.

The series is published by IOS Press, Amsterdam, and Springer, Dordrecht, in partnership with the NATO SPS Programme.

Sub-Series

A.	Chemistry and Biology	Springer
B.	Physics and Biophysics	Springer
C.	Environmental Security	Springer
D.	Information and Communication Security	IOS Press
E.	Human and Societal Dynamics	IOS Press

- http://www.nato.int/science
- http://www.springer.com
- http://www.iospress.nl

Series C: Environmental Security

Achieving Sustainability in Ukraine through Military Brownfields Redevelopment

edited by

Cezar Morar

Department of Geography, Tourism and Territorial Planning
University of Oradea
Oradea, Bihor, Romania

Laurel Berman

The Agency for Toxic Substances and Disease Registry
Atlanta, GA, USA

Serap Erdal

University of Illinois-Chicago
Chicago, IL, USA

and

Liudmyla Niemets

Department of Human Geography and Regional Studies
V.N. Karazin National University
Kharkiv, Ukraine

NATO OTAN *This publication is supported by:* The NATO Science for Peace and Security Programme

Springer

Published in Cooperation with NATO Emerging Security Challenges Division

Proceedings of the NATO Advanced Research Workshop on Achieving Sustainability in Ukraine through Military Brownfields Redevelopment
Oradea, Romania
18–21 December 2023

ISBN 978-94-024-2280-1 (PB)
ISBN 978-94-024-2277-1 (HB)
ISBN 978-94-024-2278-8 (e-Book)
https://doi.org/10.1007/978-94-024-2278-8

Published by Springer,
P.O. Box 30, 3311 GX Dordrecht, The Netherlands

www.springer.com

If disposing of this product, please recycle the paper.

Preface

Over the last decades, changes in industrial practices, geopolitical settings, and the burden of post-industrial conditions resulted in the generation of extensive areas of brownfield sites. Brownfields prevail across Europe and North America. They create blighted areas with properties that are derelict, abandoned, underused, and contaminated.

The allocation, availability, and affordability of essential environmental and natural resources shape our present-day world. However, complex environmental security challenges threaten global sustainability and public health. These challenges include human activities that contaminate and deplete clean land, air, and water. Industrialization and population growth exacerbate environmental security. As a result, we have climate change, water depletion, increased energy needs, degradation of biodiversity, loss of agriculture lands, and depletion of natural resources. Increases in population and urbanization challenges will continue to challenge world peace and environmental security, presenting pressures for urban planning reconstruction strategies to create healthy cities, green manufacturing hubs, alternative energy generators, or other sustainable land reuse practices. These pressures represent a major source of instability, conflict, unrest, and general insecurity, severely challenging global security and stability.

The NATO project, *Achieving Sustainability in Ukraine Through Military Brownfields Redevelopment* is an international collaboration promoting a unique effort to develop a framework for sustainable military brownfields redevelopment. This holistic effort incorporates geographic, urban planning, economic, technological, environmental, public health, and social improvement aspects. We focus on military brownfields regeneration as complex models and potential solutions to address threats to global environmental security.

We are embarking on a new adventure based on our combined expertise in brownfields redevelopment, public health, and sustainability. Our focus on reducing environmental security threats combines education, research, and innovation as solutions for the future's global challenges. We strive to increase awareness of the natural environment and the dire need to protect it through human solidarity, peace, and science.

The Editors would like to thank all the participants for their significant contributions and efforts in shaping a more sustainable and better world. We also express our gratitude to the *NATO Science for Peace and Security Programme, Emerging Security Challenges Division* and their research funding to University of Oradea, Department of Geography, Tourism and Territorial Planning (Romania). This funding launched our research effort and the start of an international brownfields redevelopment collaborative.

Together, we are making a difference!

Oradea, Romania — Cezar Morar

Atlanta, GA, USA — Laurel Berman

Chicago, IL, USA — Serap Erdal

Kharkiv, Ukraine — Liudmyla Niemets

Contents

About the Editors

Cezar Morar is an Associate Professor (Ph.D. Habil.) in the Department of Geography, Tourism and Territorial Planning, Faculty of Geography, Tourism and Sport, University of Oradea in Romania. Dr. Morar holds M.S. (2009) and Ph.D. (2012) degrees in Geography from University of Oradea. He is interested in and has broad experience with sustainable urban development and reuse of potentially contaminated and abandoned sites, or "brownfields" as these sites are common to many countries in Europe and North America. Dr. Morar publishes frequently on the topic of tourism, brownfields, planning, and sustainable development. He had the opportunity to work on a number of diverse international research and development projects in Hungary, Ukraine, Serbia, Canada, the United States, Poland, Albania, and France. Dr. Morar serves as a subject matter expert to evaluate project proposals submitted for funding under various programs financed by the European Union. In 2019, Dr. Morar conducted brownfields and urban planning research at the University of Illinois at Chicago, under the prestigious Fulbright U.S. Scholar Program. Dr. Morar was honored to receive the Excellence in Research Awards from University of Oradea (in 2014 and 2024). Dr. Morar is the Co-Director of the project *Achieving Sustainability in Ukraine through Military Brownfields Redevelopment, financed by the North Atlantic Treaty Organization (NATO), through the Science for Peace and Security Program (SPS)*, jointly implemented during 2023–2204 by the University of Oradea (Romania), V. N. Karazin Kharkiv National

University (Ukraine), University of Illinois at Chicago (USA), and Agency for Toxic Substances and Disease Registry (USA).

Laurel Berman is an Environmental Health Scientist with the Agency for Toxic Substances and Disease Registry (ATSDR). Dr. Berman is the agency lead on land reuse and brownfields and has coordinated the ATSDR Land Reuse Health Program for 18 years. This award-winning program integrates public health, environmental justice, and redevelopment, from the early redevelopment planning stages. Dr. Berman has created tools and resources for engaging communities in all stages of land reuse. Achievements include developing the Action Model, a tool that creates and measures indicators associated with public health and redevelopment; the Brownfields and Reuse Opportunity Working Network (BROWN), which provides free consultative brownfields assistance to communities; multiple publications; and the Environmental Health and Land Reuse training courses. Dr. Berman holds Ph.D. and M.S. degrees in Environmental and Occupational Health Sciences from University of Illinois at Chicago (UIC) School of Public Health, with a focus on industrial hygiene and toxicology. Dr. Berman holds adjunct faculty positions at UIC and DePaul University in Chicago. Dr. Berman is deeply committed to improving community health through sustainable and safe redevelopment. The Land Reuse Program and partners have won many awards. At the same time, Dr. Berman remains closely tied to UIC and is passionate about teaching. In 2023, Dr. Berman was honored to receive the Silver Circle of Excellence in Teaching Award from UIC School of Public Health.

Serap Erdal has a research program in exposure and health risk assessment for environmental and occupational hazards. Prior to her academic appointment at UIC and Rutgers University, she worked for the Health Risk Assessment Division of the EA Engineering, Science, and Technology (a for-profit consulting firm) and the Washington State Department of Ecology. Through her positions in academia, government, and industry, she has performed exposure and health risk assessments for many fortune 500 companies and governmental entities (e.g., U.S. Department of Defense, and Department of Energy). Her areas of expertise include: multi-media human exposure and risk assessment for cancer and noncancer effects; health and safety evaluation of new chemicals or products; public health impacts of renewable energy sources; multi-media (air, water, soil, sediment, fish) exposure and health risk assessment for hazardous waste sites; remediation and risk management; Brownfields or Superfund site public health evaluation and redevelopment; sustainable development; indoor and outdoor air pollution; aerosol science and technology; fine and nanoparticle exposure and risk assessment; performance assessment of low-cost sensors for air quality monitoring; community-based participatory research in partnership with environmental justice community organizations; design of community-based exposure assessment studies; environmental justice and equity analysis; industrial hygiene; retrospective occupational exposure assessment; development of exposure assessment methodologies for epidemiological investigations; technical and regulatory interpretation of environmental and occupational health and safety regulations (CAA, CWA, TSCA, CERCLA, OSHA); and regulatory science policy analysis.

Liudmyla Niemets, D.Sc. in Geography, is Professor and Head of the Department of Human Geography and Regional Studies, V.N. Karazin Kharkiv National University, the Excellent Teacher of Ukraine, and Honored Professor of V.N. Karazin Kharkiv National University. Her scientific interests are the theory and methodology of geographical research, sustainable development and global problems of humanity, methodology of social geography, territorial planning and regional management, brownfields reuse, tourist and recreational resources of Ukraine and the world. Prof. Liudmyla Niemets was the head of a number of scientific and research projects on the substantiation of contemporary processes of agglomerations, the formation of united territorial communities, sustainable regional development, formation of the competitiveness of the region, the social security of the region as a factor of its sustainable development, the peculiarities of the development of the social infrastructure of rural areas, etc. and other educational projects together with partners from Ukraine, Romania, Spain, the Czech Republic, Turkey, and other countries. She has about 300 scientific publications, including more than 10 monographs, prepared 14 candidates and 1 doctor of sciences.

Part I
Scientific Support for Redevelopment in the Environmental Security Sector

Chapter 1
Interface Between Sustainability, Public Health, and Environment: Indicators for Measuring and Tracking Impacts of Brownfields Redevelopment

Laurel Berman

Abstract This manuscript provides an overview of the Agency for Toxic Substances and Disease Registry's Action Model and Action Model Toolkit. Through a simple 4-step brainstorming process, the Action Model engages a diverse array of community members, the Development Community, in community revitalization plans. The Action Model's fourth step results in community-driven indicators associated with redevelopment. This manuscript discusses the consolidation of over 200 Action Model indicators from 40 communities, resulting in the Action Model Toolkit. The Action Model Toolkit highlights 69 redevelopment indicators common to 20% or more of the communities. The indicators are in nine categories: built environment, community involvement, economy, education, environmental improvement, environmental resources, housing, physical health, and safety and security. Example indicators from a redeveloped former United States Air Force base are featured. Any community may use the Action Model Toolkit to implement their own Action Models and encourage the selection of common indicators or to derive their own unique indicators through sustainability, public health, and environmental lenses.

Keywords Brownfields · Action Model · Public health · Sustainability indicators · Environmental indicators · Community revitalization models

1.1 Background: Brownfields and Public Health

The Agency for Toxic Substances and Disease Registry (ATSDR) includes brownfields as a type of land reuse site. These sites are potentially contaminated properties that could be revitalized or redeveloped. Land reuse sites include previous industrial sites that are now vacant; brownfields; sites permitted to treat, store, or dispose of

L. Berman (✉)
The Agency for Toxic Substances and Disease Registry, Atlanta, GA, USA
e-mail: laberman@cdc.gov

C. Morar et al. (eds.), *Achieving Sustainability in Ukraine through Military Brownfields Redevelopment*, NATO Science for Peace and Security Series C: Environmental Security, https://doi.org/10.1007/978-94-024-2278-8_1

hazardous waste that may have violated their permit (e.g., Resource Conservation and Recovery Act corrective action sites); landfills; and sites that might be on the U.S. Environmental Protection Agency (EPA) National Priorities List for remediation—commonly referred to as Superfund sites ([1], September 20). The different types of land reuse sites have different regulatory steps for redevelopment. For example, Superfund sites are prioritized and undergo rigorous investigation under the EPA Superfund program, but this step often requires years of litigation and remediation before reuse can be achieved. Some brownfield sites, such as former industrial sites, have a more complex contamination footprint than others but are not designated Superfund sites. These sites often need varied and costly remediation, requiring more resources and time than less contaminated sites. In contrast, some brownfields have less extensive contamination and can be cleaned up and redeveloped in a short amount of time. In one EPA study, approximately 30% of brownfields had contamination present that was below regulatory concern levels [2]. However, it is important to fully review all available records, such as photos and site uses, and to interview former site owners before determining that brownfields do not require cleanup. Specific to military brownfields sites, namely Formerly Used Defense Sites, the United States Government Accountability Office (GAO) [3] found that the United States Army Corps of Engineers may have underestimated the number of military sites that did not require cleanup by as much as 38 percent. This occurred due to incomplete review and investigation of potential hazards percent ([3], August).

Living with contamination from brownfields can adversely impact public health. In the U.S., brownfields tend to disproportionately impact communities with higher rates of poverty and a higher percentage of people of different races or cultures. Brownfields exist everywhere, from rural areas to indigenous lands, to urban areas. As brownfields increase in communities, so do health disparities and inequity, particularly from disinvestment and other conditions related to blight [4]. It is well documented that people living near hazardous sites may have adverse childhood or adult health effects, such as elevated blood lead levels [5], reproductive health effects [6], or higher rates of respiratory disease [7]. Mapping tools, such as ATSDR's Environmental Justice Index (EJI) [8] or EPA's Environmental Justice Screening Tool [9], indicate overall environmental and socioeconomic impacts in communities, such as proximity to hazardous waste sites, combined with factors such as poverty and race.

When people live among high numbers of brownfields, it is important to consider potential impacts of these sites on community health. As public health practitioners, our mission is protecting the public who live near or among brownfields areas to reduce potential exposures to harmful chemical contaminants while focusing on improving health outcomes overall. In 2006, ATSDR was interested in measuring changes in overall community health outcomes associated with brownfields redevelopment. We began to search the literature for brownfields models, research, or community projects that highlighted environment, sustainability, and public health indicators associated with redevelopment. We found few publications on sustainability or public health indicators associated with redevelopment.

In response, we developed the ATSDR Brownfields/Land Reuse Action Model (the Action Model). The Action Model engages the Development Community (e.g., community residents, city governments, public health departments, planners, and developers) in finding ways to make measurable health improvements part of the revitalization process.

1.1.1 Developing the Action Model to Engage the Public and Create Brownfields and Public Health Indicators Associated with Redevelopment

The Action Model is a 4-step framework that asks the following questions:

1. What are the community issues?
2. How can redevelopment address the community issues?
3. What are the associated health benefits?
4. Can we measure any changes in health status over time (indicators)?

We created the following criteria for indicators, featured in Step 4 of the Action Model, which are organized in this manuscript as "smart" and direct, or SMART-D:

- **S**pecific—provides a clear description of what will be measured, such as children's blood lead levels
- **M**easurable—can be quantified and reported numerically, such as the number of contaminated sites cleaned up or the amount of contaminated material removed
- **A**ttributable—demonstrates the connection of results to activities, such as lowering potential exposures to chemical contaminants by remediating soil
- **R**elevant—aligns with the community vision and intended outcomes, such as creation of a community garden to improve access to healthy food
- **T**angible (Practical)—can be feasibly collected, i.e., a data source exists, such as property tax values
- **D**irect—measures the intended result, such as increase in the number of jobs as a result of site reuse

To expand and vet the Action Model indicators (Step 4), we sought expert input from 11 brownfields and sustainability practitioners. We then piloted the Action Model in Milwaukee, Baraboo, and Kenosha, Wisconsin. In 2013, we published details on the three Action Model pilots in *An Indicator Framework to Measure Effects of Brownfields Redevelopment on Public Health* [10].

By 2018, 45 communities across the U.S. had used the Action Model in community revitalization projects. We consolidated their Step 4 indicators and created a data set of community-derived public health indicators associated with land reuse and redevelopment that would be available to the public through an interactive website with an "indicators scorecard" as a supplement. Ultimately, we consolidated Action Models used in 40 communities to produce 69 public health indicators associated with land reuse and redevelopment that are commonly tracked by 20% of the 40 Action Model communities. We grouped the indicators in nine community health categories:

- Built Environment: 17 indicators
- Community Involvement: 4 indicators
- Economy: 16 indicators
- Education: 4 indicators
- Environmental Improvement: 6 indicators
- Environmental Resources: 5 indicators
- Housing: 11 indicators
- Physical Health: 4 indicators
- Safety and Security: 2 indicators

The 69 indicators, organized by health categories, are available in the ATSDR Build Your Own Community Health and Land Reuse Scorecard Toolkit at www.atsdr.cdc.gov/sites/brownfields/model.html. In 2019, we described the development of the indicators and Scorecard Toolkit in the publication, *Public Health Indicators Associated with Land Reuse and Redevelopment: Results of a 40-community Analysis* [11].

1.1.2 The Action Model Indicators and Community Health and Land Reuse Scorecard

The Action Model Indicators, grouped in nine health categories, are shown in Fig. 1.1. The Development Community members can brainstorm indicators that best fit their redevelopment vision and plans. Some or most of the indicators may be selected from the nine categories shown. Once indicators are selected, the community can fill in the Community Health and Land Reuse Scorecard Toolkit, which is shown as a truncated version in Fig. 1.2.

As examples of Action Model Indicators, the Community Involvement Indicators are provided in Fig. 1.3. These include the number of city-community events and meetings; participation rates in events and meetings; number of public/outreach documents; type of public/outreach documents. The Environmental Improvement Indicators are provided in Fig. 1.4. These include an inventory to characterize land reuse sites; pre- and post-redevelopment list of site contaminants; pre- and post-redevelopment media impacts; pre- and post-redevelopment levels of contamination; number of lead abatements/remediations; and the number of asbestos abatements/remediations.

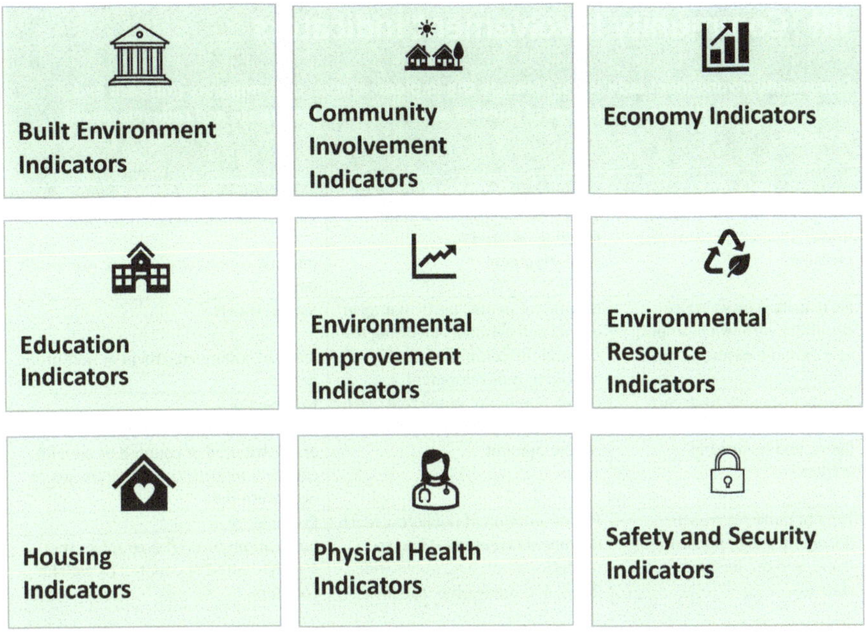

Fig. 1.1 Screenshot of the Action Model Indicator Toolkit's nine health themes [12]

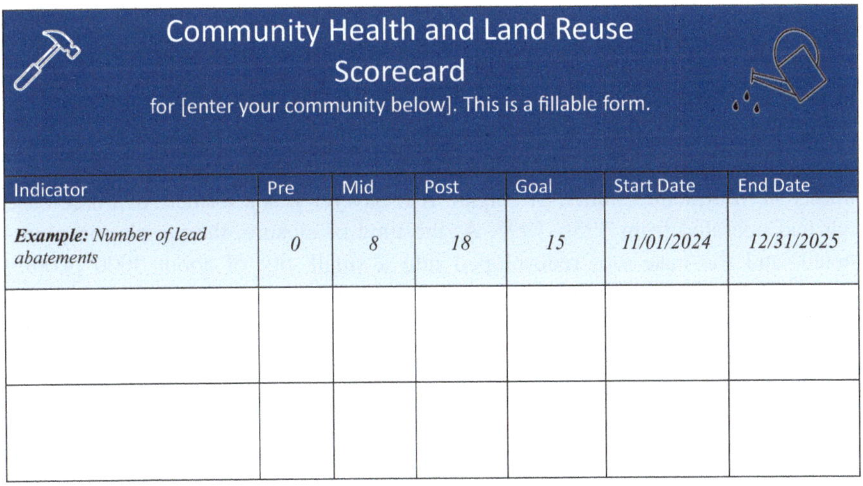
Community Health and Land Reuse Scorecard

for [enter your community below]. This is a fillable form.

Indicator	Pre	Mid	Post	Goal	Start Date	End Date
Example: *Number of lead abatements*	*0*	*8*	*18*	*15*	*11/01/2024*	*12/31/2025*

Fig. 1.2 Excerpt of the Action Model Indicators scorecard [12]

Community Involvement Indicators

Community involvement indicators: Measure the implementation of community outreach and involvement activities to populations of interest associated with land reuse and redevelopment activities.

Indicator	Definition	Data Source
Number of city-community [redevelopment] events and meetings	Count/number of events and meetings related to redevelopment.	Event/meeting flyers or agendas Internal county/city/town or non-profit data
Participation rates in city-community [redevelopment] events and meetings	Number of people participating in events and meetings involving the city and the community and related to redevelopment.	Sign-in sheets Internal county/city/town or non-profit data
Number of public/outreach documents, e.g., publications, flyers, and readership statistics	Count/number of public/outreach documents created related to redevelopment.	Numbers of publications/flyers/Infographics that are distributed or counted by web hit counters-Internal county/city/town or non-profit data
Type of public/outreach documents, e.g., publications, flyers, and readership statistics	The description of public/outreach documents created related to redevelopment, e.g., publications, flyers, and readership statistics.	Examples of publications/flyers/Infographics that are distributed or counted by web hit counters Internal county/city/town or non-profit data

Fig. 1.3 Community involvement indicators [12]

1.1.3 Action Model Example: K.I. Sawyer, a Former U.S. Air Force Base

ATSDR facilitated an Action Model process in the K.I. (Kevin Ingalls) Sawyer community in Marquette County, Michigan. K.I. Sawyer was a former Air Force base that had operated from 1956–1995. At the time of closure, the population plummeted, and the base was redeveloped into a small city of about 3000 people. However, the city still looked like a military base. K.I. Sawyer's local community group, Sawyer Community Alliance (SCA), had a vision, "From Base to Place: Building a Healthy Sawyer." Their primary goals were to improve health, economy, and social disparities by establishing a stronger sense of community; remediating blighted buildings; providing community services; and improving public services (water, transportation, etc.) [13].

ATSDR partnered with SCA from 2013–2019 to envision community revitalization, hosting their first Action Model workshop in April of 2013. The SCA focused on addressing community concerns related to former base activities that still impacted the redevelopment. These included topics of beautification; community services; governance; recreation; and food access. An excerpt of SCA's Action Model showing the topics of Recreation and Food Access is shown in Table 1.1. As shown in the table, expanding recreation was important to the community. Indicators

Environmental Improvement Indicators

Environmental improvement indicators: Measure the implementation of activities and achievement of outcomes related to hazardous chemical reduction.

Indicator	Definition	Data Source
Inventory to characterize land reuse sites (Brownfields, cleanup sites, and under-used or polluted sites) through a public health lens (e.g., size, sensitive populations, frequency of access, etc.)	The description of land reuse sites in a community from a public health/exposure perspective	ATSDR Site Screening Tool Internally-generated inventory, such as a spreadsheet maintained by a city
Pre- and post-redevelopment list of contaminants at sites	The list of contaminants suspected or confirmed to be at sites before and after redevelopment	Internal county/city/town or non-profit data
Pre- and post-redevelopment media impacts (e.g. air, soil, water, etc.)	The environmental media, such as soil, air, water, or sediment that is suspected or confirmed to be impacted by chemical contamination before and after redevelopment	Internal county/city/town or non-profit data
Pre- and post-redevelopment levels of contamination	The quantitative levels of contamination measured in environmental media before and after redevelopment (e.g. soil lead levels in parts per million)	Internal county/city/town or non-profit data
Number of lead abatements/remediations	Number of lead abatements or remediations that occur during redevelopment, such as the removal of lead-based paint from a structure slated for reuse	Internal county/city/town or non-profit data
Number of asbestos abatements/remediations	Number of asbestos abatements or remediations that occur during redevelopment, such as the removal of asbestos containing material from a structure slated for reuse	Internal county/city/town or non-profit data

Fig. 1.4 Environmental improvement indicators [12]

to measure success included the number and types of trail projects; a sidewalk survey and the number of efforts to maintain and repair sidewalks; and the number and types of community events. Increasing access to quality food and grocery stores was also important to the community. Indicators related to Food Access included the number and types of food services and the size and location of the community garden.

1.2 Conclusion

The Action Model has proven to be a useful resource for communities with concerns about brownfields and hazardous exposures. The Action Model can be used to start a conversation about community revitalization and emphasizes the importance of putting the community needs and vision first. The expertise of the community in

Table 1.1 An excerpt of Sawyer Community Alliance's action model

Community issues/ redevelopment needs	Revitalization approach	Community health benefit	Measurement indicator to track change
Recreation			
Walking trails and trailhead development.	Locate, identify, and renovate trails, including fitness trails and walking trails. Kids/volunteers can help develop trails. Look into using snowshoe and cross-country trails and coordinate with snowmobile association(s).	Access to trails may increase physical activity and draw people to Sawyer for recreation.	Number and types of trail projects
Walkability	Address walkability and sidewalks by assessing the condition of sidewalks, cleaning up and trimming sidewalks, remove snow in winter, and investigate the repair of poor condition sidewalks.	Walking may improve overall health.	Sidewalks survey and the number of efforts to maintain and repair sidewalks
Community events	Have community events, such as a soap box (airplane) derby, food fairs, a book exchange, movie night, and local concerts.	Events may promote Sawyer as a place to be, creating a sense of place for existing and potentially new residents.	Number and types of community events
Food access			
Quality food	Encourage mobile vendors, such as a taco truck, ice cream, mobile grocery buses.	Access to healthy food may increase overall health.	Number and types of food services
Grocery store/ convenience services	Create a grocery store/co-op/deli that is community run/not-for-profit and may draw in the business community.	Access to healthy food may increase overall health.	Number and types of food services
Food coop, farming	Expand the community garden to a 15-acre site.	Access to healthy food may increase overall health. More people may join the community garden.	Size and location of community garden

terms of knowing the history and locations of brownfields and yet-to-be-discovered brownfields, is also invaluable. Using the Action Model as part of redevelopment planning can lead to a health-focused, sustainable redevelopment project, supported by an informed community, which may increase stakeholder participation and sharing of resources. In addition, the Action Model's fourth step encourages the Development Community to develop and track indicators to measure changes in environmental and overall health of the community, which can maintain community motivation to stay involved in redevelopment planning.

Any community may use the Action Model Toolkit to implement their own Action Models and encourage the selection of common indicators or to derive their own unique indicators through sustainability, public health, and environmental lenses.

References

1. Agency for Toxic Substances and Disease Registry. ATSDR Land Reuse Health Program (2023, September 20). https://www.atsdr.cdc.gov/sites/brownfields/. Last accessed 2024/02/01
2. Jenkins R, Kopits E, Simpson D (2006) Measuring the social benefits of EPA land cleanup and reuse programs. U.S. Environmental Protection Agency, National Center for Environmental Economics. https://www.epa.gov/sites/default/files/2014-12/doc. Last accessed 2024/02/01
3. United States Government Accountability Office, Environmental Contamination: Corps Needs to Reassess Its Determinations That Many Former Defense Sites Do Not Need Cleanup, GAO (2002, August). https://www.denix.osd.mil/mmrp/denix-files/sites/46/2016/03/02_Environmental-Contamination-Corps-Needs-to-Reassess.pdf. Last accessed 2024/02/02
4. de Leon E, Schilling J (2017) Urban blight and public health: addressing the impact of substandard housing, abandoned buildings, and vacant lots. The Urban Institute, Washington, D.C.
5. Missouri Department of Health and Senior Services. Health Consultation: Herculaneum Lead Smelter Site 9 2001). Agency for Toxic Substances and Disease Registry, Atlanta, Georgia, United States
6. Austin AF (2011) Reproductive outcomes among former Love Canal residents, Niagara Falls. New York Environ Res 111:693–701
7. Liu XL (2012) Association between residential proximity to fuel-fired power plants and hospitalization rate for respiratory diseases. Environ Health Perspect 120(6):807–810
8. Agency for Toxic Substances and Disease Registry. Environmental Justice Index (2023), https://www.atsdr.cdc.gov/placeandhealth/eji/index.html. Last accessed 2024/01/25
9. United States Environmental Protection Agency (2024) EJScreen: environmental justice screening and mapping tool. https://www.epa.gov/ejscreen. Last accessed 2024/01/25
10. Berman L, Forrester T (2013) An indicator framework to measure effects of brownfields redevelopment on public health. J Environ Health 76(10):50–55
11. Berman L, Hall E, Omoh E (2019) Integrating public health in land reuse and redevelopment: part 3: public health indicators associated with land reuse and redevelopment: results of a 40-community analysis. J Environ Health 82(5):34–37
12. Agency for Toxic Substances and Disease Registry. Build Your Own Community Health and land Reuse Scorecard Toolkit, ATSDR Land Reuse Health Program (2019, December 2). https://www.atsdr.cdc.gov/sites/brownfields/reusescorecard.html. Last accessed 2024/02/01
13. Agency for Toxic Substances and Disease Registry. KI Sawyer Community Partnership, Upper Peninsula, Michigan, ATSDR Land Reuse Health Program (2021). https://www.atsdr.cdc.gov/sites/brownfields/factsheets/sawyer_community_partnership_factsheets.html. Last accessed 2024/02/01

Chapter 2
Public Health Considerations of Typical Contaminants at Military Brownfields Sites: Overview

Laurel Berman

Abstract This manuscript provides an overview of military brownfields and contamination associated with these sites. Military brownfields are a subset of brownfields and share similar contamination concerns as brownfields in general, such as lead, metals, petroleum, per- and poly-fluoroalkyl substances, and other hazardous chemicals. Military brownfields may also contain other chemical contaminants, including volatile organic compounds and jet fuel, which can impact drinking water, as well as unexploded ordnance, and radioactive waste. Exposure to contaminants at military brownfields, whether active or inactive sites, may lead to adverse health effects among military personnel, their families, and civilians.

Keywords Brownfields · Military brownfields · Contamination · Perfluoroalkyl substances · Polyfluoroalkyl substances · PFAS · PFOS · Volatile organic compounds · Health effects

2.1 Overview of Brownfields in the United States

In the United States, the U.S. Environmental Protection Agency (EPA) is the main federal agency that oversees brownfields. The EPA summarized the contaminants found most frequently at brownfields, which include arsenic, asbestos, lead, petroleum and hydrocarbons, polycyclic aromatic hydrocarbons (PAHs), polychlorinated biphenyls (PCBs), and volatile organic compounds (VOCs). Additional brownfields contaminants that are reported less frequently include cadmium, chromium, dioxin, mercury, and pesticides ([16], July 6).

The EPA has the authority to regulate hazardous waste sites under the 1980 Comprehensive Environmental Response, Compensation, and Liability Act (CERCLA). CERCLA enabled broad Federal authority to respond directly to

L. Berman (✉)
The Agency for Toxic Substances and Disease Registry, Atlanta, GA, USA
e-mail: laberman@cdc.gov

C. Morar et al. (eds.), *Achieving Sustainability in Ukraine through Military Brownfields Redevelopment*, NATO Science for Peace and Security Series C: Environmental Security, https://doi.org/10.1007/978-94-024-2278-8_2

releases or threatened releases of hazardous substances that may endanger public health or the environment. The act also created a tax on chemical and petroleum industries, which established a trust, called the Superfund, for cleaning up abandoned or uncontrolled hazardous waste sites. CERCLA was amended in 1986 ([18], October 30). In 2001, CERCLA was again amended, which provided authority for EPA's Brownfields Program ([19], November 6).

The Agency for Toxic Substances and Disease Registry (ATSDR) includes brownfields in its definition of land reuse sites, which are potentially contaminant properties that may be redeveloped. Brownfields account for the largest proportion of land reuse sites. The EPA estimates that there are 450,000 brownfields in the United States ([20], October 4). Figure 2.1 shows the extent and location of nearly 30,000 known brownfields in 2020 [7]. ATSDR compiled the data for this figure from publicly available information in the EPA's Assessment, Cleanup, and Redevelopment Exchange System ([21], November 10). As a federal public health agency, ATSDR's brownfields activities are typically in response to community concerns about exposures to contamination at or originating from brownfields. Figure 2.1 also highlights ATSDR's public health activities in brownfields, which include performing public health assessments, implementing community health projects, providing technical assistance, and conducting special initiatives such as soil sampling.

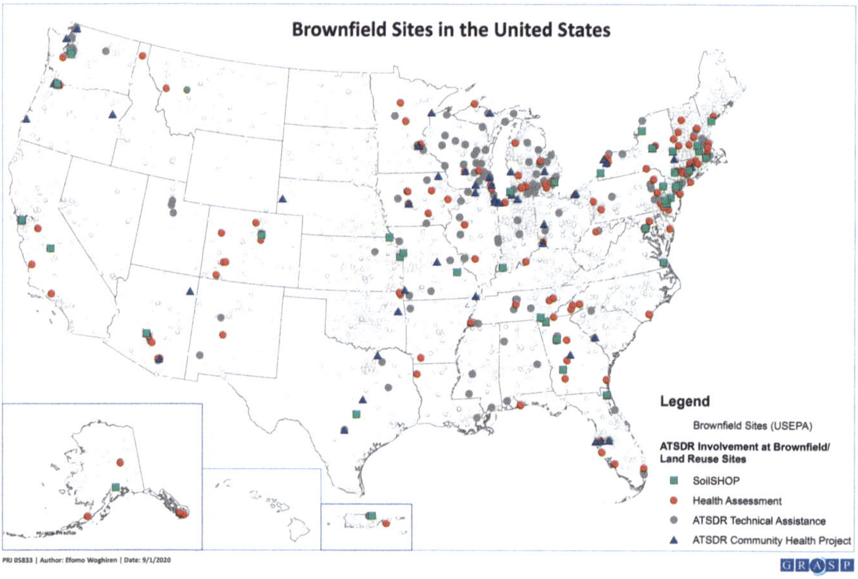

Fig. 2.1 Map of Brownfield Sites in the United States, which includes Agency for Toxic Substances and Disease Registry (ATSDR) Projects on Brownfield and Land Reuse Sites. *Note*. GRASP = Geospatial Research, Analysis, and Services Program; SoilSHOP = Soil Screening, Health, Outreach, and Partnership; U.S. EPA = U.S. Environmental Protection Agency. ([7], citing Woghiren, E. (9/1/2020, PRJ 05833))

The majority of sites shown in Fig. 2.1 have undergone site assessment activities and, as of 2020, had not been cleaned up. Using the estimate of 450,000 brownfields in the U.S. and the 30,000 brownfields represented in Fig. 2.1, less than seven percent of the estimated brownfields have undergone assessment, and only a fraction of these have been cleaned up with federal EPA funding. The low percentage of sites that received clean-up funding illuminates the challenges associated with brownfield sites from economic, regulatory, environmental, and public health perspectives. An additional challenge is the increase in non-seasonal (e.g., migrant workers) vacant properties from 2000 to 2010, increasing from 7,000,000 to nearly 10,000,000 [24]. Vacant properties built prior to 1970 are likely to contain lead-based paint or asbestos-containing material. Other vacant properties could have underground or above-ground fuel or heating oil tanks which could have leaked into soil or groundwater. These properties could be considered brownfields and could significantly increase the true number of brownfields.

Brownfields have contamination profiles that vary based on past or current use and type of chemicals used or generated on a site. For example, former metal smelters and foundries may leave behind soils contaminated with metals, whereas former manufactured gas plant sites and petroleum refineries may exhibit groundwater, soil, and air contamination from petroleum use and manufacture. Petroleum contaminants, like other VOCs, may spread through soil pores and groundwater, resulting in vapor intrusion and private well issues years later. Historic sites may have contamination that dates back several decades and occurred prior to the introduction of environmental regulations. Some brownfield sites may be currently in use or only recently closed, and some brownfields may be active or inactive sites that are incompatibly located in residential areas, such as a small food production plant or dry-cleaning plant in a residential neighborhood. Regardless of their origin, brownfields range from sites that are suspected to have minimal to no contamination to sites with more complicated contamination profiles based on different site uses over many years.

2.1.1 Military Brownfields in the United States

Military brownfields are a subset of brownfields. In the U.S., military brownfields may be identified during the Base Realignment and Closure Process (BRAC), which was implemented in 1988 and allocated federal funds for realignment and closure activities, including costs of response actions for environmental contaminants. The U.S. Department of Defense and Congress implemented five BRAC rounds in 1988, 1991, 1993, 1995, and 2005. The EPA assists in the transfer of both BRAC and National Priorities List (NPL) sites (i.e., Superfund sites) to entities such as states, counties, tribes, or territories or for business or industrial purposes. The EPA promotes cleanup and reuse of hundreds of acres of properties for economic development that can capitalize on existing infrastructure, including buildings and roads [26]. The most recent BRAC budget request in fiscal year 2023 included a new request for \$284,687,000 for environmental restoration and caretaker costs. The budget request also included funding for the emerging contamination issues associated with per- and polyfluoroalkyl substances (PFAS) [11]. PFAS are commonly found at military brownfields as PFAS was frequently used in fire-fighting foams in training or response activities.

The EPA maintains a website with listings of 136 BRAC sites across the 10 EPA regions in the United States ([26], January 31). The listing is interactive and includes links to known site information, such as whether a site is on the NPL for remediation and known contaminants of concern. For example, the U.S. Navy Avionics Center in Indianapolis, Indiana lists five contaminants of concern in groundwater: arsenic, chloroethene (vinyl chloride), manganese, trichloroethene, and cis-1,2-dichloroethene ([23], February 2).

In addition to BRAC sites, there are thousands of Formerly Used Defense Sites (FUDS) in the United States. Military bases in the U.S. often function like small cities with housing, schools, filling stations, small industries, and other activities occurring on base and related to base or military functions, such as lighthouses or arsenal storage sites. FUDS properties are more numerous than BRAC sites. Potentially contaminated BRAC and FUDS properties comprise military brownfields in the United States. These properties may be contaminated with the typical chemicals found at brownfields as well as contaminants specific to military sites, particularly munitions, including unexploded ordnance (UXO). Contaminants at military brownfields that may be eligible for cleanup under the FUDS program are referred to as "projects." Projects include unsafe buildings, structures, or debris; hazardous, toxic, and radioactive waste, which may include arsenic, certain paints, some solvents, petroleum and related products; or toxic pollutants from landfills. Projects also include containerized hazardous, toxic, and radioactive wastes, including wastes in underground and aboveground storage tanks [27].

The U.S. Army Corps of Engineers (USACE) administers the FUDS program on behalf of the U.S. Army and the Department of Defense. The FUDS program cleans up former defense sites, under CERCLA, that were transferred outside of Department of Defense control prior to October 1986 (the year that CERCLA was amended). The goal of the program is to reduce risks to human health and the environment by implementing effective, legally compliant, and cost-effective environmental response actions [12].

The USACE reports that there are about 9300 FUDS in the U.S. and its territories [15]. As of September 30, 2021, there were 5400 FUDS sites in the U.S. that required investigation and cleanup. Of these sites, 3800 had undergone clean-up and 1600 still required a CERCLA response [14]. The USACE has created an interactive map of FUDS across the U.S. and its territories. A screenshot of the map is shown in Fig. 2.2. The legend indicates properties with projects, properties with all site projects closed out, and properties without projects.

According to the U.S. Army Corps of Engineers, FUDS properties range from less than one acre to thousands of acres and are located in all areas of the U.S., from residential to industrial to tribal areas and territories. The USACE estimates that it will cost approximately \$11 billion over many decades to clean up FUDS properties [15].

Fig. 2.2 Map of Formerly Used Defense Sites (FUDS) Viewer—FY21 Data [14]

2.1.2 Contaminants and Health Risks of Exposure at Military Brownfields

As discussed above, contaminants associated with military brownfields include UXO, hazardous, toxic, and radioactive waste, including arsenic, certain paints, some solvents, petroleum and related products; toxic pollutants from landfills; and containerized hazardous, toxic, and radioactive wastes, including wastes in underground and aboveground storage tanks. In recent years, concern about PFAS chemicals associated with military brownfields and military sites has been rising. Many military bases used fire-fighting foam, which contained PFAS. These chemicals are man-made chemicals that have been widely used in industry and consumer products since the 1940s and remain in the environment for a long time ([1], January 18).

ATSDR supports and maintains the Toxic Substances Portal, which provides toxicological resources. These include information for a variety of audiences, including health effects of exposure to toxic substances; chemical classifications; information for community members, emergency responders, toxicological and health professionals, and health care provider education ([2], February 10). Using the Toxic Substances Portal's Toxicological Profiles and ToxFAQs, health effects information related to exposure to the most common brownfields and military brownfields contaminants can be obtained. This information is summarized in Table 2.1.

Table 2.1 Common brownfields and military brownfields contaminants and associated health effects of exposure

Chemical of concern	Health effects of exposure	Web source
Arsenic	Inhalation of high levels of inorganic arsenic: sore throat or irritated lungs. Inhalation of inorganic arsenic in general: increased risk of lung cancer. Ingestion of very high levels: death. Ingestion of low levels of inorganic arsenic: nausea and vomiting; decreased production of red and white blood cells; abnormal heart rhythm; damage to blood vessels; a sensation of "pins and needles" in hands and feet; and increase in the risk of skin, liver, bladder, and lung cancer. Ingesting or inhalation of low levels of inorganic arsenic for a long time: darkening of the skin and the appearance of small "corns" or "warts" on the palms, soles, and torso. Dermal (skin) contact with inorganic arsenic: redness and swelling.	ATSDR ToxFAQs[a]
Asbestos	Asbestos mainly affects the lungs and the membrane that surrounds the lungs. Inhalation of high levels of asbestos fibers for a long time among exposed workers: asbestosis. Inhalation of asbestos: can increase the risk of lung cancer and mesothelioma in people. Cigarette smoke and asbestos together significantly increase the risk of lung cancer.	ATSDR ToxFAQs[b]
Lead	Lead can affect almost every organ and system in the body. The nervous system is the main target for lead poisoning in children and adults. Long-term lead exposure: decreased learning, memory, and attention, and weakness in fingers, wrists, or ankles. Lead exposure in general: anemia, kidney damage, and increased blood pressure. Lead exposure at high levels: brain and kidney damage, and death. May cause miscarriage in pregnant women and damage to male reproductive organs. Children are more vulnerable to lead poisoning than adults. Lead is a probable human carcinogen.	ATSDR ToxFAQs[c]
Petroleum (total petroleum hydrocarbons (TPH))	TPH compounds can affect the central nervous system. Some TPH compounds at high levels: headaches and dizziness; peripheral neuropathy; and effects on the blood, immune system, lungs, skin, and eyes. Some TPH compounds in animals: affect reproduction and the developing fetus. One TPH compound (benzene) is carcinogenic to humans.	ATSDR ToxFAQs[d]

(continued)

Table 2.1 (continued)

Chemical of concern	Health effects of exposure	Web source
Per- and polyfluoroalkyls (PFAS)	High levels of certain PFAS may lead to increased cholesterol levels; changes in liver enzymes; decreased vaccine response in children; increased risk of high blood pressure or pre-eclampsia in pregnant women; small decreases in infant birth weights. Animal studies of high levels of PFAS: damage to the liver and the immune system, birth defects, delayed development, and newborn deaths in lab animals.	ATSDR ToxFAQs[e]
Polychlorinated biphenyls (PCBs)	High levels of PCBs: skin conditions such as acne and rashes. PCBs in exposed workers: changes in blood and urine that may indicate liver damage. High levels of PCBs in mothers: Babies had slightly lower birth weights, some issues with motor skills, and decreases in short-term memory, which often lasted for several years.	ATSDR ToxFAQs[f]
Polycyclic aromatic hydrocarbons (PAH)	High levels of one PAH in mice during pregnancy: difficulty reproducing in parent and in offspring. The offspring also had higher rates of birth defects and lower body weights. Animal studies: harmful effects on the skin, body fluids, and ability to fight disease after both short- and long-term exposure. These effects have not been seen in people. Some PAHs are confirmed carcinogens and others are probable carcinogens.	ATSDR ToxFAQs[g]
Radiation—Example used: Uranium	Most of the uranium inhaled or ingested is not absorbed and is excreted in the feces. Absorbed uranium: highest levels are found in the bones (66%), liver, and kidneys. Inhaled or ingested natural uranium and depleted uranium: kidney damage in humans and animals Ingesting water-soluble uranium compounds will result in kidney effects at lower doses than following exposure to insoluble uranium compounds. Uranium has not been classified as carcinogenic.	ATSDR ToxFAQs[h]
Unexploded ordnance (UXO).—example used, explosives: 2,4,6-trinitrotoluene (TNT), a common contaminant of UXO	Workers involved in the production of explosives who were exposed to high concentrations of 2,4,6-trinitrotoluene: including anemia and abnormal liver function. Ingestion or inhalation in animals: blood and liver effects, spleen enlargement, harmful effects on the immune system. Prolonged skin contact in humans: skin irritation and cataract development after long-term (365 days or longer) exposure. High doses in male animals: serious reproductive system effects. 2,4,6-trinitrotoluene is a possible human carcinogen.	ATSDR ToxFAQs[i]

(continued)

Table 2.1 (continued)

Chemical of concern	Health effects of exposure	Web source
Volatile organic compounds (VOC): Example used: trichloroethylene, a common VOC.	Moderate levels of trichloroethylene: headaches, dizziness, and sleepiness. Large amounts of trichloroethylene: coma and death. Ingesting or inhaling high levels: damage of nerves in the face; changes in the rhythm of the heartbeat, liver damage, and evidence of kidney damage. Skin contact with concentrated solutions of trichloroethylene: skin rashes. Exposure to trichloroethylene in the workplace may cause scleroderma. Some men occupationally exposed to trichloroethylene and other chemicals showed decreases in sex drive, sperm quality, and reproductive hormone levels. Trichloroethylene is classified as carcinogenic to humans.	ATSDR ToxFAQs[j]

[a] https://wwwn.cdc.gov/TSP/ToxFAQs/ToxFAQsDetails.aspx?faqid=19&toxid=3
[b] https://wwwn.cdc.gov/TSP/ToxFAQs/ToxFAQsDetails.aspx?faqid=29&toxid=4
[c] https://wwwn.cdc.gov/TSP/ToxFAQs/ToxFAQsDetails.aspx?faqid=93&toxid=22
[d] https://wwwn.cdc.gov/TSP/ToxFAQs/ToxFAQsDetails.aspx?faqid=423&toxid=75
[e] https://wwwn.cdc.gov/TSP/substances/ToxSubstance.aspx?toxid=237
[f] https://wwwn.cdc.gov/TSP/ToxFAQs/ToxFAQsDetails.aspx?faqid=140&toxid=26
[g] https://wwwn.cdc.gov/TSP/substances/ToxSubstance.aspx?toxid=25
[h] https://wwwn.cdc.gov/TSP/ToxFAQs/ToxFAQsDetails.aspx?faqid=439&toxid=77
[i] https://wwwn.cdc.gov/TSP/ToxFAQs/ToxFAQsDetails.aspx?faqid=676&toxid=125
[j] https://wwwn.cdc.gov/TSP/ToxFAQs/ToxFAQsDetails.aspx?faqid=172&toxid=30

2.1.3 *PFAS and VOCs at Military Brownfields*

The Department of Defense (DOD) has identified 687 military bases that require remedial action for PFAS groundwater contamination that originated with the use of firefighting foam. When PFAS in drinking water exceeded federal health advisory levels, DOD implemented actions, such as providing bottled water and installing water treatment systems. The DOD completed preliminary assessments/site inspections at 181 military installations and began remedial investigation/feasibility studies at 78 installations. As of 2021, the DOD had not begun cleanup at any of the 687 installations [28].

The DOD incurred $1.1 billion in actual PFAS costs in fiscal year 2020. The DOD estimated that PFAS investigation and cleanup costs in fiscal year 2021 would total more than $2.1 billion. [28].

Ruyle et al. [10] studied PFAS that originated from the use of aqueous film-forming foams (AFFF) and impacts on groundwater. The authors cited severe contamination of freshwater sources and drinking water hotspots from PFAS near hundreds of U.S. military bases that used AFFF [10]. The authors found that PFAS above regulatory guidelines will persist for centuries without remediation. Darlington, Barth, and McKernan [9] also examined impacts of PFAS from the use of AFFF at military bases and the need for effective remediation. The authors cited the persistence and mobility of PFAS in the environment as a long-term concern for military bases. The authors suggest that soil sorption technologies may be a solution for remediation of PFAS at military bases and may prevent their migration into groundwater or drinking water. However, the authors emphasize that finding ways to remediate PFAS from military bases is urgent. They report an estimated 26,000 PFAS contaminated sites (as of 2018) across the U.S. and six million people impacted by PFAS-contaminated drinking water ([9], citing Hu, Xindi C., 2016). The U.S. Environmental Protection Agency estimates that most people in the U.S. have been exposed to some PFAS ([25], June 7).

Volatile organic compounds (VOCs) are common chemical classes of concern at active military sites, FUDS, or military brownfields. ATSDR has investigated VOCs and potential adverse health outcomes among populations at hundreds of brownfields and other hazardous waste sites, including military sites. For example, ATSDR published a fact sheet in 2018 that summarized an internal morbidity study of former Marines, employees, and dependents potentially exposed to VOC-contaminated drinking water at the U.S. Marine Corps Base of Camp Lejeune, North Carolina. The contamination of drinking water started in the early 1950s, and the most contaminated wells were shut down in 1985. The Marine Corps discovered specific VOCs in the drinking water provided by two of the eight water treatment plants on base. As many as one million military and civilian staff and their families may have been exposed to the contaminated drinking water. The specific VOCs of concern included trichlorethylene (TCE), tetrachloroethylene (PCE), vinyl chloride, benzene, and trans-1,2-dichlorethylene. The maximum level of TCE found in one of the water systems was 280 times higher than the U.S. maximum contaminant level of 5 micrograms per liter of drinking water ([4], September 25, [5], January 16). Based on the results of health surveys completed by 76,000 former marines, employees, and dependents, exposure to both TCE and PCE was associated with an increased risk for kidney cancer in both Marines and civilian employees. In addition, exposure to both TCE and PCE was associated with an increased risk for bladder cancer and kidney disease in civilian employees, and exposure to PCE was associated with an increased risk for bladder cancer and kidney disease in Marines ([5, 6], April 25).

In 2021, VOCs in the form of jet fuel leaked from the Red Hill underground fuel storage facility into the aquifer that supplies water to the Joint Base Pearl Harbor-Hickam Water Distribution System in Oahu, Hawaii. The aquifer supplies water to 9700 civilian and military households, as well as schools and workplaces ([3], December 11). The jet fuels were JP-5, JP-8, and Jet A fuels. ATSDR had previously developed a toxicological profile and a shorter Tox FAQ to summarize jet fuels toxicity information for the public. ATSDR reported that little is known about the health effects of these fuels. A few studies of military personnel suggest that exposure to JP-8 can affect the nervous system. Some effects observed include changes in reaction time and changes in tests of neurological function. People who accidentally ingested kerosene, which is similar in composition to JP-5, JP-8, and Jet A fuels, suffered harmful effects to the respiratory tract, gastrointestinal tract, and nervous system. JP-5, JP-8, and Jet A fuels have not been classified as to their carcinogenicity ([7], April 06). Ultimately, the United States Navy submitted the Red Hill Bulk Fuel Storage Facility Closure Plan to the EPA and Hawaii Department of Health on November 1, 2022 ([26], January 31).

2.2 Conclusion

While only a small percentage of military brownfields and potential exposures to contaminants at these sites have been highlighted in this manuscript, brownfields in general are not benign. Contaminants at brownfield sites, including military brownfield sites, can and do impact public health. Understanding contamination associated with brownfields of all types can increase global awareness about public health hazards and impacts potentially associated with these sites.

References

1. Agency for Toxic Substances and Disease Registry, Per- and Polyfluoroalkyl Substances (PFAS) and Your Health (2024, January 18). https://www.atsdr.cdc.gov/pfas/. Last accessed 2024/02/02
2. Agency for Toxic Substances and Disease Registry, Toxicological Profiles, Toxic Substances Portal, ATSDR (2021, February 10). https://wwwn.cdc.gov/TSP/index.aspx. Last accessed 2024/02/02
3. Agency for Toxic Substances and Disease Registry, Red Hill Reponse (2023, December 11). https://www.atsdr.cdc.gov/sites/red-hill/index.html. Last accessed 2024/02/02
4. Agency for Toxic Substances and Disease Registry, Camp Lejeune, North Carolina (2018, September 25). https://www.atsdr.cdc.gov/sites/lejeune/index.html. Last accessed 2024/02/02
5. Agency for Toxic Substances and Disease Registry, Health Concerns at Camp Lejeune (2014, January 16). https://www.atsdr.cdc.gov/sites/lejeune/health_concerns.html. Last accessed 2024/02/02
6. Agency for Toxic Substances and Disease Registry, Morbidity Study of Former Marines, Employees, and Dependents Potentially Exposed to Contaminated Drinking Water at USMC Base Camp Lejeune: A Summary of Agency for Toxic Substances and Disease Registry (ATSDR) Study Design and Results (2018, April 25). https://www.atsdr.cdc.gov/sites/lejeune/morbiditystudy_factsheet.html. Last accessed 2024/02/02
7. Agency for Toxic Substances and Disease Registry, Tox FAQs™ JP-5, JP-8, and Jet A (2017, April 6). https://wwwn.cdc.gov/TSP/ToxFAQs/ToxFAQsDetails.aspx?faqid=772&toxid=150. Last accessed 2024/02/02
8. Berman L, Morar C, Unkart S, Erdal S (2022) An overview of brownfields redevelopment in the United States through regulatory, public health, and sustainability lenses. J Environ Health 84(9):8–14. https://www.ncbi.nlm.nih.gov/pmc/articles/PMC10191173/, citing Woghiren, E. (2020/09/01, PRJ 05833, Geospatial Research, Analysis, and Services Program, Agency for Toxic Substances and Disease Registry). Last accessed 2024/02/02

9. Darlington R, Barth E, McKernan J (2018) The challenges of PFAS remediation. Mil Eng 110(712):58–60. PMID: 29780177; PMCID: PMC5954436
10. Ruyle BJ, Thackray CP, Butt CM, LeBlanc DR, Tokranov AK, Vecitis CD, Sunderland EM (2023) Centurial persistence of forever chemicals at military fire training sites. Environ Sci Technol. 57(21):8096–8106. https://doi.org/10.1021/acs.est.3c00675. Epub 2023 May 15. PMID: 37184088; PMCID: PMC10233753
11. United States Department of Defense, DoD Base Realignment and Closure: BRAC Rounds (BRAC 1988, 1991, 1993, 1995 & 2005) Executive Summary. Fiscal Year (FY) 2023 Budget Estimates. Program Year 2023 (2022). https://comptroller.defense.gov/Portals/45/Documents/defbudget/fy2023/budget_justification/pdfs/05_BRAC/FY2023_BRAC_Overview.pdf
12. United States Army Corps of Engineers, Formerly Used Defense Sites Program: Program Overview (2021, January). https://usace.contentdm.oclc.org/utils/getfile/collection/p16021coll11/id/4897. Last accessed 2023/12/03
13. United States Army Corps of Engineers, Formerly Used Defense Sites Program, US Army Corps of Engineers., https://www.usace.army.mil/missions/environmental/formerly-used-defense-sites/. Last accessed 2024/02/02
14. United States Army Corps of Engineers, Formerly Used Defense Sites (FUDS) Viewer – FY21 Data, US Army Corps of Engineers (n.d.). https://geospatial.sec.usace.army.mil/arcgis/apps/webappviewer/index.html?id=fe3efb63b00e45b5857d3b9e1542092. Last accessed 2024/02/02
15. United States Army Corps of Engineers FUDS Frequently Asked Questions, US Army Corps of Engineers (n.d.). https://www.usace.army.mil/Missions/Environmental/Formerly-Used-Defense-Sites/Frequently-Asked-Questions/. Last accessed 2024/02/02
16. United States Environmental Protection Agency, Common Types of Brownfields and their Contaminants, Brownfields (2023, July 6). https://www.epa.gov/brownfields/common-types-brownfields-and-their-contaminants. Last accessed 2024/12/03
17. United States Environmental Protection Agency, About, Brownfields (2023a, October 4). https://www.epa.gov/brownfields/about. Last accessed 2024/12/03
18. United States Environmental Protection Agency, Superfund: CERCLA Overview, (2023, October 30). https://www.epa.gov/superfund/superfund-cercla-overview. Last accessed 2024/12/03
19. United States Environmental Protection Agency, Summary of the Small Business Liability Relief and Brownfields Revitalization Act, Brownfields (2023, November 6). https://www.epa.gov/brownfields/summary-small-business-liability-relief-and-brownfields-revitalization-act. Last accessed 2024/02/02
20. United States Environmental Protection Agency, About, Brownfields (2023b, October 4). https://www.epa.gov/brownfields/about. Last accessed 2024/02/02
21. United States Environmental Protection Agency, Grant Recipient Reporting, Brownfields (2023, November 10). https://www.epa.gov/brownfields/grant-recipient-reporting. Last accessed 2024/02/02/
22. United States Environmental Protection Agency, Base Realignment and closure Act Sites, FEDFacts (2024, January 31). https://www.epa.gov/fedfacts/base-realignment-and-closure-act-brac-sites. Last accessed 2024/02/02
23. United States Environmental Protection Agency, Superfund Site Information, Superfund (2024, February 2). https://cumulis.epa.gov/supercpad/cursites/csitinfo.cfm?id=0506496. Last accessed 2024/02/02
24. United State Government Accountability Office, Vacant Properties. Growing Number Increases Communities' Costs and Challenge, GAO (2011, November 11). https://www.gao.gov/assets/gao-12-34.pdf. Last accessed 2024/12/03
25. United States Environmental Protection Agency, Our Current Understanding of the Human Health and Environmental Risks of PFAS (2023, June 7). https://www.epa.gov/pfas/our-current-understanding-human health and environmental-risks-pfas. Last accessed 2024/02/23

26. United States Environmental Protection Agency, Defueling Plan and Closure of the Red Hill Bulk Fuel Storage Facility (2024, January 31). https://www.epa.gov/red-hill/defueling-plan-and-closure-red-hill-bulk-fuel-storage-facility#closure. Last accessed 2024/02/02
27. United States Government Accountability Office, Environmental Contamination: Corps Needs to Reassess Its Determinations That Many Former Defense Sites Do Not Need Cleanup, GAO (2002, August), https://www.denix.osd.mil/mmrp/denix-files/sites/46/2016/03/02_Environmental-Contamination-Corps-Needs-to-Reassess.pdf. Last accessed 2024/02/02
28. United States Government Accountability Office, Firefighting Foam Chemicals: DOD Is Investigating PFAS and Responding to Contamination, but Should Report More Cost Information (2021, June 22). https://www.gao.gov/products/gao-21-421. Last accessed 2024/02/06

Chapter 3
Risk Communication: Lessons Learned from Stakeholders in the United States

Laurel Berman

Abstract This manuscript describes risk communication in brownfields communities and features two case studies. The first case study highlights a former air force base, K. I. (Kevin Ingalls) Sawyer, in the Upper Peninsula of Michigan in the Lake Superior basin. The base closed in 1995 and the re-branded town of K. I. Sawyer struggles to find its image and purpose as a community. The primarily low-income community was concerned about a plume of per- and polyfluoroalkyl substances (PFAS) in groundwater, potentially contaminated soil in their baseball fields, and lead paint on the interior and exterior of vacant residential brownfields. The second case study features the City of Benton Harbor, Michigan. Once a thriving harbor city, it currently is navigating a burden of multiple brownfields, vacant properties, and the aftermath of a lead water crisis. In 2016, lead-lined water service lines began to fail, releasing lead into drinking water, leading to community outrage and concern from 2019–2023. The city's lead water crisis made national news. By 2023, the community felt the need to build their capacity to navigate conversations with local government and to obtain funding for community revitalization.

Keywords Brownfields · Environmental health · Risk · Risk communication · Contamination · Communities · Lead · Per- and Polyfluoroalkyl substances

3.1 Background: Risk Communication in Brownfields Communities

The Agency for Toxic Substances and Disease Registry (ATSDR) describes risk communication as how we talk to the public about substances or behaviors that can be harmful. In communities with brownfields that may have known or unknown contamination, it is especially important to listen to community concerns. Such

L. Berman (✉)
The Agency for Toxic Substances and Disease Registry, Atlanta, GA, USA
e-mail: laberman@cdc.gov

C. Morar et al. (eds.), *Achieving Sustainability in Ukraine through Military Brownfields Redevelopment*, NATO Science for Peace and Security Series C: Environmental Security, https://doi.org/10.1007/978-94-024-2278-8_3

concerns include contaminated groundwater, drinking water, or air; landfill leakage; and soil or groundwater plume spread or movement. In addition, communities may have concerns about blighted properties in general, such as mold, methamphetamine residues, lead exposures, and safety and security. Communities typically want to know how contamination may affect themselves and their families.

How we talk to the public about risk may be critical to building trust and transparent, multi-directional communication. As public and environmental health professionals, our role in communicating risks is to be open, honest, and calm to help build trust. It is critical to put the community's needs first, to be compassionate, and to be genuine. These qualities make us human, believable, and approachable. If we don't know an answer to a community member's question, we say so and offer to find out the answer. This is part of being accountable and meaningfully listening to community concerns.

In addition, we try to learn as much as possible about the community before our first visit. Here are some examples of information we collect in advance:

- Demographic information, such as education and income level, age, and languages spoken/read.
- Religious/faith beliefs, diversity of the community, geography, and cultural background and norms.
- Community organizations and local leaders that can or will be invited to the meeting and may be willing to partner on information communication and understanding.
- Room set-up for the meeting. For example, we may prefer to be sitting in a large round formation with the audience instead of up above them on a stage or at a podium. This encourages a "level playing field" and may increase trust and comfort.
- Language translators and written language translations (if needed).

An important aspect of risk communication is how we get our messages about environmental and health risks to our communities. To communicate risks, we can create written, spoken, or visual statements and use them to create messages for our community of concern. Messages are generally simple, one-sentence statements, such as:

- Reduce, reuse, recycle
- Three tests before age three (for blood lead screening)
- Boil water for five minutes
- Think green
- Shop local

If we know the environmental issues of concern, we can prepare key messages to frame our discussion and to alleviate fears or concerns the community may have specific to one or more issues. We may use a message map to convey key messages, such as the one shown in Fig. 3.1. This message map was adapted from the ATSDR Communication Toolkit [1]. You can download the message map template for your

Message Map Template

Stakeholder: Question or concern:		
Key Message 1	**Key Message 2**	**Key Message 3**
Supporting Fact 1.1	**Supporting Fact 2.1**	**Supporting Fact 3.1**
Supporting Fact 1.2	**Supporting Fact 2.2**	**Supporting Fact 3.2**
Supporting Fact 1.3	**Supporting Fact 2.3**	**Supporting Fact 3.3**

Fig. 3.1 Message map template. (Adapted from ATSDR Communication Toolkit [1])

own use here: https://www.atsdr.cdc.gov/communications-toolkit/documents/12_Message-Mapping-Tool-final-111015_508.pdf

The message map includes the community or partner (the "stakeholder") and the key question of concern, such as, will my children get sick from playing in soil with high levels of arsenic? Then, we may use 1, 2, or 3 supporting facts. A completed message map is provided in the Benton Harbor case example, below.

In the two case examples featured below, ATSDR practiced risk communication that was tailored to each community. In both communities, ATSDR was an invited guest to provide technical assistance on environmental issues.

3.1.1 Case Example: K. I. Sawyer Airforce Base, Marquette County, Upper Peninsula of Michigan

K.I. (Kenneth Ingalls) Sawyer Air Force Base was located in the state of Michigan's Upper Peninsula, a peninsula of land in the Great Lakes. The Upper Peninsula is bordered to the south by Lake Michigan, to the north by Lake Superior, and to the east by Lake Huron. Originally, the base began as the K.I. Sawyer County Airport in the mid-1930s and operated until 1955, with the establishment of the K.I. Sawyer Air Force Base. The airport then functioned primarily as a jet base with about 300 civilian flight landings per month. The K.I. Sawyer Air Force Base operated from 1955 until its closure in 1995 [8]. The base closure devastated the Marquette County area. The base closure led to the loss of as many as 3000 jobs, loss of businesses, and impacts on local schools and Northern Michigan University. The town of K.I. Sawyer suffered an economic decline, from which it did not recover. By 2016, the poverty rate was as high as 46% and unemployment reached as high as 24%. In addition, the area was perceived as a "drug den" [8].

From 2013 to 2019, ATSDR engaged with a local community group, Sawyer Community Alliance (SCA). For several years, the Marquette County Commissioner, planner, and brownfields redevelopment experts had been discussing environmental concerns and redevelopment plans with Sawyer Community Alliance. The SCA asked ATSDR to expand community engagement and risk communication activities. The SCA vision was, "From Base to Place: Building a Healthy Sawyer." Their primary goals are to improve health, economy, and social disparities by establishing a stronger sense of community. They focus on remediating blighted buildings, providing community services, and improving public services (water, transportation, etc.) [2]. By 2018, SCA had enjoyed several successes:

- Two former military members established a volunteer "community watch," which grew to over 30 members within 2–3 years of its inception. The community watch:
  - coordinated their volunteer services with the local police
  - hosted several community events with bicycle and school supplies raffles
  - worked with the state environmental agency to limit illegal garbage dumping by supplying dumpsters for community clean-up
  - worked with the Sault [pronounced "Soo"] Tribe, who owned many rental properties on-site, to improve garbage service
- The K.I. Sawyer population began to increase, and large businesses began to open on-site
- The County used brownfields funding to safely demolish 200 residential apartment units, removing lead and asbestos hazards
- The local school hosted a community needs assessment survey to focus on assets and improvements
- Local residents worked with SCA to obtain funding to clean up and re-open two baseball fields connected to the elementary school by an all-terrain trail. The local health clinic helped expand the community garden, which relocated to the school grounds.

While focusing on small wins and positive successes, the SCA and partners continuously address and communicate risks during regularly scheduled SCA meetings (see Fig. 3.2). For example, the baseball fields were filled with stamp sands, a coarse sand formed from processing metal ore in a stamping mill, which crushes material to remove ore. Residents were concerned about metal contamination in the stamp sands. The ball fields were in disrepair and not usable without the correct fill. Local residents obtained funding to have the stamp sands sampled and then removed. Ultimately, metal contamination was not a concern. The ball fields were updated with clean fill and reopened. In addition, 200 apartment housing units were vacant

Fig. 3.2 Left: SCA members received funding for community projects. (Image taken by Lloyd DeGrane, 2014. Right: SCA members used funding to clean up and re-open two baseball fields. Image taken by ATSDR, 2018)

and abutted the school property. The harsh winter conditions had led to peeling, lead-based paint hanging in long strips on the exterior of the units. The units were an attractive nuisance to local teens, who trespassed in the units. The deteriorated buildings posed a risk of injury to the teens. The teens also risked bringing lead dust home on their clothing, potentially impacting younger siblings. The units were demolished and the soil that encapsulated the demolished remains was tested for lead. While lead was present, it was below regulatory limits. The County continuously discussed with SCA and the community the demolition risks as well as the levels of lead in soil.

One ongoing concern at K.I. Sawyer is a plume of per- and polyfluoroalkyl substances (PFAS) in groundwater. PFAS were often used on military bases in firefighting foam used for training and response. The County coordinated continuous delineation and monitoring of the plume area. In 2016, the Air Force detected PFAS at 13 of 15 sites. A former fire training area has the highest levels of PFAS. The drinking water at K. I. Sawyer is continuously monitored. Testing of 28 residential wells were non-detect for PFAS. Multiple agencies and partners inform the community about monitoring and PFAS risks. There is ongoing risk communication, using plain language and a website designated to provide information about PFAS and monitoring activities [4].

3.1.2 *Case Example: Benton Harbor, Berrien County, Michigan*

Benton Harbor is in Berrien County, Michigan, a largely agricultural area of rural towns and small cities. The City of Benton Harbor was once a thriving harbor city on the St. Joseph River, which empties into nearby Lake Michigan. The City of Benton Harbor is 87% African American (Black), with a median household income in the city of $26,592 and 44% of the population in poverty. The United States

Census Bureau defines poverty as a status when the family's total income is less than the family's money income threshold (pre-determined by the Census Bureau). Then the family and every individual in it is considered in poverty ([6], June 15). In comparison, Berrien County overall is 80% White and 14% African American, and the poverty level is 15% [5]. Figure 3.3 indicates 43 brownfields in Berrien County that have received federal funding, primarily for site assessment, A high density of these sites are clustered in the lower income areas, including Benton Harbor ([7], January 19). State-funded brownfields projects in Benton Harbor include 10 site assessments and more than 10 redevelopment grants and loans [3].

In September of 2019, ATSDR was invited by local organizations to hold a soilSHOP—soil screening, health outreach, and partnerships—event during a weekly farmers market in downtown Benton Harbor. The community had been going through a lead water crisis attributed to failing, lead-lined water services lines. Partners from the local healthcare organization offered blood lead screening for children and environmental health education. Community members were concerned that soil lead levels may have increased from watering lawns and gardens with household water. Using an Xray Fluorescence machine, ATSDR screened over 80 soil samples for lead. None of the samples were of concern in terms of exceeding regulatory levels of lead in soil. However, lead pipe replacement at the time was only slowly progressing.

In 2019, Benton Harbor residents were using bottled water, largely distributed by churches and volunteer organizations. The Berrien County Health Department was distributing faucet water filters. While the County Health Department had a

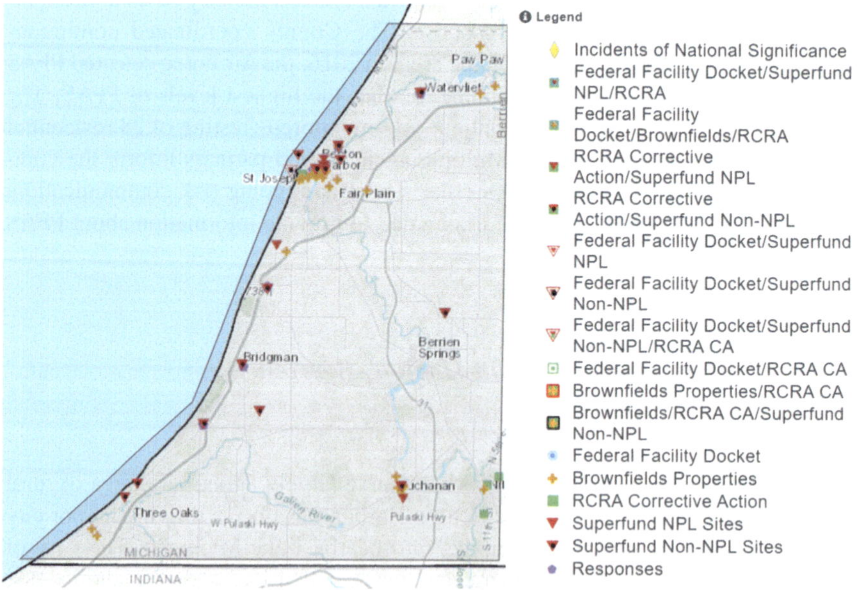

Fig. 3.3 Land reuse sites in Berrien county: map and legend ([7], January 19)

statement on their website regarding distribution of water filters, they had not implemented concentrated risk communication efforts, such as partnering with the local faith-based leaders or going door-to-door to conduct outreach. In addition, if lead levels are very high, water filters may not function effectively. The water crisis made national news, and residents, the Mayor, and one City Commissioner (i.e., a role within the Mayor's office that is part of the legislative and policy-making body) were interviewed. People were, in general, confused and outraged. By 2023, over 99% of lead service lines (about 4500) in Benton Harbor were replaced, and the remaining lines were scheduled for replacement. In addition, the state established a lead home assessment program in which inspectors would identify any sources of lead in homes, such as windows or paint, and repair and replace the items at no cost to the home owner.

In early 2023, a City Commissioner reached out to ATSDR with a specific request to help the City and residents increase their capacity to understand environmental and health risks. The Commissioner relayed that residents had feelings of distrust in their own city government and a need to understand chemical contamination. In response, we offered a 4-hour training in Communicating Environmental and Health Risks. The training offerered four continuing education units and a certificate of completion from our partner, National Environmental Health Association (NEHA). In March, we facilitated a 2-day event, with the training on Day 1 and a soilSHOP and community environmental education event on Day 2.

Knowing that the community might still be outraged and distrustful of government, our first goal was to establish trust early. We did so by holding the event at a local Boys and Girls Club. We ensured that the City Commissioner and a strong, well-known community advocate within the City government designed flyers for the training, conducted outreach, and introduced the training team from ATSDR and NEHA. We also included a guest lecturer from the Michigan Department of Environment, Great Lakes, and Energy, who described multiple brownfields projects and successes funded by the state specific to assessment and remediation in Benton Harbor. We opened the training with, "what is a part per billion (ppb)?" demonstration, using images of one billion particles of soil with one white particle, a flask filled with one million particles of water and a drop of blue dye, and a bucket filled with one million particles of water and one red bead (i.e., one part red bead per million parts water, or 1 ppm), shown in Fig. 3.4. We tailored the training around interactive exercises associated with lead in water and soil. Practice sessions included developing key risk messages with the participants, specifically regarding lead in water and soil. The key messages included links to resources for information, such as water quality updates or to sign up for the home assessment program. An example message map for concerns about lead in water is provided in Fig. 3.5.

In total, 35 participants attended the training. The participants included residents, local non-profits, city personnel, partners from the United States Environmental Protection Agency (EPA) who also hosted the soilSHOP event on Day 2, and a class of eight students and one professor from nearby Andrews University. Several participants attended Day 2 of the event and had a chance to observe risk communication and environmental health education through the soilSHOP activity (see

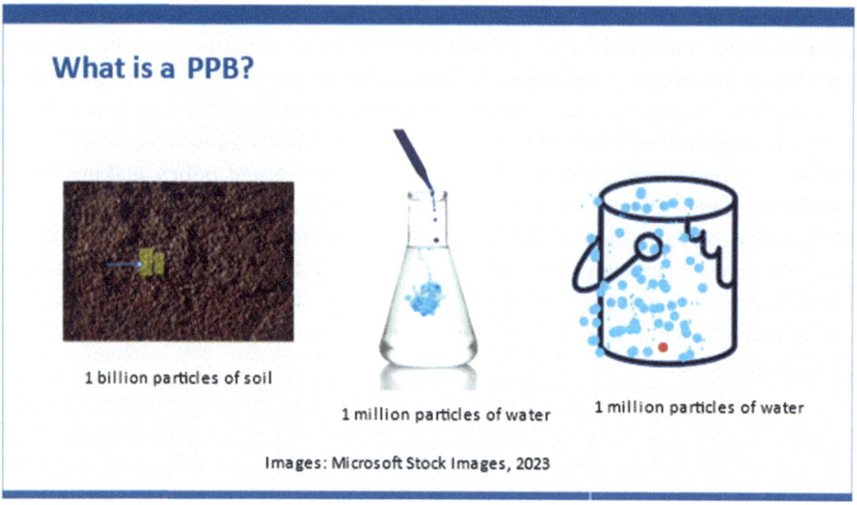

Fig. 3.4 What is a PPB? (Image created by the Agency for Toxic Substances and Disease Registry Land Reuse Program)

Message Map 1: Water Crisis in Benton Harbor (example only)

Stakeholder: Benton Harbor Citizens Question or concern: Is there lead in our water?		
Key Message 1	**Key Message 2**	**Key Message 3**
Before the water pipes were replaced, lead in water was found to be above the action level of 15 parts lead per 1 billion parts of water (15 ppb)	The water is safe to drink	The state (EGLE) is offering free whole home lead assessments
Lead in water above the action level can lead to lead poisoning in small children and some effects in adults	Over 99% of lead water lines were replaced and water tested was below the action level in June (14 ppb) and December (9 ppb)	EGLE will replace any sources of lead in your home
Berrien County Health Department reviewed children's blood lead data from 2010 – 2021+ and did not find significant increases in lead poisoning*	If you need more time to feel safe, you can drink bottled water or obtain a free water filter. Information is here: https://www.michigan.gov/egle/newsroom/press-releases/2022/11/28/bh-water	Removal of lead sources should reduce potential lead exposures
You can get free blood lead screening for kids ages 0 - 6 at: (March 11 community event, TBD)	Every six months EGLE will post water quality updates on this website: https://www.michigan.gov/egle/about/featured/benton-harbor	There is no charge for the home assessments or repairs. You can sign up here: Lead Safe Home Program ALE application, or call 866-691-5323.

*Lead data depends on number of children tested, who is tested, and other factors

Fig. 3.5 Example training exercise message map related to concerns about lead in water

Fig. 3.6). The EPA partners screened over 20 soil samples for lead, and none of the samples exceeded regulatory limits. During the event, the Berrien County Health Department provided free water filters and radon test kits. The State of Michigan also provided information about and a sign-up sheet for free home lead assessments.

Fig. 3.6 Left: An EPA representative with a community member, scanning a soil sample for lead. Right: Risk communication and environmental health education activities during the soilSHOP event. This image was taken in 2023 by ATSDR

3.2 Conclusion

The K.I. Sawyer and Benton Harbor case studies demonstrate the importance of community engagement and effective risk communication in two very different communities with very different concerns about contamination. In K.I. Sawyer, risk communication was coordinated through the County's municipal entities and happened frequently, helping to build trust within the community. In turn, this motivated SCA to attain small successes to achieve a sense of place and community pride. In Benton Harbor, a City Commissioner led the charge to build the capacity of the community to navigate complex issues associated with lead contamination concerns in water and soil. By participating in a 2-day training and community event, Benton Harbor changed from a community in distrust of their government over a water crisis to a community with skills to understand chemical contamination. They learned how to access resources, such as water quality data or lead home assessment. In addition, the community had new partners to guide them on State-led brownfields remediation in Benton Harbor.

References

1. Agency for Toxic Substances and Disease Registry. ATSDR Communication Toolkit (2018). https://www.atsdr.cdc.gov/communications-toolkit/documents/12_Message-Mapping-Tool-final-111015_508.pdf. Last accessed 2024/02/01
2. Agency for Toxic Substances and Disease Registry. KI Sawyer Community Partnership, Upper Peninsula, Michigan, ATSDR Land Reuse Health Program (2021). https://www.atsdr.cdc.gov/sites/brownfields/factsheets/sawyer_community_partnership_factsheets.html. Last accessed 2024/02/01
3. Michigan Department of Environment, Great Lakes, and Energy (2023) Environmental Mapper, https://www.mcgi.state.mi.us/environmentalmapper/#. Last accessed 2024/02/01

4. Michigan PFAS Action Response Team. K.I. Sawyer Air Force Base (Gwinn, Marquette County) (2022). https://www.michigan.gov/pfasresponse/investigations/sites-aoi/marquette-county/ki-sawyer-air-force-base. Last accessed 2024/02/01
5. United States Census. Quick Facts (2023). https://www.census.gov/quickfacts/fact/table/US/PST045218/. Last accessed 2024/02/01
6. United States Census (2023, June 15) How the census bureau measures poverty. https://www.census.gov/topics/income-poverty/poverty/guidance/poverty-measures.html. Last accessed 2024/02/27
7. United States Environmental Protection Agency. Cleanups in My Community Map (2024, January 19). https://cimc.epa.gov/ords/cimc/f?p=cimc:MAP::::71:P71_WELSEARCH:MI|State|MI|||true|true|true|true|true|true||-1|sites|N|basic. Last accessed 2024/02/01
8. Witanen D (2017) History of the K.I. Sawyer Air Force Base, Military History of the Upper Great Lakes. https://ss.sites.mtu.edu/mhugl/2017/10/20/history-of-the-k-i-sawyer-air-force-base/. Last access 2024/02/01

Chapter 4
Training for Stakeholders: Evaluating Environmental and Health Risks

Laurel Berman and Christopher A. De Sousa

Abstract This manuscript provides an overview of a special training session offered during the Advanced Research Workshop (ARW) in Oradea, Romania. The training provided is the Agency for Toxic Substances and Disease Registry's (ATSDR) Environmental Health and Land Reuse (EHLR) Basic Training Module 2: Evaluating Environmental and Health Risks. The purpose of the training is to promote health-focused redevelopment, known as Healthfields, by safely reusing contaminated land to improve community health outcomes. One goal of offering the training among brownfields experts participating in the ARW was to expand an existing North American-Eastern European collaboration focused on sharing best practices and resources related to land reuse that incorporates public health, particularly improving health outcomes as a result of redevelopment. The training topics include an overview of toxicology, epidemiology, risk assessment; Phase I and II environmental site assessment; and ATSDR's Public Health Assessment program and process. Resources related to the evaluation of environmental and health risks are also provided in the training. Twenty-two ARW participants attended the training, 14 virtual participants from Ukraine and eight live participants from seven additional countries.

Keywords Brownfields · Public health · Environmental site assessment · Environmental health · Toxicology · Epidemiology · Risk assessment · Public health assessment

---

L. Berman (✉)
The Agency for Toxic Substances and Disease Registry, Atlanta, GA, USA
e-mail: laberman@cdc.gov

C. A. De Sousa
School of Urban and Regional Planning, Toronto Metropolitan University, Toronto, ON, Canada

C. Morar et al. (eds.), *Achieving Sustainability in Ukraine through Military Brownfields Redevelopment*, NATO Science for Peace and Security Series C: Environmental Security, https://doi.org/10.1007/978-94-024-2278-8_4

4.1 Background: Environmental Health and Land Reuse Training

The Agency for Toxic Substances and Disease Registry (ATSDR) developed the Environmental Health and Land Reuse (EHLR) Basic Training for environmental professionals; planners; municipal entities; and students in environmental science, public health, and planning. Our purpose was to promote health-focused redevelopment, known as Healthfields redevelopment. Such redevelopment safely reuses contaminated land to reduce potential exposures to chemicals and hazards while focusing on improving overall health outcomes in communities with brownfields and other potentially contaminated sites.

There are five modules of the EHLR Basic Training. Each module of the training represents one step of ATSDR's 5-step Land Reuse Model:

1. Engaging with Your Community
2. Evaluating Environmental and Health Risks
3. Communicating Environmental and Health Risks to the Community
4. Redesigning with Health in Mind
5. Measuring Success

Each module of the 5-module EHRL training (Fig. 4.1) is eligible for a Certificate of Completion and continuing education units (hours) from ATSDR's education partner, the National Environmental Health Association (NEHA). However, to be fully certified in EHLR, ATSDR encourages learners to complete all five modules. All training materials may be download from ATSDR's website, at: https://www.atsdr.cdc.gov/sites/brownfields/classroom-training-materials.html, including the supplemental course textbook, Land Reuse and Redevelopment: Creating Healthy Communities. The book was written by brownfields experts and contains a wealth of resources related to the 5-step Land Reuse Model and the EHLR training overall. The book is available for download at: https://www.atsdr.cdc.gov/sites/brownfields/classroom_training/Creating_Healthy_Communities-508.p. In addition, ATSDR provides a link to an online, asynchronous version of the EHLR training that is hosted and maintained by NEHA, which also includes a Certificate of Completion and continuing education units. The training materials are shown in Fig. 4.2.

4.1.1 Providing the EHLR Module 2 Training: Evaluating Environmental and Health Risks at the ARW

For the Advanced Research Workshop (ARW), we provided EHLR Module 2 training, Evaluating Environmental and Health Risks. We selected this module as most appropriate for addressing environmental and health risks associated with military brownfields and brownfields in general. More specifically, we felt this training would broaden an existing brownfields collaboration among North American and

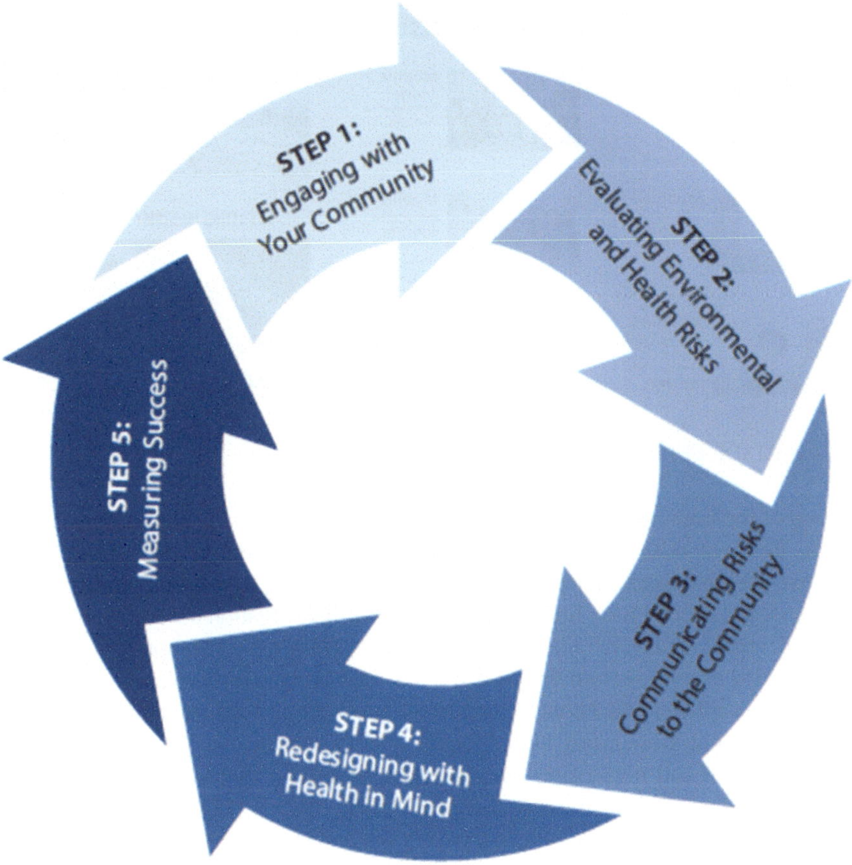

Fig. 4.1 ATSDR's 5-step Land Reuse Model

European partners that focuses on integrating and improving public health through safe land reuse and redevelopment. Dr. Berman presented in the module virtually live and Dr. DeSousa presented in-person with live ARW participants in Oradea. Twenty-two ARW participants attended the Module 2 training, of which 14 from Kharkiv University in Ukraine, attended virtual live. The other eight participants who were physically in Oradea represented seven different countries.

The EHLR Evaluating Environmental and Health Risks module incorporates a review of basic environmental health topics, including toxicology, epidemiology, and risk assessment. Knowledge of toxicology can form the basis of regulatory guidelines related to specific contaminants. The primary content we covered in the toxicology segment focused on dose (how much of a contaminant a person may be exposed to) and exposure to contaminants. In the epidemiology section, we focused on diseases related to specific toxicant exposures. During the risk assessment

Fig. 4.2 ATSDR EHLR classroom training materials

section of the training, we focused on four steps of risk assessment: hazard identification, dose-response assessment, exposure assessment, and risk characterization.

Next, we discussed environmental site assessments (ESA). We highlighted the differences between Phase I, ESA (is research and investigation based) and Phase II ESA (identifies actual contaminants and conducts sampling and laboratory testing). We provided examples from the United States and Canada to supplement the training. The examples showed how these assessments were carried out in both countries and to highlight similarities and differences between the two jurisdictions with regard to when and how ESAs are conducted, who is responsible and eligible to undertake them, and how these assessments fit into the regulations governing contaminated properties and the procedures for taking these properties from potentially contaminated to safe for reuse. For instance, in the United States, a "Licensed Environmental Professional" must conduct a Phase I or II ESA. In Ontario, Canada, only a "Qualified Person" (Engineer or Geoscientist) can conduct or supervise a Phase 1 ESA, which is only required by law if one intends to change the use of a property (i.e., from commercial/industrial to residential/park). However, in some circumstances, lenders, insurers, municipalities or other stakeholders may want a Phase 1 ESA conducted before taking certain actions (e.g., making a loan, granting a planning permission). While these rules may vary somewhat by country and local jurisdiction, the overall objective of this part of the training was to introduce participants to the relatively standard scientific rationales and procedures for assessing risk while also getting them to think about how these align with environmental cleanup, planning, and land redevelopment.

Following the ESA segment, the EHLR module content describes ATSDR's public health assessment program. This program emphasizes the identification of exposure pathways at specific sites; evaluates environmental and health data to identify potential health risks; recommends ways to prevent exposures; and educates affected communities and local health professionals, as well as others participating in the training, about site contamination and potential health effects.

We ended the EHLR Evaluating Environmental and Health Risks training by highlighting a list of free tools and resources summarized in the module that may be helpful for all participants. These included

- ATSDR Brownfield and Land Reuse Site Tool: https://www.atsdr.cdc.gov/sites/brownfields/site_inventory.html
- ATSDR Toxicological Profiles and ToxFAQs: https://www.atsdr.cdc.gov/substances/index.asp
- ATSDR Action Model: https://www.atsdr.cdc.gov/sites/brownfields/model.html
- Health Impact Assessment: https://www.cdc.gov/healthyplaces/hia.htm
- Protocol for Assessing Community Excellence in Environmental Health Guidebook: https://www.cdc.gov/nceh/ehs/docs/pace-eh-guidebook.pdf
- Choose Safe Places for Early Care and Education: https://www.atsdr.cdc.gov/safeplacesforECE/index.html)
- Community Mapping and Data Tools
  - EnviroAtlas: https://enviroatlas.epa.gov/enviroatlas/interactivemap/
  - Environmental Justice Screening and Mapping Tool: https://www.epa.gov/ejscreen
  - National Center for Environmental Health Tracking Program: https://www.cdc.gov/nceh/tracking/BetterInfoBetterHealth.htm
  - Community Health Rankings and Roadmaps: https://www.countyhealthrankings.org/explore-health-rankings#county-select-38

4.1.2 Next Steps: Building a Training Collaboration

As a result of the Module 2 training during the ARW, participants from Kharkiv University in Ukraine and one participant from University of Tirana in Albania requested the full EHLR 5-module training for doctoral and Master of Science students in Geography and Tourism. ATSDR is planning to provide the full training in 2025.

4.2 Conclusion

One of the great challenges and joys associated with managing brownfields and recycling land is that it brings together many subjects, professions, and stakeholders. Public health has always been central to brownfields practice and outcomes, and as a result, it is important to introduce all stakeholders to public health concepts. ATSDR's training modules have been designed to do just that. The delivery of this module at the ARW provided an opportunity not just to train participants but also to introduce them to an important tool that can be used to prepare prospective students and others keen to address brownfield challenges in Ukraine and to work toward more sustainable solutions that manage risk and provide a better quality of life to its citizens. The request for the full, 5-module training, helps us reach our goal to expand our North American and Eastern European collaboration while increasing the focus on improving public health through safe land reuse and redevelopment.

The EHLR training is free of charge, and we encourage anyone interested in using it in their curriculum to do so. ATSDR can provide technical assistance, as needed. For more information, email atsdr.landreuse@cdc.gov.

Chapter 5
Infusing Sustainability into Brownfields Planning and Redevelopment: Lessons from Canada and the United States

Christopher A. De Sousa

Abstract The current war in Ukraine and ongoing Russian aggression are leaving behind a legacy of polluted, damaged, and destroyed property that the country will need to address once the tyrants are defeated. This paper provides a brief overview of brownfield policy and activity in the United States and Canada over the past three decades to help Ukraine understand how the two countries have laid the groundwork for assessing, remediating, and redeveloping their potentially contaminated lands, including those damaged by military activities. The paper then reviews brownfield redevelopment fundamentals and explores best practices in infusing sustainability into brownfield redevelopment activities and measuring sustainability outcomes. The hope is that future brownfield redevelopment activities in Ukraine will help repair the past, promote sustainable practices, and lay the foundation for the future that Ukrainians deserve.

Keywords Brownfields · Sustainability · Policy · Contaminated lands · United States · Canada

5.1 Introduction

Although it might seem premature to advise Ukraine on brownfield planning and redevelopment while they are in the throws of fighting for their freedom and survival, the intention here is to help lay a hopeful path for the future that Ukrainians are fighting for. While the scale and scope of the urban environmental damage that Ukraine is facing due to Russia's relentless aggression is far beyond the scope of what the typical planner and policymaker in the United States or Canada are challenged with, it might find value in learning about the experiences and efforts that

C. A. De Sousa (✉)
School of Urban and Regional Planning, Toronto Metropolitan University, Toronto, ON, Canada
e-mail: chris.desousa@torontomu.ca

C. Morar et al. (eds.), *Achieving Sustainability in Ukraine through Military Brownfields Redevelopment*, NATO Science for Peace and Security Series C: Environmental Security, https://doi.org/10.1007/978-94-024-2278-8_5

both countries have made in turning polluted and derelict industrial, commercial, and military properties, often referred to as brownfields, into more sustainable and resilient places. Based on past research, this paper reviews the evolution of brownfield policy in the two countries, brownfield redevelopment fundamentals, and trends and best practices in infusing sustainability into brownfield redevelopment and measuring sustainability outcomes.

5.2 The Early Evolution of Brownfields Policy in the United States and Canada

Despite an uptick in manufacturing during the Second World War, many cities in the United States and Canada set off on their post-industrial transition from manufacturing to service-based economies soon after. While some manufacturing remained in the central city and along waterfronts, some moved to more spacious and modern suburban locations (along with affluent populations) or entirely offshore. Left behind was a growing inventory of underused, vacant, and derelict industrial properties, along with a plethora of employment and economic challenges. Adding to the problem was an emerging awareness of the potential risks to human health and the environment caused by historic industrial pollution and waste management practices, which further limited interest in the reuse of these sites. Well-publicized disasters in the US, such as a fire at a chemical-waste treatment facility in Bridgeport, New Jersey, that left six dead and 35 in hospital in 1977, and Love Canal, which led President Jimmy Carter to announce a federal health emergency on August 7, 1978, intensified public concern and led to government actions like the Comprehensive Environmental Response, Compensation, and Liability Act (CERCLA aka Superfund) passed by the US Congress in 1980 (Public Law 96-510, HR 7020, SEC 104(a)) [1]. Superfund gave the US Environmental Protection Agency (EPA) and other federal agencies the regulatory authority to respond to a release, or threat of a release, of a hazardous substance or "any pollutant or contaminant which may present an immediate and substantial danger to public health or welfare." The legislation enabled the federal government to recover the costs of cleanup actions from responsible parties or to clean up contaminated sites at their own expense.

Similar efforts were also taking place in Canada, where environmental calamities like the Sydney Tar Ponds in Nova Scotia, where hundreds of thousands of tons of toxic chemical contaminants lingered from over a century of steel making, galvanized national attention and demanded government action. However, as scientists and regulators on both sides of the border worked to understand the science of contamination and to track down those responsible for it, the number of known and potentially contaminated sites in both countries piled up into the hundreds of thousands as deindustrialization fed the inventory. All the while, purchasers, investors, and developers seemed content to leave these previously used and potentially contaminated sites to idle and rust and build a cleaner and greener future in the suburbs.

By the early 1990s, mayors and supporters of older industrialized communities in both countries grew impatient and pushed for reform. In response, governments at all levels and other affected stakeholders began to shift the focus from regulating the risks and liabilities associated with contaminated sites to facilitating their remediation, reuse, and redevelopment. Most potentially contaminated sites were re-branded as "brownfields," and policies, programs, and procedures were developed, tested, and implemented to get people and markets to shift their perception of these sites from liabilities to opportunities. Early work by Simons [2] and the US Government Accountability Office [3] estimated the number of brownfield sites across the US at 450,000 or more, representing about 5% of the land area in 31 large cities studied. Canadian estimates put the number of brownfields at over 64,000 [4], representing 3.3% [5] to as much as 25% of land urban land [6]. Canada's Federal Contaminated Sites Inventory alone lists over 23,078 sites (including 5337 active, 2355 suspected, and 15,386 closed sites) under the custodianship of federal departments, agencies, and consolidated Crown corporations, including military properties [7].

Despite similar goals and trajectories, there were a few differences in the approaches taken in the United States and Canada that may be instructive. In the US, political interest, will, and commitment to brownfield redevelopment intensified in the 1990s, driven by all levels of government and led by the EPA with strong federal agency partnerships [8]. Federal attention and investment initially emphasized economic development projects that aimed to preserve and create businesses, jobs, and taxation structures in urban, suburban, and exurban areas throughout the country, but with a particular emphasis on the inner cities of the Northeast and Midwest most affected by deindustrialization [9]. State and local governments also favoured directing their considerable technical assistance and financial resources to brownfields that had some economic development potential or taxable end use to make up for the losses from post-industrialization, as opposed to just cleaning up sites or using them for alternative activities like housing development [8].

In Canada, even though stakeholders decried the challenges associated with remediation and redevelopment, governments at all levels gave the issue less attention than in the US. While the federal and provincial governments worked cooperatively to develop a common approach to solving the contaminated sites problem in the early 1990s, this relationship began to fizzle as the federal government transferred environmental management responsibilities to provincial governments and focused their efforts on managing federally-owned property and other high-risk sites, and continue to refer to them as contaminated sites as opposed to brownfields [10]. Most provincial governments began to take a more independent approach to formulating legislation and management policies, resulting in variability across the country due to the absence of federal leadership [11]. While political fragmentation did not help advance or sustain brownfield efforts, support did come fortuitously from the fact that nearly every province pursued some form of local restructuring and growth management over the last two decades. Efforts in the large urban centers of Ontario, British Columbia, and Quebec, in particular, to restrict urban sprawl and force growth inward (via infill of previously developed land) and upward (via density) have facilitated the reenvisioning of old industrial districts by local governments and the cleanup and redevelopment of those districts through private-sector redevelopment or public-private partnerships [12, 13]. Directing regional growth is more feasible in Canada than in the US because provincial governments maintain more authority over municipal autonomy. However, while this approach has worked in urban and suburban areas with fast growth and strong real estate markets, smaller peripheral communities with industrial legacies, weaker real estate markets, limited resources, and a plentiful supply of competing agricultural greenfield land continue to face challenges [8, 14].

Therefore, an optimal approach for Ukraine is to marry the two approaches, tightly regulating land supply and restricting urban growth to brownfields, as in Canada, while providing solid and strategic national policy direction and support for brownfields, as in the United States.

5.3 Taking a Brownfield from Assessment to Redevelopment

Given the plethora of potentially contaminated sites in both countries, it became clear early on that identifying, assessing, and remediating the vast inventory would require more resources than could be mustered from governments and evasive responsible parties alone. Government efforts, particularly those of environmental agencies like the EPA, to acknowledge, understand, facilitate, and partner in "redeveloping brownfields" with the private market were pivotal in changing the narrative and approach. At its core, this required a legally secure process for interested parties to take a potentially contaminated brownfield property from assessment to remediation to redevelopment. Several chapters in this special issue describe the Environmental Site Assessment (ESA) procedures used in the US (which are similar to those used in Canada and other nations) to assess and remediate sites to make them safe for reuse. This ESA process is aligned with those governing planning and development and has since become well-understood and normalized. The main challenge to brownfield redevelopment that has remained, particularly when compared with the development of clean property, is how to deal with the additional costs of reclamation and redevelopment (i.e., assessment, cleanup, demolition, time, finance, legal, etc.) and how governments could support a project's overall economic viability.

The European CABERNET Network's A-B-C model [15] is perhaps the most straightforward conceptual model for characterizing the economic viability of different types of brownfields and informing government intervention (see Fig. 5.1). Its basic premise is that brownfield redevelopments are fundamentally affected by land value (vertical y-axis) and reclamation costs (horizontal x-axis). As outlined in Fig. 5.1 below, sites with high land values and low reclamation costs can be profitably developed by the private sector (A), those on the borderline of profitability with

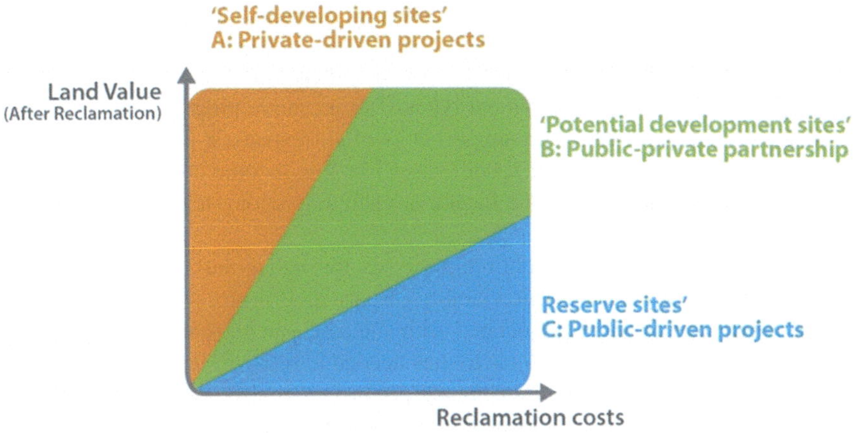

Fig. 5.1 European CABERNET Network A-B-C Model [15]

lesser land value and higher reclamation costs can be reclaimed via public-private partnerships (B), and those with low land values and high reclamation costs are not profitable and can only be redeveloped by the public sector or with significant public funding (C). The straightforward conceptualization summarizes how government agencies began to characterize and categorize brownfield sites, as well as understand the non-financial and financial strategies necessary for supporting their redevelopment.

As mentioned, governments in fast-growing regions have been able to raise the land value of urban brownfields and increase the number of sites that the private market can manage (in Category A) by rezoning lower-value industrial land to higher-value residential and commercial uses, increasing the building density permitted, and, particularly in Canada, blocking sprawl development into cheaper agricultural greenfields in the periphery. All levels of government, particularly in the US, have also been very creative in tools and approaches for supporting public-private partnerships and public-led brownfield redevelopment to address sites in Categories B and C [16]. Research by Paull for the Northeast-Midwest Institute estimates that most brownfield projects in the US (between 55%–80%) involve some form of public subsidy, although the main goal is to use tools that incur the least public expense to achieve the highest returns in employment and municipal property taxes [17].

To assist with economic feasibility, governments can offer incentives that the EPA has grouped into three broad categories: (1) offsets to brownfields financing needs, (2) tax incentives, and (3) direct financing [18].

1. Offsets refer to financially indirect measures, such as technical assistance, process facilitation, and project support. Popular among developers are single points of contact for government assistance that bring together representatives from multiple agencies to work in a coordinated manner. Municipal governments can

also rewrite local plans, target infrastructure spending, rezone, and increase density.
2. Government tax incentives can steer investment capital to weaker market areas and promote specific development types. Tax incentive programs (i.e., credits, abatements, forgiveness) can support brownfield expenses, attract business to particular locations, or foster specific uses like low-income housing. For example, the US Opportunity Zone incentive encourages long-term investments in high-poverty census tracts.
3. Direct financing of brownfield-related costs through grants, loans, and other means improves the economic viability of particular projects. These funds can cover front-end expenses associated with site assessment, remediation, demolition, and other site-preparation activities needed to ready a site for construction. These tools and other public insurance products can also increase the lender's comfort by limiting the risk of potential losses or defaults.

While the discussion above pertains to all brownfields, military brownfields typically have additional layers of oversight, complexity and support due to their public ownership. In the US, the Department of Defence (DoD) generally is the sole owner of military installations. When a military base is no longer needed, its realignment and closure are facilitated through the Base Realignment and Closure (BRAC) program, chartered in 1988. The Defense Base Closure and Realignment Act of 1990 established an independent commission to ensure a timely, independent, and fair process for closing and realigning US military installations [19]. The law included numerous provisions, including the Base Closure Account, designated as the exclusive source of funds for environmental restoration projects. Almost $10 billion has been requested for cleanup, restoration, and caretaker costs for past rounds of base closures.

A comprehensive but concise comparison of the BRAC and brownfield statutes and programs by the EPA identifies their key similarities and differences concerning a variety of factors, including property identification, ownership and history, the extent of environmental contamination, addressing contamination, site assessment, cleanup standards, institutional controls, funding availability and sources, future liability, environmental insurance, community involvement, reuse plans, assistance for the homeless, property transaction processes, and environmental considerations in property transactions [20]. While there are many similarities between military and non-military brownfields, military properties tend to be larger and more contaminated and owned by the DoD, which is responsible for identifying excess property, undertaking cleanup following federal procedures outlined in the National Contingency Plan (NCP), maintaining institutional controls, funding cleanup, and indemnifying purchasers. Military brownfield activities also have heightened expectations regarding public benefit and sustainability given federal requirements for greater public involvement in cleanup and reuse decisions and a federal screening process that considers other federal and public interests (e.g., public benefit, economic development, homelessness assistance) before a site can be made available to the private market. Indeed, in comparing BRAC and state brownfield programs, renowned US legal scholar Joel Eisen praises the BRAC over state programs for its insistence on comprehensive remedial action, emphasis on public participation, and oversight [21].

In Canada, the planning, acquisition, use, and disposal of military properties managed by the Canadian Armed Forces is guided by the federal government's Directive on the Management of Real Property [22]. The disposal of property requires undertaking due diligence concerning its environmental condition, consulting with other federal agencies, Indigenous peoples, the Canada Lands Company (to identify surplus property), and the public, and granting priority to acquire the property to federal, provincial, municipal, and Indigenous groups for beneficial public use before any sale to the private market. Mitigating risks associated with the disposal of federal contaminated sites is guided by the Federal Approach to Contaminated Sites and the Decision-Making Framework (DMF) required by the Federal Contaminated Sites Action Plan (FCSAP), established in 2005 with $3.5 billion in federal funding [23]. The DMF's 10-step process guides federal custodians on (1) identifying suspect sites, (2) historical review, (3) initial testing, (4) site classification, (5) detailed testing, (6) site reclassification, (7) development of a remediation/risk management strategy, (8) implementation of the remediation/risk management strategy, (9) confirmatory sampling and final reporting, and (10) long term monitoring (if required). As in the US, federal efforts in Canada on military sites require more comprehensive remedial action, oversight, and public participation. The most recent DMF also incorporates FCSAP's Sustainability Strategy with guidance on improving the sustainability of site assessment and remediation/risk management activities, integrating climate change matters in the contaminated site management plan, considering sustainable options, and weighing the impacts of contaminated site management on social, financial and environmental factors. Therefore, military brownfields in both countries are required to achieve more significant public benefit and sustainability outcomes than conventional development projects.

5.4 Advancement of Sustainable Brownfield Redevelopment in the United States and Canada

As mentioned, governments in the US and Canada began in the 1990s to shift from enforcement-driven regulators of contaminated site pollution to facilitators of brownfields remediation, reuse, and redevelopment. While managing brownfield risks remained a public health and environmental priority, broader sustainability goals associated with recycling land, removing blight, restoring employment, revitalizing communities, combatting sprawl, and strengthening the urban real estate market were added as critical factors motivating public, private, and community partners. This more collaborative and facilitation-oriented approach immediately began to bear fruit in shifting market interest to brownfields, increasing redevelopment, and supporting community revitalization and local economic development. As such, by the early to mid-2000s, governments, communities, and increasingly confident partners hoped to build on their success and advance brownfield projects even further to achieve an even more comprehensive array of sustainable features and beneficial outcomes. Though this has been commendable, it is important to note that some in the private and public sectors worry that sustainability objectives will add costs to the brownfield reclamation axis of the A-B-C model while adding little to land and project value, thus making it more challenging.

5.4.1 Sustainable Brownfield Redevelopment: Priorities, Realities, and Lessons

Since its introduction in the mid-1980s, sustainability has become a widely embraced approach for integrating environmental needs with economic and social ones in human development. The model widely used to depict sustainable development is the United Nations' three pillars, popularly portrayed by three overlapping circles representing economic, social, and environmental factors, with sustainability achieved when all three are considered [1]. Adopted in 2015, one of the UN's seventeen Sustainable Development Goals relevant to brownfields includes making cities and human settlements inclusive, safe, resilient and sustainable (Goal 11) [24]. In addition, the New Urban Agenda report generated by the Habitat 3 conference explicitly recommends the promotion of planned urban extensions and infill, prioritizing renewal, regeneration and retrofitting of urban areas [25].

In the US, sustainability has always been at the core of the government's environmental mission and brownfield efforts. In the late 1990s, the EPA published two comprehensive reports outlining the Characteristics of Sustainable Brownfield Projects [26] and a Sustainable Brownfields Model Framework [27] to help communities advance sustainability by reusing brownfield assets. It has since produced numerous reports on sustainability programs and partnerships funded [28–30] and other topics for improving the sustainability of brownfield projects through ecological revitalization [31], green remediation techniques [32], urban agriculture [33], green infrastructure [34], environmental justice [35, 36], renewable energy [37], and climate resiliency (see the EPA's 2021 Climate Smart Brownfields Manual for tools, strategies, and a comprehensive list of resources and references) [38].

As a researcher who had already spent over a decade studying this transformation in attitudes, policies, and practices in both the US and Canada, I was keenly interested in examining brownfield projects that aimed to achieve more sustainable forms of redevelopment. In 2009, with support from the EPA, I started tracking two dozen projects across the US, nine of which were already considered sustainability trailblazers in the brownfields arena, along with over a dozen others that the EPA funded to help incorporate sustainability as part of a Sustainable Brownfields Pilot program [28]. I continued to track these projects for over a decade and published a book of case studies in 2021 entitled *Sustainable Brownfield Development: Building a Sustainable Future on Sites of our Polluting Past* [1]. The discussion below draws on the key findings and lessons from the book, with Fig. 5.2 providing a visual summary of the sustainability-oriented features and elements incorporated into the two-dozen case study projects analyzed.

Despite some early resistance to environmental sustainability features at brownfields due to their perceived costs and low returns, most sustainable brownfield projects studied incorporate the development of green space or habitat (21 case study projects) and green infrastructure (i.e., stormwater management) (19). These have become popular in private, public, and nonprofit-led projects because they provide a green aesthetic to erase past blight and attract use while performing ecological and environmental infrastructure functions. More than half of the projects studied include green building design elements (16), energy efficiency, conservation or generation (16), and alternative transportation infrastructure for public transit, walking, and cycling (15) aimed at alleviating costs and improving access. Other standard

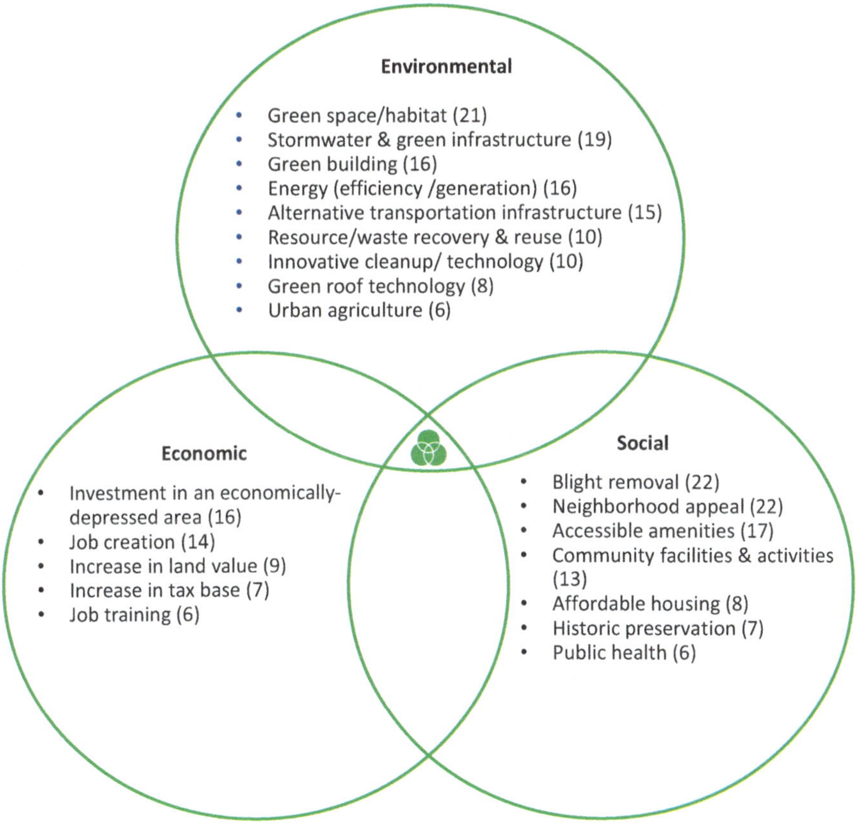

Fig. 5.2 Summary of sustainable brownfield redevelopment features and amenities (number of case study projects in brackets) [1]

green practices include green roofs (8), resource/waste recovery and reuse (10), innovative cleanup/technology (10), and urban agriculture (6), the latter three having important considerations for dealing with buildings and communities affected by military activities. An important lesson learned from the projects examined is the importance of sustainability and green design guidelines, like the LEED (Leadership in Energy and Environmental Design) rating system, which provides helpful direction on rebuilding sustainably versus conventionally. Indeed, more and more municipalities like Toronto and Boston have established voluntary and mandatory green building standards that apply to all development activities. Some challenges experienced include conflict of sustainable practices with planning and building policies, high technology costs, difficulties incorporating green components into building design, and issues related to procurement and financing. Many of those involved in these projects emphasized governments' critical role in promoting, supporting, and helping fund green practices.

From a social and community development perspective, most projects strove to remove blight (22), enhance neighbourhood appeal (22), and improve the availability and access to amenities (e.g., parks, trails, paths) (17). It is here where planning and public health can help lay out a hopeful vision for an area devastated by industrial decline or war. Other social features included the provision of community-oriented facilities and activities (e.g., community centers, educational workshops) (13), affordable housing (8), historical preservation (7), and public health improvements (6). Projects led by public and nonprofit groups are particularly effective at emphasizing cultural goals that celebrate local history, increase opportunities for social interaction and exchange, and tackle complex social and environmental justice issues related to demographics (i.e., race, income, crime), housing, and health.

As for economic sustainability, about two-thirds of the projects studied stressed investment in economically depressed areas (16) and job creation (14), and about a third also noted using redevelopment to raise land values (9), increase the local tax base (7), and add job training opportunities (6). Emphasizing tax and employment generation helps justify public investments, although linking jobs to local residents is also important. Challenges typically include limited funding, high contamination costs, and low buyer demand. Overall, the economic value of sustainability is still tied mainly to conventional economic objectives. While there seems to be a hope that addressing conventional economic objectives will tackle the more complex problems affecting communities surrounding many brownfields, more projects in the US, Canada, and Ukraine would gain from explicit sustainable economic targets for the surrounding community (e.g., raising minimum wage levels, local employment and ownership, health care, employment density, and workforce training and development).

In addition to their sustainability features, other aspects of the redevelopment process that offer lessons for Ukraine are as follows. Regarding motivation, visioning, and leadership, the research undertaken in the US reveals that sustainable brownfield redevelopment projects continue to be driven mainly by a focus on stimulating local development and making brownfield sites more attractive through blight removal, new construction, adaptive reuse, and greening. Sustainability, green building, and environmental justice considerations on their own are not primary motivators as of yet, so it is still necessary to integrate them. Public leadership was, and continues to be, essential for advancing sustainability in brownfield redevelopment. In the projects examined led by private and public partners, sustainability was a response to a government objective or requirement, or government incentives seeded the sustainability ideas. Sustainability actions are also facilitated when early visioning activities address the goals of multiple stakeholders and community members, which also helps form a coalition of supporters. That said, projects face concerns and barriers that are standard to brownfield reclamation, so proponents of sustainability need to show awareness of and support for the extra costs and any loss of opportunity associated with sustainability measures.

With regard to location, market, and reuse, understanding the local real estate market is essential for sustainable brownfield redevelopment, particularly in secondary markets near and outside the downtown core. Projects in these markets may have more difficulty attracting tenants/development, leading some to be delayed and others to change use. Redevelopment projects must be mindful of the market viability of a location and the planned new uses (e.g., industrial, retail, residential, office) and not bow down to political pressure and interference. Unless the proposed use is public, the government should target their support to private and nonprofit projects that are economically and operationally viable to realize both brownfields and sustainability objectives.

Government, particularly at the local level, plays a vital role in land acquisition and assembly of privately and publicly-owned parcels, which provides leverage for encouraging sustainable reuse. Sustainable brownfield redevelopment benefits immensely from being accompanied by a comprehensive vision and plan, a detailed roadmap for undertaking it, and a viable catalytic project to get it started. Visions and plans work best when they incorporate community input, tie into a master plan for the area, and contribute to achieving multiple stakeholder and community objectives. Planning to advance sustainable innovation can also benefit from the support and technical expertise of upper levels of government, national/international consulting firms, universities, and professional associations.

With regard to site preparation (remediation, building rehab, infrastructure), earlier sections of this chapter have emphasized the importance of clear regulatory standards and procedures for developers and environmental consultants to follow to take sites from polluted to suitable for reuse. It is also crucial that governments make funding available for infrastructure and the cleanup of public and private spaces. Cleanup must employ conventional approaches and technologies, as well as consider more sustainable methods that require less material removal (e.g., through capping or using institutional controls to manage them in place) or use innovative techniques (e.g., phytoremediation and bioremediation) that allow nature to attenuate low-market areas over time. Resource recovery and adaptive reuse are options for many older buildings, not just those with heritage value, but the decision to reuse, deconstruct, or demolish must still be based on sound economic principles.

A lesson from the sustainable brownfield projects examined, particularly those that would be placed in CABERNET's B category, is that public funding must be made available, pursued, and creatively assembled from numerous sources to address the many cost barriers that impede brownfield redevelopment and sustainability elements related to green development, affordable housing, and other community programs and activities. Municipal governments will likely need to assume considerable financial risk to advance brownfield projects in weaker markets but can use this to leverage environmental and social benefits. Besides greening, sustainable visions and elements are most often seeded, mandated, or core to the mission of public/nonprofit proponents instead of being in high demand by the private market and consumers. As learned from the analysis of US projects, redevelopment projects should not be given sustainability funding in the hopes of becoming financially viable. Instead, economically viable projects should be given sustainability funding in the hopes of becoming redeveloped sustainably.

Overall, strong inter and intra-governmental cooperation, public-private partnerships, and public-non-profit partnerships are vital for the success of sustainable brownfield redevelopment efforts. Robust stakeholder participation, which includes representation from affected communities, is an essential element in all of the projects studied, along with opportunities for public consultation in general.

5.4.2 Post-project Evaluation of Sustainable Brownfield Redevelopment

While most policymakers and developers typically move on to their next project once construction and sales are complete, it is vital to share the stories of successful projects once they are completed, celebrate best practices, and track long-term progress toward sustainability over time to ensure environmental, social, and economic objectives and interventions are achieved and impactful. Award programs like the US Phoenix Awards and Canadian Brownie Awards have recognized extraordinary practitioners and sustainability-leading brownfield projects in both countries for over 25 years. Numerous case studies of award winners and other best-practice projects have also been published by relevant government, university, non-profit, and building industry partners. However, despite their importance, longer-term and more in-depth assessments of sustainable brownfield projects post-redevelopment are few [39].

One example of such a project is the Menomonee Valley Benchmarking Initiative (MVBI), which I co-initiated and co-led for over a decade as a professor at the University of Wisconsin-Milwaukee with the Sixteenth Street Community Health Center. The core objectives of the MVBI were: (1) to raise awareness in the community regarding the current condition of the 565-hectare Menomonee Valley in Milwaukee, Wisconsin (the state's largest brownfield) and the progress made towards its revitalization, which started in the late 1990s and continues to this day; (2) to create an information clearinghouse on data related to environmental, economic, and social indicators; (3) to promote the principles of sustainability in an urban context by exploring issues and assembling data in a more holistic manner that considers economic, environmental and social concerns; (4) to generate a practical synthesis of the raw data for the benefit of a wide variety of users; and (5) to stimulate research interest in the Valley as a complex laboratory for studying urban environments (for a comprehensive overview of the MVBI process and history, see De Sousa 2016) [40].

In 2001, Indicator Work Group meetings focusing on social, environmental, and economic performance were held with stakeholders to determine key "issues of concern" for the Valley and to select specific "indicators" for investigating sustainability issues. Sustainability reports were generated in 2003, 2005, and, most recently, in 2013. The latest report includes almost 40 benchmarks (Economic 4 issues and 19 benchmarks, Community 4 issues and 11 benchmarks, and Environmental 4 issues and 9 benchmarks). The 2013 report commences with an overall introduction to the Valley and the MVBI and includes maps of the study areas. Indicator analyses are then sorted into Community, Environment, and Economy sections, each commencing with an introductory page that highlights the most important results from the section and presents an index of the issues and indicators [41]. The analysis of each indicator addresses three fundamental questions: (1) what has been measured? (Measurement) (i.e., benchmark, sources of data, and methodological approach); (2) why is it important? (Importance) ((i.e., explains the indicator's role in achieving sustainability); and (3) how are we doing? (Analysis) (i.e., describes the performance of each indicator). The analysis of each indicator is summarized on a single page, while tables, figures, and/or maps illustrate the results. Following the indicator analyses is a section describing catalytic projects built in the Valley, photos of buildings, landmarks, and other Valley features, and an Appendix with various demographic data tables.

Measuring and tracking sustainability outcomes allows a government or organization to assess and highlight the outcomes of their redevelopment and identify issues that still require attention and investment. For example, below are some key sustainability highlights from the 2013 Menomonee Valley study tracking a decade of activity.

- Environmental indicators reveal that the natural environment in the Valley improved with better air quality, an increase in porous ground and vegetation cover, growing bird populations, and a slight increase in woody plant species. Unfortunately, water quality and aquatic IBI have not improved despite the creation of new stream and environmental corridors.
- Community indicators reveal increases in population, housing units, public art, recreation venues, events, and a reduction in child lead poisoning rates and teenage pregnancy. While there has been an overall increase in owner-occupied homes, the study area has witnessed a decrease in the owner-occupancy rate. Unfortunately, there has been a slight decline (−<1%) in household income and a slight rise in crime.
- Economic indicators reveal that the Valley has witnessed an increase in occupied industrial space, accessibility, and labour market growth, but a decline in the

proportion of employees residing in the surrounding area and the provision of health insurance. There has been little change in local ownership, sales and expenditures, and employer interest in the Valley.

While the MVBI data gathering, analysis, and writing process were very extensive, complex and somewhat costly, the joint coordination of the project by a university and a local nonprofit made it possible to access researchers, students, and nonprofit stakeholders involved in specific issues (e.g., birding, crime, greenspace), in addition to various government departments at the local and state level. Overall, efforts to track sustainability outcomes have added to the holistic efforts by multiple stakeholders to support the Menomonee Valley's sustainable redevelopment, helping to make it one of the US's most widely recognized, awarded, and celebrated sustainable brownfield projects [42].

5.5 Concluding Remarks

The incremental advancement of brownfield redevelopment efforts in the US and Canada offers practical examples and lessons for those in Ukraine struggling to envision a better future for their war-ravaged communities. While the country is vast, and one could build new cities on clean land, repairing and redeveloping brownfields affected by war and pollution will be a testament to Ukraine's resolve, resilience, and commitment to its people and history. As with the phoenix in ancient Greek mythology, for which the US brownfield awards are named, the reuse of these sites symbolizes immortality, resurrection and life after death. The hope presented in this chapter is that the new redevelopment projects, like the new phoenix that springs from the ashes of its predecessor, will epitomize sustainable best practices and symbolize the modern future that Ukrainians long for and deserve.

References

1. De Sousa C (2021) Sustainable brownfield development: building a sustainable future on sites of our polluting past. Routledge, New York
2. Simons R (1998) How many urban brownfields are out there?: an economic base contraction analysis of 31 US cities. Public Works Manag Policy 2(3):267–273
3. US Government Accountability Office: Economic Development Administration (2005) Remediation activities account for a small percentage of total brownfields grant funding. US Government Accountability Office, Report GAO-06-7, Washington, DC
4. ECO Canada (2007) Who will do the cleanup? Canadian labour requirements for remediation and reclamation of contaminated sites 2006–2009. Environmental Careers Organization Canada, Environmental Labour Market Research, Calgary
5. De Sousa C (2006) Urban brownfields redevelopment in Canada: the role of local government. Can Geogr 50(3):392–407
6. Benazon N (1995) Soil remediation: a practical overview of Canadian cleanup strategies and commercially available technology. Hazard Mater Manag Fall:10–26

7. Government of Canada (2024) Inventory of Federal Contaminated Sites Homepage. https://www.canada.ca/en/environment-climate-change/services/federal-contaminated-sites/inventory.html. Last accessed 2024/04/24
8. Adams D, De Sousa C, Tiesdell S (2010) Brownfield development: a comparison of North American and British approaches. Urban Stud 47(1):75–104
9. Bartsch C (1996) Paying for our industrial past. Commentary Winter:14–24
10. Harrison K (1996) Passing the buck: federalism and Canadian environmental policy. University of British Columbia Press, Vancouver
11. De Sousa C, Ridsdale R, Lima I, Easton M (2018) The state of brownfields in Canada: renewing Canada's National Brownfield Redevelopment Strategy. Report prepared for the Canadian Brownfields Network. Brownfields Research Lab at Ryerson University, Toronto
12. De Sousa C (2002) Brownfield redevelopment in Toronto: an examination of past trends and future prospects. Land Use Policy 19(4):297–309
13. De Sousa C (2017) Trying to smart-in-up and cleanup our act by linking regional growth planning, brownfields remediation, and urban infill in Southern Ontario cities. Urban Plan 2(3):5–17
14. De Sousa C, Ridsdale DR (2021) An examination of municipal efforts to manage brownfields redevelopment in Ontario, Canada. Can J Urban Res 30(1):99–114
15. Ferber U, Grimski D, Millar K, Nathanail P (2006) Sustainable brownfield regeneration: CABERNET Network Report. Land Quality Management Group, University of Nottingham, Nottingham
16. De Sousa C, Carroll AMM, Whitehead S, Coffin S, Heberle L, Hettiarachchi G, Martin S, Sullivan K, Van Der Kloot J (2023) A review of brownfields revitalisation and reuse research in the US over three decades. Local Environ 28(12):1629–1654
17. Paull E (2006) Using tax increment financing for brownfields redevelopment. Northeast Midwest Economic Review, pp 5–8
18. US EPA (2007) Financing brownfields: State program highlights. US EPA Office of Solid Waste and Emergency Response Report EPA 560-F-07-252, prepared by ICF International, Washington, DC
19. US Department of Defense (2022) DoD Base Realignment and Closure, BRAC Rounds (BRAC 1988, 1991, 1993, 1995 & 2005), Executive Summary, Fiscal Year (FY) 2023 Budget Estimates Program Year 2023. Department of Defense, Justification Data Submitted to Congress, April
20. ICMA (International City/County Management Association) (2006) BRAC and Brownfields: Different Means to the Same End. https://icma.org/sites/default/files/8593.pdf
21. Eisen JB (2008) Brownfields and BRAC: a surprising compatibility. William Mary Environ Law Policy Rev 32:403–425
22. Government of Canada (2012) Directive on the management of real property. Treasury Board of Canada, Ottawa
23. Government of Canada (2018) Federal Contaminated Sites Action Plan (FCSAP): decision-making framework (DMF). Environment and Climate Change Canada, Ottawa
24. United Nations General Assembly (2015) Transforming our world: the 2030 agenda for sustainable development. United Nations, New York
25. United Nations General Assembly (2016) New urban agenda: Quito declaration on sustainable cities and human settlements for all. United Nations, New York
26. US EPA (1998) Characteristics of sustainable brownfields projects. Office of Solid Waste and Emergency Response report EPA500-R-98-001, Washington, DC
27. US EPA (1999) Sustainable brownfields model framework. Office of Solid Waste and Emergency Response report EPA500-R-99-001, Washington, DC
28. US EPA (2009) Building a sustainable future: a report on the Environmental Protection Agency's Brownfield's Sustainability Pilots. Office of Brownfields and Land Revitalization report EPA-560-F-09-500, Washington, DC

29. US EPA (2012) Partnership for sustainable communities: three years of helping communities achieve their visions for growth and prosperity. Office of Sustainable Communities report EPA 231-R-12-001, Washington, DC
30. US EPA (2014) Partnership for sustainable communities: five years of learning from communities and coordinating federal investments. Office of Sustainable Communities report EPA 231-R-14-004, Washington, DC
31. US EPA (2009) Ecological revitalization: turning contaminated properties into community assets. Office of Solid Waste and Emergency Response report EPA-542-R-08-003, Washington, DC
32. US EPA (2008) Green remediation: incorporating sustainable environmental practices into remediation of contaminated sites. Office of Solid Waste and Emergency Response report EPA 542-R-08-002, Washington, DC
33. US EPA (2011) Brownfields and urban agriculture: interim guidelines for safe gardening practices. Region 5 Superfund Division report EPA 560/S-11/001, Washington, DC
34. US EPA (2013) Implementing stormwater infiltration practices at vacant parcels and brownfield sites. Office of Solid Waste and Emergency Response, Office of Water report EPA 905-F-13-001, Washington, DC
35. US EPA (1996) Environmental justice, urban revitalization, and brownfields: the search for authentic signs of hope. National Environmental Justice Advisory Council Waste and Facility Siting Subcommittee report EPA 500-R-96-002, Washington, DC
36. US EPA (2016) National Environmental Justice Advisory Council 20-year retrospective report (1994–2014). Office of Environmental Justice report EPA-300-R-16-002, Washington, DC
37. US EPA (2012) Handbook on siting renewable energy projects while addressing environmental issues. Office of Solid Waste and Emergency Response, RE-Powering America's Land Initiative, Washington, DC
38. US EPA (2021) Climate Smart brownfields manual. Office of Land and Emergency Management report EPA 560-F-21-002, Washington, DC
39. Berman L, De Sousa C, Linder T, Misky D (2012) From blighted brownfields to healthy and sustainable communities: tracking performance and measuring outcomes. In: Hula RC, Reese LA, Jackson-Elmoore C (eds) Reclaiming brownfields. Ashgate, London, pp 311–340
40. De Sousa C (2015) Evaluating progress towards sustainable development: a critical examination of a decade-long university-non-profit collaboration. Curr Opin Environ Sustain 17:10–21
41. Tilidetzke, J., De Sousa, C. and Gramling, B.: The Menomonee Valley state of the Valley report: Ten years benchmarking change in the Valley 2003–2013. Report prepared for Menomonee Valley Benchmarking Initiative, UW-Milwaukee Center for Urban Initiatives Research, Ryerson University, and the Sixteenth Street Community Health Center, Milwaukee (2013)
42. Woodward C (2016) How Milwaukee shook off the rust. Politico Magazine, August 18

Chapter 6
Public Health Considerations of Typical Contaminants at Military Brownfields Sites: Part 2, Estimation of Health Benefits of Brownfield Remediation in Case Studies

Serap Erdal

Abstract This manuscript provides an overview of the health benefit estimates related to Brownfield development as reported in scientific literature. Although available literature provides evidence that there are public health benefits for cancer and non-cancer effects and health risk reduction associated with Brownfield remediation, there is uncertainty due to availability of scant information on numerical estimates of benefits or benchmarks associated with Brownfield cleanups based on risk-based estimates utilizing pre- and post-cleanup site investigation data or epidemiological data or community health data. Based on this limitation, we advocate that the regulatory agencies (e.g., Environmental Protection Agency (EPA) in the U.S.) begins to lay the foundation and to establish the infrastructure needed to be able to assess public health implications of Brownfield redevelopment. We also call for funding for research at the local and national level in North America, Europe, and elsewhere that leads to better understanding of the estimates of health benefits and measurable outcomes of community-level public health related to Brownfield cleanups to reduce the knowledge gap in this area and stimulate brownfield development in socio-economically disadvantaged neighborhoods.

Editors' Note: This two-part manuscript represents two presentations on the public health considerations of contaminants at military Brownfield sites. Part 1 was written by Dr. Laurel Berman, who provides an overview of military Brownfields in the United States, highlighting contaminants found at military Brownfields, and adverse health effects associated with exposure to contaminants. In Part 2, Dr. Serap Erdal provides a summary of estimates of public health benefits associated Brownfield remediation through case examples and studies reported in the literature and by Dr. Erdal's research team. This presentation was made by Dr. Serap Erdal during the NATO Advanced Research Workshop (ARW) on "Achieving Sustainability in Ukraine through Military Brownfields Redevelopment" in Oradea, Romania on December 18–21, 2023.

S. Erdal (✉)
University of Illinois-Chicago, School of Public Health, Chicago, IL, USA
e-mail: erdal@uic.edu

C. Morar et al. (eds.), *Achieving Sustainability in Ukraine through Military Brownfields Redevelopment*, NATO Science for Peace and Security Series C: Environmental Security, https://doi.org/10.1007/978-94-024-2278-8_6

Keywords Brownfields · Military brownfields · Public health · Risk · Clean-up · Remediation · RBCA

6.1 Background: Estimation of Health Benefits Associated with Brownfield Remediation

There has so far been very limited focus on health outcomes associated with Brownfield redevelopment/remediation in the literature. The scant information available in terms of numerical benchmarks that provide comparable standard benchmarks or measures that signify public health benefits related to Brownfield redevelopment serves as an impediment to advocating for resources and funds for Brownfield remediation, particularly in environmental justice communities. Here, we summarized available literature and our research in this area to perform a state-of-the-science assessment in this topic.

Litt et al. [1] and Litt and Burke [2] attempted to link contamination in Brownfield sites with adverse health outcomes. The authors examined the characteristics of urban Brownfield properties in southeast Baltimore, Maryland, and screened sites for their potential environmental hazards. In addition, demographic and health data were evaluated to profile the social and health status of those in Brownfield communities. The results showed that Brownfields in southeast Baltimore represented a range of historic operations, including metal smelting, oil refining, warehousing, and transportation, as well as paints, plastics, and metals manufacturing. The screening method identified a range of substances associated with these properties, including heavy metals, chlorinated hydrocarbons, and polycyclic aromatic hydrocarbons, all of which are suspected or recognized toxicants, and many of which are persistent in the environment. Spatially, these sites are concentrated in white, working-class neighborhoods in which poverty levels exceed and educational attainment lags behind state and national averages. Moreover, these sites are concentrated in communities in which excess mortality rates due to respiratory disease, cancer, and heart disease exist when compared to the city, state, and national averages. The health analysis showed existence of disparities across Southeast Baltimore communities, including excess deaths from respiratory illness (lung cancer, chronic obstructive pulmonary disease, influenza, and pneumonia), total cancers, and a "leading cause of death" index. Furthermore, there was a spatial and statistical relationship between environmentally degraded Brownfields areas and at-risk communities.

Another study by Wang [3] studied public health impacts of Brownfields redevelopment in Charlotte, North Carolina. Using qualitative and quantitative data, this study found that living close to a Brownfield was significantly related to having a higher incidence of low birth weights, stillbirths, and infant mortality problems, but the relationship disappeared once the effects of socioeconomic status were controlled for statistically. Because Brownfields cleanup did not have a measurable epidemiological effect, the author used an alternative method to measure public health impact by which it was estimated that the Brownfield cleanup program averted approximately 1.87 statistical cancer deaths [3].

Although there has been limited information regarding the increased health risks from Brownfield sites, there has been a fair amount of research conducted to assess the ecological association between a variety of health outcomes and living near hazardous waste sites using GIS analysis. Buzelli and Jerrett [4] interpolated air monitoring data for total suspended particulate matter measured in 20 stations in Hamilton, Canada using GIS techniques. Furthermore, the authors used cumulative proximal exposure (CPE) measure, which integrates distances to exposure sources from a multitude of hazards (including hazardous waste sites) in a geographic location. The monitored data were compared against the proxy CPE estimates. The authors concluded that distance from the sites was a reliable measure for health hazard in environmental justice research and proxy measures instead of actual monitoring data are useful indicators [4].

Miranda and Dolinay [5] employed GIS to target areas of high risk and linked address level childhood lead poisonings in six counties in North Carolina to individual address location. The authors developed a GIS analysis methodology to validate predictive risk models. They concluded that spatial analytic approaches are very useful in elucidating relationships among genetic, environmental, and social factors and are important tools for advancing scientific understanding of neurotoxins and other children's environmental health threats [5]. The GIS method has also been used to link neural tube defects to proximity to hazardous waste sites as well as proximity to toxic release inventory facilities. Using a case control study, Suarez et al. [6] reported that proximity to living near hazardous waste sites (National Priority List Superfund sites) did not lead to increased neural tube defects but living in close proximity to a Toxic Release Inventory (TRI) facility did significantly increase neural tube defects experienced in the study population [6].

In another study, increased serum concentrations of polychlorinated dibenzofurans (PCDD/F) were observed in a cohort of individuals who may have been affected by contamination attributable to a Superfund site in Florida (i.e., wood treating company in Pensacola, Florida) when compared to the general U.S. concentrations as reported by the CDC [7]. The implications of increased blood levels of PCDD in individuals who are exposed to PCDDs/Fs at a Superfund site could also be relevant to communities exposed to the same compounds in Brownfield sites. This study provides biological evidence that communities surrounding Superfund sites may experience increased blood level of chemicals, which are demonstrated to be carcinogenic to humans (IARC 1997), but the increased cancer incidence might not yet be observed in communities.

Erdal and O'Hara [8] estimated reduction in excess cancer risks associated with Brownfield remediation using data from ten Brownfield sites in Cook County, IL. The authors found an excess cancer risk reduction varying from 9.2×10^{-5} to 3.9×10^{-3} across ten sites for adult and children exposures combined throughout lifetime. The most benefit in terms of risk reduction was estimated for a Chicago Transportation Agency site, yielding an excess cancer risk reduction of 5 out of 10,000 exposed for adults. For children, the estimated risk reduction was higher by an order of magnitude, i.e., 3 cases out of 1000 exposed. This site was contaminated with metals, Volatile Organic Chemicals, Benzene, Toluene, Ethyl Benzene, and Xylenes, and Poly Aromatic Nuclear Hydrocarbons. The authors noted that "Although these risk reduction benefit estimates for a given site appear small, their impact for a large urban geographical area with thousands of sites should not be underestimated. If we assume that the risk reduction is 1×10^{-3} (i.e., mean across all sites studied and across adults and children) due to site clean-up and Brownfield redevelopment and exposed population size is 10,000, this will result in 10 avoided cancer cases."

The study by Chilton et al. [9] used the risk-based method employed by Erdal and O'Hara [8] and attempted to estimate the avoided cancer incidences because of Brownfield clean-up. In this study, clean-up of Brownfield sites was found to avert 3 expected cancer incidences in an affected population of 8066 in Charlotte, North Carolina. The authors further adjusted their estimate based on a 5-year relative survival rate of 63% for all cancers combined, resulting in 1.1 statistical deaths averted by Brownfield clean-up. They also estimated that the Brownfield clean-up in Charlotte, NC generated an economic benefit of $7.59 million in public health improvements based on the EPA-derived value of a statistical life of $6.9 million in 2008 dollars [9].

In a more recent study, Sinnett et al. [10] performed an assessment of the evidence for the effectiveness of remediation and redevelopment of contaminated sites in reducing environmental and health risks. The study focused on the redevelopment of contaminated sites to new residential neighborhoods and public or recreational functions. The authors concluded that there is good evidence that remediation via removal, capping, and replacing soil, and planting of bare soils, can reduce concentrations of lead and chromium in blood and urine in children. However, they found more limited evidence (from one study) that remediation of soils can improve cognitive performance in children living near waste dumps. There was also some evidence (also from one study) that sediment dredging could reduce PCB concentrations in umbilical cords in infants. The authors noted that study designs are relatively weak, highlighting the uncertainties in their assessment.

6.2 Conclusions and Recommendations for Assessing Public Health Benefits of Brownfield Cleanup

The summary of the literature provides evidence that there are public health benefits for cancer and non-cancer effects and health risk reduction associated with Brownfield cleanup. However, there are limited numerical estimates of health benefits or health outcomes or benchmarks associated with Brownfield cleanups based on risk-based estimates utilizing pre- and post-cleanup site investigation data or epidemiological data or community health data.

Based on this, we advocate that the regulatory agencies (e.g., Environmental Protection Agency (EPA) in the U.S.) begins to lay the foundation and to establish the infrastructure needed to be able to assess public health implications of Brownfield redevelopment. As the limited data presented above demonstrate, neither the federal EPA nor the state/local environmental agencies in the U.S. currently have a system in place to perform quantitative assessment of public health effects related to site clean-up and redevelopment. Although CDC's Environmental Health Tracking Program and a recent emphasis of environmental justice research have improved our understanding of community level effects, the emphasis on Brownfield redevelopment is lacking.

We propose that an element of the infrastructure development effort should be a short electronic data entry form to be filled in by the site owner and reviewed and authorized by the responsible agency overseeing the site clean-up. This data entry form should include the following information:

- Site full address and its geo-code
- Short site history
- Source of contamination
- Exposure media and pathways of concern
- Maximum and mean concentrations of chemicals detected in each exposure medium of concern prior to site remediation
- Site clean-up levels (e.g., Tier 1 Tiered Approach to Corrective Action Objectives (TACO) in Illinois or similar Tier 1 risk-based cleanup levels developed in the U.S. by other States)
- Identification of chemicals exceeding Tier 1 risk-based clean-up levels for each exposure pathway studied
- Type of remediation performed at the site
- Volume of soil excavated/removed and/or volume of groundwater pumped from the site
- Presence of engineered barriers at the site as "No Further Remediation" (NFR) condition
- Presence of institutional controls at the site as NFR condition
- Site current and future zoning and proximity to residential areas
- Neighborhood characteristics in the form of access to health care, transportation, schools, healthy food sources, parks, recreation activities, walking trails, sidewalks

If a data bank containing the information documented above for, at least, publicly funded sites across the U.S. were to be created and maintained, more robust estimates of risk reduction associated with site clean-up could be developed. This information is likely to inform site developers, federal/state/local agencies, researchers, policy makers about optimum conditions for maximum health risk reduction at Brownfield sites and at other hazardous waste sites. The intersection between energy-efficient and sustainable approaches to Brownfield redevelopment and health risk reduction would be further examined and understood with the availability of these data.

We should emphasize that there are indicators or benchmarks that provide social, public health and biological conditions at a given site (e.g., access to parks, sidewalks, healthcare facilities, fresh food sources; concentrations of biomarkers of chemicals of concern in biological tissues) before and after Brownfield redevelopment. The health or risk-based benefit estimates should be utilized in conjunction with other social and public health-related neighborhood-level indicators to obtain a more reliable and comprehensive assessment of improvement of community health in Brownfield impacted areas.

Lastly, we strongly advocate that regulatory agencies continue to secure funding for the Brownfield Program to clean-up Brownfield sites to revitalize the economic and health well-being of the communities. At the same time, it is essential that research that leads to better understanding of the estimation of health benefits and measurable outcomes of community-level public health impacts should be funded at the local and national level in North America, in Europe, and elsewhere to reduce the knowledge gap in this area and stimulate brownfield development in socio-economically disadvantaged neighborhoods.

References

1. Litt JS, Tran NL, Burke TA (2002) Examining urban Brownfields through the public health "Macroscope". Environ Health Perspect 110(Supplement 2):183–193
2. Litt JS, Burke TA (2002) Uncovering the historic environmental hazards of urban Brownfields. J Urban Health 79(4):464–481
3. Wang J (2008) Public health impacts of brownfield redevelopment in Charlotte, NC. Doctoral dissertation submitted to the Public Policy Program, University of North Carolina Charlotte
4. Buzelli M, Jerrett M (2003) Proximity measures of exposure to geostatistical estimates in environmental justice research. Environ Hazards 5:13–21
5. Miranda ML, Dolinoy DC (2005) Using GIS-based approaches to support research on Neurotoxicants and other children's environmental health threats. Neurotoxicology 26:223–228
6. Suarez L, Brender JD, Langlois PH, Zhan FB, Moody K (2007) Maternal exposures to hazardous waste sites and industrial facilities and risk of neural tube defects in offspring. Ann Epidemiol 17:772–777
7. Karouna-Renier N, Ranga Rao K, Lanza JJ, Davis DA, Wilson PA (2007) Serum profiles of PCDDs and PCDFs, in individuals near the Escambia Wood Treating Company Superfund site in Pensacola, FL. Chemosphere 67:1312–1319
8. Erdal S, O'Hara M (2012) Public health evaluation of Brownfield sites using risk indicators: case study of ten sites in Cook County, IL. Report submitted to U.S. Environmental Protection Agency, Washington, D.C.
9. Chilton K, Schwarz P, Godwin K (2009) Verifying the social, environmental and economic promise of Brownfield programs. Final Report to EPA's Brownfields Training, research, and Technical Assistance Grants and Cooperative Agreements Program, BFRES-04-02
10. Sinnett D, Bray I, Baranyi G, Braubach M, Netanyanhu S (2022) Systematic review of the health and equity impacts of remediation and redevelopment of contaminated sites. Int J Environ Res Public Health 19:5278. https://doi.org/10.3390/ijerph19095278

Chapter 7
Risk Assessment Approaches to Site Remediation: Development of Health-Protective Cleanup Standards

Serap Erdal

Abstract This manuscript provides an overview of the development of risk or health-based cleanup levels or concentrations for chemical contaminants present at contaminated exposure media i.e., air, water, soil, sediment, fish) at Brownfield contaminated sites in general and also in military Brownfields. Military Brownfields are a subset of Brownfields, and they are subject to the same regulatory requirements for site-cleanup including the clean-up to federal (i.e., U.S. Environmental Protection Agency in the U.S.) or state-specific (e.g.., Illinois Environmental Protection Agency in the U.S.) cleanup standards developed using the approaches presented in this manuscript. It is critically important for the sites to be remediated based of health/risk-based clean-up standards for cancer and non-cancer effects to ensure that the health of future residents of the site, including sensitive subpopulations such as children, pregnant women, and elderly, are protected and the site does not pose a health risk to residential, recreational, or occupational users of the site. For military Brownfield sites, the site development per health or risk-based cleanup levels ensures the protection of the health of military personnel, their families, and civilians.

Keywords Brownfields · Military brownfields · Exposure pathways · Health risk · Clean-up · Remediation · RBCA

Editors' Note: This manuscript represents a presentation made by Dr. Serap Erdal Serap Erdal during the NATO Advanced Research Workshop (ARW) on "Achieving Sustainability in Ukraine through Military Brownfields Redevelopment" in Oradea, Romania on December 18-21, 2023, on the risk assessment approaches to site remediation, particularly focusing on development of health-protective cleanup standards for contaminants present at Brownfield sites, including the military Brownfield sites.

S. Erdal (✉)
University of Illinois-Chicago, School of Public Health, Chicago, IL, USA
e-mail: erdal@uic.edu

C. Morar et al. (eds.), *Achieving Sustainability in Ukraine through Military Brownfields Redevelopment*, NATO Science for Peace and Security Series C: Environmental Security, https://doi.org/10.1007/978-94-024-2278-8_7

7.1 Overview of Tiered Risk Assessment Process for Brownfields Remediation in the U.S.

The risk-based approach to site remediation gained momentum in the U.S. after the publication of ASTM Standard E1739-95 guidance for Risk-based Corrective Action (RBCA) at the Petroleum Release Sites in 1995 [1]. The RBCA process is a three-tiered approach where assessment and remediation activities are tailored to site-specific conditions and risks with increasing site-specificity and sophistication. At the end of each tier, the user has the option of continuing to the next tier or pursuing remediation activities based on the tier just completed. The increased costs of performing site assessment activities at the next tier are compared against the potential savings in remediation costs that the less conservative, more site-specific assessment in the next tier may provide. Under this process, if the concentration of the contaminant is below the generic Tier 1 risk-based cleanup concentration, that contaminant is not of concern and no further evaluation is needed for that specific contaminant. However, if the Site concentration is greater than the Tier 1 value, a Tier 2 or Tier 3 site-specific evaluation may be pursued or remediation to Tier 1 value must occur to protect potentially exposed individuals. Since the generic Tier 1 levels determine which, if any, contaminants and environmental media may warrant further evaluation or remediation to protect public health and the environment, they are termed as "Tier 1 Risk-Based Cleanup Levels (RBCLs)" in this manuscript, although the terminology used for generic, or default cleanup concentrations vary from state to state in the U.S. [7].

Each state in the U.S. develops its own state-specific Tier 1 RBCLs for Brownfield site cleanup or remediation within their jurisdiction. For example, the Illinois EPA established and tabulated risk-based Tier 1 Tiered Approach to Corrective Action Objectives (TACO) for contaminated soil and groundwater at Brownfield sites to protect human health and take into account site conditions and land use. This signifies that the site contaminants above the Illinois EPA Tier 1 TACO remediation objectives for soil and groundwater are harmful to human health and the Site must be cleaned up to comply with Tier 1 TACO remediation objectives and protect health of all human receptors at Brownfield sites in Illinois [9]. The U.S. Environmental Protection Agency (EPA) also publishes, and updates Generic Risk-based Clean-up Levels called "Regional Screening Levels" to support site clean-up and remediation across the U.S. [6].

7.2 Overview of Regulatory Health Risk Assessment Steps for Brownfield Remediation

According to the U.S. Environmental Protection Agency (EPA), "Human health risk assessment is an estimation of the nature and probability of adverse health effects in humans who may be exposed to chemicals in contaminated environments, and it is

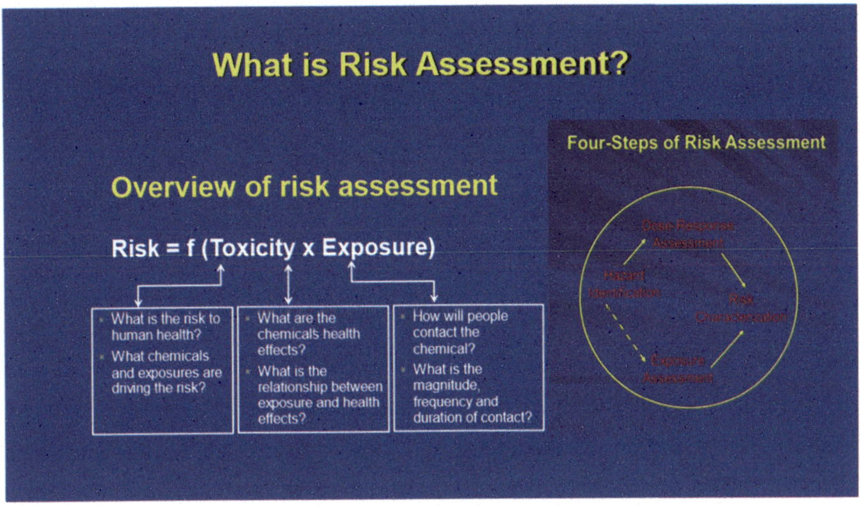

Fig. 7.1 Overview and Steps of Risk Assessment-Basis of Brownfield Clean-up Process

the process to estimate the nature and probability of adverse health effects in humans who may be exposed to chemicals in contaminated environmental media, now or in the future." Human health risk assessment includes the following four basic steps as shown in Fig. 7.1 [5]:

- *Hazard Identification*, which examines whether a stressor has the potential to cause harm to humans and/or ecological systems and, if so, under what circumstances,
- *Dose-Response Assessment*, which examines the numerical relationship between exposure/dose and effects,
- *Exposure Assessment*, which examines what is known about the frequency, timing, and levels of contact with a stressor, and
- *Risk Characterization*, which examines how well the data support conclusions about the nature and extent of the risk from exposure to environmental stressors" (see.

Although risk assessment has been in use by the federal government in the U.S. since the mid-1970s, in 1983, the National Academy of Science (NAS) formalized the risk assessment process by instituting the aforementioned four-step process (consisting of hazard identification, dose-response (or toxicity) assessment, exposure assessment, and risk characterization) and published "Risk Assessment in the Federal Government: Managing the Process" [10]. EPA has integrated the principles of risk assessment from this groundbreaking report into its practices to this day for environmental risk management and public health protection.

As the Venn diagram connecting risk assessment and risk management shows in Fig. 7.2, risk assessment is an integral part of risk management used in Brownfield remediation. Risk assessment scientists communicate the scientific/technical results

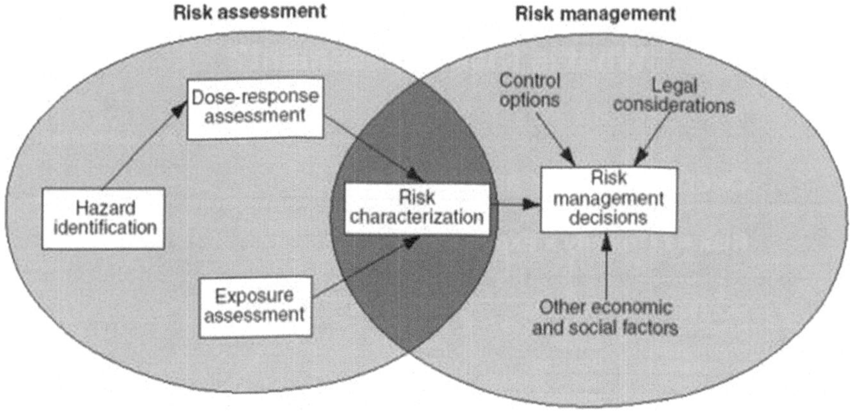

Fig. 7.2 The Interface between Risk Assessment and Risk Management at Brownfield Sites

of the health risk assessment to risk managers, who consider the technical information (i.e., output from risk assessment) along with other non-technical information (cost, economic/technologic feasibility of risk management alternatives, legal considerations, community concerns, etc.) in adopting a risk management solution(s) to mitigate and/or eliminate the exposure/health risk to the exposed population at the Brownfield sites. Any risk assessment and management must accompany effective risk communication to be successful in the real world.

7.3 Development of State-Specific Tier 1 Risk-Based Cleanup Standards in the U.S.

Risk is a function of exposure and toxicity (see Fig. 7.1). While the exposure component of the risk equation provides a measure of average dose (mg/kg-day) received by an exposed individual and is a function of concentration of chemical in exposure medium (C), exposure frequency (EF), and exposure duration (ED), and contact rate (CR), e.g., ingestion rate or inhalation rate) of the exposed individual with the exposure medium, toxicity is a quantitative measure of the ability of a substance to cause harmful health effects (for both cancer and non-cancer health end-points) (see Fig. 7.3). The toxicity component of the risk equation is a toxicity value, i.e., Cancer Slope Factor (CSF) (mg/kg-day)$^{-1}$ for oral (ingestion) pathway or Inhalation Unit Risk (IUR) (mg/m^3)$^{-1}$ for inhalation pathway for cancer effects; and Reference Dose (RfD) (mg/kg-day) for oral (ingestion) pathway or Reference Concentration (RfC) (mg/m^3) for inhalation pathway for non-cancer effects (see Fig. 7.3, [8]). Cancer Slope Factor (CSF) for ingestion pathway (or Inhalation Unit Risk (IUR) for inhalation pathway) is defined as a plausible upper-bound estimate of the probability of a cancer risk per unit dose (or concentration) of chemical over a lifetime and

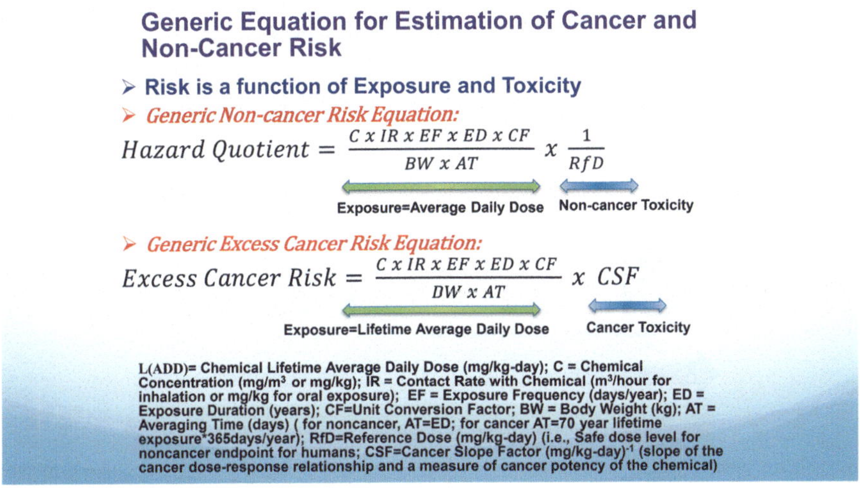

Fig. 7.3 Generic Equation for Estimation of Excess Cancer Risks and Non-Cancer Risks at Brownfield Sites

they are used to estimate lifetime excess cancer risks associated with ingestion (or inhalation) exposure to a carcinogen. The cancer toxicity values, CSFs or IURs, are presented with the scientific weight-of-evidence classification to indicate the strength of evidence that the chemical is a human carcinogen [2–4]. On the other hand, the toxicity values for non-carcinogens, Reference Dose (RfD) for ingestion pathway or Reference Concentration (RfC) for inhalation pathway, represent safe dose (or concentration) level for humans, i.e., an estimate (with uncertainty spanning perhaps an order of magnitude) of a daily exposure to the human population (including sensitive subgroups) that is likely to be without an appreciable risk of deleterious effects during a lifetime. While the non-cancer risk for a single Contaminant of Concern (COC) is termed Hazard Quotient (HQ) (unitless), the cumulative non-cancer risk for multiple noncarcinogenic COCs is termed Hazard Index (HI) (unitless), which is estimated by dividing average daily dose received by an individual by the safe dose (or concentration) [i.e., RfD (or RfC)].

In documenting development of the state-specific risk -based cleanup levels, Illinois EPA methodology is presented here as the U.S. example since the same approach is used across 50 states in the U.S. In estimating Illinois EPA Tier 1 remedial objectives (or clean-up standards or RBCLs), the risk equation is used to solve for the concentration term (rather than the cancer or non-cancer risks in forward risk assessment) by assigning a target risk level to protect exposed individuals/population from cancer and non-cancer health effects (see Fig. 7.4; [8]). In deriving Tier 1 cleanup levels, Illinois EPA selected the following target risk (TR) levels for carcinogens and non-carcinogens to protect public health: TR = 1×10^{-6} (one excess case of cancer over one million exposed) at the point of human exposure for carcinogens; and Target Hazard Quotient (THQ) = 1) for noncarcinogens (i.e., average daily dose

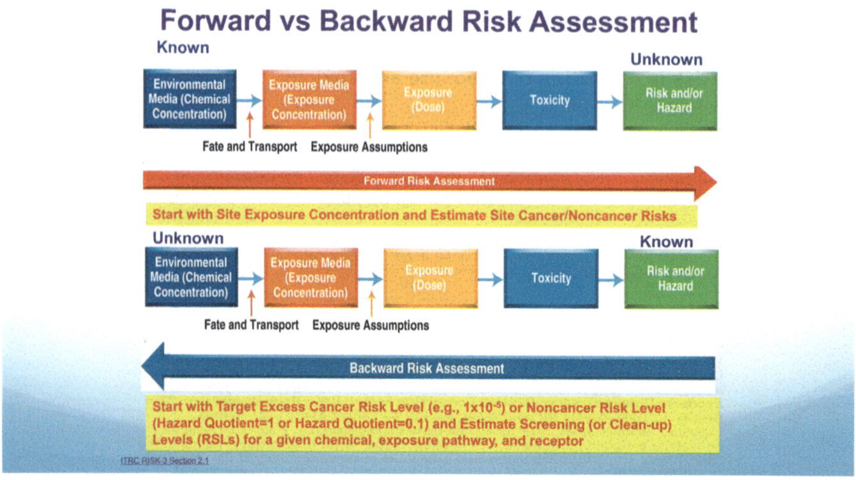

Fig. 7.4 Forward Risk Assessment for Calculation of Cancer and Non-Cancer Risks and Backward Risk Assessment for Calculation Risk-Based Cleanup Concentrations/Levels

is equal to safe dose level). Specifically, the Illinois IEPA's risk-based remediation objectives (or cleanup levels) are derived using the risk equations for different exposure pathways listed in "IL Administrative Code Section 742.APPENDIX C Tier 2 Illustrations and Tables, Section 742.TABLE A SSL Equations". In these equations, the parameters listed in "Section 742.TABLE B SSL Parameters" are used to back-calculate health-protective Tier 1 cleanup (remediation) levels (objectives) for exposure pathways considered under Tier using Illinois EPA's target risk criteria of (for carcinogens, TR = 1×10^{-6} at the point of human exposure; for non-carcinogens, Target Hazard Quotient (THQ = 1). The Illinois EPA Soil and Groundwater Tier 1 remediation objectives for contaminants (e.g., Volatile Organic Compounds, Semi-Volatile Organic Compounds, Heavy Metals) derived using this approach are tabulated under "IL Administrative Code. Part 742-Tiered Approach to Corrective Action Objectives" [9].

This means that if the carcinogenic COCs at the site found to be greater than the risk-based Illinois EPA Tier 1 remediation objectives, the Illinois EPA's target risk criterion of one in one million is exceeded and potential excess cancer risks for the exposed population is of concern at the site. Similarly, for COCs that cause non-cancer effects, if the exposed receptors have average daily dose levels above the safe dose levels for a given COC and an exposure pathway, then the target risk criterion of THQ = 1 is exceeded, signifying that potential non-cancer risks for exposed human receptors are of concern at the site. In both cases, site conditions pose health risks above the regulatory health-protective criteria (i.e., target risk level) set by the Illinois EPA to protect public health. This forms the basis of site clean-up/remediation so that unhealthy exposure conditions at the Site are eliminated.

7.4 Conclusions

Although basing Brownfield remediation, including military Brownfield remediation, on risk-based clean-up levels to protect public health standardizes the clean-up process using scientific principles for exposure and toxicity assessment, the varying risk-based clean-up standards from State to State across 50 States in the U.S. complicates site-cleanup and raises questions on the integrity of the process. For protection of public health, the ultimate goal of this approach is simply to cost-effectively reduce the actual or potential risk posed by specific hazardous contaminants at Brownfield sites to humans who may reside at the remediated site to acceptable risk levels for cancer and non-cancer effects. However, the application of this simple concept to risk-based remediation has become quite complex when implemented by the States, especially in the definition of what level constitutes as the safe exposure concentration that can remain at a site without causing harm to exposed individuals. This is the critical question of "How clean is clean enough?" The answer to this question invariably varies from site to site due to the differing complexity of risk-based remediation from one site to another and also due to differing land-use options (e.g., residential, industrial/commercial, recreational). However, in the absence of national policy and guidance from the EPA, each State has answered this critical question differently for similar sites and exposure conditions and developed its own cleanup criteria using different target cancer and non-cancer criteria, compounding the complexity of Brownfield site remediation across the United States. This question becomes more complex for the clean-up of military sites globally in the absence of policies or standardized approaches used by agencies in Europe and elsewhere in the world. Thus, there is a critical need for the development of standardized risk-based protocol for Brownfield clean-up across the globe to protect public health as the agencies around the world tackle climate change as a serious threat to the environment and public health.

References

1. ASTM (American Society of Testing and Materials) (1995) Standard Guide for Risk-Based Corrective Action Applied at Petroleum Release Sites. E 1739-95. Philadelphia, PA
2. EPA (U.S. Environmental Protection Agency) (1986) Guidelines for Carcinogen Risk Assessment. 51 FR 33992-34003. Available at: https://cfpub.epa.gov/ncea/risk/recordisplay.cfm?deid=54933
3. EPA (U.S. Environmental Protection Agency) (2005) Guidelines for Carcinogen Risk Assessment. 70 FR 17765-17817. Available at: https://www.epa.gov/risk/guidelines-carcinogen-risk-assessment
4. EPA (U.S. Environmental Protection Agency) (2023) Risk Assessment for Carcinogenic Effects. Available at: https://www.epa.gov/fera/risk-assessment-carcinogenic-effects
5. EPA (U.S. Environmental Protection Agency) (2024a) Human health risk assessment. Available at: https://www.epa.gov/risk/human-health-risk-assessment

6. EPA (U.S. Environmental Protection Agency) (2024b) Regional screening levels (RSLs) – generic tables, Tables as of: May 2024. Available at: https://www.epa.gov/risk/regional-screening-levels-rsls-generic-tables
7. Erdal S, Carollo AN (2007) Assessment of health protectiveness of the risk-based soil remediation standards of the Midwestern states. Hum Ecol Risk Assess Int J 13(3):543–573. https://doi.org/10.1080/10807030701341126
8. Erdal S (2023) Risk Assessment Approaches to Site Remediation: Development of Health-Protective Cleanup Standards. Presented at the NATO Conference on Achieving Sustainability in Ukraine through Military Brownfields Redevelopment held at University of Oradea, Romania, December 18
9. IEPA (Illinois Environmental Protection Agency) (2002) Title 35: Environmental Protection Subtitle G: Waste Disposal Chapter I: Pollution Control Board Subchapter F: Risk Based Cleanup Objectives Part 742-Tiered Approach to Corrective Action Objectives (amended in R00-19(C) at 26 Ill. Reg. 2683, effective February 5, 2002). Available at: https://www.ilga.gov/commission/jcar/admincode/035/03500742sections.html
10. NRC (National Research Council) (1983) Risk assessment in the Federal Government: managing the process. The National Academies Press, Washington, DC. https://doi.org/10.17226/366

Chapter 8
Urban Reconstruction in Post-War Ukraine: Development Strategies

Mircea-Sebastian Mancia , Gabriela A. Popoviciu , and Aurora Mancia

Abstract The current paper, based on current events, aims to describe some possible development and reconstruction strategies regarding the restoration of Ukraine's heritage, which was severely damaged by the ruthless attack of the Russian Federation starting two years ago. Russia's aggression against Ukraine resulted in the loss of human lives, the destruction of strategic, industrial, residential, cultural objectives, the contamination of land, the destruction of agriculture. The current war uses both the military use of drones and satellites, but also the fight in trenches, in bunkers and, implicitly, excavations that lead to the destruction of cultural heritage. Ukraine is affected both in the active war zones and in the areas near Kiev. The true extent of the destruction can be determined by the systematic research of the affected locations, which can last a long time after the war ends. Ukraine and Russia are signatories of the Hague Convention of 1954, with the obligation to protect heritage in case of armed conflict. For example, 7 Ukrainian sites can be found in the UNESCO world heritage.

Keywords Systematic research · Cultural heritage · Strategic objectives · Sustainable development · War · Destruction

M.-S. Mancia (✉) · A. Mancia
University of Oradea, Faculty of Construction, Cadastre and Architecture, Oradea, Bihor, Romania
e-mail: mircea.mancia@uoradea.ro

G. A. Popoviciu
University of Oradea, Faculty of Environmental Protection, Oradea, Bihor, Romania
e-mail: gpopoviciu@uoradea.ro

C. Morar et al. (eds.), *Achieving Sustainability in Ukraine through Military Brownfields Redevelopment*, NATO Science for Peace and Security Series C: Environmental Security, https://doi.org/10.1007/978-94-024-2278-8_8

8.1 Introduction

"Ukraine is currently the country with the most mines and unexploded ordnance remnants in the whole world, with a third of the area affected, which is equivalent to seven Switzerlands" [1]. The mined areas are mainly those in the vicinity of the Zaporoje power plant—the largest nuclear power plant in Europe. Agricultural and marshy areas are largely mined, Ukraine being a large agricultural producer. The European Union finances international demining agencies by offering Ukraine vehicles, drones, robots, special suits for these operations. "UN agencies, such as the World Food Program (WFP), the United Nations Organization for Agriculture (FAO), also contribute to demining efforts" [1].

According to the World Bank, the estimated costs for these operations would be over 34.5 billion Euros, at the end of 2022.

8.1.1 Some Statistical Data Regarding the Financial Aid Granted to Ukraine

The war in Ukraine affected cultural heritage, art and tourism. It is estimated that 1.4 million houses in which about 18 million people lived were damaged or destroyed. Water supply systems, electrical networks, services, health and education systems, industrial and infrastructure activities were affected. According to the UN and the Government of Kiev, the necessary sums, through specialized agencies, are around 486 billion dollars, in March 2024. It is estimated that of these sums, 17% should be allocated for housing reconstruction, 15% for transport, 14% for trade and industry, 12% for agriculture, 10% for energy, 9% for social assistance and means of subsistence and 7% for the management of explosion hazards.

The Estimated Reconstruction of Ukraine The reconstruction of Ukraine is and will be a long process estimated by the UN at min. 10 years. The costs change permanently depending on the time of the estimate. In 2023, the European Union launched the intra-judgment program of Ukraine called "Ukraine Facility", for the period 2024–2027.

"The reconstruction of Ukraine is a process that must begin now, not from the moment when the war started by Russia ends," says Werner Hoyer, president of the European Investment Bank (EIB), considered "the financial arm of the E.U." [2]

The EU states will be able to get involved in the reconstruction of Ukraine with the help of local and central Ukrainian authorities; by concluding public-private partnerships, transparency in all projects, financial support and specialized personnel will be ensured.

At the opening of the "Reconstruction of Ukraine and the strategic role of Romania" conference, held in December 2023 in Sinaia, the Prime Minister of Romania said: "I emphasize that Romania acts and will continue to be the promoter of the concerted measures of the Euro-Atlantic community for the resistance of Ukraine and the reconstruction of this countries in the European integration process."

> An essential component of the partnership will be the logistic hub for the reconstruction and assistance of Ukraine. [3]

At the same conference, Ukrainian Prime Minister Denys Shmyhal said: "We suffered unprecedented damage: 257,000 civilian objects were destroyed or damaged, including schools, hospitals, over 30% of buildings and 50% of the electricity network. What was destroyed must be rebuilt. The reconstruction of Ukraine is based on the principle of rebuilding again, better" (see Fig. 8.1).

The head of the Government in Kyiv mentioned that the reconstruction of Ukraine is based on six priorities. The first priority is in the energy field, where the Ukrainian Government wants to ensure and develop the production capacities of nuclear energy, energy from renewable sources and repair and strengthen the transport networks in this area.

The second priority refers to the military field, to the construction of capacities in the defense industry.

The third priority concerns the development of translation markets, the fourth refers to the development of natural resources, the fifth to infrastructure and the development of construction equipment, and the sixth refers to the resilience of public authorities and institutions.

In addition to the authorities from Romania, delegates from Germany, France, and the USA offered their support in the reconstruction of Ukraine, along with 50 other states.

Fig. 8.1 Map of Ukraine on March 25, 2024. (Source: https://en.wikipedia.org/wiki/Russian_invasion_of_Ukraine#/media/File:2022_Russian_invasion_of_Ukraine.svg)

Within the framework of the Eastern Partnership and the Energy Community, the EU promotes ecological reforms, the reconstruction and modernization of Ukraine, materially supporting it together with European financial institutions.

Russia's war of aggression is ongoing: the sums needed for the reconstruction of Ukraine cannot be precisely calculated, but the basic elements for reconstruction and recovery can be designed and conceived.

In January 2023, the multi-institutional platform for the coordination of international donors was launched to support the recovery and reconstruction of Ukraine, which is carried out in a transparent, continuous and responsible manner.

In October 2023, the European Parliament adopted the amendments related to the proposed regulation of the European Parliament and of the Council establishing the Mechanism for Ukraine.

The Regulation of the European Parliament and the Council establishing the mechanism for Ukraine also provides: "(18) Within the new mechanism, investments in the recovery, reconstruction and sustainable modernization of Ukraine should begin urgently to ensure decent living conditions for the Ukrainian population and the reconstruction of critical infrastructure, at the same time ensuring the mitigation of serious environmental damage, as far as possible in a country devastated by war, and the support of Ukraine in the double green and digital transition." [4]

> (20) The medium-term perspective offered by the Plan for Ukraine through a unique instrument should also favor building better by encouraging and helping Ukraine to channel investments and reforms towards a sustainable green, digital economy, contribute to the mobilization to some donors, including from the private sector, where investments should be aligned as much as possible with the Union acquis in the field of climate and environment and should contribute to the implementation of Ukraine's national climate and energy plan. [4]

> (21) Recovery, reconstruction and modernization efforts should be based on the assumption of responsibility – by Ukraine on close cooperation and coordination with countries and organizations that provide support. It is also expected that local organizations and experts from Ukraine will play an important role in this process, fully participating in its design and evaluation. Peer-to-peer cooperation (*inter pares*) and programs integrated into partnerships between cities and regions of the Union and those of Ukraine have already facilitated the delivery of humanitarian aid to Ukraine and thus provide a basis for enrich and accelerate the process of recovery, reconstruction and modernization. [4]

> (32) The mechanism should contribute to the accession to the Paris Agreement and the United Nations Framework Convention on Climate Change, the UN Convention on Biological Diversity and the UN Convention to Combat Desertification and should not contribute to environmental degradation or damage the environment or climate. In particular, the funding allocated in the context of the mechanism should be consistent with the long-term objective of maintaining the global average temperature increase to 2 degrees Celsius above pre-industrial levels and continuing efforts to limit the temperature increase to 1,5 degrees Celsius. This should be consistent with the objective of increasing the capacity to adapt to the negative effects of climate change and to promote resilience to climate change, as well as with the support for the conservation of biodiversity, for the circular economy and the reduction to zero of pollution. Special attention should be paid to actions that create related benefits and fulfill multiple objectives, including for the climate, biodiversity and environment.

> The mechanisms should contribute to mitigating and repairing the enormous environmental damage caused by the Russian aggressors, including the floods caused by the blowing up of the Kakhovka hydroelectric dam and the bombing and mining exploration of a large part of the territory of Ukraine. To this end, the measures supported by the Plan for Ukraine within the mechanism should, as far as possible, contribute to mitigating climate change and adapting to it, to environmental protection, to the green transition, including biodiversity, to addressing the challenges generated by them, and aim to provide an amount that represents at least 20% of the total allocation of the Plan for Ukraine based on the methodology for tracking climate-related expenses provided for in this Regulation. [4]

Urban planning means inventorying the territories in order to adopt the most effective with which to obtain territorial performances with effects in sustainable development.

The effects of the war in Ukraine are felt in the degradation and even destruction of urban spaces. The reconstruction must be carried out in harmony with the architectural style, with the requirements of environmental protection and those of society.

During the period in which the war is still ongoing, analyzes can be made, projects drawn up to outline optimal directions for sustainable development.

The detailed analysis will be possible when all the sites will be inventoried, and the problems of a physical nature, pollution, contamination and the related necessary measures will be taken into account.

The strategic urban development and the revitalization of war-affected areas in Ukraine can be achieved based on preliminary analyzes by processing the existing statistical data in the official databases. The results of such analyzes make up the diagnosis of the economic, social, sanitary, educational, administrative, digital, touristic, environmental situation in administrative territories.

The preliminary analysis will be integrated into the regional, national and European plan detailing the proposals based on technical projects respecting the minimization of the impact on environmental factors.

The president of the parliamentary committee on regional development and urban planning in Ukraine, Olena Shulyak, estimated on December 1, 2023 that more than 170,000 buildings were damaged or destroyed.

Of these, about 20,000 buildings are residential, 3500 are educational institutions, 420 are large and medium-sized companies, 344 are bridges and 25,000 are roads.

Three months after the start of the war in Ukraine, a group of specialists is trying to save the cultural heritage of Ukraine. According to euronews.ro from 31.05.2022, French and Ukrainian specialists are trying to reconstruct the images of some destroyed historical monuments in Kiev, Lvov, Chernihiv and Kharkov [5]. In this sense, 3D scans were made that will be useful to architects and experts in historical constructions for the reconstruction of buildings based on scanned images.

On September 5, 2022, UNESCO published a report on the verification of the destruction of cultural buildings. 186 Ukrainian buildings were checked and it was found that 79 religious buildings, 13 museums, 36 historical buildings, 17 monuments, 9 libraries and 32 buildings for cultural activities were destroyed.

Ukraine has seven monuments inscribed on the UNESCO Heritage List. According to Le Figaro, UNESCO reported in August 2023, a number of 625 partially destroyed cultural sites, 23 completely destroyed sites and 115 destroyed sites without knowing the extent of the damage [6]. By the end of 2023, 5000 cultural sites were destroyed and 341 were damaged 2 UNESCO heritage sites were affected by the Russian bombings: the historical center of the city of Lvov and the Cathedral of the Transfiguration in Odessa (see Fig. 8.2).

UNESCO estimates in February 2024 that 9 billion Euros would be needed just for the restoration of cultural heritage [7, 8].

In order to achieve the objectives related to the reconstruction in the field of residential, administrative, cultural and industrial buildings, a set of mechanisms must be developed to protect foreign investments in Ukraine.

At the level of professional associations, the European Federation of the Construction Industry (FIEC) and the European Federation of European Builders (EIC) have established Task Force Ukraine [9].

According to the press release sent after the establishment "The main purpose of the group is determined by the reconstruction project of Ukraine, and the main points on which the group's activity is focused are coherent actions regarding the implementation of the recovery plan of Ukraine, the establishment of a permanent dialogue with the EU and other structures involved in the analysis/evaluation and establishment of guidelines to be followed, centralization of collected information." [9]

Fig. 8.2 The Transfiguration Cathedral in Odessa damaged. (Source: Oleksandr GIMANOV / AFP / Profimedia)

An element of first necessity is the assessment of needs and costs in order to establish the solutions that will be applied in design and execution.

The collaboration with the local authorities in Ukraine is an important point in these actions. There were meetings with the local authorities in Irpin who presented a plan for the reconstruction of the city.

FIEC also met with the representatives of the Confederation of Builders of Ukraine (CBU) who presented the requirements for personnel specialized in demolitions, construction equipment for dismantling and waste disposal, construction material factories.

We are working on the basis of the local plans of the Ukrainian authorities of the Recovery Plan (UPC) 2022, developed by the Government of Ukraine.

Ukraine's recovery and development plan is designed in 3 stages [10]:

1. Short-term resilience/recovery – until the end of 2022 – for humanitarian aid, restoration of water supply, winter heating capacity, bridges ($60–65 billion).
2. Medium-term recovery 2023–2025 would mean the reconstruction of infrastructure, schools, hospitals, homes (300 billion USD).
3. Modernization and sustainable growth 2026–2032, related to long-term investments and reforms, leading to eventual accession to the EU (400 billion USD).

A large part of these estimated investments is represented by infrastructure, housing, energy, and the environment.

So, environmental recovery means USD 20 billion and implies:

- Energy independence and the transition to the green economy (130 billion USD).
- Investment in logistics infrastructure (120–160 billion USD).
- Reconstruction of housing and regional infrastructure (150–250 billion USD).
- Investments in social infrastructure (35 billion USD).

These documents developed in 2022 are updated with the reconstruction needs of 2024.

The Council of Europe has published documents in which it says that the EU and its citizens are fully in solidarity with Ukraine and the Ukrainian people [11]. The EU has taken unprecedented measures to support Ukraine and its people [11]. These measures include:

- economic and financial assistance,
- humanitarian aid,
- support in matters of civil protection,
- military support,
- receiving refugees through the EU's temporary protection mechanism,
- support for the investigation and prosecution of war crimes,
- support for the reconstruction of a democratic Ukraine [11].

As a result of the war launched by Russia against Ukraine, soil, water and air pollution have devastating effects with long-term consequences [12, 13].

According to the international analytical center Globsec, 174,000 km^2 of land is contaminated. They represent about 30% of the territory of Ukraine, being mined in a larger area than in Syria and Afghanistan [5].

The mines were placed both on the front line and in agricultural lands, forests or roadsides. The most mined seem to be the Kharkov and Kherson regions recaptured by the Ukrainians.

The explosions in Ukraine and the fires that follow them are factors of air pollution (see Fig. 8.3). The air in Romania, especially in Moldova, can be affected if the movement of air currents is in the western direction, towards our country [13].

In March 2024, the number of pyrotechnic units is 340 with 1900, and continues to grow. The information was taken from the statement of Serhiy Reva, head of the department of humanitarian organization and demining of the State Emergency Service of Ukraine.

For the reconstruction of the localities destroyed by the war, the inventory and mapping of the affected buildings and surfaces must be carried out in safe conditions without endangering the lives of the civilians and specialists involved [14].

Companies from Ukraine get involved with those from abroad for the preparation and start, when possible, of these activities.

An architecture firm has created an online platform for the reconstruction of buildings damaged or destroyed by the war [15]. The goal is to preserve the organic image of Ukrainian villages, according to the architect Balbek Bureau.

Citizens are advised how to rebuild their houses by receiving on the platform plans, drawings in .pdf file format, with isometry, facade plan, technical description, but also recommendations regarding the use of construction materials [15].

The annual report of the EU for 2023 regarding the situation of regions and cities has clear specifications regarding Ukraine and its reconstruction [14, 16].

Fig. 8.3 The airfield in Vasilkiv (Kiev region) was completely destroyed by Russian troops. (Source: https://www.radiooltenia.ro/razboi-in-ucraina-aerodromul-din-vasilkiv-regiunea-kiev-a-fost-distrus-complet-de-trupele-ruse/)

The EU should use the potential, the expertise of its regions and cities to anchor the reconstruction, reform and European integration of Ukraine at the local level and continue to support decentralization. The Mechanism for Ukraine provides resources and funding for this purpose. [16]

The European Alliance of Cities and Regions for the Reconstruction of Ukraine, coordinated by the CoR, contributes to the strengthening of local and regional capacities in Ukraine, stimulating *inter pares* cooperation, helping local authorities in Ukraine to familiarize themselves with the Community acquis and providing advice on putting in application of the decentralization process. [17]

The EU should provide regions and cities with a coherent and consolidated framework for sustainable development, which aligns with global climate and sustainability agendas. This should take into account the transformative challenges facing regions and cities, because putting in application to reflect more the local realities.

EU institutions must work with all levels of government to close financial gaps, providing direct funding and concrete guidance, introducing green budgets and better linking the EU Green Deal with EU economic governance. [18].

8.2 Conclusions

The strategies for regeneration and sustainable development reflect the stage in which the studied region is.

In the conditions of globalization and EU expansion, the urban development of Ukraine is conditioned by the completion of the war of aggression caused by Russia, the mobilization of financial resources, the coherence of urban planning projects, and the professionalism of specialists in the affected fields.

Urban regeneration in Ukraine means the reconsideration of spaces for a superior conformation, a positive aesthetic change with the preservation of identity, of cultural dimensions, respecting the European norms of environmental protection and urban development.

It is the development of economic, social, health, and relaxation, urban remodeling opportunities with the conclusion of administrations, investors, architects and all the factors involved.

Reconstruction in the case of Ukraine means measures that will have to be related to its gradual accession to the EU, to the participation in the initial "Green Deal" by the EU, with the achievement of environmental, energy efficiency and sustainable development objectives.

The war in Ukraine produced effects not only in this country, but all over the world.

So:

- The EU mobilized and responded firmly to the geopolitical realities resulting from this conflict: the cohesion of the EU member states and the USA was proven.
- There were disruptions in the energy supply, resulting in increased inflation.
- The post-pandemic recovery was slowed down.
- The "Green Economy" projects have been slowed down.

- The bombing of some strategic objectives in Ukraine, such as: the Zaporoje nuclear power plant or warehouses with dangerous substances can affect the climate of the whole of Europe.

The EU must develop a strong eco-region policy, based on the reality of the area—to be able to help places and citizens; the efforts to reconstruct Ukraine together with its citizens will be intensified.

Summing up, national programs must focus on solving the most pressing environmental problems. National and regional and why not, transnational authorities, must formulate policies that provide economic incentives for environmental activities. National Environmental Plans must be developed, which include mandatory quantitative protection targets, but also qualitative objectives. The new situation requires new types of regulations, new administrative and enforcement procedures. The role of non-governmental actors and local and regional administrations in the application of the new environmental policy must also be strengthened.

Environmental impact assessment must be mandatory for large development projects, as they provide clear signals for state, local and regional administrations regarding how to proceed.

The capacity of national, regional, transnational and local administrations must be strengthened to be able to implement EU environmental policies and standards. Therefore, it is considered necessary to modernize the administrative structures, so that they are able to cope with the assistance offered by the EU. In the same spirit, a great educational effort and local initiatives are needed to change people's attitude towards their environment [19].

The highest priority in the field of environmental protection, as the primary objective of cohesion policies, must be given to the ecological rehabilitation and reconstruction of degraded areas. In turn, the elimination of the biggest sources of pollution must be a priority objective, a context in which the development and expansion of treatment plants in large cities is one of the most important tasks.

In certain situations, privatization can be coupled with environmental rehabilitation by including pollution reduction obligations in contracts, corresponding to some facilities in the price of privatization. At the same time, the attitude of voluntary protection and attention to the environment must be strengthened.

In this sense, the environmental protection programs can be organized and carried out effectively only on the basis of a set of priorities arising from the integration of scientific, technical, economic and social information and their influences on the resources of the natural environment.

As can be seen, almost all the recommendations in this work contain almost the same characteristics as those of economic, social or territorial cohesion. Including, extrapolating the area of analysis, we can frame the recommendations as being part of a specific field, as usually are various fields, especially the new ones, focused primarily on the problems affected by the law, opposed to the traditional fields. This is why, we can say that the inclusion of the environment in the process of territorial cohesion has its own conceptual unity, a unity that was developed by the environment itself, through its own needs [19].

References

1. Vălică C, Avram A (2024) 24 februarie 2024: doi ani de război. Drumul de la Slava Ukraini la ajutoare blocate și depozite de muniție goale. O cronologie; https://romania.europalibera.org/a/doi-ani-de-razboi-rusia-ucraina/32829382.html. Last accessed 2024/03/26
2. Popescu P (2024) Ucraina, războiul pentru reconstrucție; https://karadeniz-press.ro/ucraina-razboiul-pentru-reconstructie/. Last accessed 2024/03/26
3. Marinescu G (2023) România – hub logistic pentru reconstrucția Ucrainei; in Bursa Newspaper / 14.12.2023; https://www.bursa.ro/romania-hub-logistic-pentru-reconstructia-ucrainei-86174154. Last accessed 2024/02/20
4. https://eur-lex.europa.eu/legal-content/RO/TXT/PDF/?uri=OJ:L_202400792. Last accessed 2024/03/26
5. Euronews.ro: Ucraina, cea mai minată țară din lume. Aproximativ 30% din teritoriul țării, contaminat de obiecte explozive; in Euronews.Romania / 28.04.2023; https://www.euronews.ro/articole/ucraina-cea-mai-minata-tara-din-lume-aproximativ-30-din-teritoriul-tarii-contamin. Last accessed 2024/03/20
6. Marinescu A (2023) Război în Ucraina - Siturile culturale sunt țintite din ce în ce mai mult, potrivit UNESCO. 23 de situri au fost complet distruse iar 625 au fost parțial deteriorate; in Newsro / 09.08.2023; https://www.news.ro/cultura-media/razboi-in-ucraina-siturile-culturale-sunt-tintite-din-ce-in-ce-mai-mult-potrivit-unesco-23-de-situri-au-fost-complet-distruse-iar-625-au-fost-partial-deteriorate-1922400809492023081621279818. Last accessed 2024/03/20
7. https://www.hotnews.ro/stiri-razboi_ucraina-26903167-unesco-razboiul-provocat-distrugeri-patrimoniu-valoare-3-5-miliarde-dolari-ucraina.htm. Last accessed 2024/03/26
8. Andrei V (2023) Războiul din Ucraina a distrus moștenirea culturală a țării la o scară nemaivăzută de la ultima conflagrație mondială; in Ziare.com / 01.12.2023 / https://ziare.com/ucraina/ucraina-razboi-rusia-invazie-distrugeri-opere-de-arta-cladiri-istorice-1839203. Last accessed 2024/03/27
9. Federația Patronatelor Societăților din Construcții (FPSC) (2022) Task Force Ukraine / 08.07.2022; https://federatiaconstructorilor.ro/agenda/task-force-ucraina. Last accessed 2024/03/20
10. Umland A (2022) Cum poate Occidentul să ajute Ucraina. Trei strategii pentru o victorie și o revenire a Ucrainei; in Contributors.ro / 31.12.2022; https://www.contributors.ro/cum-poate-occidentul-sa-ajute-ucraina-trei-strategii-pentru-o-victorie-si-o-revenire-a-ucrainei/. Last accessed 2024/02/20
11. Popescu P (2024) Ucraina, războiul pentru reconstrucție; in Karadeniz Press / 16.02.2024; https://karadeniz-press.ro/ucraina-razboiul-pentru-reconstructie/. Last accessed 2024/03/20
12. Agerpres: Ucraina-doi ani de război. Bilanțul uman al războiului din Ucraina, un subiect sensibil / 24 febr.2024; https://covasnamedia.ro/actualitate/bilantul-uman-al-razboiului-din-ucraina-un-subiect-sensibil. Last accessed 2024/03/27
13. Victoria Kopenko: Colții nevăzuți ai conflictului din Ucraina și cum pot fi afectate apele, aerul și solul din România, ani de acum încolo; in Panorama – Business of tomorrow / 15.10.2023; https://panorama.ro/razboi-ucraina-ecologie-poluare-impact-mediu-romania/. Last accessed 2024/03/20
14. https://cor.europa.eu/en/engage/brochures/Documents/EU%20Annual%20Report%20on%20the%20State%20of%20Regions%20and%20Cities%202023/4892%20-%202023%20Annual%20Report%20EN.pdf. Last accessed 2024/03/27
15. zonele eliberate de ocupația rusă; https://www.veridica.ro/stiri/un-birou-de-arhitectura-din-ucraina-vrea-sa-simplifice-reconstructia-in-zonele-eliberate-de-ocupatia-rusa. Last accessed 2024/03/27
16. https://www.europarl.europa.eu/doceo/document/A-9-2023-0286_EN.html. Last accessed 2024/03/27

17. https://eu-solidarity-ukraine.ec.europa.eu/index_en. Last accessed 2024/03/27
18. https://www.consilium.europa.eu/en/policies/eu-response-ukraine-invasion/eu-solidarity-ukraine/. Last accessed 2024/03/27
19. Popoviciu GA (2011) Coeziune teritorială şi environmentală (politici şi implicaţii legislative). Universitatii of Oradea Publishing House. pp 166–167, 187, ISBN 978-606-10-0543-7

Chapter 9
Issues of the Military Brownfields in Ukraine: A Sustainable Development Perspective

Liudmyla Niemets, Kateryna Sehida, Anatoliy Babichev, Josep Vila Subiros, Cezar Morar, Vasile Grama, Kateryna Kravchenko, and Ievgeniia Telebienieva

Abstract The turbulent history of Ukraine, particularly its post-Soviet legacy and geopolitical changes, along with the ongoing war with Russia since 2014, which escalated into full-scale dimensions from 2022, has left an imprint on its landscape in the form of abandoned former military facilities scattered throughout Ukraine. The intensity of hostilities since 2014 and their spatial scale since 2022 have led to the emergence of numerous new military centers in practice. Some of these centers are directly linked to urban, industrial, and natural areas that have been damaged. A new challenge and open question arise: how to manage the recovery of Ukraine's territories after achieving victory and restoring territorial integrity. The purpose of this article is to identify the challenges posed by the war in Ukraine and explore the potential of repurposing military brownfields to contribute to the social, economic, and environmental reconstruction (development) of Ukraine in the post-war period. The paper elucidates the definition of military brownfields and clarifies the essence of post-military and post-war brownfields. Consequences of war, military operations and attacks on frontline and rear cities and terror of the civilian population are outlined. Taking into account the principles of the concept of sustainable development, the possibilities of reusing military brownfields in Ukraine were considered.

L. Niemets (✉) · K. Sehida · A. Babichev · K. Kravchenko · I. Telebienieva
V.N. Karazin Kharkiv National University, Kharkiv, Ukraine
e-mail: ludmila.nemets@karazin.ua; kateryna.sehida@karazin.ua; babichev@karazin.ua; kateryna.kravchenko@karazin.ua; eugene.telebeneva@karazin.ua

J. V. Subiros
Department of Geography, University of Girona, Catalonia, Spain
e-mail: josep.vila@udg.edu

C. Morar · V. Grama
Department of Geography, Tourism and Territorial Planning, University of Oradea, Oradea, Bihor, Romania
e-mail: cezar.morar@gmail.com; vasile.grama2014@gmail.com

C. Morar et al. (eds.), *Achieving Sustainability in Ukraine through Military Brownfields Redevelopment*, NATO Science for Peace and Security Series C: Environmental Security, https://doi.org/10.1007/978-94-024-2278-8_9

Keywords Military brownfields · Post-military brownfields · Post-war brownfields · Sustainable development · Brownfields sustainable regeneration · Ukraine

9.1 Introduction

Ukraine is the largest country in Europe in terms of area and the sixth-largest in terms of population. It is currently embroiled in the largest war of the twenty-first century, amidst significant global geopolitical changes [18]. Russia's war against Ukraine began in February 2014. However, in 2014, Russia tried to portray the war as an internal conflict. The full-scale invasion of the Russian army, under Russian flags, into the northern, eastern, and southern regions of Ukraine in February 2024 marked the beginning of a new active phase of the war [18]. All regions of Ukraine, without exception, are under enemy fire and shelling. Russia's war against Ukraine has been going on for more than 10 years, and a full-scale war has been going on for more than 800 days. The result of it, and maybe the goal, is genocide, urbicide, ecocide.

The turbulent history of Ukraine, particularly its post-Soviet heritage, and the war with Russia, ongoing since 2014, have left an imprint on the landscape of the country in the form of abandoned former military facilities scattered throughout its territory. A number of our previous studies were devoted to the study of the experience of reuse of brownfields, including military brownfields, in Ukraine and other European countries [20, 21, 24]. The intensity of hostilities since 2014 and their spatial scale since 2022 have led to the emergence of numerous new military centers. Some of these areas are directly linked to damaged urban, industrial, and natural zones. A new challenge and open question arise: how to manage the recovery of Ukraine's territories after achieving victory and restoring territorial integrity.

The purpose of this article is to identify the challenges posed by the war in Ukraine and explore the potential of repurposing military brownfields to contribute to the social, economic, and environmental reconstruction (development) of Ukraine in the post-war period. Achieving the set goal is provided through the implementation of three research tasks: to reveal the definition of military brownfields and to define the definitions of "post-military brownfields" and "post-war brownfields"; outline the consequences of war, military actions and attacks on frontline and rear cities and terror of the civilian population; consider the possibilities of reusing military brownfields in view of the principles of the concept of sustainable development.

9.2 Military, Post-military, Post-war Brownfields

Interest in the study of military territories has increased in European countries over the past 20 years, attracting attention from both scientists and practitioners. Urban planners, architects, urban designers, economists, geographers, ecologists, historians, political scientists and others deal with brownfields issues. A significant portion of research on former military territories is conducted from the perspective of human geography [25], focusing on the economic opportunities and challenges associated with the redevelopment and reuse of brownfields [32]. The possibilities of sustainable reconstruction and rehabilitation of brownfields for urban planning, ecology, and improving the quality of life are gaining new relevance, especially in the context of sustainable development, urban renewal, and the revitalization of urban centers [9, 23].

When discussing the concept of brownfields, it's important to note that there is still no universally accepted definition. The term "brownfield site" is commonly used to refer to "any land or premises which has previously been used or developed and is not currently fully in use, although it may be partially occupied or utilized. It may also be vacant, derelict or contaminated" [1]; brownfields as "properties (lands, objects, areas) that are underused, neglected, and can be contaminated; they are relics of industrial, agricultural, residential, military or other activities" [14]. Due to the complex interplay between a specific area, its history, social, legislative, and environmental processes, brownfields provide valuable insights into the state of society and the environment [26]. Another concept, "drosscapes" is also used to describe "waste and wasted land abandoned as a result of the socio- and spatio-economic processes of deindustrialization, post-Fordism and technological innovation" [3].

Brownfields are classified based on the type of previous use of the site, including military, rail and transport, agricultural, institutional, commercial, cultural, or leisure activities [5, 30]. Military brownfields comprise built or utilized areas where military structures or facilities were present, such as barracks, military towns, airfields, training grounds, and other related premises, often including remnants of fortification systems [11]. Considering the broad range of military brownfields, they are further differentiated based on categorization criteria. The first group of military brownfields consists of visible areas retaining traces of their original military use, including ruins of buildings and signs of military presence, such as pyrotechnic contamination or distinctive architecture. The second group comprises barracks and housing facilities. The third group encompasses unused land and partially developed brownfields on the outskirts [16]. Post-military/post-war brownfields refer to sites where two or more types of activities were closely linked in the past and underwent significant transformation during wartime [30].

Military brownfields encompass a wide array of former military sites, including bases, arsenals, ammunition depots, fuel and lubricant storage facilities, tank training sites, landfills, and hazardous waste disposal sites [22]. Military brownfields exhibit various features, including the functional diversity of buildings on the sites and their locations [9]. For example, military training grounds represent a distinct type of territory. They are often entirely inaccessible to the general public, and military practices create unique geographical conditions where military activities can significantly impact the physical and social landscape of the surrounding area [12]. Additionally, human military activity impacts not only the appearance but also the structure and functions of the landscape. This is particularly evident in the case of fortification lines. During wartime, the landscape acquires meanings imbued with military significance, including battlefields, bunkers, defensive batteries, restricted reserves, military industries, bomb shelters and air-raid shelter [32]. The post-war landscape also serves as a testament to historical development and, accordingly, constitutes part of the cultural heritage of the area [15]. The possibility of conversion of military brownfields is considered in the context of attractiveness and renewed dynamism [16]. The term "post-military landscape" is used when the landscape has "imprint of a former military function [that] remains too pervasive to enable the erasure of their military origins" [33].

Since the origins of post-military/post-war brownfields are fraught with tense and tumultuous geopolitical and historical events characterized by military action, they evoke complex and often negative memories, particularly concerning military attacks and occupation. Such memories linger in the landscape and must be considered both physically, in terms of reconstructing the remains, and psychologically, in the context of rethinking and re-evaluating heritage [32] searching for new meanings, and forming a collective memory.

"The literature presents various approaches to the reuse and reconstruction of brownfields, including post-agricultural brownfields [13], post-military [33], post-conflict and post-war brownfields [6]. However, it remains an open question whether these abandoned sites will remain deserted for an extended period or whether the war/conflict will persist or spread further. Additionally, it's uncertain which strategies are the most optimal for addressing post-conflict/post-military/post-war brownfields and which of the potential options are the most sustainable [13].

When discussing Ukraine, we should focus on post-war brownfields, which encompass not only military sites but also various other types of brownfields. These post-war brownfields can range from small sites in mixed urban or rural areas to large military complexes, public spaces, industrial facilities, airports, and more, all of which emerged as a result of military operations. Currently, Ukraine faces numerous post-war brownfields that resulted from shelling and military attacks in central areas, on the outskirts of cities, in historical districts, traditional industrial zones, metropolitan areas, and rural areas. The full-scale war in Ukraine has caused significant destruction and unprecedented consequences, leading to the emergence of new types and conditions of brownfields.

9.3 Background of Post-military Brownfields in Ukraine: The 10 Years in War

Ukraine has been at war for more than 10 years. As of May 3, 2024, a full-scale war has been ongoing for 800 days. What is happening nowadays? Fierce battles continue on the front lines, and confrontation persists along the entire front line. Shelling of Ukrainian cities, rocket attacks, guided aerial bombs, drones, etc., continue to terrorize the civilian population. There are constant attacks on energy and critical infrastructure facilities. Ukraine's economy is subsidized and dependent on foreign investments. Despite successful examples of territory restoration, large-scale reconstruction has not yet begun and may not start until Ukraine achieves victory.

Almost 18% of Ukraine's territory is currently under temporary occupation. Over the course of 2 years of war, the Ukrainian army has liberated more than 7% of the country's territory, including areas in the regions of Kyiv, Chernihiv, Sumy, Kharkiv, Mykolaiv, and Kherson. Front-line territories are under constant artillery fire, affecting cities, towns, and villages. The consequences of Russia's war in Ukraine are evident in the form of genocide, ecocide, and urbicide.

Despite the concentration of military confrontation in the south and east of Ukraine, the entire territory of the country is under shelling and missile strikes, targeting the civilian population, destroying infrastructure, and claiming lives. The active phase of the war has lasted for 800 days. Within the first 100 days of this active phase, 147 cities, or roughly a third of all cities in Ukraine, were directly and deliberately destroyed, either partially or completely [18].

There is an interactive map publicly available on the Internet with a recording of shelling since February 22, 2024 by type of shelling: artillery, missile or air (Fig. 9.1), and information about the date of shelling, its location, coordinates, etc. A number of areas (indicated by a blue rectangle) contain satellite images showing the destruction caused by shelling (Fig. 9.1). A number of attacks were directed against the civilian population, destroyed cities, formed abandoned territories, have social, economic, environmental impacts. The Russians are attacking Ukrainian cities with missiles, guided aerial bombs, and drones, destroying infrastructure and claiming lives. In frontline areas, artillery fire compounds the missile attacks. For instance, Saltivka (Kharkiv) was the largest residential area in Ukraine, with nearly 300,000 residents as of February 24, 2022, situated along the immediate border with the Russian Federation. Today, the district lies almost entirely in ruins and deserted. Another tragic incident occurred in the small village of Hroza in Kharkiv Oblast, where rocket fire killed more than half of the villagers in the afternoon of October 5, 2023 (59 people killed, 7 wounded). Numerous similar examples abound. Several settlements and districts, including the cities of Mariupol, Izyum, Bakhmut, and Avdiivka, have been devastated. The physical destruction extends beyond housing to include communication networks, transport infrastructure (roads, bridges),

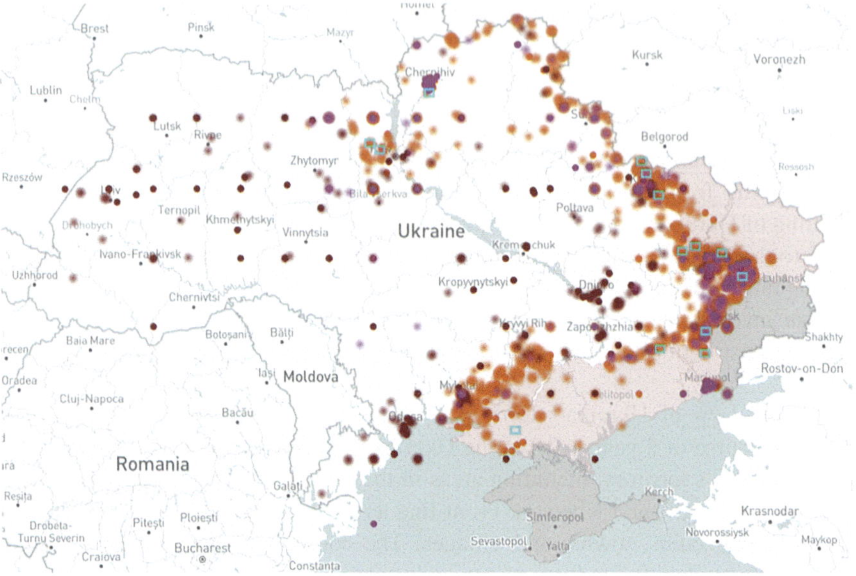

Fig. 9.1 Map of shelling of Ukraine by Russian troops [10]

industrial facilities, administrative buildings, universities, schools, kindergartens, churches, hotels, shopping centers, and historical landmarks. The appearance of regions affected by active hostilities has been irreversibly altered and continues to change. A true understanding of the scale of the disaster will only emerge post-occupation. Therefore, the issue of post-war restoration of these objects is deeply challenging but crucial, especially considering its novelty in the European space in recent decades.

Russia is purposefully destroying Ukrainian cities, towns, and villages. The radical destruction of Mariupol and Popasna, the baseless massacre in Bucha, and the daily destruction in Kharkiv, Mykolaiv, and Kherson and other cities, along with the imposition of the "Russian world" in occupied cities [18] represent yet another planned crime by Russia. There is a daily killing of ordinary residents (civilians), public spaces, and entire city blocks, with the aim of destabilizing security, disrupting public order, displacing residents, and either completely destroying or weakening cities. Despite this devastation, most of the cities that are being destroyed continue to function as fully-fledged cities, with public transport, garbage collection, street cleaning, businesses operating, and salaries and pensions being paid [18].

The next important issue is ecocide—the irreversible damage to the environment. Due to Russia's aggression, not only the cities but also the natural landscapes of Ukraine are suffering tremendous harm [4]. Nuclear power plants, seaports, and storage facilities for hazardous waste (such as mineral fertilizers, polyurethane foam, paint, and lubricants), as well as industries including chemical and metallurgical enterprises, have been within the active conflict zones. Incidents of fires have been reported at oil depots, gas stations, and landfills. Moreover, there are instances of damage to heat and water supply facilities, including sewage pumping stations and water pipes. Ecosystems have also been directly impacted by fires and the destruction of rare species and habitats, affecting natural reserves and ecosystems such as forests, steppes, and seas [28]. One tragic example is the disaster at The Kakhovka Hydroelectric Power Plant. The issue of demining the territory is extremely acute and large-scale. Currently, mined areas include not only zones of direct combat operations but also forests, parks, fields, infrastructure areas, locations near bridges and crossings, bodies of water such as ponds, rivers, and wells, abandoned combat positions, trenches, roadblocks, and checkpoints, military and strategic facilities, as well as abandoned equipment and buildings [19]. Mine terror is a form of Russian terror that Ukraine will have to face for many years. Regrettably, many civilians, including children, have already perished due to landmines in cities and villages.

The list of terrible crimes committed by Russia and the consequences of the war could go on indefinitely. Furthermore, it should be emphasized once again that the issue of post-war reconstruction is entirely new and necessitates detailed study and the development of effective strategies tailored to each situation, based on principles of sustainable development.

9.4 Sustainable Development and Post-military Brownfields Regeneration

Redevelopment of post-military brownfields is a complex and collaborative process driven by the goal of transforming a once abandoned site into a thriving community asset. Such a transformation requires an interdisciplinary approach that combines environmental expertise, community engagement and sustainable development practices to create positive impacts on both the environment and local populations in a sustainable development context [17]. Post-military brownfields regeneration requires a high degree of integration of tools and strategies across economic sectors and management institutions, considering both spatial and temporal dimensions [8, 25]. Studies on the brownfields sustainable regeneration encompass integrated approaches to spatial and environmental planning [22]. Sustainable development necessitates balancing social, economic, environmental, and institutional dimensions to achieve environmental performance, quality of life, social justice, etc. This involves engaging all relevant stakeholders [5].

Sustainable communities/places represent the principles of sustainable development by balancing and integrating the social, economic, and environmental components of their community. They aim to meet the needs of both present and future generations while also respecting the needs of other communities in the wider region or internationally to achieve sustainability [17]. All of these principles should be considered during of post-military brownfields regeneration to ensure their sustainability. Military Brownfields Regeneration project deals with site assessment, remediation and cleanup, environmental restoration, infrastructure and facilities community engagement, economic redevelopment, monitoring and maintenance. Military Brownfields Regeneration project is comprehensive and integrated, providing a solution for the existing problems, a long-term and lasting development, aim to improve economic, physical, social and environmental aspects with equal weight.

The concept of sustainable regeneration of brownfields is viewed as an ideal that may be realized in the future. However, achieving absolute sustainability is not feasible, as different aspects of sustainability will inevitably compete with each other to some extent. Therefore, rather than focusing on situations deemed optimal from today's perspective, emphasis should be placed on adaptability and high flexibility when considering approaches to sustainable development. "Sustainable brownfield regeneration is the management, rehabilitation and return to beneficial use of brownfield land in such a manner as to ensure the attainment and continued satisfaction of human needs for present and future generations in environmentally sensitive, economically viable, institutionally robust and socially acceptable ways within the particular regional context" [29]. Sustainability is not static over time and does not imply a fixed spatial perspective. It is a dynamic process that encompasses not only specified parameters but also a time dimension and spatial dimension, adapting to changing situations [7]. After a long period of exposure to negative factors, the sustainable development of brownfields or contaminated land begins with an assessment of the consequences and real impact on territorial development practices [34]. Post-military brownfields regeneration offers the opportunity for a new economic boost, demonstrated through increased residential and economic attractiveness, as well as an enhanced image effect [16].

The reuse of military brownfields should consider actively involving all interested parties in joint planning to ensure project sustainability. This involves incorporating the information and experiences of various stakeholders to reach agreement on different objectives [3, 12, 34]. Among the limitations of effective participation and the involvement of a wide range of interested parties are the risks of insufficient and untimely participation, the lack of authority to enhance sustainability options, the challenges in employing sustainable technologies and materials [2], the presence of vested interests within certain segments of the local community, and resistance to change [11].

The reconstruction of military brownfields has a significant positive impact on regional development and the enhancement of urban aesthetics. Military sites serve as landmarks, particularly due to the distinctiveness of their cultural heritage [22]. Each object is unique, many of them can be symbolic of the history of the state as a whole or a specific region [30]. Many of these former military sites and buildings hold not only historical significance but also possess considerable architectural and emotional value [22]. Landmarks cannot be preserved without transformation and adoption of new functions. Spatial identity and characteristics are crucial elements for the spatial quality of a sustainable future, as emphasized by the thesis that "no new development should be done without considering its past" [31]. Redevelopment plans often include the preservation of old buildings and infrastructure to preserve the historic character of the area and collective memory. But sometimes a complete transformation is foreseen, then it becomes difficult to recognize the historical appearance in the new one [5].

The issue of reusing military brownfields based on sustainability is fundamental [9], as sustainable development remains the primary goal of the European Union [22]. Specific sustainable development policy measures are introduced and implemented through national grant schemes of local spatial development policy [9] and should be incorporated into (re)development practices aimed at preserving and protecting the cultural environment, thereby minimizing the negative impact of brownfields [22].

There are several examples of successful practices in reusing military brownfields, including their utilization as landfills for the destruction of obsolete weapons and tools [23], as sites for new tourist destinations [23], as designated world heritage sites [3], and for various purposes such as administrative and residential buildings, warehouses, garages, etc. [23]. Reuse projects are primarily oriented towards the local community and secondarily towards tourists [31].

For Ukraine, which has been at war for 10 years, with the active phase lasting for more than 2 years, the development of the defense sector is crucial, both during and after the end of the war. The experience of the Association of Defense Communities, which brings together various interests of communities, state governments, the private sector, and the military, can be invaluable as it considers the necessity of military partnerships between communities [3].

In today's context, the sustainable reconstruction of military brownfields becomes increasingly relevant, given ongoing research on environmental pollution mitigation, the challenges and principles of brownfield reconstruction, and repurposing brownfields for sustainable development [34]. The organization of the process for planning the regeneration of military brownfields in a society oriented towards democracy, but with an inherited system of rational territory planning, remains an important issue [27].

Successful post-war rehabilitation of former military facilities based on sustainable development can be achieved through a combination of key components: cooperation between various stakeholders, including local communities, government bodies, and private investors; availability of financial resources, international support, and funding from governmental and non-governmental organizations, along with rational resource distribution and strategic planning; adoption of the approach of adaptive reuse. This will contribute to the post-military / post-war brownfields regeneration, economic growth, and the development of communities, as well as the preservation of the nation's history and the memory of Ukrainians' struggle for independence and sovereignty.

9.5 Conclusions

Questions related to the study of military brownfields in Ukraine have been a focus for scientists, particularly in light of the legacy of the Soviet era and the search for opportunities for successful reuse of such sites. The ongoing war between Russia and Ukraine since 2014 has introduced new challenges. Given the escalation of the war and its expanded scale since February 2022, the definition of "military brownfields" as established in scientific research may not fully correspond to the realities in Ukraine. In our opinion, the definitions of "post-war brownfields" and "post-military brownfields", which will include territories of various purposes that have become neglected and abandoned as a result of military operations, are more complete. The substantiation of the specified definitions on the basis of the available scientific foundation will be implemented in the following works of the authors.

During the active phase of Russia's war against Ukraine, daily attacks target both frontline and rear cities, instilling terror among the civilian population and leading to the genocide of the Ukrainian people. These attacks deliberately aim at infrastructure and residential buildings, rendering normal and safe life impossible. The aggressor purposefully carries out urbicide in Ukrainian cities. Additionally, ecocide occurs as natural objects suffer significant losses and damage. Many areas affected by hostilities have been physically destroyed, impacting not only housing but also communication networks, transport infrastructure, industrial enterprises, and other facilities. Consequently, the territories where active hostilities are taking place have undergone irreversible changes and continue to experience ongoing transformations.

Full comprehension of the scale of this disaster will only emerge after the territory is liberated from occupation. The issue of post-war recovery of post-war brownfields is both painful and incredibly important. It presents an entirely new challenge for the global community, as Ukraine is currently embroiled in the largest conflict of the twenty-first century. The sustainable regeneration of post-war brownfields in Ukraine involves not only physical revitalization in accordance with the principles of sustainable development but also the preservation of collective memory, the strengthening of national self-identification, and the development of a prosperous, united, independent, and sovereign Ukraine. The development and implementation of strategies, programs and plans for the sustainable regeneration of post-war brownfields with the involvement of financial investments and the participation of local communities will contribute to the economic growth and development of territorial communities, the preservation of the nation's history and the strengthening of Ukrainian identity.

The established challenges caused by the war in Ukraine and the possibilities of re-using military brownfields, given their potential to contribute to the social, economic and ecological post-war recovery (development) of Ukraine, indicate that post-war sustainable regeneration of post-war brownfields in our country will be significantly different from experience and practices of other countries. Given the realities of today, post-war sustainable regeneration of post-war brownfields is possible only after the Victory. It will require scientific research of all territorial communities of Ukraine, establishment of existing problems and opportunities, development of appropriate measures, which will be the focus of the authors' further research.

References

1. Alker S, Joy V, Roberts P, Smith N (2000) The definition of brownfield. J Environ Plan Manag 43(1):49–69. https://doi.org/10.1080/09640560010766. https://www.researchgate.net/publication/227619224_The_Definition_of_Brownfield
2. Čiháková AS, Černíková M, Dubová M (2011) Historical contamination and brownfield management: what does economic theory say about it? In: Ecosystems and Sustainable Development VIII, p 339. https://doi.org/10.2495/ECO110301
3. Clark CM (2009) Drosscapes or brownfields? Differing processes to bring redundant industrial land, including military sites, back into productive use. In: Sustainable Development and Planning IV, vol 1, p 175. https://doi.org/10.2495/SDP090171
4. Ecoaction. https://en.ecoaction.org.ua/
5. Elrahman A (2016) Redevelopment Aspects for Brownfields sites in Egypt. https://www.sciencedirect.com/science/article/pii/S1878029616300251?ref=pdf_download&fr=RR-2&rr=8348796f0de6247c
6. Far MS, Duarte CM (2014) Geopolitical Brownfields: introducing a new terminology for brownfields within conflict zones. In 21st Annual European Real Estate Society Conference. ERES: Conference. Bucharest, Romania, 2014. https://eres.architexturez.net/doc/oai-eres-id-eres2014-158
7. Franz M, Pahlen G, Nathanail P, Okuniek N, Koj A (2006) Sustainable development and brownfield regeneration. What defines the quality of derelict land recycling? Environ Sci 3(2):135–151. https://doi.org/10.1080/15693430600800873
8. Garb Y, Jackson J (2009) Brownfields in The Czech Republic 1989–2009: the long path to integrated land management. J Urban Regen Renew 3(3):263–276. https://doi.org/10.13140/2.1.2099.7767. https://www.researchgate.net/publication/264157896_Brownfields_in_the_Czech_Republic_1989-2009_The_long_path_to_integrated_land_management
9. Hercik J, Simacek P, Szczyrba Z, Smolová I (2014) Military brownfields in The Czech Republic and the potential for their revitalisation, focused on their residential function. Quaestiones Geographicae 33(2):127–138. https://doi.org/10.2478/quageo-2014-0021. https://www.researchgate.net/publication/263469262_Military_Brownfields_in_the_Czech_Republic_and_the_Potential_for_their_Revitalisation_Focused_on_their_Residential_Function
10. Interactive map of locations and dates of shelling. https://texty.org.ua/projects/107440/pid-udarom-sho-i-koly-obstrilyuvala-rosiya/
11. JakovčićJ M, Kajinič J, Cašparovič S (2013) Redevelopment of military brownfield sites: example of redevelopment of military complex Karol Rojc in Pula. ANNALES Ser hist sociol 23:487–500. https://www.researchgate.net/publication/288793814
12. Klusáček P, Martinát S, Krejčí T, Bartke S (2019) Re-development of a former military training area—the case of Brdy told from a local actors' perspective. Land Use Pol 82:147–157. https://doi.org/10.1016/j.landusepol.2018.11.045
13. Klusáčeka P, Navrátila J, Martináta S, Charvátováb K, Krejčía T (2021) From large-scale communist agricultural premise through abandoned contaminated ruin to organic farming production: The story of successful post-agricultural brownfield regeneration. https://www.deturope.eu/artkey/det-202103-0002_from-large-scale-communist-agricultural-premise-through-abandoned-contaminated-ruin-to-organic-farming-producti.php

14. Kunc J, Martinat S, Tonev P, Frantál B (2014) Destiny of urban brownfields: spatial patterns and perceived consequences of post-socialistic deindustrialization. Transylvanian Review of Administrative Sciences, March 2014. https://www.researchgate.net/publication/267864127
15. Kupka J, Brázdová A, Vodová J (2022) Units of military fortification complex as phenomenon elements of the Czech borderlands landscape. Land 11:79. https://doi.org/10.3390/land11010079
16. Lotz-Coll S (2019) The military brownfields, an opportunity for the urban redevelopment of small mountain towns? J Alpine Res. http://journals.openedition.org/rga/5198; https://doi.org/10.4000/rga.5198
17. McDonald S, Malys N, Maliene V (2009) Urban regeneration for sustainable communities: a case study. https://www.researchgate.net/publication/225071288_Urban_regeneration_for_sustainable_communities_A_case_study
18. Mezentsev K, Mezentsev O (2022) War and the city: Lessons from urbicide in Ukraine. https://doi.org/10.12657/czageo-93-20.; https://www.researchgate.net/publication/366639879_War_and_the_city_Lessons_from_urbicide_in_Ukraine
19. Mining territory caused by Russian aggression in Ukraine. https://freewayua.com/korusna-informacija/karta-zaminovanykh-terytoriy-ukrayiny/
20. Morar C, Nagy G, Dulca M, Boros L, Sehida K (2019) Aspects regarding the military cultural-historical heritage in the city of Oradea (Romania). ANNALES—Ann Istrian Mediterranean Stud Series Historia et Sociologia 29(2):303–322. https://doi.org/10.19233/ASHS.2019.21. https://www.researchgate.net/publication/336720209_ASPECTS_REGARDING_THE_MILITARY_CULTURAL-HISTORICAL_HERITAGE_IN_THE_CITY_OF_ORADEA_ROMANIA
21. Morar C, Nagy G, Boros L, Gozner M, Niemets L, Sehida K (2021) Heritage, culture and regeneration of the former military areas in the City of Oradea, Romania. Architektúra & urbanizmus 55(1–2). https://doi.org/10.31577/archandurb.2021.55.1-2.6. https://www.researchgate.net/publication/353554192_Heritage_Culture_and_Regeneration_of_the_Former_Military_Areas_in_the_City_of_Oradea_Romania
22. Morar C, Nagy G, Nagy G (2016) Brownfields regeneration, between sustainable urban development and cultural heritage. The former military sites in Oradea, Romania. J Urban Reg Anal. https://doi.org/10.37043/JURA.2016.8.1.6. https://www.researchgate.net/publication/306137002
23. Musa S, Šakić D (2019) Military brownfields—potential for tourist revitalization in Bosnia and Herzegovina. Geo J Tourism Geosites 26(3):1021–1032. https://doi.org/10.30892/gtg.26326-414. https://www.researchgate.net/publication/337652299
24. Niemets L, Husieva N, Suptelo O, Sehida K, Kravchenko K (2018) Research of brownfields and greyfields of the city: theory and practice. Proceedings of the 32nd International Business Information Management Association Conference (IBIMA) (15–16 Nov. 2018, Seville, Spain). Vision 2020: Sustainable Economic Development and Application of Innovation Management from Regional expansion to Global Growth, 2018, pp 409–436. https://www.scopus.com/record/display.uri?eid=2-s2.0-85063055212&origin=inward&txGid=c31a1b1bcc2a4286989fb5df8e318cbc
25. Niemets L, Sehida K, Kobylin P, Şahin T, M., Bartosh, O., Telebienieva, I., Kravchenko, K., & Kliuchko, L. (2023) Integrative role of human geography in solving global problems of modernity. Visnyk of V N Karazin Kharkiv National University, Series. Geol Geograp Ecol 58:171–187. https://doi.org/10.26565/2410-7360-2023-58-14; https://periodicals.karazin.ua/geoeco/article/view/22199
26. Osman R, Frantál B, Klusácek P, Kunc J, Martinat S (2015) Factors affecting brownfield regeneration in post-socialist space: the case of the Czech Republic. https://www.sciencedirect.com/science/article/abs/pii/S0264837715001702
27. Peric A, Miljus M (2021) The regeneration of military brownfields in Serbia: moving towards deliberative planning practice? Land Use Policy 102:105222. https://doi.org/10.1016/j.landusepol.2020.105222

28. Potential environmental impacts caused by Russian aggression in Ukraine. https://en.ecoaction.org.ua/warmap.html?gad_source=1&gclid=EAIaIQobChMIwZfp6uzyggMVQaWDBx2o7wY-EAAYAyAAEgKk8vD_BwE
29. RESCUE Work Package 1, Development of an Analytical Sustainability Framework for the Context of Brownfield Regeneration in France, Germany, Poland and the United Kingdom, Project Deliverable, 2003
30. Rey E et al (2022) Urban brownfields: origin, definition, and diversity. Chapter 2. Neighbourhoods in Transition, The Urban Book Series, https://doi.org/10.1007/978-3-030-82208-8_2
31. Verschuure-Stuip G (2016) Military brownfields in the Netherlands. The revitalisation of the New Dutch Waterline (1980–2014) Sustainable regeneration of former military sites. Chapter: 9. In: Bagaeen S, Clark C (eds) Research in planning and urban design. Routlegde. https://www.researchgate.net/publication/304253530_Military_brownfields_in_the_Netherlands_The_revitalisation_of_the_New_Dutch_Waterline_1980-2014
32. van Vliet W (2016) The future of relics from a military past. A study to the history, contemporary role and future development of 20th century military heritage at the waterfront of Tallinn, Estonia. https://www.rug.nl/research/kenniscentrum-landschap/voor-studenten/masterscripties/ma_scr_wvanvliet_2016.pdf
33. Woodward R (2014) Military landscapes: agendas and approaches for future research. Prog Hum Geogr 38(1):40–61. https://www.researchgate.net/publication/270623926_Military_landscapes_Agendas_and_approaches_for_future_research
34. Zheng B, Masrabaye F (2023) Sustainable brownfield redevelopment and planning: Bibliometric and visual analysis. https://www.sciencedirect.com/science/article/pii/S2405844023004875

Chapter 10
Transformation of Greenfields in Ukraine

Taras Pohrebskyi, Sergiy Kostrikov, Kateryna Sehida, and Oleksii Niemets

Abstract Russian-Ukrainian war has become the main catalyst for alterations in the state national economy, which is currently affected by the recession. Therefore, the financial contributions, especially foreign investment, drastically necessary for economic recovery.

Greenfield projects are those projects, the main purpose of which is a continuous process of using the lands, that have been vacant or have been used in agriculture or forestry for a long period of time. The key feature of greenfield investment strategy is building residential or industrial facilities from scratch. They will not be restricted by existing material and technical base, and the investor will be completely independent in his construct solutions. It should be mentioned that it is the "green projects", which are the most relevant for Ukraine's reconstruction, as the costs for green project implementation are much cheaper, than brownfield projects, which may include liquidation or partial reconstruction of industrial facilities, if the latter have been destroyed by the hostilities.

Over the past 10 years, hundreds of greenfield projects, which are characterized by its industrial, agricultural, and innovative directions, have been implemented in Ukraine. The development of greenfield areas can be an important element for ensuring the spatial development of cities and regions, as well as for creating new workplaces and attracting investments. However, it is also important to consider the impact of this process on the environment and the need for balanced development to avoid negative environmental or social consequences.

Keywords Greenfield projects · Foreign investment · Ukraine

T. Pohrebskyi (✉)
Lesya Ukrainka Volyn National University, Lutsk, Ukraine
e-mail: pogrebskyi.taras@vnu.edu.ua

S. Kostrikov · K. Sehida · O. Niemets
V.N. Karazin Kharkiv National University, Kharkiv, Ukraine
e-mail: sergiy.kostrikov@karazin.ua; kateryna.sehida@karazin.ua; oleksii.niemets@student.karazin.ua

C. Morar et al. (eds.), *Achieving Sustainability in Ukraine through Military Brownfields Redevelopment*, NATO Science for Peace and Security Series C: Environmental Security, https://doi.org/10.1007/978-94-024-2278-8_10

Russian-Ukrainian war has become the main catalyst for change in the state national economy, which is currently affected by the recession. Therefore, the financial flows, especially foreign investment, will be subsequently needed for economic recovery.

At the current stage of social development a variety of approaches to definition of concept of "investment" are springing up. According to the research subject, we consider the term "investment" as the set of economic resources which are aimed at increase of the socio-economic and environmental benefits in the different spheres of economic activity.

Depending on the type of lands which can be used in Ukraine's post-war reconstruction process, foreign direct investments are divided into:

- Brownfield projects include investments into lands with already existing facilities of the enterprises and infrastructure facilities. The clear advantage of a brownfield investment strategy is the opportunity to get rapid economic effect, as the investor's time devoted to construction can be avoided.
- Greenfield projects are the projects, the main purpose of which is a continuous process of using the lands which have been vacant or have been used in agriculture or forestry for a long period of time. The key feature of greenfield investment strategy is building residential or industrial facilities from scratch. They are not restricted by existing material and technical base, and the investor is completely independent in his construction solutions. It should be mentioned that it is the "green projects" that are the most relevant for Ukraine's reconstruction, as the costs for green project implementation are much lower than brownfield projects, which may include liquidation or partial reconstruction of industrial facilities that have been destroyed by the hostilities [4].

The term "greenfield projects" was originally used to describe constructive processes of production complexes within suburban or agricultural areas in the early XX century in Great Britain. In 1966 above 10% of all England's manpower worked in industrial enterprises. All this contributed to territorial differentiation stability of country's economic development [3].

Apart from that, world economy recession in 1930s and further economic decline indicators as a result of World War II events encouraged developed countries to adopt regional development policy. This policy attracts foreign investments in the establishment of branch networks in less developed regions emphasizing benefits of greenfield projects as the catalyst for depressed areas' economic growth. Particularly, during 1954–1982 French government developed a new regional development policy of the country. This policy provided tax benefits for enterprises which would be relocated from the Paris region urbanized areas into less developed agricultural territories in the West and South-West of the country. This has led to jobs growth in industrial sector and number of jobs reached for about 368 thousand vacancies.

During this period greenfield projects started to be used in urban planning. Particularly, in 1980s for the first time in the history of the USA independence financial grants and tax benefits were granted to support industrial and residential buildings construction in the underdeveloped southern part of the country.

These areas were used as a location for enterprises in the final stage of production which needed a large number of labor resources and were granted with special tax benefits by local authorities in the early XXI century. At that time, greenfield projects became the primary tool for electrical engineering industries. During the XXI century foreign direct investments from multinational companies were rapidly growing. The money were invested into innovative technologies of industrial production, which changed the concept of "greenfield" into investments as an economic category that includes any investment into new operations regardless of geographical reference (no matter is it urban or rural area). At the same time, the analysis of the latest scientific works by David Harvey, Jan Gehl, Ernst Glaser, K. Seto, and other outstanding scientists, demonstrated that modern concept of "greenfield" includes three important fields not of the economy research, but of human geography [1, 8]. Particularly, regional development of the territories, its industrial specialization, and investment provision, which are the basis for country's national economy growth.

As of January 1, 2023, international community announced the development of investment projects, the total number of which is over 16 thousands. These projects were worth more than 1155 trillion dollars, and their implementing would provide over 2,2 millions jobs. Extractive industry, IT sector, and urban planning still remain the key areas for implementing these projects. Today, unlike previous years, in the territorial structure of foreign direct investments 3 regions of the recipient countries can be clearly distinguished. Particularly, these are North American, European, and Asian regions. Let's take the closer look at each of them.

At the time of the study, in the North American region represented by the United States of America about 2330 greenfield projects with a total cost of more than 178, 2 million dollars were implemented in 2022. The cost is 29% more than in 2019. The key areas for implementing the projects were mechanical and technical engineering. The most successful greenfield project in this region was semiconductor manufacturing plants construction by "Taiwan Semiconductor Manufacturing", the Taiwanese company, in Arizona state. The total cost of the construction was 40 billion dollars. Currently, there is a positive trend of using not only the agricultural lands which have been mastered for a long time, but developing offshore fields as well. Thus, the Dutch company "Royal Dutch Shell" in partnership with the Norwegian company "Equinor" plans to implement the project worth 4.3 billion dollars, which will specialize in oil production in the Gulf of Mexico off the USA coast at a depth of over two thousand meters [2].

European region is the second largest source of foreign direct investment. Particularly, 10 leading countries in FDI include Great Britain, Germany, Spain, France, Poland, Netherlands, Ireland, Belgium, Turkey, and Portugal. The key feature of this area is not only the high level of greenfield projects innovativeness, but also customer-focus. During 2022, 5250 projects were implemented in Europe and provided more than 383 thousand workplaces. Among the leading areas in investment attraction is the energy sector. Particularly, 13 of the 15 mega-projects (those which are worth more than 1 billion dollars) were aimed at construction of territories for renewable energy facilities. The latest ScotWind leasing round also enabled developers to apply for seabed rights to plan and build major offshore wind projects in Scottish waters. One of such projects is project of international energy conglomerate Total energies, Green Investment Group, and RIDG, which is aimed at offshore wind farm which can value up to 4 billion pound sterling. This project will reduce the dependence of industrial capacities in Orkney on the capabilities of the country's central energy system.

During 2022, there has been a tendency to a significant increase of foreign direct investments in industrial use of Asia's countries agricultural lands. At the time of the study, the total number of FDI is 279, 7 billion dollars, which is 65% more than in 2021. India is one of the leading region's countries in territorial distribution of greenfield projects. India now accounts for 994 projects, the main specializations of which are engineering, metallurgy, and renewable energy. It should be mentioned that unlike previous regions, this area is characterized by the high level of free lands use into urban planning. Particularly, Bin Zayed International company established in the United Arab Emirates is going to invest more than 9, 6 billions in the hotel complex "Langkasuka" located in Malaysia, which involves more than 2 thousand acres land use, where resort buildings, living quarters, villas, shopping centers, and golf course will be placed. From our point of view, a truly unique example of greenfield projects implementation is construction of industrial complex for the production of green hydrogen. This project is valued to 40 billion dollars, and in the future it can become the foundation of new era in metallurgy field, specially era of green steel.

African and Latin American regions are characterized by a small number of investment projects. More exactly, 2132 projects fall on African region, and 1267 fall on Latin American region. Above 76% of free lands in these regions will be used for industrial enterprises construction specialized in electricity and oil production, as well as wood processing. It should be noted that South-African countries are characterized by the use of green territories for urban planning purposes. Arabian company URB specializing in sustainable cities development is going to invest more than 20 billion dollars in construction of the city from scratch, which will represent 12 residential areas with a total area of 1700 hectares. The implementation of this project will give an opportunity to provide housing for more than 150 thousand people.

At the moment of the study, the leading global companies investing in greenfield projects are International Workplace Group, Amazon, Deutsche Post, Panattoni, Volkswagen, Stellantis and others. Also, one of the major players in international arena is American automobile company startup Tesla, which invested in these territories over 93, 7 billion dollars. In addition, this company is one of the leaders in mega-projects implementation. Particularly, from 2016 till 2022 company's total number of projects exceeded the mark of 59. Those are presented not only by mega factories in Shanghai, Brandenburg, and Montevideo, which specialize in electric vehicles and automotive components, but a range of enterprises for the batteries, solar panels, and robotics production in Sydney, Berlin, and Texas.

Over the past 10 years, hundreds of greenfield projects, which are characterized by its industrial, agricultural, and innovative directions, have been implemented in

Ukraine. At the moment of the study, one of the most ambitious projects of Ukraine's agro-industrial complex is the construction of the new industrial facility "Allseeds Black Sea", which is one of the largest manufacturers of vegetable oils and meals. The total project cost is more than 200 million US dollars. The plant is located near "Pivdennyi" seaport, and plant's total area is 28 hectares. The key feature of this complex is its proximity to the open water, as well as the availability of its own independent laboratory, which carries out the control over the raw materials quality and provides the breeding of new seed products varieties.

Favorable economic and geographical position, cheap labor, and excellent material and technical base have become the main factors of Japanese company "Fujikura" facilities construction in Lviv, which produce cable networks of the European standard for the "Volkswagen" automotive company. The complex though is not large by the area, but currently provides jobs for about 4 thousand people, and has total production capacity of 143 thousand cables per day.

It should be mentioned that during 2015 in Ukraine the first and the biggest greenfield project "Biopharma" has been implemented. The area of the project is pharmaceutics and the total cost is 40 million US dollars. Biopharma company is the leading Ukrainian pharmaceutical developer and manufacturer of innovative blood-derived medicines, recombinant preparations, and spore-forming bacteria drugs. Currently, the enterprise is under expansion that provides for fractionator plant, in which the company invested more than 30 million US dollars. In the nearest future, this plant will greatly increase our blood-derived medicines production capacity, which is currently 90 tons per year. In addition, it will contribute to implementation of reforms, particularly, active development of blood service [5].

At the moment of study, the most successful greenfield project in the field of innovations is Vinnytsia Industrial Park with a total area of 37, 5 hectares. This park is the center of high-technology industries, which are consolidated into clusters. Namely, the park priority industries are repair and installation of machines and equipment, engineering, furniture production, sports equipment, logistics, and service. The favorable conditions of the industrial enterprises provide an opportunity to employ 6 thousand Vinnytsia residents [6].

After February 24, 2022, the society perception of modern geopolitical vision of the world has completely changed. This date has led to the search of ways how to minimize economic and environmental consequences of Russia-Ukraine war for our state and how to enable the development of agriculture, industrial system, and urban planning. Let's take a closer look at possible ways of greenfield lands use depending on their geographical position and types of the damages.

According to Zoï Environment Network and Regional Eastern European Fire Monitoring Center, more than 3 million hectares of forest plantations and 471 thousand hectares of grassland have been burned since a full-scale invasion started, and long-time lands restoration is needed. At the time of the study, Russian military scorched earth tactics dealt a crushing blow on Chernihiv, Kyiv, Sumy, Zhytomyr, Kharkiv, Donetsk, and Luhansk regions' forests, where the total area of destroyed plantations is over 2.1 million hectares [9]. It should be noted that more than 36% of these lands cannot be rehabilitated in several decades due to the changes of tree growth conditions. Particularly, this applies to the pine and mixed forests, as their full restoration will take from 50 to 100 years.

The proximity of these territories to settlements contributes to the intense searches for the involvement of affected regions in the development of urban landscapes. Particularly, the international experience of Estonia, Finland, and Germany regarding the involvement of beligerative landscapes in post-war urban planning will be used in the restoration of Ukrainian cities in Kharkiv region. This will expand boundaries by building sleeping areas and suburbs with an emphasis on public spaces, safe pedestrian areas, and transport accessibility of residential areas to the residents' work premises.

An example of greenfield territories' involvement in Ukraine's post-war reconstruction is the project of the new Oleksiivka City microdistrict in Kharkiv, which provides for the construction of 16 residential blocks with a total area of 22 hectares with own kindergartens, schools, public spaces, commercial premises, underground parking lots, stadiums, etc. This concept is new in Ukraine's urban planning and provides for the creation of a "city within a city". According to the analysis of project documentation, in case of implementation of the plan, the area of residential development will be 201 thousand square kilometers and it will provide permanent residence for more than 10 thousand people. Depending on the level of investment inflows, the duration of microdistrict construction can vary from 5 to 10 years.

Another type of free lands that can be used in the reconstruction process are the territories of fertile agricultural lands affected by explosions of ammunition, military maneuvers with heavy equipment, construction of fortifications, etc.

As of 2022, the total cost of losses from damage to lands and soil disturbance is 9.8 billion dollars. Besides, more than 113 thousand square kilometers located in the 10-kilometer frontline zone need three-stage demining, which cost 1.2 billion dollars and can be used in the post-war reconstruction process only for industrial purposes, particularly, for the implementation of "greenfield" projects.

A vivid example of these lands is the surrounding areas of Izium city in the Kharkiv region, which in the pre-war period were empty or actively used in agriculture. According to satellite images, at the time of the study, there were an average of 480 to 540 craters from 82 millimeters shells, 580 to 760 craters from 120 millimeters shells per 1 square kilometer of the region. The largest number of craters that was recorded from 150 millimeters shells is more than 1500. There are also separate areas with an area of soil disturbances of more than 375 square meters with funnels with a diameter of about 8 meters, resulting from the explosion of 250-kilogram aerial bombs. As a result of shells detonation in the examined area, more than 90,000 tons of soil cover were damaged, which potentially could not be used for agricultural purposes for 10–25 years, depending on the degree of pollution by heavy metals, which exude into the soil from the missiles and bombs explosions.

At the same time, the favorable economic, geographical, and transport position of the region, in particular, its proximity to the raw material bases in Donetsk region and Kharkiv, as the innovation and technical center of the East, contributes to the growing interest of foreign investors in using these territories for industrial purposes. At the time of the study, 34 Italian companies which specialize in energy, engineering and light industry are ready to participate not only in the reconstruction of Izium city, but also to join the construction of an industrial hub in the above-described territories to ensure their comprehensive development. This project provides for the allocation of more than 1.2 million dollars to Ukraine in the form of foreign direct investment, which in the long term will enable to attract from 27 to 58% of the war-affected lands.

Besides, satellite images of the cities' affected lands indicate the artificial lakes occurrence in place of deep craters, which is a kind of method of beligerative landscapes self-renewal. In the future, they can be used as recreational areas in case if the regions do not pose a threat to human life. For the first time in Ukraine, the idea of such a transformation is planned to be used in the restoration of Borodyanka city, where the craters of the urban fabric will become the "epicenters of urban biotopes" of squares, recreation areas, cultural and sports grounds in the future.

Ukraine's post-war reconstruction provides the involvement of another category of "greenfield" lands, namely agricultural areas affected by anthropogenic disasters or have been artificially flooded for a long time. Let's take a closer look at the possibility of using these territories in the example of the Kakhovka Hydroelectric Power Station.

According to our calculations, based on processed satellite images in the period from June 5 (before the explosion) till June 9 (after the explosion), the destruction of the dam led to temporary flooding of 612 square kilometers, of which 554.6 square kilometers of land are located within Kherson region and 57.8 square kilometers within Mykolaiv region. This delimitation has become the main problem on the way to rapid elimination of the disaster's consequences because more than 62% of the entire territory is currently under the temporary occupation of Russian troops.

In the process of Ukraine's post-war reconstruction, the full use of flooded lands is possible only if the Kakhovka HPP is rebuilt, which at this stage of scientific and technological development progress is inexpedient. At the time of the study, it is impossible to predict the approximate duration of hostilities in these territories, and the analysis of the available design documentation for the construction of the "Kakhovska 2.0" dam showed that its total cost will be more than 1.5 billion dollars, and the construction itself will last 5 years. In addition, the stable functioning of this ecosystem involves additional costs for regulating surface runoff, counteracting flooding and erosion, improving air quality, and regulating the outflow of pollutants. The total cost will be 172 billion hryvnas per year. From our point of view, an alternative way to solve this problem is the local irrigation systems construction that will minimize the territorial dependence of the right-bank regions on the development of hostilities, and what is most important, it will reduce the time of putting the system into operation and attracting funds from international donors. At the same time, the implementation of these projects within the temporary left-bank occupied territories in the post-war period will lead to the transformation of the agro-industrial complex by more than 400 thousand hectares of Ukrainian black soils and will attract more than 17 million dollars of foreign investments [7].

Our research makes it possible to assert the need to change the vision regarding the use of lands that have been flooded for a long time by the water column.

According to the qualitative characteristics of soil horizons, the value of the destroyed reservoir territory can be estimated at a total value of 0.25 billion dollars with a growth potential of up to 0.5 billion dollars in the post-war period, which is a positive economic factor for foreign investors. The most rational option for using 2155 square kilometers of lands is their involvement in the development of the country's energy sector, which will eliminate the existing deficit in Ukraine's unified energy system that arose as a result of massive strikes of the Russian Federation on infrastructure and dam destruction. As of 2021, the average annual electricity production at the Kakhovka Hydroelectric Power Plant was 1489 million kilowatt hours (1.5 terawatt hours). Our calculations demonstrate that solar panels installation with a capacity of 330 watts on the territory of the reservoir, which area is 150 square kilometers, with a combined loss of 15% of electricity will generate about 37.8 terawatt hours per year, which is 25 times more than the generation of the Kakhovka Hydroelectric Power Plant. At the same time, alternative energy sources can become a catalyst for the development of the energy sector, as well as reduce the anthropogenic load on the territory's soil cover, because the accumulation of water masses is the main factor in the long-term soil salinization and the development of soil erosion.

From our research, it can be concluded that currently existed and perspective greenfield projects are an important factor in the growth of national economies of the world highly developed countries, as well as of depressed regions that use "green" projects as a catalyst for their production and economic complex transformation. "New" projects remain an important subject for further research of the social geography of Ukraine because the modern concept of "greenfield" includes three important areas of its study. Particularly, territories' regional development, corresponding industrial specialization, and investment support, which are the basis for the Ukraine's successful post-war reconstruction, together with international experience provides an opportunity to assert the possibility of beligerative landscapes further use for agricultural, industrial, and urban planning purposes.

References

1. Adams D, Watkins C (2002) Greenfields, brownfields and housing development. Blackwell Science, Oxford
2. Hennart JF, Park YR (1993) Greenfield vs. acquisition: the strategy of Japanese investors in the United States. Manag Sci 39(9):1054–1070
3. Nagengast A, Hendrickson C, Lange D (2011) Commuting from U.S. brownfield and greenfield residential development neighborhoods. J Urban Plan Develop:298–304. https://doi.org/10.1061/(ASCE)UP.1943-5444.0000072
4. Niemets L, Husieva N, Suptelo O, Sehida K, Kravchenko K (2018) Research of brownfields and greyfields of the city: theory and practice. стаття Proceedings of the 32nd International Business Information Management Association Conference (IBIMA) (15–16 Nov. 2018, Seville, Spain). Vision 2020: Sustainable Economic Development and Application of Innovation Management from Regional expansion to Global Growth, pp 409–436

5. Report of the World Commission on Environment and Development: Our Common Future. URL: https://www.britannica.com/topic/Brundtland-Report
6. The World Bank. World Bank's Peace Project Supports Key Government Programs in Ukraine. URL: https://www.worldbank.org/en/news/feature/2023/07/10/?cid=eca_fb_ukraine_uk_ext
7. Ukraine's National Recovery Plan. National Recovery Council. URL: https://uploadsssl.webflow.com/Ukraine_Recovery_Plan_blueprint_ENG.pdf
8. Walden DD. Brownfield systems development: moving from the Vee Model to the N Model for Legacy Systems. INCOSE International Symposium, Orlando, FL (US), 29(1), pp 1084–1092
9. Zoï Environment Network. URL: https://zoinet.org/

Chapter 11
A Metamodel-Based Scoping Tool for the Assessment of Cultural Significance in the Redevelopment of Brownfield Sites (STASH)

Matthias Ripp, Christopher A. De Sousa, Christer Gustafsson, and Sebastian Daniel

Abstract The redevelopment of brownfield sites is a multi-layered and complex process in which a large number of various aspects and stakeholders interact with each other. If cultural heritage is viewed from a systemic perspective, it is often a vital component of such redevelopment processes. The aim of this paper is to develop a prototype for a tool that can be used for screening and scoping brownfield sites to understand if and what kind of cultural heritage is relevant in a specific case. Based on a metamodel for heritage-based urban development that reflects a holistic new heritage approach, the tool is developed using a Design Research Methodology (DRM) approach. The tool can also be used to collect information, assess the role of cultural heritage in specific brownfield projects and support the related decision-making and prioritisation processes. Potential use cases are also briefly explained.

Keywords Brownfield redevelopment · Cultural heritage · Metamodeling · Assessment · Evaluation

---

M. Ripp (✉) · S. Daniel
World Heritage Coordination, Department of Cultural Heritage, City of Regensburg, Regensburg, Germany
e-mail: matthiasripp@posteo.de; sebastianfjdaniel@web.de

C. A. De Sousa
School of Urban and Regional Planning, Toronto Metropolitan University, Toronto, ON, Canada
e-mail: chris.desousa@torontomu.ca

C. Gustafsson
Conservation Unit, Department of Art History, Uppsala University, Uppsala, Sweden
e-mail: christer.gustafsson@konstvet.uu.se

C. Morar et al. (eds.), *Achieving Sustainability in Ukraine through Military Brownfields Redevelopment*, NATO Science for Peace and Security Series C: Environmental Security, https://doi.org/10.1007/978-94-024-2278-8_11

11.1 Introduction

Industrialisation was accompanied by a change in the way people made things. In contrast to production by hand, the industrial age is still characterised by machine-based manufacturing processes that require large factory buildings and industrial complexes. Industrialisation is one of the most crucial factors in human prosperity and a significant step in the development and modernisation of societies. Nevertheless, it is often anchored in the collective memory as the starting point for the alienation of man and nature, which is currently culminating in the creation of artificial intelligence, social networks and digital worlds. Harmful emissions, toxic substances, greenhouse gases, hazardous goods, etc. also pollute the environment on a specific level, making industrialisation the most important trigger factor for climate change (cf. [2]).

The decontamination of former industrial sites is often associated with excessive costs. Today, brownfield redevelopment is important in many regions around the world where industrial or commercial activities have been subject to change, as well as in regions where new uses have been introduced due to destruction caused by wars or rapid-onset natural events (earthquakes, fires, floods) (cf. [19]).

Although the definition of brownfields has evolved over time and varies slightly by jurisdiction, at its core it refers to the reuse of land ranging from previously developed land to commercial and industrial sites to those complicated by the presence or potential presence of hazardous substances, pollutants or contaminants. The CABERNET research project defines Brownfields as "sites that:

- have been affected by the former uses of the site and surrounding land,
- are derelict and underused,
- may have real or perceived contamination problems,
- are in developed urban areas,
- and require intervention to bring them back to beneficial use." [5]

In many brownfield sites, some kind of cultural heritage is prevalent, especially when cultural heritage is understood as a system and a process that belongs to local communities (cf. [28]) and that includes both tangible and intangible cultural heritage; not only objects (monuments) but also processes, functions, memories, people, etc. This broad understanding of cultural heritage could be categorised as the New Heritage Approach, which includes the 'users' of cultural heritage, but also values, functions etc., and goes far beyond the physical aspects of cultural heritage, which Laurajane Smith has referred to as the 'authorised heritage discourse' (cf. [34]). This understanding of cultural heritage serves as a starting point for the development of a prototype for a cultural heritage assessment tool for specific brownfield sites to better understand and capture their role (potential, barriers, challenges, etc.) in brownfield redevelopment.

The problem identified is that there are no fully developed instruments with which (potential) brownfield sites can be analysed and evaluated in terms of their cultural-historical qualities. This makes it difficult to uniformly assess cultural significance and thus to efficiently and sustainably reuse space that is already scarce in industrialised countries (cf. [5]).

The research question we start with is: What can a tool for understanding, scoping and assessing cultural heritage on brownfield sites look like that is based on a holistic understanding of cultural heritage? The authors do not intend this article to be an in-depth analysis or inventory of cultural heritage, but rather an initial screening and inventory of existing heritage-related elements that can serve as a starting point for subsequent processes such as documentation, in-depth analysis, adaptive reuse of places etc. The tool developed should, therefore, be easy to understand, user-friendly and selective, meaning that no relevant categories can be omitted. The means of data collection for the tool should be variable and as open as possible, without limiting the use of the tool to specific contexts, data formats etc. The tool should be more of a communication tool to support strategic decisions and promote the design of successful brownfield regeneration processes that incorporate or are based on cultural heritage. Furthermore, it should be able to support decisions on prioritisation between different brownfield projects, structure in-depth analysis processes and support related decisions. It should enable preliminary assessments of existing heritage values, functions, relevance etc. (Fig. 11.1)

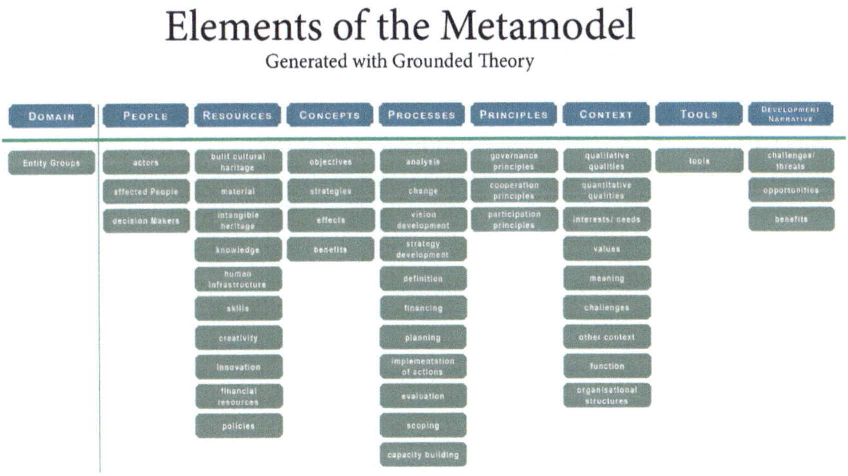

Fig. 11.1 The elements of the metamodel were developed through the application of grounded theory by Matthias Ripp as part of a comprehensive research project for his doctoral degree. (© Matthias Ripp)

11.2 Brownfield Regeneration and Cultural Heritage

Brownfield redevelopment has evolved significantly since the late 1970s, with an initial focus on understanding and managing the health and environmental risks posed by contaminated land incidents such as the Love Canal chemical spill in the United States and the explosion at the Seveso chemical plant near Milan, Italy. This risk-oriented focus on the assessment, remediation and removal of contaminated land and buildings was replaced in the early 1990s by a broader focus on the regeneration and redevelopment of the newly labelled 'brownfield sites', with governments working with a wider range of stakeholders to develop and implement mechanisms to attract and support investment. Some of these mechanisms focussed on funding and financial support, while others involved comprehensive planning and policy efforts to reimagine brownfields and communities and turn these visions into reality.

At this time, cities began to better understand the value of built and cultural heritage in developing and driving the future vision for brownfield sites. As opportunities to address pollution risks and remediation costs improved in the new millennium, the broader set of stakeholders involved in and affected by brownfield regeneration continued to pursue a wider range of environmental, social and economic objectives. This progress in sustainability objectives, including those associated with the preservation and appreciation (including commercialisation) of cultural heritage, has impacted all types of brownfield redevelopment projects, from high-value redevelopment projects undertaken by the private sector to public-private partnerships or purely public sector-led projects.

Brownfields are not alienated areas separated from communities or the social context. Rather, they are products of human activity, be it industrial production, destruction (war), activities related to mobility (railway land), mining, housing, political movements etc. Whilst the definition of brownfield land has evolved over time and differs slightly from country to country, at its core it generally refers to the re-use of land and property, which in the UK ranges from "previously developed land (PDL) that is or has been occupied by a permanent structure" [38] to "property the expansion, redevelopment, or reuse of which may be complicated by the presence or potential presence of a hazardous substance, pollutant, or contaminant" in the United States [42].

Having a history does not necessarily mean that there is a cultural heritage. Only when 'something' is somehow valuable and efforts are made to preserve it, it is generally referred to as cultural heritage (Fig. 11.2):

> There are three conditions that are necessary for something to be valued as heritage. First, it needs to have come from the past to the present, accompanied by a desire that it continues into the future. This makes the idea of inheritance a condition of heritage preservation. Second, and linked to this, there needs to be a community that values this possibility. Third, there must be the idea of what risk looks like, and how the value of the heritage might be preserved and protected. [33]

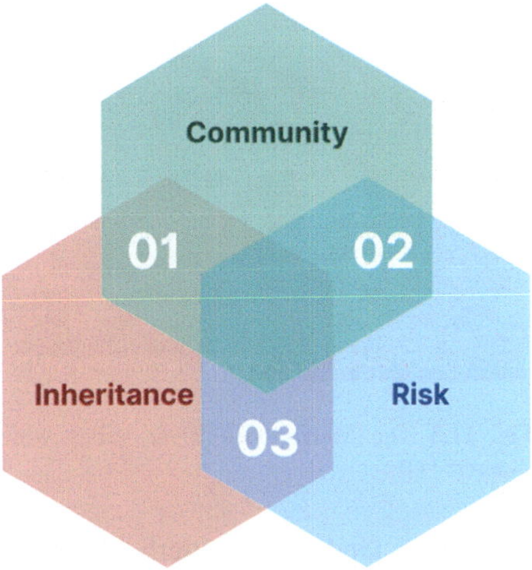

Fig. 11.2 Criteria for the valorisation of cultural heritage. (© Sebastian Daniel)

Although the Venice Charter refers to "a message from the past" with which monuments are "imbued" [20] as early as 1964, traditional approaches have mainly focussed on the physical aspects of cultural heritage, as can be seen from Article 1 of the UNESCO Convention Concerning the Protection of the World Cultural and Natural Heritage from 1972:

> For the purpose of this Convention, the following shall be considered as 'cultural heritage':
>
> – monuments: architectural works, works of monumental sculpture and painting, elements or structures of an archaeological nature, inscriptions, cave dwellings and combinations of features, which are of outstanding universal value from the point of view of history, art or science;
> – groups of buildings: groups of separate or connected buildings which, because of their architecture, their homogeneity or their place in the landscape, are of outstanding universal value from the point of view of history, art or science;
> – sites: works of man or the combined works of nature and man, and areas including archaeological sites which are of outstanding universal value from the historical, aesthetic, ethnological or anthropological point of view. [39]

In the mid-twentieth century, the concept of ensembles, which encompassed entire areas and went beyond a single-building approach, emphasised the fabric between buildings. At the beginning of the twenty-first century, the UNESCO Recommendation on the Historic Urban Landscape then included functions and different urban layers in the understanding of cultural heritage (cf. [40] and [27]). Only recently, new heritage approaches (cf. [25]) are emerging that understand cultural heritage as a system and a process that belongs to local communities (cf. [28]). With this broader holistic and systemic understanding, it becomes clear that most brownfield sites are about cultural heritage. Either with physical objects that were part of a production process (Toronto's Distillery District, Zollverein Coal Mine Industrial Complex in Essen etc.), a strong community that still identifies with former functions (Erzgebirge/Krušnohoří Mining Region), their function as iconic elements of

Fig. 11.3 The Menomonee River Valley which was filled in to create an industrial area. (Source: [1])

the (urban) landscape (Batterfield Station London etc.) or the identities of entire regions (Mühlviertel etc.) that are still very visible in street names etc.

Associations with past uses are not always positive and can be complex. For example, the Menomonee Valley in Wisconsin has a natural history as a rich environment that provided Native Americans with a canoe route inland from Lake Michigan and abundant resources, including Menomin (Algonquin for wild rice), for which the valley and river are named (cf. [7, 8]). This version of cultural heritage is often used to support efforts to restore the Valley's natural environment, honour Native history, and implement projects to support Native communities such as the Potawatomi tribe, who have played a significant role in the valley's revitalisation. At the same time, Milwaukee boasted in the early 1900s that it was the 'machine shop of the world', and the Menomonee River Valley (which was filled in to create an industrial area, see Fig. 11.3) was its engine, so manufacturing industries and restoration efforts are celebrated as both environmentally and economically beneficial (cf. [22]). Both aspects are well balanced in the city's regeneration story.

For successful brownfield redevelopment processes, it is therefore beneficial to understand the cultural heritage dimension already at the beginning of the process, which is often referred to today as the scoping or screening phase, as a kind of preliminary phase of each planning/development phase (planning phase 0) (cf. [31]). This tool is to be used if the relevant cultural heritage on brownfield sites is not already recorded in existing inventories of cultural heritage at national, regional or local levels.

11.3 Heritage as a Tool for Regeneration and Urban Development in Europe

In recent times, the cultural heritage sector has been increasingly involved in the green and digital transition. Protecting cultural heritage is mentioned in Agenda 2030 (cf. [37]) and UNESCO (cf. [41]) has presented 22 new indicators to follow it up. The European Commission has launched the creative and interdisciplinary initiative The New European Bauhaus (cf. [11]) with the aim of connecting the European Green Deal (cf. [10]) to our living spaces and experiences.

It can be understood that cultural heritage today is an important resource for sustainable development, but at the same time, it places new demands on developing methods, strategies and business models to clarify the importance of cultural heritage in new contexts (cf. [14, 16, 29, 30]). The relations of culture and cultural heritage to the surrounding society have been the starting point in studies of cultural vibrancies in western Sweden (cf. [3, 4]). In the industrial district north of Gothenburg, for example, it was found that the intangible industrial heritage still played a key role through the various organisations and decision-making processes that were created when the shipyards, car factories, paper mills, etc. were still in operation. In the studies, it was found that in some cases today the intangible industrial heritage constituted obstacles in the green and digital transition.

The EU Horizon research and innovation projects CLIC and Be.CULTOUR were partly based on these studies. In the Horizon projects, the focus has been to develop new activities in innovative, entrepreneurial and circular adaptive re-use of brownfield areas based on circular economy workshops together with the local population, policymakers, and investors (cf. [35, 36]). In the workshops, they were presented with innovative ideas that have been developed into finished services or products aiming at the three pillars of sustainable development and ready to market together with researchers from various disciplines.

In EIT Culture & Creativity one objective is to clarify the relations between culture and cultural heritage—creativity—innovation (cf. [17]). One research question raised is if it is possible to move creativity and innovation from one part of a city to another. In the Danish city Aarhus, the cultural production centre Godsbanen opened in 2012 in former train station buildings with the aim to support culture and creativity. In the old and abandoned shipyards buildings in Gothenburg, the Lindholmen Science Park has been established as a new tech arena for startups and technology-driven companies. On the other side of the river Göta älv and next to the city's creative heart in historic districts, a new business and residential area is developed now. One question for the developers and urban planners is whether and, if so, how they can relocate the creativity to these new premises.

11.4 Methodology and Scope

The tool was developed using a Design Research Methodology (DRM) comprising five different steps as described by Peffers et al. [26], Niedderer [24], Weber [43], Gregory [13], Kelly [21], Easterday et al. [9] and Schneider et al. [32]. In the 1960s, Design Research Methodology became popular as a scientific method to develop a new artifact (which can be, for example, a model or, as in this case, a prototype) according to a solid and reproducible method with scientific rigor. It is used for the development of the prototype for precisely these reasons. From 1980 onwards, the method became more popular as "creative and interdisciplinary" [24] and was used in various fields such as architecture, information science and others.

Design research is a proven method when a new "artifact, tool, etc." is to be developed based on a scientifically rigorous methodology. To develop the prototype of the STASH tool, a 5-step approach is used, based on Peffers et al. [26], Niedderer [24], Weber [43] and others, as described by Ripp [31] in *A Metamodel for Heritage-based Urban Development*. The steps can be described as:

1. **State the problem/research question** (see previous chapters and especially the introduction).
2. **Define Requirements**: The authors defined the requirements as:
   (a) Easy to use: No major explanation, introduction or training should be required to use the tool.
   (b) Considering various aspects related to cultural heritage following a holistic and systemic understanding of cultural heritage: Going beyond a traditional material-focused definition of cultural heritage.
   (c) Easy to understand: After the tool is applied the results should be easy to understand for someone who was not involved.
   (d) Able to digest different forms of data (text, pictures, etc.): Open in the way it can be used.
   (e) Able to use data derived from a wide variety of sources.
   (f) Easily accessible: not in a complicated format where you need to register first hand etc.
   (g) Assist in the evaluation and prioritisation of Brownfields based on their potential heritage value: Be able to include an additional quantitative-based level for this purpose.
3. **Design tool**: The prototype has been designed using the Example of Sporer-Quetsch and considering the defined requirements.
4. **Demonstrate/test**: The Demonstration and Testing of the tool has been done by a student that has not been involved in the design of the prototype. By this setup potential flaws and needs for improvement have been made visible.
5. **Refine/adjust according to findings**: More substantial advances of the prototype can only be done if more testing with different Brownfields have been taken place.

The development of the tool is described in Chap. 6. It was demonstrated and first tested in Chap. 7 using the Bavarian example of Sporer-Quetsch. A preliminary evaluation of the tool, based on this first test, is discussed in Chap. 8.

11.5 The Metamodel: Development and Nature

The core of the tool to be developed is a new approach that understands cultural heritage as a system and process that belongs to local communities (cf. [28]). A metamodel for heritage-based urban development (cf. [31]) will provide the structure and logical understanding of different areas and entities relevant to cultural heritage understood as a system.

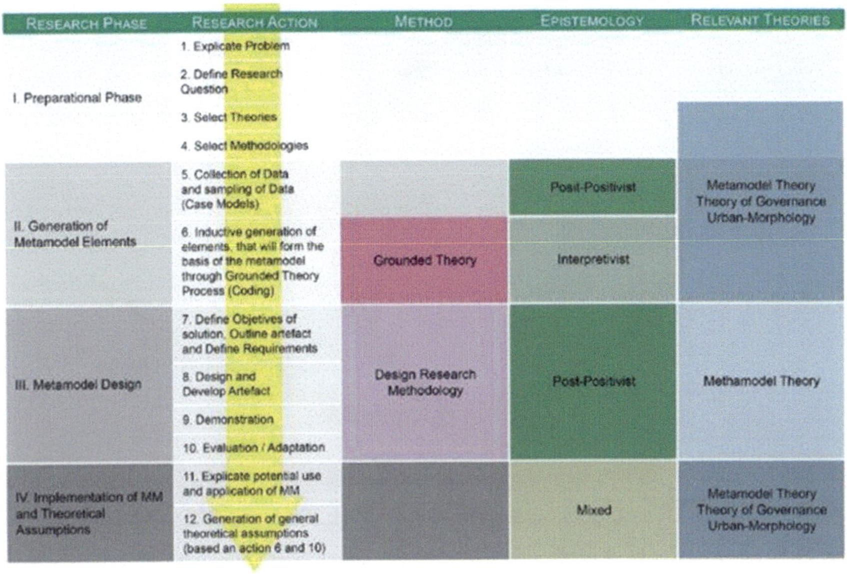

Fig. 11.4 The mixed-methods-research-design for the development of the metamodel. (© Matthias Ripp)

The metamodel was developed as part of a dissertation and is based on three successful case models of European projects where cultural heritage was the starting point for urban development processes: Hero (heritage as opportunity), Comus (community based urban development) and the Halland model (a regional development project in the Halland region in Sweden) ([15], cf. [6, 12, 29]) (Fig. 11.4).

The elements of the metamodel were developed from the three case studies, which were examined using grounded theory, and common patterns and parallels were sought. In the second step, the metamodel was designed using the design research methodology (Fig. 11.5).

A number of use cases have been developed: One of them is the design of a successful heritage-based urban recovery process, which is close to the purpose of the tool to be defined as a result of this article (cf. [18]).

For the tool to be developed here, the metamodel will serve as a basic structure to understand which elements (or entities, to stay in the language of the metamodel) are relevant and to ensure that all potential aspects of cultural heritage are part of the tool. Based on the experience with brownfield development projects of different scales and regions, the metamodel will be refined to meet the needs of brownfield sites. The challenge will be to translate the abstract elements of the metamodel into a conceptual model that is applicable to brownfields.

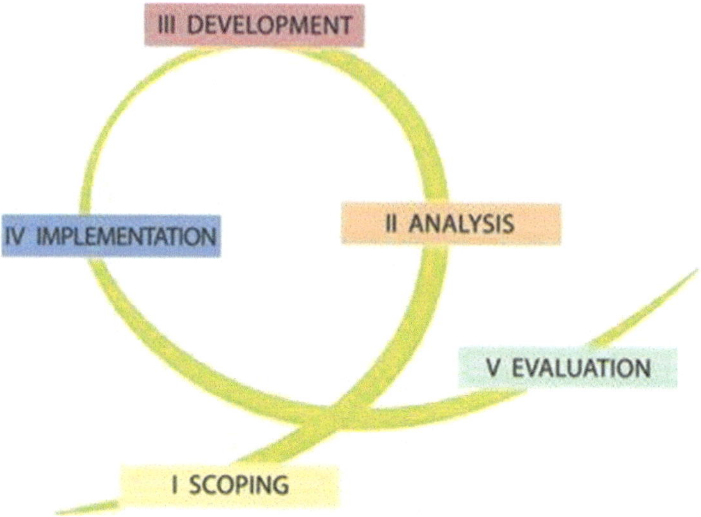

Fig. 11.5 The five phases of heritage-based urban development. (© Matthias Ripp)

11.6 The Prototype: STASH

Based on the requirements and the research question, a prototype was developed. According to the third step of the research methodology design, the following process was conducted: The metamodel had to be adapted to the reality of brownfields. To this end, the metamodel was evaluated based on the experiences of the four authors and gaps were identified. In a virtual collaboration forum, the elements of the metamodel (Fig. 11.1) were presented, and all participating authors were asked to develop specific entities relevant to brownfields by making post-it notes on the general framework of the metamodel elements and on the development process with its five phases.

The introduction of a co-creation process (cf. [23]) at this stage of prototype development provides a broader experience base and increases the universality of the to be developed prototype by integrating a more diverse context in terms of temporal and geographical scope as well as individual experience.

After the co-creation process, the results were discussed, evaluated and structured. Based on these real representations or entities, relevant questions and aspects for the prototype were elaborated; for a detailed explanation of the metamodel logic and associated terminology refer to Ripp [31]. In the next step, the prototype was designed and created from the draft (Fig. 11.6, Table 11.1).

Fig. 11.6 Exterior view of the historic building “Sporer-Quetsch” quartz stone grinding plant in Viechtach, Bavaria. (© Anna Fechter)

Table 11.1 Exemplary application of the STASH prototype to the "Sporer-Quetsch" brownfield site

Basic data and information about the brownfield site					
Title	Sporer-Quetsch	**Location**	Viechtach	**Historic uses**	Quartzstone Grinding Production
Size	?	**Year of origin**	?	**Operating years**	?
Relevance score (= R)		0 = no potential relevance		10 = extremely high relevance	
Domain	**Entity Groups**	(Possible) **Questions**	**Answers**	**R**	**Further information/ Sources**
People	Actors Affected people Decision makers	What is the history of workers, owners etc.?	Story of workers, owners etc.	6	BBV, Nr. 160 (13.7.2013) BBV, Nr. 227 (29.9.2004) BBV Nr. 102 (3.5.2019)
Resources	Built Heritage Material Heritage	What is the history/ historical relevance of the product?	Story of the product	3	BBV, Nr. 102, (3.5.2019)
	Intangible Heritage Knowledge	Were materials used that are culturally significant?	Description of the cultural significance of the used materials	2	
	Human infrastructure Skills	Were techniques/ skills or tools used at the site that represented a decisive innovative advance for a particular type of production relevant to the cultural heritage?	Description of the related skills and techniques	6	
	Creativity Innovation Finances Policies	Has the previous operation of the site contributed to significant new knowledge in a related field?	Report on the knowledge development	7	
Concepts	Objectives Strategies Effects Benefits	Did the former production contribute significantly to the development/ prosperity of the region?	Story of wealth/ development	5	BBV, Nr. 227 (29.9.2004)

(continued)

Table 11.1 (continued)

Basic data and information about the brownfield site					
Processes	Analysis	To what extent was the production facility affected by changes? What internal changes were there?	Description of the internal and external changes	5	/
	Change				
	Vision development				
	Strategy development				
	Definition				
	Financing				
	Planning				
	Implementation of actions				
	Evaluation				
	Scoping				
	Capacity building				
Principles	Governance	How did the operational management of the production facility change over time? What principles were advocated?	Description of governance principles	5	/
	Cooperation				
	Participation				
Context	Qualitative qualities	What is the history of the site?	Story of the site	6	BBV, Nr. 149 (30.6.2012) Denkmalatlas Bayern
	Quantitative qualities	What is the value of related buildings, production facilities, landmarks?	Description of the cultural significance	10	
	Interests/needs				
	Values				
	Meaning	Is the site located in vicinity or (functionally) connected to other heritage-relevant sites?	Report on the geographical context	4	
	Challenges				
	Other context				
	Function	Are or were intangible values present?	/	9	
	Organisational structures				
Tools	Tools	/	/	/	/

This questionnaire was created with the help of Julia Schimmeyer (Julia.Schimmeyer@st.oth-regensburg.de)

11.7 Data Resources and Use Cases

A. Resources: The STASH sheet can be used in conjunction with a variety of information sources. It is important to note that the tool is designed as a quick assessment tool and is not intended (although it could be used) to structure long-term research projects. Possible sources of information include:

- publicly available online information
- interviews with stakeholders, owners, experts from the heritage sector, local NGOs, workers, etc.
- documents, archive records related to the site
- existing inventories
- media sources like radio, TV, newspapers
- existing research publications
- other means of documentation

B. Use Cases

(a) **General**: The STASH sheet can be used for a quick assessment of brownfields and their heritage-related values.

- Scope: Any brownfield site that is to become part of a development or redevelopment process
- Objective: Understanding the heritage facilities and values associated with the brownfield site in order to design a successful process for development or redevelopment
- Reason: Any situation related to or embedded in a broader urban planning context (e.g. national, regional or state regional development strategies) where a decision needs to be made whether to activate or develop a brownfield site and use cultural heritage as a driver for development [31]. Or in all cases where a choice has to be made as to which brownfield sites should be selected or which should be started if several are available.
- How to use the prototype: The STASH sheet can be used for structured data collection. The use of the STASH sheet ensures that no information relevant to the monument is left out. It can also be used to determine whether information is missing on some aspects and further investigation is required to obtain a clearer picture.

(b) **Post-war-scenarios:** The STASH sheet can also be used for an assessment of brownfields and their heritage-related values in terms of post-war-scenarios.

- Scope: In a post-war situation that has resulted in many brownfield sites and in the post-war reconstruction process, decisions need to be made about which brownfield sites to start or work on.

– Objective: Assess brownfield sites on the basis of their heritage value to understand their significance and relevance and to gain an overview of the situation to inform decisions on priorities and actions.
– Reason: In an initial brownfield site assessment, where decisions need to be made about prioritization and selection, the heritage values of the site play a crucial role in making informed decisions. The STASH sheet is a tool that can be used for this purpose.
– How to use the prototype: see above.

The questionnaire can be used by one person or a team. In a first step, the participants should familiarize themselves with the questionnaire and the general idea of the instrument. Then, in a second step, information from the sources described is used to collect information and complete the questionnaire. In the third step, the relevance (R) is determined by the responsible person or team. If several brownfields are assessed, it is highly recommended to discuss the parameters and criteria for the relevance assessment in advance. In the fourth step, the results can be discussed/used for the purpose for which they were developed (e.g. prioritization, selection etc.).

11.8 Evaluation and Conclusion

Following the DRM-Approach the artefact (here the STASH sheet prototype) will be evaluated according to the defined user requirements:

1. Easy to use: In general, the STASH questionnaire is easy to use. The limitation lies in the accessibility and the effort required to access and record the information.
2. Consideration of a variety of aspects related to cultural heritage in terms of a holistic and systemic understanding of cultural heritage: As the STASH sheet is based on the meta-model, this user requirement is met.
3. Easy to understand: The STASH sheet is almost self-explanatory, and only a short instruction manual is required to work with it. Some, but not much, prior knowledge is required to use it.
4. Ability to process different forms of data (text, images etc.). Column 6 can be used to collect short descriptions/titles of brownfield sites. There is also space for links to additional information, documents, images, audio files etc. However, it has been shown that there is a conflict between the compactness of the tool and the possibility for additional information. For future refinement of the tool, possibilities should be considered, such as working with an additional file where resources can be cut or combined and the compressed text in the table can be linked to these.
5. Possibility of using data from a variety of sources: The design of the STASH questionnaire is intentionally open and can be used with a variety of different data sources.

6. Easily accessible: The overview nature of the STASH questionnaire makes it easily accessible, but at the same time allows access to further information and resources via column 6. This has proved difficult when using the tool (see point 4). Further thought needs to be given to how the STASH tool can be developed as a kind of front-end together with a file/folder in which the information collected is collated.
7. To assist in the assessment and prioritization of brownfield sites based on their potential heritage value. Column 5 makes it possible to assess the relevance (R) of the various listed properties and thereby prioritize brownfield sites based on their heritage value. However, the quantitative assessment based on numbers now contains a strong subjective element related to the person using the tool. For the future refinement of the tool, a kind of user manual for this template seems necessary.

The limitations of the STASH sheet and its use are determined by the accessibility and availability of the required data/information. In some cases, information may be difficult to obtain or may have been lost, e.g. due to war. In such cases, obtaining information may prove to be complicated and the STASH sheet may reach its limits.

This article shapes the first idea and prototype of the STASH sheet. To refine and fully develop it, it must be used in some specific cases to gather more experience and data that can be used to evaluate what worked well and where adaptations and refinements are needed. The prototype does not yet allow a standardized evaluation of the answers, so that the assessment of relevance (R) is not yet reliable, but rather subjective. The aim of further research work must be to define questions for each entity group that are linked to items according to which the answers can be assessed.

References

1. Beck & Pauli. (1882) Milwaukee, Wis. Milwaukee. [Map] Retrieved from the Library of Congress. https://www.loc.gov/item/75696717/
2. Bundeszentrale für politische Bildung (2022) Industrialisierung/industrielle Revolution Bundeszentrale für politische Bildung
3. Buscema M, Ferilli G, Gustafsson C, Sacco PL (2023) Toward a precision, complexity-informed cultural policy design: structural bottlenecks to culture-led development in Skaraborg, Sweden. Commun Nonlinear Sci Numer Simul 116:106886. https://doi.org/10.1016/j.cnsns.2022.106886
4. Buscema PM, Ferilli G, Gustafsson C, Sacco PL (2020) The complex dynamic evolution of cultural vibrancy in the region of Halland, Sweden. Int Reg Sci Rev 43(3):159–202. https://doi.org/10.1177/0160017619849633
5. CABERNET Coordination Team (2006) Sustainable brownfield regeneration: CABERNET network report. University of Nottingham, Nottingham
6. City of Regensburg, Ripp M, Scheffler N (2011) HerO – Heritage as opportunity. The road to success: integrated Management of Historic Towns-Guidebook, Regensburg
7. De Sousa C (2011) Greening the Industrial District: transforming Milwaukee's Menomonee Valley from a blighted brownfield into a sustainable place to work and play. In: Slavin MI (ed)

Sustainability in America's cities. Island Press/Center for Resource Economics, Washington, DC, pp 45–67
8. De Sousa C (2018) Menomonee Valley. In: Anderson M, Seligman A (eds) Encyclopedia of Milwaukee. Board of Regents—University of Wisconsin System, Milwaukee
9. Easterday MW, Rees Lewis DG, Gerber EM (2018) The logic of design research. Learn Res Pract 4(2):131–160. https://doi.org/10.1080/23735082.2017.1286367
10. European Commission (2024) The European Green Deal. https://commission.europa.eu/strategy-and-policy/priorities-2019-2024/european-green-deal_en. Accessed 29 Apr 2024
11. European Union (2024) New European Bauhaus: beautiful, sustainable, together. https://new-european-bauhaus.europa.eu/index_en. Accessed 29 Apr 2024
12. Ferilli G, Gustafsson C, Sacco PL (2017) Cognitive Keynesianism: heritage conservation as a platform for structural anti-cyclic policy. The case of the Halland Region, Sweden. J Cult Herit 27:10–19. https://doi.org/10.1016/j.culher.2017.02.019
13. Gregory RW (2011) Design science research and the grounded theory method: characteristics, differences, and complementary uses. In: Heinzl A (ed) Theory-guided modelling and empiricism in information systems research, 1st edn. Physica-Verl, Berlin, Heidelberg, pp 111–127
14. Gustafsson C (forthcoming) Cultural heritage and circular business models. ICOMOS Austria
15. Gustafsson C (2009) The Halland Model: a trading zone for building conservation in concert with labour market policy and the construction industry, aiming at regional sustainable development. Zugl.: Göteborg, Univ., Diss., 2009. Doktorsavhandlingar vid Chalmers Tekniska Högskola, N.S., 2990. Chalmers Univ. of Technology, Göteborg
16. Gustafsson C (2019) CONSERVATION 3.0: cultural heritage as a driver for regional growth. Sci Res Inform Technol (SCIRES-IT) 9(1). https://doi.org/10.2423/i22394303v9n1p21
17. Gustafsson C, Lazzaro E (2021) The innovative response of cultural and creative industries to major European societal challenges: toward a knowledge and competence base. Sustainability 13(23):13267. https://doi.org/10.3390/su132313267
18. Gustafsson C, Ripp M (2022) A metamodel for heritage-based urban recovery. Built Herit 6(1):29
19. Hou D (2023) NREE. 2023. Sustainable remediation and redevelopment of brownfield sites Nature reviews. Earth 4(4):271. https://doi.org/10.1038/s43017-023-00404-1
20. ICOMOS Deutschland, ICOMOS Luxemburg, ICOMOS Österreich, ICOMOS Schweiz (2012) Internationale Grundsätze und Richtlinien der Denkmalpflege. Monumenta, vol 1. Fraunhofer IRB Verlag, Stuttgart
21. Kelly AE (2010) When is design research appropriate? In: Plomp T, Nieveen N (eds) An Introduction to educational design research: proceedings of the seminar conducted at the East China Normal University, Shanghai (PR China), November 23–26, 2007. Enschede, pp 73–87
22. Menomonee Valley Partners (2024) History. https://www.thevalleymke.org/history. Accessed 29 Apr 2024
23. Michalik G (ed) (2023) Co-creation mindset: eight steps towards the future of work. Management for professionals. Springer International Publishing; Imprint Springer, Cham
24. Niedderer K (2009) Understanding methods: mapping the flow of methods, knowledge and rigour in design research methodology. Proceedings of third international conference of the International Association of Societies of design research (IASDR 2009). Seoul. http://niedderer.org/IASDR2009_niedderer.pdf
25. OurWorldHeritage (2021) New heritage approaches. OurWorldHeritage Foundation for civil society heritage partners
26. Peffers K, Tuunanen T, Rothenberger MA, Chatterjee S (2007) A design science research methodology for information systems research. J Manag Inform Syst 24(3):45–77. https://doi.org/10.2753/MIS0742-1222240302
27. Ripp M, Rodwell D (2016) The governance of urban heritage. Hist Environ: Policy Prac 7(1):81–108

28. Ripp M (2018) Heritage as a system and process that belongs to local communities. Reframing the role of local communities and stakeholders. In Faro Convention Workshop. C. o. Europe. Fontecchio
29. Ripp M, Stein P (2018) Applying the Faro convention principles to deliver heritage-based urban development: the COMUS-project. Community-led urban strategies in historic towns
30. Ripp M, Hauer S, Cavdar M (2019) Heritage-based urban development: the example of Regensburg. In: Reshaping urban conservation: the historic urban landscape approach in action. Springer, pp 435–457
31. Ripp M (2022) A Metamodel for heritage-based urban development: enabling sustainable growth through urban cultural heritage. Heritage studies. Springer, Cham
32. Schneider D, Oliveira LF, de Souza J (2017) Designing, building and evaluating a social news curation environment using the action design research methodology. Clust Comput 20(2):1731–1748. https://doi.org/10.1007/s10586-017-0781-z
33. Sloggett R, Scott M (2023) The story of climate change: narratives as influencers. In: Sloggett R, Scott M (eds) Climatic and environmental threats to cultural heritage. Routledge Taylor and Francis Group, London/New York, pp 10–26
34. Smith L (2010) Uses of heritage. Heritage studies. Routledge, London
35. Stanojev J, Gustafsson C (2019) Circular economy concepts for cultural heritage adaptive reuse implemented through smart specialisations strategies. Proceedings of the STS Conference Graz 2019, Graz, pp 415–436. https://doi.org/10.3217/978-3-85125-668-0
36. Stanojev J, Gustafsson C (2021) Smart specialisation strategies for elevating integration of cultural heritage into circular economy. Sustainability 13(7):3685. https://doi.org/10.3390/su13073685
37. The Global Goals: Goal 11 (2024) Sustainable cities and communities. https://www.globalgoals.org/goals/11-sustainable-cities-and-communities/. Accessed 29 Apr 2024
38. UK Government (2012) National planning policy framework: annex 2: Glossary. https://www.gov.uk/guidance/national-planning-policy-framework/annex-2-glossary#prev-dev-land. Accessed 29 Apr 2024
39. UNESCO (1972) Convention Concerning the Protection of the World Cultural and Natural Heritage. https://whc.unesco.org/archive/convention-en.pdf
40. UNESCO (2011) Recommendation on the Historic Urban Landscape. https://whc.unesco.org/uploads/activities/documents/activity-638-98.pdf
41. UNESCO (2024) UNESCO Culture|2030 Indicators. https://whc.unesco.org/en/culture2030indicators/. Accessed 29 Apr 2024
42. USA Public Law 107-118 (H.R. 2869) (2002) Small Business Liability Relief and Brownfields Revitalization Act
43. Weber S (2010) Design science research: paradigm or approach? AMCIS 2010 proceedings

Chapter 12
The Impact of the Russian Invasion of Ukraine on Protected Natural Areas: A Quantitative Approach and the Opportunities and Challenges of Nature Restoration

Josep Vila Subiros, Liudmyla Niemets, and Kateryna Sehida

Abstract The large-scale invasion of Ukraine by Russia as of February 24, 2022, has also been a catastrophe for its natural heritage and the country's biodiversity. Many protected natural areas have been turned into real brownfields. By April 2024, it has been estimated that 37.7% of the country's protected areas have been directly affected by the fighting in the northeast and east of the country and along the Dnieper River up to its delta. The future reconstruction of the country must take this reality into account to expand the extent of protected areas in Ukraine, considering international commitments (*Kunming-Montreal Global Biodiversity Framework*) and the new European legislative framework (*Nature Restoration Law*). Decisions that must integrate climate change scenarios to ensure a recovery of natural heritage and biodiversity that will guarantee resilient landscapes in the face of climate change and promote the well-being of the population.

Keywords Ukraine · Russian invasion · Brownfields · Protected natural area · Nature restoration · Resilient landscapes · Climate change

---

J. V. Subiros (✉)
Department of Geography, University of Girona, Catalonia, Spain
e-mail: josep.vila@udg.edu

L. Niemets · K. Sehida
Department of Human Geography and Regional Studies, Karazin Kharkiv National University, Kharkiv, Ukraine
e-mail: ludmila.nemets@karazin.ua; kateryna.sehida@karazin.ua

C. Morar et al. (eds.), *Achieving Sustainability in Ukraine through Military Brownfields Redevelopment*, NATO Science for Peace and Security Series C: Environmental Security, https://doi.org/10.1007/978-94-024-2278-8_12

12.1 Introduction

The large-scale Russian invasion of February 24, 2022, has been a humanitarian and political catastrophe for Ukraine, but also an environmental disaster [20]. A situation that has been particularly affected the country's natural heritage and nature conservation areas becoming active zones of military confrontations. The first approximations reported by the Ukrainian government itself suggest more than 20% of nature conservation areas of Ukraine are affected by the war [10]. According to the quantification of the data collected by this research, the impact has been significantly higher at around 40% (29,240,7 km^2) to be precise, with varying degrees of incidence depending on the time they were occupied or their proximity to the current front line (April 2024).

The intensity of the fighting has generated in practice many new military brownfields. Some of these areas are directly linked to damaged urban and industrial areas, but others correspond to natural areas. As against standard definitions of brownfield as for example in the Cambridge dictionary "*used to refer to an area of land in a town or city that was previously used for industry and where new buildings can be built*". This chapter focuses precisely on these greenfields that have been turned into brownfields by intense military activity. In these areas, many of the related environmental problems have been reproduced, such as soil contamination and the presence of waste and other highly dangerous and directly lethal materials such as mines and various types of explosives. In this context, we can focus our attention on a different kind of brownfields, the brownfields of war linked to urban, rural and natural areas (Photo 12.1).

Photo 12.1 Pollution of the Udy river in Kharkiv region after a Russian attack on an oil depot. (Source: Dmytro Hrebinnyk/Suspilne Kharkiv)

The few studies elaborated so far to assess the situation highlight that the main impacts would be related to habitat and wildlife destruction and behavioral changes; chemical pollution; fires; soil erosion and land cover change; and exploitation of natural resources. [6]. This situation has a very negative impact on people's health, food security and all the different types of ecosystem services, worsening air quality, increasing greenhouse gas emissions, reducing biodiversity, increasing soil contamination levels and reducing soil fertility, as well as water availability and quality [14, 16]. In the current context of war and occupation, it will be years before Ukraine can account for the full extent of the damage caused to its protected areas and natural heritage [6, 7] to such an extent that some authors have used the term like Protected Areas Genocide (PAsCide) [13] or collateral ecocide [9] to define this dramatic situation. The level of destruction, contamination and degradation of some protected areas and natural areas has turned them into real brownfields precisely because of the contamination caused by the massive use of weapons and military equipment, both in terms of their direct impact and all the indirect effects described above. The economic quantification of the damage to the natural environment in November 2022 was estimated at a total amount of 1.69 trillion UAH which is equivalent to about 40.000 million euros [19].

Therefore, to have a first approximation of the consequences on the natural heritage and protected natural areas (PNA), it is necessary to determine which areas have been most exposed to this type of process up to now. It is precisely the main objective of this chapter to make a first quantitative approximation of which protected natural areas or proposed protected natural areas have been most affected by the Russian invasion and occupation. It also provides some recommendations on future lines of action with the aim of recovering, restore and extending the network of protected areas in line with international and European Union guidelines to facilitate the process of integration into EU environmental policies (Photos 12.2 and 12.3).

The Impact of the War on Ukraine's Protected Natural Areas: A Quantitative Approach

For the elaboration of the assessment of the impact on protected areas, we have opted for an eminently cartographic work that has allowed us to map the most affected areas, also differentiating degrees of affectation. This in turn has produced quantitative information to be obtained on the degree of impact of the different categories of protection in square kilometers and percentages in relation the total area occupied by each of these protection categories.

Mapping layers on the distribution of protected sites in Ukraine have been obtained in April 2024 from *Protected Planet*. World Database on Protected Areas (WDPA) from UNEP-WCMC and IUCN. This is the world's leading platform on the status and trends of protected areas and other effective conservation measures (OECMs). The data it offers as an open source is designed to facilitate trend assessment, decision making and reporting [21]. It is therefore perfectly aligned with the objectives of this publication by providing evidence-based information to assess the impact of the war on the structure of protected areas and OECMs in Ukraine. It is a key tool for monitoring and reporting on progress towards international

Photo 12.2 Floading & draining of the Kakhovka Dam after it was blown up by the Russian army. Zaporizhzhia region, June 2023. (Source: Photo: Valentyna Hurova/Suspilne Mykolaiv)

Photo 12.3 Floading & draining of the Kakhovka Dam after it was blown up by the Russian army. Zaporizhzhia region, June 2023. (Source: State Agency of Fisheries of Ukraine)

environmental targets such as Target 3 of the Kunming-Montreal Global Biodiversity Framework (GBF) and the 2030 Sustainable Development Goals.

According to *Protected Planet* the terrestrial and inland waters protected area coverage is 12.96% and the marine 9.24%. The percentage of territory under some form of protection in Ukraine is low and there is a long way to go to meet international conservation targets at global and EU level. In this respect, it is worth recalling the commitments made. In the map the extension of natural areas that can be seen is higher than the percentage indicated as officially protected, as it also includes those considered by the Emerald Network. This network focused in Areas of Special Conservation Interest highlights precisely those areas that should be protected because of the natural values they contain and offers a clear distinction of priority natural areas for conservation and the achievement of international commitments. This was an initiative launched by the Council of Europe with the aim of helping to implement in the European framework the commitments arising from the Berne Convention on the Conservation of European Wildlife and Natural Habitats (1979). Emerald network has played a positive role in the conservation for example of the forest heritage under Ukrainian control since the beginning of the conflict in 2014 by preventing dramatic deforestation processes experienced in the sites under Russian control [18].

Map 12.1 shows a cartography of all the protected natural areas together with those forming part of the Emerald network in Ukraine, which occupy a total area of 77,474,9 km^2 (12,96% of the total area of the country). In the first place, a distinction is made between those which have been directly affected by the Russian invasion, which in total represent an area 29,240,7 km^2 and 37,7%. Therefore, approximately more than a third of them have been directly affected by the fighting. Evidently, another part of the natural areas has also suffered serious effects, but they

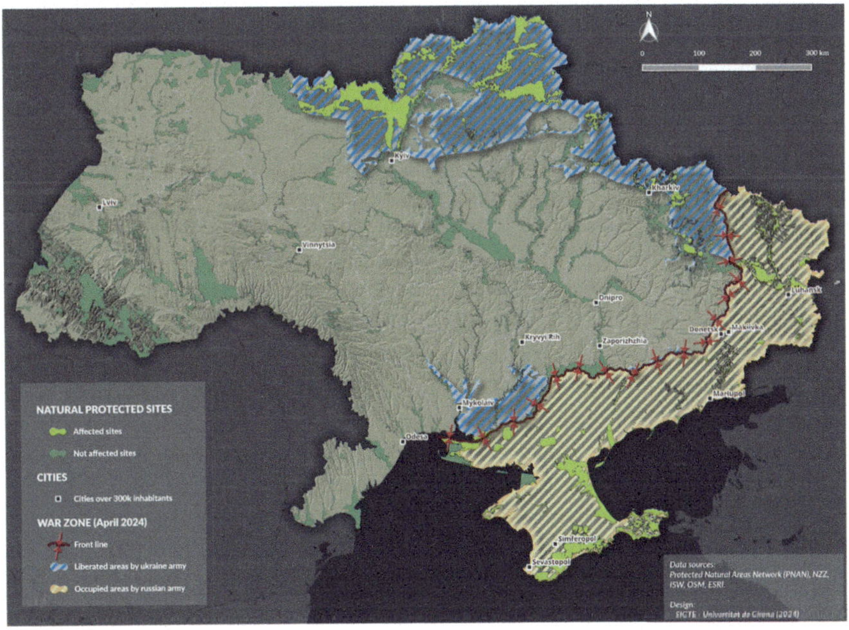

Map 12.1 Protected natural areas in the occupied and liberated areas (April 2024)

Table 12.1 Affected territorial size of protected natural areas in the occupied and liberated areas (April 2024)

Natural protected sites	Affected area (km²)	Percent (%)
Biosphere reserve (Chornobyl Radiation and Ecological Biosphere Reserve)	2265,6	100
Biosphere reserve (Chernomorskiy / Askaniya—Nova Zapovednik)	1067,3	57
Emerald network	34795,5	49,6
Nature zapovednik	675,7	42
Regional nature monuments	15,5	40,7
State zakaznik	1449,2	39,6
Ramsar sites	1404,8	39
Regional landscape parks	1197,4	37,2
State nature monuments	5,9	36,2
Zapovedne urotchische	108	32,2
Regional zakaznik	1258,1	31
National parks	516,8	11
National nature parks	12	2,4

Source: Own elaboration based on *Protected Planet*. World Database on Protected Areas (WDPA) from UNEP-WCMC and IUCN

are located outside the area directly liberated or that remains under the invader's control.

Table 12.1 shows the extent in km^2 and the percentage of invasion and fighting in protected natural areas in eastern Ukraine. First of all, it is striking to note the significant degree to which most of the protected sites have been affected, with a percentage of more than 30% of their area. In this regard, it is worth highlighting the serious impact in the first months of the invasion of the Chornobyl Radiation and Ecological Biosphere Reserve, which was affected by 100% of its area. This is very relevant considering the environmental importance of this Biosphere Reserve on a European scale [15] and especially the impact suffered, for example, by the forest fires linked to the invasion, which affected more than 62,000 ha. The burned areas have a good capacity to recover but the fires have also facilitated the spread of some invasive species in this early stage of vegetation succession [25].

The second most affected protected area is the Chernomorskiy / Askaniya—Nova Zapovednik Biosphere Reserve, with a degree of war damage of almost 50%. In this case, it should be noted that the intensity and extension of the front line in the vicinity of the reserve undoubtedly had a critical impact on the many natural values of this area, which is also of international importance.

In the second map, the focus is on the front line, which remains fairly stable after the Ukrainian army's successful offensive in September 2022, largely following the course of the Dnieper River. Since that date, there has been little change in the war line in either direction. This means that for almost 20 months the fighting has been concentrated in this part of the country, with a devastating effect on the natural heritage of the surrounding areas.

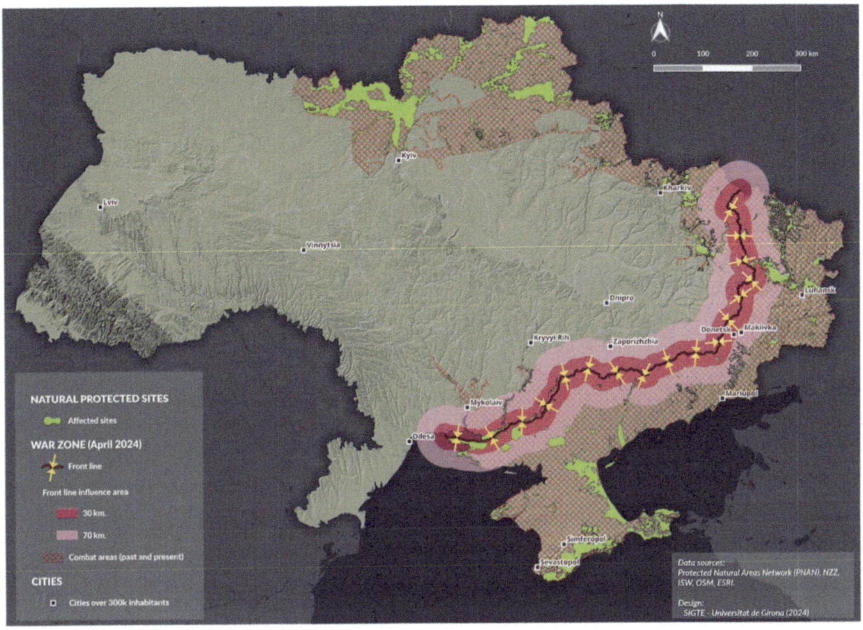

Map 12.2 Protected natural areas in front line (April 2024)—Buffer 30 km & 70 km

To quantitatively assess which protected natural areas are most affected on this frontline, two buffers of 30 km and 70 km were mapped along the boundaries of the active war zone. The boundaries of the buffers were determined considering the firing capacity of the ordinary artillery units located on the front line and depending also on the ammunition (Map 12.2).

Taking into consideration the operational range of the artillery located on the frontline we find a first buffer conditioned by the severe impact of the fighting. It is the 70 km buffer, which is also exposed to some of the ordinary artillery with the use of appropriate ammunition. The area affected by this situation includes an area of 4.098 km^2, what it represents 5,29% of PNA of Ukraine. In this case, the impact on the 331,8 km^2 that form part of the Biosphere Reserves (BR) of *Chernomorskiy / Askaniya—Nova Zapovednik* which represents the virtual loss of 17.7% of Ukraine's BRs. A percentile figure that is followed by the 450,8 km^2 of Ramsar Sites by affecting the entire Dnipro Delta River which means the destruction of a large part of the natural values of 12.5% of the country's Ramsar areas. Finally, the impact on the Emerald network deserves a special mention, as 5.7% of it lies within the 70 km buffer. This seems to be a relatively small percentage, but if we talk about absolute data, we also become aware of the large area seriously damaged, namely 3.991 km^2.

On the other hand, we observe that a total extension 2075,9 km^2 (2,68% of PNA of Ukraine) constitutes the 30 km buffer which is the most exposed to the effects of the fighting (Table 12.2). These are the natural conservation areas most affected by the armed conflict and undoubtedly in a situation of critical impacts on their

Table 12.2 Affected territorial size of protected natural areas in front line (April 2024)—Buffer 70 km

Natural protected sites	Affected area (km^2)	Percent (%)
Biosphere reserve (Chernomorskiy / Askaniya - Nova Zapovednik)	331,8	17,7
Emerald network	3991	5,7
Nature zapovednik	1,2	0,073
Regional nature monument	0,8	2,1
State zakaznik	61,2	1,7
Ramsar site	450,8	12,5
Regional zakaznik	46,8	1,2
National nature park	11,6	2,4

Source: Own elaboration based on *Protected Planet*. World Database on Protected Areas (WDPA) from UNEP-WCMC and IUCN

Table 12.3 Affected territorial size of protected natural areas in front line (April 2024)—Buffer 30 km

Natural protected sites	Affected area (km^2)	Percent (%)
Biosphere reserve (Chernomorskiy)	63,8	3,4
Emerald network	2016,4	2,9
Regional nature monuments	0,5	1,4
State zakaznik	52	1,4
Ramsar sites	361,9	10,1
Regional zakaznik	31	0,8
National nature parks	2,4	0,5

Source: Own elaboration based on *Protected Planet*. World Database on Protected Areas (WDPA) from UNEP-WCMC and IUCN

biological integrity and the natural and cultural heritage values they sustain. The areas with the highest percentage of incidence are the Ramsar sites, which, with a total of 361,9 km^2, represent the loss of 10% of this type of area in all Ukraine. This percentage is followed in second place by the 63,8 km^2 of the Chernomorskiy Biosphere Reserve (3.4% of the total number of BRs in Ukraine) and the more than 2.016 km^2 of sites included in the Emerald network (2.9% of the total number of the network) (Table 12.3).

Another effect is the loss of ecological connectivity between the different protected natural areas along the war front. This situation has continued for the last 20 months and is also a very negative factor for the maintenance of biodiversity and the capacity of ecosystems to recover. It should be recalled that the United Nations Global Biodiversity Framework recognizes the importance and the need for *maintain, enhance & restore the integrity, connectivity and resilience of all ecosystems* by 2050. A new challenge for the future will therefore be to design a network of well-connected protected areas and to increase the connectivity also of green and blue spaces in urban areas to ensure the recovery of ecosystem services and the implementation of more resilient landscapes, facilitating their recovery and their better adaptation to new climatic and environmental conditions [11].

Can the War Brownfields Be New Greens Opportunities for Ukraine? The Nature Restauration of Military Brownfields in Eastern Ukraine
Brownfields have demonstrated on many occasions their capacity to function as true reservoirs of biodiversity conservation even in situations of high environmental degradation [1]. As well as their high resilience after extreme alterations of environmental conditions. The brownfields linked to warfare have the same potential. Likewise, the renaturalization of areas close to the borders with Russia and Belarus can also be a strategic defense decision, as the restoration of forest areas, marshlands and rivers can act as a preventive defense system against potential future military conflicts [8] (Photo 12.4).

The new challenge is how to manage the recovery and restoration of Ukraine's nature conservation areas when the war is over and the country can regain its territorial integrity. In this context, it is necessary to promote the implementation of international recommendations and commitments. In this framework we have on the one hand the international considerations linked to the last COP 15 and on the other hand the more innovative European policies focused especially on the *Nature Restoration Law*.

Photo 12.4 Rocket in Sokolnyky-Pomirky. Kharkiv forest park. (Source: (c) malva.tv)

In the COP 15 on biodiversity was held in Montreal. The result was the establishment of the so-called *Kunming-Montreal Global Biodiversity Framework* [2] which establishes a broad set of milestones to be implemented by 2030. In relation to natural areas and the conservation of natural heritage and biodiversity, the following are particularly relevant:

- Promote the conservation of 30% of the planet, both on land and in marine and coastal areas. 10% under strict conservation figures.
- Restore 20% of degraded ecosystems.
- Promoting nature-based solutions to achieve these objectives.

On the other hand, we have the European Union dimension, which bases its roadmap on the *European Green Deal & the EU Biodiversity Strategy*. One of the objectives of the European Green Deal in the context of transforming the EU economy for a sustainable future is precisely the preservation and restoration of ecosystems and biodiversity. In this context, the European Union's big bet is the so-called *European Nature Restoration Law* [3], which was the old continent's contribution to COP 15 in December 2022. The law lost some of its ambition as it moved along the long road to approval. The motivation for this lesser ambition has been to seek political consensus to overcome the reluctance of the main opposition groups, such as the right-wing and extreme right-wing groups and the agricultural and livestock lobby. After a long and winding legislative path in the framework of the European institutions, the law was adopted by the Parliament on 27 February 2024 [4], but is still awaiting ratification by many EU member states. An unprecedented situation that demonstrates the difficulties to implement restoration and conservation policies in this context of high social conflict with the farming sector and a greater influence of right-wing and extreme right-wing parties.

Main objectives of *European Nature Restoration Law*:

- 2030. Recover 20% of degraded ecosystems in the European Union. Starting with areas included in the Natura 2000 network. Based on specific plans.
- 2050. Recover all degraded ecosystems in the EU.
- Agricultural areas. Achieve a "positive trend" for at least two of these three indicators: butterfly index, % of agricultural land with high landscape diversity, and organic carbon stock in the mineral soil of arable land.
- Trees: Plant 3 million trees.
- Rivers: Restore 25,000 km.
- Urban green spaces: Do not lose and increase urban green spaces.

Considering these initiatives in combination, we can establish a few key priorities for nature restoration in Ukraine base on the implementation of the principles of the most innovative policies at the international level and aligned with European Union legislation:

1. Increasing the legally protected area from 12% of the territory to 30% in the 2030 scenario. A first action would be the incorporation of the areas integrated in the Emerald network and secondly, extending the protected natural areas

affected by the Russian invasion as a measure to favour their recovery. Special mention should also be made of increasing the number of strictly protected natural areas (from IUCN category I to IV), which currently occupy a practically irrelevant percentage of the country to reach the 10% target set by the COP 15 agreements on biological diversity.
2. Focus efforts to restore 20% of degraded ecosystems in the 2030 scenario on war-affected areas with a special focus on river ecosystems and steppe restoration. These areas offer opportunities for such restoration through the prioritized implementation of nature-based solutions. This aligns restoration efforts with international policy (COP 15) and the guiding principles of the EU Nature Restoration Law.
3. Consider restoring the natural functioning of the rivers affected by the war. At this point, the maintenance of river connectivity is a key aspect that recommends not rebuilding large river barriers such as mega dams like the Kakhovka dam on the Dnieper River.
4. Increase urban green spaces with abundant woodland to reduce the impacts of climate change linked to the expected increase in temperatures especially in summer. The expansion of streets and green spaces with trees can favour the creation of climatic refuges for the benefit of people's health, with special attention to the most disadvantaged social groups.

Ukraine in the Context of the Climate and Environmental Change: Resilient Landscapes and Future Scenarios

In the process of recovery and reconstruction of the country, the new climatic and environmental context in which the country is immersed must be taken into consideration. The effects of climate change in which we are immersed, which is an element that must condition our approach to the country's reconstruction policies in general and military brownfields redevelopment in particular. The emission reduction targets of not exceeding 1.5 or 2 degrees Celsius following the Paris agreements (2015) are, unfortunately, only a dream that we will not achieve. It is worth remembering and emphasizing that 2023 was the year with the highest emissions since records have been kept and the warmest, with an average temperature increase of about 1.2 degrees Celsius compared to the pre-industrial era (1850). War is also a very important net contributor to greenhouse gas emissions. During the first year of war, CO2 emissions equivalent to 109 million tones have been estimated by Ukraine's Ministry of Environmental Protection and Natural Resources [5]. It is therefore easy to estimate that this is a figure that has now more than doubled, also influencing the acceleration of global climate change.

We must therefore think about planning our actions in a scenario of at least a 3 degree increase in average temperature by the end of this century together with a significant reduction in precipitation. Both temperatures and precipitation are particularly marked during the summer, which implies a very significant synergistic impact on agricultural production as well as on the health of people and ecosystems [24]. This is a very different climatic and environmental reality from the one we know, but one that we must take into consideration to facilitate adaptation processes and resilient landscapes. In the context of the reconstruction of the country, measures must be incorporated in advance to deal successfully and in the most efficient way with investments and decisions in the imminent future.

In this context, the key concept that emerges is the concept of Resilience. And a derivative is the concept of Resilient Landscapes adapted to new environmental conditions that have a high capacity to recover and maintain ecosystem services to the maximum. This is a great challenge, but there is no other way to conceptualize future scenarios from the perspective of complexity [17]. Conservation and Sustainability are no longer sufficient concepts, they do not allow us to address the biodiversity and climate crisis. More proactive action is needed. We must make way for Regeneration and Restoration become the new key concepts for the future management of protected natural areas in Ukraine.

A new approach to the management and restoration of natural heritage to be implemented in a context of increasing international collaboration. The contexts of geopolitical conflict must give way to new contexts of cooperation as the only way to face the great environmental and climatic challenges that the international community must confront in the framework of the Anthropocene [22]. In this context, geography can play a key role in facilitating the implementation of solutions that consider and integrate complexity and the search for broad consensus [12]. A perspective that has also been emphasized by the latest World Economic Forum report highlights that environmental and climate issues are the focus of public concern in the short term (2 years) but much more clearly in the medium term (10 years). Some of the most severe risk in a new era characterized "*As we stand on the edge of a low-growth and low-cooperation era, tougher trade-offs risk eroding climate action, human development and future resilience*" [23].

12.2 Conclusion

The impact of the Russian invasion on Ukraine's protected areas is significantly higher than the data available to date, amounting to 37.7% of the total, which represents a very significant statistic. Evidently, this impact does not respond to a homogeneous intensity but with different degrees of incidence depending on the time during which they were occupied or their proximity to the current front line of the war (April 2024).

A second conclusion is that most of them are in the northern part of the country, areas that were badly affected during the first months of the invasion. As of September 2022, the most negatively affected areas so far (April 2024) have been concentrated towards the east of the country, following especially the course of the Dnieper River up to its delta. It also highlights the importance of the rivers, true refuges of biodiversity in the country. Here too, the situation on the Dnieper River and the war front is dramatic. For example, in areas such as Dnipro Delta River, National Nature Park—Lower Dnieper National Nature Park include in the Ramsar Convention.

These concepts take on even more interest, meaning and strength in the case of Ukraine and the impacts of the Russian invasion. In addition, there is another element of great importance and interest for Ukraine, the convergence with European policies, which are in the process of incorporating this conceptualization to their legal framework with the approval soon of the nature restoration law.

The redevelopment of military brownfields and the post-war reconstruction of the Ukraine must take into consideration the new climatic and environmental context, especially the new risks associated with climate change: higher temperatures and lower rainfall especially in summer. On the other hand, it is also an opportunity to address the biodiversity crisis taking implementing the basic principles of the international & European commitments to biodiversity conservation and restoration.

It possible to provide Ukrainian society with an environment that favors the quality of life of its inhabitants, better adaptation to and mitigation of climate change, and ensure a model of progress and well-being. The restoration of the natural areas most affected by the war is the opportunity to turn these objectives into reality.

References

1. Brown HP (2014) Is the new green: brownfield sites often harbour a surprisingly large amount of biodiversity. EMBO 15(12):1238–1242. https://doi.org/10.15252/embr.201439736
2. CDB (2022). Kunming-Montreal global biodiversity framework. Conference of the parties to the Convention on Biological Diversity. United Nations. https://www.cbd.int/doc/decisions/cop-15/cop-15-dec-04-en.pdf
3. European Commission (2022) Nature restoration law—for people, climate, and planet, Directorate-General for Environment. Publications Office of the European Union. https://data.europa.eu/doi/10.2779/86148
4. European Parliament (2024) Legislative resolution of 27 February 2024 on the proposal for a regulation of the European Parliament and of the Council on nature restoration (COM(2022)0304-C9-0208/2022–2022/0195(COD)). https://www.europarl.europa.eu/doceo/document/TA-9-2024-0089_EN.pdf
5. de Klerk L, Shlapak M, Shmurak A, Mykhalenko O, Gassanzade O et al (2023) Climate damage caused by Russia's war in Ukraine. UNFCCC: Initiative on GHG Accounting of War. https://climatefocus.com/wp-content/uploads/2022/11/clim-damage-by-russia-war-12months.pdf
6. Timmins HL, Petrovych O, Drapaliuk A, Polianska K, Vasyliuk O, Bragger J, Kuzemko A, Vishnevsky D (2023) The state of Ukraine's protected areas: an interim update on damages from the full-scale invasion. PARKS, Int J Prot Areas Conservation 29(2):52–63. https://doi.org/10.2305/WUER4443
7. Hartmane I, Biyashev B, Getman AP, Yaroshenko OM, Anisimova HV (2024) Impacts of war on Ukrainian nature. Int J Environ Stud 81(1):455–462. https://doi.org/10.1080/00207233.2024.2314856
8. Khvesyk M, Ilina M (2023) After the war recovery of the environment of Ukraine: the vision and provision. Business Econ Sustainab Leader Innov Ukrainian-American Sci J 10:30–36. https://doi.org/10.37659/2663-5070-2023-10-30-36
9. Mammadov S, Luhovyi S, Starodubets O, Kalynychenko H, Trybrat R (2024) Collateral ecocide. The impact of war on Ukrainian flora and fauna. Int J Environ Stud 81(1):446–454. https://doi.org/10.1080/00207233.2024.2314851

10. Ministry of Environmental Protection and Natural Resources of Ukraine (2023) Міндовкілля: Війною уражено понад 20% природоохоронних територій України. https://www.kmu.gov.ua/news/mindovkillia-viinoiu-urazheno-ponad-20-pryrodookhoronnykh-terytorii-ukrainy. Accessed 29 Apr 2024
11. NaturaConnect (2024) Guidelines for connectivity conservation and planning in Europe. International Institute for Applied Systems Analysis (IIASA) & Martin-Luther Universität Halle-Wittenberg (MLU). Provisional version 2024.04.29
12. Niemets L, Sehida K, Kobylin P, Şahin TM, Bartosh O, Telebienieva I, Kravchenko K, Kliuchko L (2023) Integrative role of human geography in solving global problems of modernity. Visnyk of V N Karazin Kharkiv National University, Series. Geol Geograp Ecol 58:171–187. https://periodicals.karazin.ua/geoeco/article/view/22199, https://doi.org/10.26565/2410-7360-2023-58-14
13. Nikolaychuk T (2023) Protected area genocide in Ukraine: an aspect of genocide. Soc Legal Stud 6(4):132–144. https://doi.org/10.32518/sals4.2023.132
14. Pereira P, Ferdo BF, Bogunovic I, Barcelo D (2022) Russian-Ukrainian war impacts the total environment. Sci Total Environ 83:155865. https://doi.org/10.1016/j.scitotenv.2022.155865
15. Perino A et al (2019) Rewilding complex ecosystems. Science 364. https://doi.org/10.1126/science.aav5570
16. Rawtani D, Gupta G, Khatri N, Rao PK, Hussain CM (2022) Environmental damages due to war in Ukraine: a perspective. Sci Total Environ 850:157932. https://doi.org/10.1016/j.scitotenv.2022.157932
17. Schmidt C (2022) Landscape resilience. Basics, case studies, practical recommendations. Springer-Verlag, Berlin, p 245
18. Shumilo L, Skakun S, Gore ML et al (2023) Conservation policies and management in the Ukrainian Emerald Network have maintained reforestation rate despite the war. Commun Earth Environ 4:443. https://doi.org/10.1038/s43247-023-01099-4
19. Trotsiuk N, Hrabovsky H (2023) Legal protection of the environment in Ukraine: current state and prospects. Sci Works Natl Aviat Univ Ser Law J Air Space Law 1(66):156–164. https://doi.org/10.18372/2307-9061.66.17431
20. Tsaryk L, Ihor Kuzyk I (2022). Russian-Ukrainian war: Environmental aspect. // Наукові записки Тернопільського національного педагогічного університету ім. Володимира Гнатюка. Сер. Географія. Тернопіль : Тайп, 2022. Вип. 2(53):С. 100–106. https://doi.org/10.25128/2519-4577.22.2.13
21. UNEP-WCMC and IUCN (2021) Protected Planet Report 2020. UNEP-WCMC and IUCN: Cambridge UK; Gland, Switzerland
22. Vila-Subirós J, Salhi A (2022) El Antropoceno. Una interpretación desde la geografía. Documents Anàlisi Geogràfica 68:493–503. https://doi.org/10.5565/rev/dag.746
23. WEF (2023) The global RisksReport 2023, 18th edn. World Economic Forum, p 96
24. World Bank Group (2021) Climate Change Knowledge Portal (CCKP). For Development Practitioners and Policy Makers. https://climateknowledgeportal.worldbank.org/country/ukraine. Accessed 29 Apr 2024
25. Zymaroieva A, Kolomiychuk V, Fedoniuk T, Goncharenko I, Borsuk O, Melnychuk T, Svenning JC (2024) Post-fire recovery of vegetation in the Chornobyl radiation and ecological biosphere reserve. Int J Environ Stud 81(1):489–509. https://doi.org/10.1080/00207233.2023.2287345

Chapter 13
Lessons of the Regeneration of Former Soviet Military Sites in Hungary

Bence Szabó, Tamás Kovalcsik, and Zoltán Kovács

Abstract In the final phase of World War II Hungary was occupied by the Soviet troops who remained in the country until June 1991. After the withdrawal of the Soviet army most of their former barracks and other military sites were abandoned, became dilapidated and were considered as brownfields. Using a complex methodology this study aims to capture the location and actual use of former Soviet military sites in Hungary with special attention to the level and circumstances of regeneration of such sites. Our findings confirmed that, on the one hand, next to the size of the settlement (i.e. population number) and its position in the urban hierarchy, the socio-economic dynamism of the municipality has an impact on the level of regeneration and the reuse of these sites. On the other hand, local conditions also matter, like the relative location of the site within the city (distance to the centre), its accessibility, quality of the building stock, heritage preservation, ownership situation and the attitude of the municipality regarding regeneration. Centrally located sites, in cities with higher position in the urban hierarchy, with relatively good accessibility and in good economic conditions have higher chances for complete regeneration, especially if they are owned by a single owner (preferably by the state), the local government is pro-active in the field of urban regeneration and it has the power (through ownership) to attract investors.

Keywords Military brownfields · Soviet military sites · Cold war · Brownfield regeneration · Hungary

B. Szabó · T. Kovalcsik
University of Szeged, Szeged, Hungary

Z. Kovács (✉)
University of Szeged, Szeged, Hungary

HUN-REN Research Centre for Astronomy and Earth Sciences, Budapest, Hungary
e-mail: zkovacs@iif.hu

C. Morar et al. (eds.), *Achieving Sustainability in Ukraine through Military Brownfields Redevelopment*, NATO Science for Peace and Security Series C: Environmental Security, https://doi.org/10.1007/978-94-024-2278-8_13

13.1 Introduction

Under-utilized, vacant or derelict former military sites in metropolitan areas are commonly classified as 'military brownfields' in the literature [15]. Such brownfield sites usually include downtown command buildings, military garrisons with barracks, warehouses, hospitals, or military airbases often covering significant surfaces and creating urban enclaves that may pose several challenges for urban development and regeneration [5, 10]. Disused military areas stand out from the wide spectrum of brownfield sites due to their contamination problems (caused by e.g. bomb residues, heavy metals, residues from waste incineration, fuel contamination) and disadvantageous locations, often being far from the densely built urban cores with poor accessibility. These factors altogether make them less attractive for real estate developers and hampers their regeneration.

The end of the Cold War, and the subsequent decrease of military budgets as well as enhancing military technology led in the 1990s to the emergence of excessive military brownfields around the world in general, and in Europe in particular [6]. This phenomenon seems to be especially an important issue in Central Europe on either side of the former Iron Curtain where troops form the NATO and the Warsaw Pact had been stationed before the end of the Cold War [9, 15]. To this group of countries belongs also Hungary where a growing awareness towards the challenges caused by military brownfields could be observed in recent literature [2–4, 8].

Soviet troops invaded Hungary and other Central European countries in the final stage of World War II as a liberation from Nazi occupation in autumn 1944 and they withdrew nearly half a century later in summer 1991. Due to confidentiality issues the number of Soviet military troops and the exact location of sites remained hidden for a long time, and in this respect relevant data differ also considerably in the—otherwise—scarce literature [8]. According to Csapody [3] the Soviet Army stationed ca. 55,000 soldiers and 10,000 civilians on 288 sites in 104 municipalities in Hungary before they were withdrawn. Irrespective of the number of military personal, a study on the National Environmental Remediation Programme identified 171 former Soviet military sites in the country [12], whereas Kádár [7] found during her research 342 sites scattered in 108 municipalities. These fluctuating numbers raise the question about the exact geographical location, size and current use of former Soviet military sites in Hungary, and especially about their regeneration after three decades of the withdrawal of the Soviet troops. In this paper, based on a complex methodology, we try to scrutinize the actual use of Soviet military brownfields in Hungary. The main questions of the research are as follows:

- Where are the former Soviet military sites located in Hungary and what are their actual use?
- To what extent have former Soviet military sites gone through regeneration and what are their new functions?
- Is it possible to make a typology of sites according to the level of regeneration?
- What are the factors of success and/or failure of regeneration?

The chapter is structured as follows. In the following section we briefly introduce our research methodology and data collection. It is then followed by the analysis of the data and our main research findings. In the final section of the chapter we discuss the key findings and draw the conclusions.

13.2 Data and Methods

In the first phase of the research we tried to explore the exact location of former Soviet military brownfield sites in Hungary and to identify their current functions. For this purpose, desk research was carried out analyzing the existing literature and archive materials. As a starting point we used the date set comprised by Kádár [7] in her PhD dissertation, the most profound collection of data in the field. As it turned out a significant part of the 349 Soviet military objects identified by Kádár [7] could either not be found anymore (e.g. former military training grounds that became green areas in the meantime) or their size was too small, and they could not be considered as real military brownfields (e.g., smaller transmission towers). We also noted that Kádár handled some of them separately, but in certain cases some of these military infrastructures were located on the same plot (e.g., shooting range and ammunition storage), therefore, they had to be handled together (Table 13.1). Based on various sources like academic publications [2, 4], satellite images and aerial photos, we could finally identify 135 former Soviet military brownfield sites, that were digitized in a GIS database.

After the identification, geolocation of sites, and collection of descriptive information, a qualitative evaluation of the current use of the Soviet military brownfields was carried out. This phase of the research included also site visits and expert interviews in selected municipalities. Obviously, it was not possible to visit all the sites covered by the study, but we could gain useful information crosscheck them from multiple sources. With the applied research methodology we aimed to make a typology of the sites and analyze their relative location within the country and in the urban hierarchy.

Table 13.1 Soviet military brownfield sites in Hungary identified on the basis of Kádár [7]

Type	Number of sites
Geolocated, digitized sites considered for the analysis	135
Geolocation cannot be performed within the settlement	54
Sites belonging to other identified areas	58
Areas that are irrelevant for the research (not brownfields)	54
Sites that had only proprietary relationship with the soviet Army	48
Total	*349*

Source: own survey based on Kádár [7]

13.3 Analysis of Data and Main Research Findings

13.3.1 Spatial Distribution of Former Soviet Military Sites in Hungary and Their Actual Use

Altogether, more than half (54.1%) of the 135 identified Soviet military sites could be classified as former military barracks (i.e. buildings lodging soldiers and commanders extended by other facilities), this group was followed by warehouses (storing explosives, fuel etc.), airfields, training fields, radio stations, hospitals and residential enclaves. Regarding the spatial distribution of military brownfields, we could detect high concentration of Soviet garrisons and sites in the central part of the country, in and around Budapest, and north of Lake Balaton (Fig. 13.1). At the same time the border zones with former Czechoslovakia in the North, Soviet Union (Ukraine) and Romania in the East and Yugoslavia in the South were sparsely settled by the Soviet army. The map also shows that next to Budapest, the primary strategic location in the country, cities like Győr, Veszprém and Székesfehérvár in the western part of Hungary, as well as Kecskemét, Szolnok and Debrecen in the East, were more frequented by Soviet military objects. This spatial pattern can be explained by military strategic factors of the Soviet army occupying the central part of the country and concentrating its forces in major transport (rail, road and air) hubs providing easy communication among the troops and the homeland [8].

Fig. 13.1 Spatial distribution of former Soviet military sites in Hungary

In the next phase of the analysis, we elaborated a typology of the Soviet military brownfields according to the extent and form of regeneration. We based the typology on observations, field survey and available media news and the classification was based on intuition rather than metrics. According to the available information the Soviet military brownfields were divided into three main categories. We classified 'renewed', if the site has gone through large-scale or complete regeneration in the past with new emerging functions (e.g., businesses, housing). We used the term 'partially renewed', if only some parts of the site have been renovated, and new business functions did not dominate. In later case some parts of the former military buildings and infrastructure were used for lower quality activities (e.g. car repair, storage facilities) or they remained empty. Finally, we classified the site 'intact', if no or hardly any renovation has taken place since the withdrawal of the Soviet army, and the site completely lacked new business or residential functions (Fig. 13.2).

Figure 13.2 shows the spatial distribution of different types of Soviet military sites in Hungary and reflects a random pattern. The otherwise prevailing East-West or Centre-Periphery dichotomy within the country [13, 14] cannot be identified. Therefore, the map raises an intriguing question about the main explanatory factors for success and/or failure of regeneration of Soviet military sites. Answers to this question can be manifold. On the one hand, we assumed that next to the size of the city (i.e. population number) and its actual position in the urban hierarchy, the socio-economic dynamism of the municipality may have an impact. On the other hand, we also assumed that the relative location of the military site (i.e. distance from the core of the municipality and its accessibility), its physical and

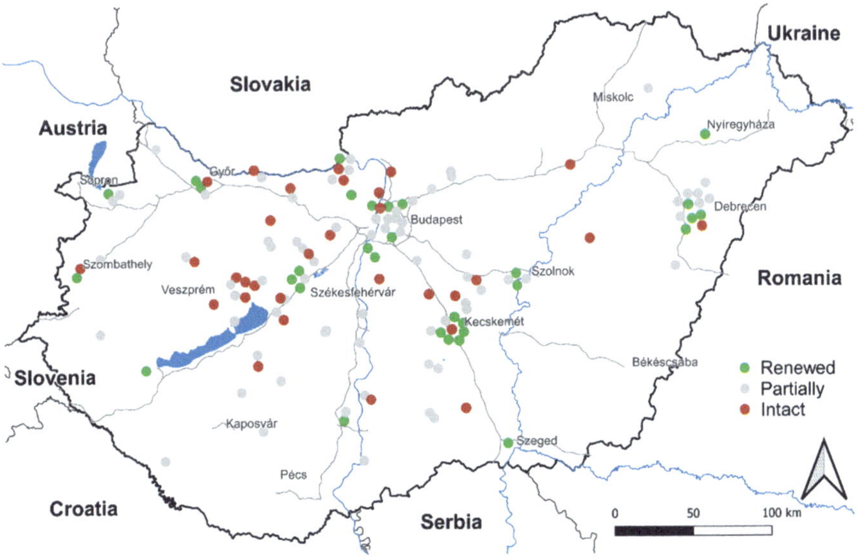

Fig. 13.2 Spatial distribution of different types of former Soviet military sites in Hungary

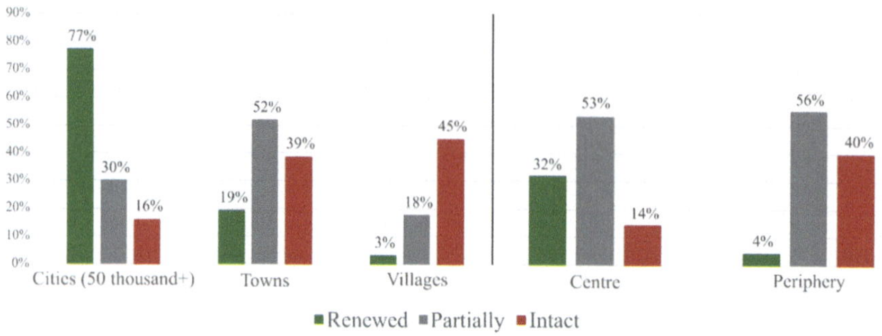

Fig. 13.3 Distribution of different types of military sites in the urban hierarchy and according to the socio-economic dynamism of cities

environmental conditions, as well as local political factors (i.e. abilities of the town hall to attract investors, long-term development strategies) play a role here.

To test our preliminary hypotheses, first we investigated the spatial distribution of different types of former Soviet military sites according to the urban hierarchy (Fig. 13.3). As demonstrated by our data, the overwhelming majority of renewed Soviet military sites are located in larger cities, belonging to the upper levels of the urban hierarchy, whereas the intact military brownfields are mainly concentrated in smaller towns and villages, at lower levels of the settlement hierarchy. This implies that the size of the municipality and local budgetary conditions really matter.

As a next step we created an index to check the centre-periphery relations in absolute and relative terms. In absolute term, we considered the distance of the municipality where the site is located from Budapest, the capital city of the country, on road (in km). In relative term, we considered per capita income and the level of unemployment as proxies for economic success. The integration of these three variables in one index provided us an indicator which indirectly reflects the socio-economic dynamism of the municipality. As Fig. 13.3 shows, renewed Soviet military sites tend to concentrate in economically thriving places, located closer to Budapest the dominant centre of Hungary, whereas intact military objects are more dominant in peripheral locations.

13.3.2 Influencing Factors of Regeneration

During the site visits and interviews we could identify several causes that can explain the success of regeneration of Soviet military brownfield sites, and if regeneration at all has been initiated in any form. Our findings showed that the success or failure of regeneration projects of Soviet military brownfields depends on several factors. The most important among them are as follows:

- location within municipal area;
- origin of the buildings;

- heritage preservation;
- quality of the infrastructure and buildings;
- ownership situation;
- local urban development and regeneration strategies;

Our study gave evidence that *relative location* and physical accessibility within municipal area matters. Military brownfields that are located far from the city centre, isolated from the urban fabric, are very much disadvantaged regarding the potentials of regeneration compared to those lying close to the city centre. This is in line with the findings of Kádár [8]. A good example is the Frigyes barracks in the economically prosperous city of Győr, that were built in 1897 for the Hungarian army. After four decades the Soviet troops left behind the buildings in very dilapidated conditions that became, however, subjects of heritage protection in 2001. Finally, in 2005 the site was bought by the Austrian owned Leier International company specialised in building materials manufacturing. In the subsequent 15 years the six buildings located on 3 hectares in the centre of Győr were completely renovated and converted to office (giving home to the Hungarian headquarter of the company) and high quality residential spaces. The project can be considered very successful and it has contributed to the extension of the city centre of Győr (Photo 13.1).

The *origin of the buildings* refers to their inception, whether they had been built already before the arrival of the Soviet troops, or they were built by the Russians on site. Soviet army confiscated several buildings from the Hungarian army after World War II, that were built predominantly during the Austro-Hungarian Monarchy (i.e. before the First World War). Despite their age such buildings have usually preserved their architectural quality and their modernization (i.e. instalment of new amenities) have been relatively easy. On the other hand, buildings constructed by the Soviet army followed architectural standards different from the European norms (e.g. size of windows and doors), therefore, it has been difficult to renovate them. Also, buildings erected by the Soviet army were often made of pre-fab concrete panels representing poor quality and their renovation was costly and technologically difficult. Therefore, buildings built by the Soviet army have often been demolished completely after they left.

Photo 13.1 The former Soviet barracks (left) and the Leier City Centre project (right) in Győr

Photo 13.2 The main building of the former Hajmáskér barracks (left) is protected but in very poor conditions, and a nicely renovated rear wing (right)

The role of *heritage preservation* is also important when it comes to regeneration [11]. On the one hand, if a military brownfield building is officially listed as a historic monument it makes the renovation cumbersome and costly. On the other hand, such buildings cannot be torn down and the potentials of full renovation by a new investor (i.e. new owner) remain high. An example for this is the Hajmáskér barracks, which were built during the Austro-Hungarian Monarchy and are under heritage protection, and the full renovation of the military brownfield would require a large-scale investment (Photo 13.2).

The *quality of the infrastructure, buildings* and the environment of the site is also important regarding the potentials of regeneration projects. If a site is poorly supplied with basic infrastructure, the buildings are dilapidated and the environment is degraded (e.g. lacking green surfaces), then it is difficult to find investors. On the contrary, sites supplied by great variety of facilities and infrastructures, retaining well preserved buildings and attractive environment usually encourage potential investors.

Regarding *ownership situation*, it is advantageous if there is a single landowner who owns the land and controls the reuse and refurbishment of the buildings, and especially if the landlord is the state (central or local government). In the case of multiple ownership and especially if the land is owned by a multitude of private landowners the successful regeneration of military brownfield sites is often difficult. Such ownership situation tends to complexify land management issues and makes long-term planning impossible.

The existence of clear *local urban development and brownfield regeneration strategies* are also important [1]. Most of the successful regeneration projects of Soviet military brownfield sites are located in municipalities where the municipal government realized the importance of local brownfield regeneration programmes, such programmes were elaborated shortly after EU accession, and additional funding for regeneration from national and EU budgets could be raised.

The above listed and discussed factors have been of paramount importance for the success or failure of the regeneration of Soviet military brownfield sites in Hungary, and we think they also highlight the heterogeneity of situations.

13.4 Discussion and Conclusions

The state of the art of the regeneration of Soviet military brownfield sites in Hungary provides us many lessons. On the one hand, as supported by our data, the effectiveness of regeneration of Soviet military brownfields depends very much on the size of the municipality and its socio-economic dynamism where they are located. On the other hand, local conditions also matter, like the relative location of the site (distance to city centre), its accessibility, building quality, heritage preservation, ownership situation and the attitude of the municipality towards regeneration. Centrally located sites with good accessibility, higher up in the urban hierarchy and in relatively good conditions have higher chances for complete renovation, especially if the local government is pro-active in the field of urban regeneration and it has the power (through ownership) to attract investors. Next to this general wisdom our research results also show that there is no general recipe. We found Soviet military brownfields successfully renovated in locations that do not correspond the above criteria, but they represent a tiny minority.

Acknowledgments This research was supported by project no. TKP2021-NVA-09 implemented with the support of the Ministry of Culture and Innovation of Hungary from the National Research Development and Innovation Fund, financed under the TKP2021-NVA funding scheme.

References

1. Bagaeen S (2006) Redeveloping former military sites: competitiveness, urban sustainability and public participation. Cities 23(5):339–352. Available at https://doi.org/https://doi.org/10.1016/j.cities.2006.05.002
2. Bársony R (2018) Bezárt katonai- és határőrlaktanyák hasznosításának helyzete Magyarországon [The utilization of closed military and border guard barracks in Hungary]. Területfejlesztés és Innováció 12(2):40–52
3. Csapody T (2000) A gyalogsági aknák Magyarországon. Országjelentés a Landmine Monitor részére [Landmines in Hungary. Country report for Landmine Monitor]. MTA, Politikai Tudományok Intézete, Budapest, p 53
4. Dannert É, Pirisi G (2016) Demilitarizálódó kisvárosok – a katonai barnamezők hasznosításának helyi sikerei és kudarcai [Demilitarizing small towns – local successes and failures of military brownfield utilization]. Területfejlesztés és Innováció 10(2):27–44
5. Hercik J, Šimáček P, Szczyrba Z, Smolová I (2014) Military brownfields in The Czech Republic and the potential for their revitalisation, focused on their residential function. Quaestiones Geographicae 33(2):127–138
6. Jauhiainen J (2007) Conversion of military brownfields in Oulu. In: Rebuilding the City. Managing the built environment and remediation of brownfields. The Baltic University Press
7. Kádár K (2013) Funkcióváltás szovjet katonai objektumok helyén. Barnamezős katonai területek újrahasznosítása hat magyar megyeszékhelyen [Functional change at the location of Soviet military sites. Reuse of military brownfields in six Hungarian county centers]. PhD-dissertation, University of Pécs, Pécs
8. Kádár K (2014) The rehabilitation of former Soviet military sites in Hungary. Hungarian Geographical Bulletin 63(4):437–456

9. Kunc J, Martinát S, Tonev P, Frantál B (2014) Destiny of urban brownfields: spatial patterns and perceived consequences of post-socialistic deindustrialization. Transylv Rev Adm Sci 41:109–128
10. Morar C, Nagy G, Boros L, Gozner M, Niemets L, Sehida K (2021) Heritage, culture and regeneration of the former military areas in the city of Oradea, Romania. Architektura & Urbanizmus 55(1–2):76–87
11. Morar C, Nagy G, Dulca M, Boros L, Sehida K (2019) Aspects regarding the military cultural-historical heritage in the City of Oradea (Romania). Ann Istrian Mediterranean Stud Ser Historia et Sociol 29(2):303–322. Available at https://doi.org/10.19233/ASHS.2019.21
12. Nagy Á (ed) (2003) Az EU strukturális alapok keretében barnamezős rehabilitációra kiírandó pályázatok szakmai megalapozása (professional support of applications to be submitted for brownfi eld rehabilitations stemming from the structural funds of the EU). VÁTI Kht, Budapest, p 44
13. Pénzes J, Demeter G (2021) Peripheral areas and their distinctive characteristics: the case of Hungary. Moravian Geograph Rep 29(3):217–230. Available at https://doi.org/10.2478/mgr-2021-0016
14. Pósfai Z, Nagy G (2017) Crisis and the reproduction of core-periphery relations on the Hungarian housing market. Eur Spat Res Policy 24(2):17–38
15. Rey E, Laprise M, Lufkin S (2022) Neighborhoods in transition: brownfield regeneration in European metropolitan areas. Springer, Cham

Chapter 14
GIS Based Decision Support System for Brownfield Redevelopment: The Case Study Municipality of Prokuplje, Serbia

Aleksandar Valjarević, Kh. Md Nahiduzzaman, and Filip Vujović

Abstract Over the past 30 years, wastelands have increased in Serbia, particularly in the south and southeast. Prokuplje and surrounding municipalities lack protected areas and face significant depopulation and contamination issues. The Toplica region, with 88,900 inhabitants, is strategically located with key transportation routes, but former military sites in Prokuplje present challenges due to contaminated soils and proximity to urban areas, necessitating remediation and redevelopment. Three zones are ecologically suitable for conversion in Prokuplje, while one is moderately suitable. Two brownfields near the Niš-Prokuplje-Merdare-Priština road pose remediation challenges due to contamination with mercury(I) oxide, aluminum, and boric acid. Shallow contamination near densely populated areas makes remediation difficult. Legal regulation, homeowner assessment, and in-depth chemical analysis are essential for effective remediation. Essentially, Prokuplje has three areas suitable for conversion and one moderately suitable. However, the two brownfields near the Niš-Prokuplje-Merdare-Priština road face significant remediation challenges due to contamination with mercury oxide, aluminum, and boric acid. Addressing these issues requires legal regulations, homeowner assessments, and detailed chemical analysis to ensure safe redevelopment and boost investor confidence. Future efforts may include spatial analysis of former military sites, expanding to the Toplica region.

Keywords Brownfield · Redevelopment · Depopulation · Military facilities · Contaminated soils · Decision support system

The original version of the chapter has been revised. A correction to this chapter can be found at https://doi.org/10.1007/978-94-024-2278-8_36

A. Valjarević (✉)
Faculty of Geography, University of Belgrade, Belgrade, Serbia
e-mail: aleksandar.valjarevic@gef.bg.ac.rs

K. M. Nahiduzzaman
Mohammed VI Polytechnic University, Ben Guerir, Morocco
e-mail: Khmd.NAHIDUZZAMAN@um6p.ma

F. Vujović
University of Montenegro, Podgorica, Montenegro

C. Morar et al. (eds.), *Achieving Sustainability in Ukraine through Military Brownfields Redevelopment*, NATO Science for Peace and Security Series C: Environmental Security, https://doi.org/10.1007/978-94-024-2278-8_14

14.1 Introduction

Over the last thirty years, the number of wastelands has steadily increased in the Republic of Serbia which has an area of 88,361 Sq. km. The situation is particularly precarious in the south and south-east of the country. The municipalities in south and southeast, such as Prokuplje, Kuršumlija, Žitorađa, Blace, Aleksinac, Niš, Negotin, Kladovo, etc., do not have any protected areas that were affected by the first, second and most recent conflicts in the former Yugoslavia. The municipality of Prokuplje, together with the municipalities of Žitorađa, Blace and Kuršumlija, belongs to the Toplica region. The region has a total population of 88,900 inhabitants with an average population density of 39.1 inhabitants per km^2 i.e., lower than the European average, 105 inhabitants per km^2 (Census, 2022). The Toplica region lies between the mountains of Jastrebac in the north and the mountains in the south, including Sokolovica, Vidojevica and Radan [1]. The Toplica region and the municipality of Prokuplje have a good geographical position and belong to the south-eastern part of Serbia. The municipality of Prokuplje covers an area of 759 km^2, of which 20.96 km^2 are urban areas. The average population density of the municipality is 38.054, while the average density of urban and rural areas is 50.

It is interesting that the municipality of Prokuplje had a downsize of population from 59,292 (1948) to 38,291 in 2024. The main river of the municipality, the Toplica, flows for 45.5 km. The Toplica River flows through the central and western part of the valley. The Toplica region has a very favorable geographical position and terrain. The most important railroad lines and highways of the Balkan Peninsula pass through the region and represent the shortest connection between the Vardar and Morava valleys and the Adriatic Sea. This route also passes through the central part of the municipality. Most of the inhabitants live in the town of Prokuplje (23,435), the fewest in the rural settlements, which fluctuate between 0 and 100 [2]. The depopulation of the municipality is partly due to economic and social reasons. The war in the former Yugoslavia in the early 1990s had a strong influence on the fact that some parts of the municipality were uninhabited. The strong army in the former Yugoslavia built military facilities throughout the municipality. Most of the facilities were built in the large capital Prokuplje. All the buildings are located in the eastern suburbs of the city and one building in the west. The area of these three main objects is 12,000,000 m^2. There were total five projects in the city of Prokuplje, the other three military objects are smaller and are spread out in the mountains and hills [3]. After the year 2000, the three military facilities, one of which was located in the hills and the other in the mountains, were decommissioned. The first object in the northern part of the city has not been sufficiently rehabilitated and the land is now being restructured into rural areas. The main problem of the first object in the northern part of the city is the deep intrusion into the urban areas of the city. The area of this property is 23 hectares, and it is surrounded by 120 houses and more than 1000 households. Another problem is that it is a former military object and some parts of the area of the object have poisoned soils with a high content of mercury (I) oxide Hg2O, aluminum (Al) and boric acid (BH3O3). The northern and eastern parts of this site are now covered with fruit and vegetables such as plums, apples, peas, wheat, corn, potatoes and cabbage. The situation on the second former military site is much more complex. This site has an area of 25 hectares and is home to more than 200 houses and 5000 homeowners. The situation could be even worse as it is in close proximity to the springs and two schools. In the past, there was a landfill with contaminated waste on this site (military land). Some recent investigations have revealed high levels of heavy metals and hydrochloric acid (HCl) [4]. After 2000 years, the city's industry was dispersed in the parts of the city where three military sites exist today. This can also be a problem as daily migration (workers) takes place very close to the three sites.

14.2 Methodology

The integration of the Geographic Information System with Multi-Criteria Analysis (GIS-MCDA) represents an important approach for transforming and combining geospatial data to obtain new information of great importance for decision making. The process of mutual comparison and evaluation of the influence of different criteria makes it possible to assess the suitability of a particular space for a specific phenomenon or process [5–7]. Geographic information systems in combination with numerical analyzes can accelerate the process of analyzing spatial data. The common and advanced GIS methods are well suited to determine spatial problems. The balance of space is one of the main issues in the modern world. In today's world, there are many problems such as climate change, land degradation, major floods, extreme weather conditions, urban sprawl, industrial pollution and, of course, brownfields. Three analysis steps were approved and defined in the study. The first step was the search for suitable documents and data such as old topographical maps, spatial plans, historical documents, old photos, etc.

The second step was the digitization of data available in analog formats. The digitization process was carried out in three main phases. The first phase consisted of georeferencing old topographic maps at scales of 1:25,000, 1:50,000 and 1:100,000. The second phase consisted of vectorizing all spatial data from the analogical topographic maps. The third phase consisted of identifying old military objects and analyzing their characteristics in space and time.

The third step consisted in the digitization and vectorization of all data and the better analysis of old photographs and satellite images [8]. The main methods used in this research are inreptolation, kriging, zonal statistics and two types of zonal statistics. The zonal statistical methods together with AHP and modified AHP methods provided satisfactory scientific results. These results can be divided into statistical and numerical results. The statistical methods are used to produce better analyzes of brownfields and former military sites. The AHP methods are well suited to determine the process of selecting the remediation method and the process of brownfield adaptation [9].

One of the main difficulties associated with brownfields is the uncertainty about the cost and time required to conduct site investigation and remediation activities. In addition, the risk to human health and the environment should be one of the most important factors in prioritizing brownfield sites for investigation, remediation and redevelopment. To incorporate this dimension into our indexing scheme, we considered parameters that can be easily quantified with available data sources and that provide a rough representation of the three elements that are typically part of environmental risk assessment: Source, pathway and receiver. Six criteria were selected (Table 14.1). The input data for the criteria were collected on the basis of available spatial planning documents.

The GIS-MCDA procedure implemented in the QGIS software in this study is based on six steps: Determination of the objective of the multi-criteria analysis/problem definition; determination of the criteria; standardization of the criteria values; determination of the weighting of the criteria; the results of the analysis (standardization of the standardized values and weights of the criteria) and the validation of the results [10, 11]. The weighting coefficients for the criteria in terms of importance are determined by applying the Analytic Hierarchy Process (AHP) (Fig. 14.3).

The AHP methods are very important for the classification of land and soil degradation. Another important tool is the distance matrix, which can be used in GIS software. The distance matrix software is very useful in combination with the buffer zone statistics. The buffer zone statistics are well suited for creating a real distance to brownfields. The distances (buffer zones) are subdivided in this study according to the respective problems of the municipality.

The minimum distance is 10 m, the average 500 m and the maximum 1500 m. The criteria for the distance are divided into (low, medium, high). The criteria for the type of industrial site Industrial, commercial and residential area. The criteria for determining the distance from surface waters. The criteria for determining the distance from protected areas. The criteria for the distance from forests. The criteria for distance from the last spatial level of the municipality (see Table 14.1).

The other AHP criteria are defined by the type of land, zoning and distance to the last spatial level of the municipality. Distance to surface water. Distance to the municipality's protected areas. Distance to the municipality's forest areas.

Table 14.1 Criteria for environmental indices within this research

Criteria	Low	Medium	High
Soil type (C1)	Impermeable	Semi-permeable	Permeable
Zoning according spatial plan (C2)	Industrial	Commercial	Residential
Distance from surface water bodies (C3)	<100 m	100–300 m	>900 m
Distance from protected areas (C4)	<500 m	500–1000 m	>1500 m
Slope (C5)	>10	2–10	<2
Distance from forests (C6)	<100 m	100–300 m	>300 m

14.3 Results (Fig. 14.1)

The second brownfield is located in the hilly area (Borovnjak), only 750 m from the city center. The main problem of this former military site is the high concentration of contaminated soil. The soils have a high concentration of mercury (I) oxide Hg2O, aluminum (Al) and boric acid (BH3O3). The soils in this brownfield site are very close to the built-up areas. The average distance is 100 m. Another potentially major problem is the short distance to the urban areas of the city. The urban zones of the city have encroached into some areas of this brownfield (Fig. 14.2).

The average height of this brownfield is 500 m. The soil type of this fallow land is permeable. According to the zonal statistics, it is industrial and commercial land. The distance to the protected areas is 300 m. The average slope is over 10°. The main problem is the very close location of the urban areas, which have an average density of 1000 inhabitants per km^2. In some zones, the density is almost 2000 inhabitants.

The third brownfield is located in the southern part of the Toplica River. The average distance between the third brownfield and the Toplica River is 200 meters. The area of this brownfield is 1,000,000 m^2. The type of soil is semi-permeable and permeable. According to the last spatial level, it is commercial and residential land. The average distance to surface waters is 200 m. The average distance to forests in the municipality is 300 m. This brownfield on the former military site is a military landfill and the concentration of chemical substances is not high. This former

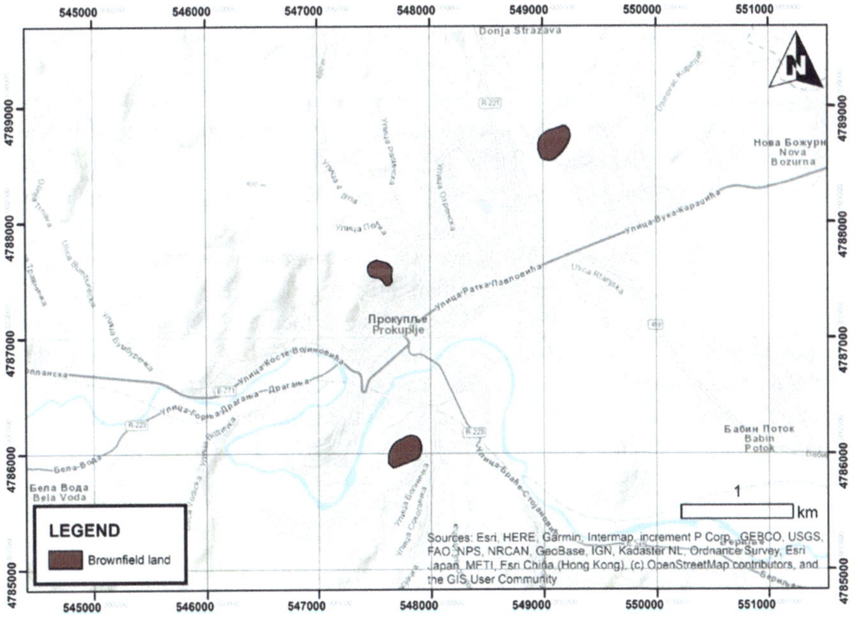

Fig. 14.1 Location of the actual brownfield in the municipality of Prokuplje

Fig. 14.2 The second brownfield in the city of Prokuplje. (Photo by the author Aleksandar Valjarević)

military site was only used for physical waste and not for chemical content and is therefore not dangerous for life.

The main problem associated with this site is its proximity to the main bus stations. The municipality's new spatial planning program envisages that the redevelopment of the brownfield site will take up part of the bus site. The ambitious concept also includes the creation of green spaces between the bus station and the brownfield site (Fig. 14.3).

After the overall analysis of the environmental characteristics and spatial location of the three brownfields in the municipality of Prokuplje, the main conclusion is that the potential for brownfield redevelopment is medium (Fig. 14.4). The environmental index was between 5 and 6 on a scale of 0–10. The reason for these results is this the location of the current brownfields is not good, as the urban areas are very close to each other. The second brownfield site examined also has a high concentration of chemical constituents (Fig. 14.4; Tables 14.1 and 14.2).

Fig. 14.3 The central bus station and the former military landfill. (Photo by the author Aleksandar Valjarević)

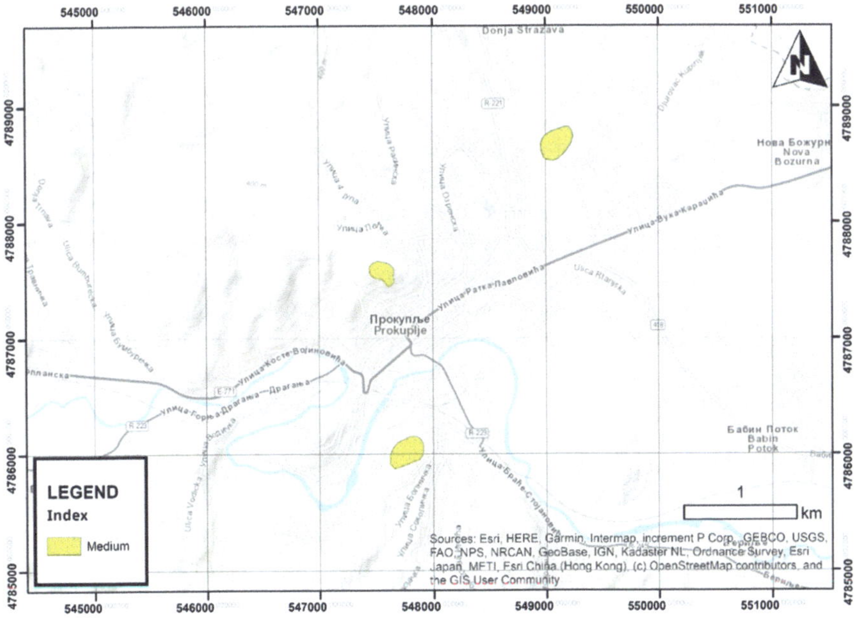

Fig. 14.4 Suitability map for brownfields according to the Environmental Index

Table 14.2 AHP matrix for environmental indices within this research

Criteria	C1	C2	C3	C4	C5	C6	Weight
C1	1.00	2.00	3.00	3.00	3.00	3.00	0.328
C2	0.50	1.00	2.00	3.00	3.00	2.00	0.226
C3	0.33	0.50	1.00	2.00	3.00	2.00	0.159
C4	0.33	0.33	0.50	1.00	3.00	2.00	0.125
C5	0.33	0.33	0.33	0.33	1.00	2.00	0.085
C6	0.33	0.50	0.50	0.50	0.50	1.00	0.077

14.4 Discussion

According to the results obtained in the Prokuplje area, three areas are environmentally suitable for conversion for various purposes, while one area is moderate. The environmental impact assessment (EIA) is of key importance for brownfields and their redevelopment for several reasons. Brownfields are sites that were previously used for industrial or commercial purposes and now need to be revitalized or redeveloped due to potential contamination. Here are some reasons why the environmental index is important in this context:

Identification of contamination: The Environmental Index allows for a systematic examination of the soil to identify potential hazards and contaminants that may have resulted from past industrial activities. This is key to understanding the extent of the problem and determining the extent of contamination [12].

Risk assessment: Based on the data obtained through the environmental index, a risk assessment for human health and the environment can be carried out. This assessment helps to make decisions about the need for remediation and to set priorities.

Planning and implementation of remediation measures: The Environmental Index forms the basis for developing a brownfields remediation plan. These plans include strategies to eliminate or reduce contamination to ensure safe land use in the future.

Legal and regulatory requirements: In many cases, an environmental index is required to meet legal and regulatory requirements. Authorities and regulators often require an environmental impact assessment as part of the approval process for work on brownfield sites. Investor confidence: The existence of an environmental index can increase investor and community confidence in the brownfield revitalization process. Transparency about the extent of pollution and the remediation plan can facilitate acceptance of the project by various stakeholders. An environmental index is crucial for brownfields as it identifies contamination, assesses risks to health and the environment and enables the planning and implementation of remediation measures. In addition, it meets legal requirements, strengthens investor confidence and ensures transparency in the revitalization of contaminated sites. According to the results obtained in the Prokuplje area, three areas are ecologically suitable for conversion for various purposes, while one area is moderately suitable. The environmental impact assessment (EIA) is of key importance for brownfields and their redevelopment for several reasons. Brownfields are sites that were previously used for industrial or commercial purposes and now need to be revitalized or redeveloped due to potential contamination. Here are some reasons why the environmental index is important in this context: Identification of contamination: The Environmental Index allows for a systematic investigation of properties to identify potential hazards and contamination that may stem from past industrial activities. This is crucial to understand the extent of the problem and determine the level of contamination. Risk assessment: Based on the data obtained through the environmental index, a risk assessment for human health and the environment can be carried out. This assessment helps to make decisions about the need for remediation and the prioritization of actions. Planning and implementation of remediation measures: The Environmental Index forms the basis for developing a brownfields remediation plan. These plans include strategies to eliminate or reduce contamination to ensure safe future land use. Legal and regulatory requirements: In many cases, an environmental index is required to meet legal and regulatory requirements. Authorities and regulators often require an environmental impact assessment as part of the approval process for work on brownfield sites. Investor confidence: The existence of an environmental index can increase the confidence of investors and communities in the brownfield revitalization process. Transparency regarding the extent of the pollution and the plan to solve the problem can facilitate acceptance of the project by the various stakeholders. The environmental index is crucial for brownfields as it identifies contamination, assesses risks to health and the environment and enables the planning and implementation of remediation measures. In addition, it meets legal requirements, strengthens investor confidence and ensures transparency in the revitalization of contaminated sites.

14.5 Conclusion

According to the results obtained in the Prokuplje area, three areas are ecologically suitable for conversion for various purposes, while one area is moderate. The two areas (brownfields) are very close to the international road connection Niš-Prokuplje-Merdare-Priština. This could be a potential problem for better remediation of these sites. The first brownfields is located near built-up areas and protected forest areas. The second brownfield site is located in the hilly areas that have strictly encroached into the city. These brownfields have contaminated contents and a high concentration of mercury (I) oxide Hg2O, aluminum (Al) and boric acid (BH3O3). These chemical constituents are located at very shallow depths and close to urban areas with a high population. The remediation of this site is very difficult to authorize. The height of this brownfield is 600 m and the slope is higher than 10°. The slope of the second brownfield site is about 2°, that of the second with the area. The third brownfield site is the smallest in the area and has a slope of 3°. Planning and implementation of remediation measures including the next levels. Legal regulation of brownfields in the municipality in accordance with local and national laws. The assessment of a number of homeowners especially in the second brownfield. The in-depth analysis of the chemical and environmental problems associated with brownfields. It also meets legal requirements, boosts investor confidence and ensures transparency in the revitalization of brownfield sites. Future work may include spatial analysis of all former military sites and brownfields. The investigation can be extended to the area of the Toplica region.

References

1. Valjarević A, Djekić T, Stevanović V, Ivanović R, Jandziković B (2016) GIS numerical and remote sensing analyses of forest changes in the Toplica region for the period of 1953–2013. Appl Geogr 92:131–139
2. Predojević-Despić J, Penev G (2016) Population of Serbia abroad by destination countries: regional approach. Bull Serbian Geogr Soc 96(2):83–106
3. Valjarević A, Živković D, Valjarević D, Stevanović V, Golijanin J (2014) GIS analysis of land cover changes on the territory of the Prokuplje municipality. Sci World J 2014:1–8
4. Gržetić I, Ghariani RHA (2008) Potential health risk assessment for soil heavy metal contamination in the central zone of Belgrade (Serbia). J Serb Chem Soc 73(8–9):923–934
5. Kpiebaya P, Amuah EEY, Shaibu AG, Baatuuwie BN, Avornyo VK, Dekongmen BW (2022) Spatial assessment of groundwater potential using quantum GIS and multi-criteria decision analysis (QGIS-AHP) in the Sawla-Tuna-Kalba district of Ghana. J Hydrol Reg Stud 43:101197
6. Ferrando A, Faccini F, Paliaga G, Coratza P (2021) A quantitative GIS and AHP based analysis for geodiversity assessment and mapping. Sustain For 13(18):10376
7. Siddayao GP, Valdez SE, Fernandez PL (2015) Modeling flood risk for an urban CBD using AHP and GIS. Int J Infor Educ Technol 5(10):748–753
8. Valjarević A (2024) Long-term remote sensing-based methods for monitoring air pollution and cloud cover in the Balkan countries. Environ Sci Pollut Res 31:27155–27171. https://doi.org/10.1007/s11356-024-32982-y
9. Lange D, McNeil S (2004) Clean it and they will come? Defining successful brownfield development. J Urban Plan Dev 130(2):101–108
10. Jiang H, Eastman JR (2000) Application of fuzzy measures in multi-criteria evaluation in GIS. Int J Geogr Inf Sci 14(2):173–184
11. Domazetović F, Šiljeg A, Lončar N, Marić I (2019) Development of automated multicriteria GIS analysis of gully erosion susceptibility. Appl Geogr 112:102083
12. Nikolić G, Vujović F, Grozdanić G, Valjarević A (2024) Geomorphological characteristics of Montenegro. In: Speleology of Montenegro. Springer International Publishing, Cham, pp 67–78

Part II
Social, Economic, and Environmental Aspects of Reusing Former Military Lands

Chapter 15
Sustainable Use of Materials Found in Military Brownfield Sites: Repurposing and Recycling Opportunities for Industrial Growth

Lucien N. Brush

Abstract The abandoned Brownfield military installations may provide resources that may provide regional economic benefits. Materials that are recovered may be re-utilized in the local economy and remaining structures and real estate also re-purposed to help local economies. A community engaged approach for Brownfield development is outlined, emphasizing a materials science and engineering perspective that will be important in deciding appropriate pathways for resource utilization and re-purposing. An example of a possible economic and environmentally sustainable industrial application as a development pathway is also provided.

Keywords Military Brownfields · Material repurposing · Materials processing · Economic opportunity

15.1 Background and Approaches for Sustainable Materials Recovered in Brownfield Sites

The abandonment of military installations leads to possible opportunities for the redevelopment of the Brownfield sites to benefit the regional economy. Examples of Brownfield sites are shown in Fig. 15.1, revealing a variety of recyclable materials, equipment and structures including buildings that typically reside on land areas that are large enough to support new industries.

NATO Advance Research Workshop (ARW), Achieving Sustainability in Ukraine through Military Brownfields Redevelopment University of Oradea, Romania, December 18, 2023.

L. N. Brush (✉)
Department of Materials Science and Engineering, University of Washington, Seattle, WA, USA
e-mail: brush@uw.edu

C. Morar et al. (eds.), *Achieving Sustainability in Ukraine through Military Brownfields Redevelopment*, NATO Science for Peace and Security Series C: Environmental Security, https://doi.org/10.1007/978-94-024-2278-8_15

Fig. 15.1 Examples of abandoned Brownfield sites illustrating materials, equipment and structures remaining. (a) From Ref. [5]. (b) From Ref. [6]

In general, multiple approaches to the development of Brownfield sites are applicable. Approaches may be unsustainable, in which resources are irreplaceably consumed; or sustainable, in which the resources can be repurposed for continuous use. There have been several detailed studies that focus on the redevelopment of military Brownfield sites [1–4]. To aid in understanding best practice pathways for Brownfield re-development, the materials science and engineering perspective can be followed. This perspective involves creating an inventory by categorizing and characterizing materials, equipment and structures, as well as processing methods, and developing a framework to allow optimization of benefits of resource utilization for the local region.

15.1.1 First Stage: Community-Engaged Approach to Materials Recovery

Figure 15.1. Shows examples of abandoned Brownfield sites, illustrating that materials, equipment and structures that may remain at the site. Categorization and characterization of the materials and structures that remain needs to be carried out to understand how these resources provide benefit to the local economy. Categorization of materials is carried out according to standard definitions of material classes. Material classes include (1) metals and alloys, with examples being aluminum alloys, copper alloys and other non-ferrous alloys, steels or ferrous alloys, (2) ceramic materials such as alumina, silica, glasses, perovskites, rock salt, (3) polymers including Teflon, PVC, PE, plastics and (4) composites such as carbon fiber composites, metal matrix composites, ceramic matrix composites and fiberglass. These represent only a small fraction of the materials that are associated with each material class. There could also be advanced materials such as semiconductors, biomaterials, nanomaterials and quantum materials present at Brownfield sites.

Materials harvested from the Brownfield sites may be characterized or analyzed in order identify the material class. Different industries utilize materials from the different material classes and so once the materials recovered from the Brownfield site are classified, the materials may then be identified with the industry or product for which it will be useful. For example, engineering industries that use metals and alloys in their products are automotive, aerospace, energy and infrastructure to name a few; ceramics are used in the production of glassware, refractories, electronic devices, superconductors and magnets; polymers produce plastics, coatings, adhesives and packaging; composites are heavily used in aerospace, other transportation industries, storage and containers. These are many more such examples of material usage in a variety of industries. In this way, the materials recovered may be mapped to the specific local/regional industries that will be able to acquire materials useful for their operation and production, leading to beneficial use of materials. By identifying the class of material and linking the material resource with specific products, the best local destination for the material resource can be identified.

To summarize, possible steps involved in classification of materials for re-utilization include:

1. Creating an inventory of recoverable/reusable materials,
2. Classifying the materials by material class,
3. Collaborate with the local community and policymakers (e.g., governmental agencies and economic development council) to identify potential needs or applications in the community for these materials,
4. Match materials with local/regional resources and economic needs,
5. Estimate sustainable time frame of the material use and repurposing. If it is long time frame, then there can be consideration of the resource use in long range economic planning. If it is only a short time frame, then there can be additional collaboration and planning with stakeholders for repurposing these materials from the Brownfield site.

15.1.2 Second Stage: Community-Engaged Approach to Sustainable Brownfield Development—Materials Processing

To identify additional potential pathways for Brownfield resource utilization, insight can be gained by examining a fundamental tenet of materials science and engineering, illustrated by the materials tetrahedron shown in Fig. 15.2. This materials science and engineering perspective is relevant for all materials applications including Brownfield material resource repurposing. The tetrahedron shows the links between material properties, required for any application, with the internal structure of the material, referring to the (micro)structure and molecular level structure, which dictates the material properties, with the processing methods or synthesis methods needed to create the structure that leads to the properties of the material, and ultimately to the performance of the material in application. Characterization refers to the methods used to determine the properties and structure of the materials and to analyze the processing techniques [7].

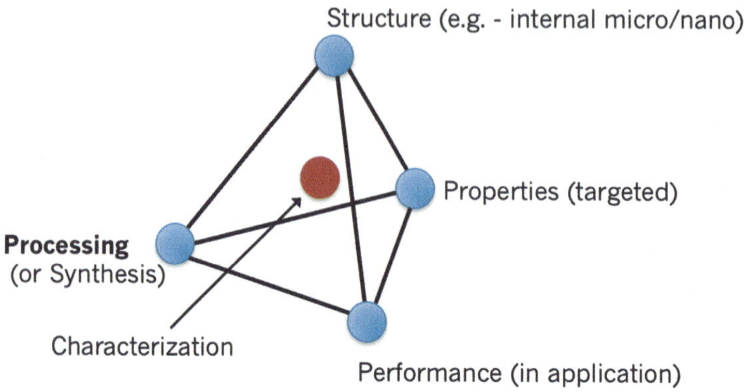

Fig. 15.2 Materials Tetrahedron. (For a discussion see, for example [8])

A particular component of the materials engineering tetrahedron that could be helpful in sustainably repurposing the abandoned Brownfield sites to help the local economy is the processing component. Processing refers to the engineering methods used to make any product. A summary of characteristics of materials processing is listed below:

1. Materials processing transforms raw/recyclable/input materials into useful products.
2. The large-scale manufacturing of semiconductor devices, carbon fiber, steel beams and rebar, optoelectronic devices, and a wide variety of other material products all are the end products of materials processing methods.
3. Materials processing provides materials with designed shapes, with dictated internal structures to provide targeted material properties.
4. Materials processing can be a sustainable activity and there are an enormous number of opportunities for repurposing materials found in Brownfield sites

An example of a sustainable material processing facility that could serve as a model for repurposing a military Brownfield site with benefit to the local economy is the Nucor steelmaking facility located in Seattle, WA (Fig. 15.3). This is a facility in which 99% of the material used to make steel product is recycled scrap steel. Scrap steel is available from a variety of sources world-wide, and its use eliminates the need for furnaces to reduce iron oxide, ovens to produce coke from coal, and furnaces to make steel from the carbon saturated iron, among other processes. The main inputs for this facility include the recyclable scrap metal, electricity used to melt the scrap in an electric furnace and alloying elements such as lime, silicon, manganese and carbon, needed to refine the steel composition [9]. Appropriate compositions for different products can be achieved by introducing scrap metal of different compositions. Once molten, the steel is typically continuously cast, and the solidified ingots can be formed into product.

The processing method here also addresses environmental sustainability in addition to a economic benefit. Per Nucor Corporation estimates, "the U.S. is expected

Fig. 15.3 Example: NUCOR Steel Seattle—sustainable steel and steel products manufacturing, Seattle, WA [9]

to build 30 GW of offshore wind energy by 2030, which will require an estimated 7 million tons of steel. If that steel is made using the extractive method, there will be 16.2 million tons of CO_2 generated. However, if the steel is made using the circular method, as Nucor, it will only produce 5.4 million tons of CO_2 – a differential of 10.9 million tons of CO_2" [10]. By focusing on the development of a materials processing facility on the Brownfield site, a sustainable local economic benefit can be obtained.

15.2 Conclusions

In summary, among the possible uses of materials found in military Brownfield Sites are sustainable approaches for re-purposing and recycling leading to opportunities for industrial growth. These may be implemented in a multiple-tiered plan with community collaboration, to ensure that appropriate local/regional distribution is available for reusable/recyclable materials; sustainable processes for local/regional community benefit are established; and environmentally responsible and just choices are applied to materials recovery, processing, and repurposing for environmental sustainability and economic benefit.

References

1. Jauhiainen JS (2007) Chapter 6: "Conversion of Military Brownfields in Oulu". In: Vestbro DU (ed) Rebuilding the city. Managing the built environment and remediation of Brownfields. Baltic University Press, Hamburg, p 27
2. Christopher Wedding G, Crawford-Brown D (2007) Measuring site-level success in brownfield redevelopments: a focus on sustainability and green building. J Environ Manag 85:483–495
3. Peric A, Miljus M (2021) The regeneration of military brownfields in Serbia: moving towards deliberative planning practice? Land Use Policy 102:105222
4. Musa S, Šakić D (2019) Military Brownfields–Potential for tourist revitalization in Bosnia and Herzegovina. Geo J Tour Geosites 26(3):1021–1032
5. Benzon B, Brownfields Photograph (https://www.flickr.com/photos/stc4blues/7899232142/) cited by Laura Taylor at the Center for Environmental and Resource Economic Policy at North Carolina State University in an article entitled "Remediated Brownfield Sites: Environmental and Property Value Benefits". Available at: https://cenrep.ncsu.edu/2016/07/14/remediated-brownfield-sites-environmental-and-property-value-benefits/
6. Gorman P, "Student corner: flexible opportunities in North Carolina's brownfields program environmental and property value benefits". Brownfield Photograph cited, University of North Carolina. Available at: https://ced.sog.unc.edu/2016/10/flexible-opportunities-in-north-carolinas-brownfields-program/
7. "Materials science and engineering for the 1990s." National Academies Press, Washington, DC, 1998, p 27
8. Callister WD, "Materials science and engineering: an introduction", 10e edition, Chapter 1, Section 1.2. Wiley Publishers, p 4
9. Nucor Steel Seattle Photographs. Available at https://app.leg.wa.gov/committeeschedules/Home/Document/162766. Accessed on 16 Dec 2023
10. Nucor Corporation, "Scopes 1, 2 and 3 emissions for steel purchasers: how to reduce your carbon footprint". Available at: https://nucor.com/newsroom/scopes-emissions-for-steel-purchasers-how-to-reduce-your-carbon-footprint

Chapter 16
Military Brownfields in City Peripheries: Emerging Areas for Recreation and Landscape Protection, a Case Study of Ostrava and Milovice (Czech Republic) and Kharkiv (Ukraine)

Pavel Bednář, Lenka Smékalová, Pavlo Kobylin, Serhii Chekhov, and Yevhen Shpak

Abstract Military brownfields, unique in their extensive areas and significant environmental contamination, present a paradox. Despite their burdens, these areas on the outskirts of cities often boast substantial forest cover or vegetation reminiscent of steppe landscapes. This unique juxtaposition offers potential for their utilisation, after remediation of environmental burdens, in landscape management and public accessibility for recreational purposes. To explore this potential, a narrative analysis with data triangulation from various sources was employed. The results reveal that selected military brownfields in the Czech Republic have already advanced in their transformation, being repurposed into recreational and landscape protection areas accessible to the public. Meanwhile, in the case study area of Ukraine, such activities are still in their early stages, with the potential to draw from experiences in transforming industrial brownfields. The chapter underscores the importance of a strategic and coordinated approach involving all stakeholders in land transformation, public land ownership, and harmonising commercial activities with landscape preservation.

Keywords City periphery · Czech Republic · Kharkiv · Landscape protection · Military brownfields · Recreation · Ukraine

P. Bednář (✉) · L. Smékalová · P. Kobylin
Tomas Bata University in Zlín, Zlín, Czech Republic
e-mail: bednar@utb.cz; smekalova@utb.cz; pavlo.kobylin@karazin.ua

S. Chekhov · Y. Shpak
V.N. Karazin Kharkiv National University, Khrakiv, Ukraine
e-mail: serhii.chekhov@student.karazin.ua; yevhen.shpak@student.karazin.ua

C. Morar et al. (eds.), *Achieving Sustainability in Ukraine through Military Brownfields Redevelopment*, NATO Science for Peace and Security Series C: Environmental Security, https://doi.org/10.1007/978-94-024-2278-8_16

16.1 Introduction

Military brownfields in city peripheries have the potential to provide useful and even valuable spaces for urban residents, offering opportunities for recreation and landscape protection, among others. In this case study of Ostrava and Milovice in the Czech Republic, and Kharkiv in Ukraine, we explore how these former military sites can be transformed into vibrant and sustainable community assets, enlivening the peripheral locations of cities. Through carefully selected reuse and thoughtful planning of such, military brownfields can contribute to the overall quality of life in urban areas by creating new public spaces, public greenery, housing estates, fostering entrepreneurial activity, and, among other uses, providing opportunities for landscape protection and recreational opportunities for local residents.

Ukraine has many common features in socio-economic and political relations with the Czech Republic since both countries were under the influence of the Soviet Union to varying degrees. Many military bases were located on its territory. After the collapse of the Soviet Union and gaining the independence due to a decrease in the funding of the army, their network began to be optimized. Accordingly, some bases became abandoned. These territories require attention from the point of view of assessing the degree of impact on the environment and their further integration into the territorial planning system. Today, Ukraine is in a state of war as a result of Russian aggression. There are destructions of infrastructural and residential facilities, and deaths of people. The issue of infrastructure restoration, attraction of foreign investments and international financial assistance is becoming more and more urgent. In this aspect, the issue of involving military brownfields in economic activity can also be of great importance.

The objective of the paper is to present and understand the unique characteristics of selected military brownfields in the Czech Republic and explore their contribution to the development of their respective regions. By examining the specific strategies and challenges faced in Ostrava and Milovice (Czech Republic), as well as in Kharkiv (Ukraine) urban planners, policymakers, and other stakeholders can gain insights into successful practices for revitalising military brownfields to inform their decisions concerning military brownfield regenerations in other cities around the world by using narrative analysis in constructive perspective with data triangulation and applying local knowledge intrinsic illustrative/descriptive multiple parallel cases approach for qualitative case studies research. Considering the European integration aspirations of Ukraine, it is very important to know the successful cases of reuse of military brownfields in the Czech Republic for the purpose of their use during the post-war reconstruction of Ukraine.

The chapter is structured as follows. First, the roles of military brownfields in urban areas, especially on their outskirts, including the possibilities of their transformation for new use, are discussed. This is followed by a description of the causes of their emergence and an overview of the current state of military brownfields in the Czech Republic and Ukraine. The next section of the chapter focuses on presenting the state and transformation of military brownfields in the outskirts of cities such as Ostrava, Milovice, and Kharkiv for the recreation of residents, including landscape protection. The conclusion of the chapter provides a summary of the results and lessons learned through policy recommendations.

16.2 Theoretical Background

16.2.1 Brownfields: From Unused Land to New Activities

Alker et al. [1] define a brownfield as previously developed land that is currently unused and may be contaminated, although contamination is not an integral part of their definition. This kind of definition fits many types of brownfields, including military ones, based on their characteristics from their previous use by the armed forces. These types of brownfields do have a variety of prior uses. Winkelmann [2] names military training facilities, airports, warehouses, barracks, bunkers and fortifications, battlefields, explosives and munitions factories and depots, and facilities for the production of chemical warfare agents. As with other types of brownfields, some are contaminated [3], although even the contamination might be specific due to their previous use, as few other types of brownfields might include abandoned munition or military equipment. The main difference from other types of brownfields, though, is that they were abandoned by the military, which poses unique difficulties and, in some cases, even legal restraints [4]. In their classification of brownfields according to former functions and economic sectors, Turečková [5] sets the military brownfields among others, separate from brownfields created by agriculture, industry, or services, and among blackfields, as military brownfields often carry significant environmental burdens.

The presence of any brownfield within an urban environment is as much an obstacle as it is an opportunity. That much is true also about military brownfields that, too, have the potential for redevelopment inside urban areas [6]. Frequently, abandoned military sites in urban areas were once part of the urban fabric. Once unused, they may be an obstacle to the development of a wider area. Challenges that are frequently observed include the presence of a hazard to human health, whether from abandoned military equipment and structures or pollution, and they too represent a substantial financial burden with regards to any renewed use [7]. Their previous use does not always necessitate reconversion to entirely another one (i.e. previous military housing can become civil housing), although it is a significant factor affecting the new use of any brownfield, as is the ownership and location [8–10]. However, not all military brownfields are within urban areas. Many military complexes were secreted away from the public; others were known but inaccessible.

The abandonment of these sites led to the emergence of rural military brownfields that face some similar issues as urban ones, further exacerbated by their peripheral location [11].

The potential of military brownfields to contribute to the development of urban areas is as multifaceted as their previous use; however, much like other brownfields, the cost of redevelopment is occasionally staggering and requires an influx of public resources, a common public expense throughout Europe [12]. The practical examples show that military brownfields have been given new life and successfully utilised. Preston et al. [6] present brownfields in general as valuable resources of greenery, and indeed, there are cases of even military brownfields that were redeveloped as public green spaces, thus improving the sustainability of the urban environment. Certain military brownfields in urban areas have the opportunity to become combined-use spaces. They are often too large to remain dedicated to a single use. Bagaeen [13] gives examples that span the United Kingdom, Germany, France, or Jordan. Careful land use planning is a necessity during the redevelopment of brownfields [9] and allows for the combination of complementary functions in their new use [13]. Hercik et al. [14] give the example of such a former military facility in the Czech Republic and emphasise the activity of local public administration in the successful reconversion.

Redevelopment of any brownfield seeks to utilise the land again for a variety of new activities. These depend largely on the existing characteristics of a brownfield, such as previous use, location, presence of environmental burden, ownership, etc. [9, 10]. Brownfields located in the cores of urban areas are frequently planned to become commercial and housing estates, such as the plans for post-military complexes in the Tel-Aviv region [15]. Brownfields, which are less desirable on the real estate market, present an opportunity to create sites dedicated to leisure, recreational activities, and environmental protection [16]. Some military brownfields attract tourism based on the curiosity of visitors to see previously inaccessible sites [4] or their historical significance [17, 18] thus becoming a centre of the leisure-related tourism industry. Others have been deliberately redeveloped into leisure activity areas. Alker and Stone [19] emphasise the benefits for local inhabitants and the economy. Examples of such conversions are abundant. Navrátil et al. [20] give an example of uses in the Czech Republic, De Sousa [21] examines cases of US cities, and Derek [22] presents examples from multiple Polish cities. The general methodological problems of the military brownfields were considered in the study of the integrative role of human geography in solving the global problems of modernity [53], the analysis of the informational and synergistic component of human-geographic research [52], modern approaches to the methodology of human-geographic research [51], in the research of theory and practice of brownfields and greyfields of the city [49] and research of the heritage, culture and regeneration of the former military areas in the city of Oradea (Romania) [47].

16.3 Emergence and Redevelopment of Military Brownfields in the Czech Republic and Ukraine

16.3.1 Emergence of Military Brownfields in the Czech Republic

The existence of military brownfields is widespread. The Czech Republic has had its share of brownfields distributed in no particular pattern across the country [23], including historic and more recent military sites that are no longer used by any army. The rather recent and substantial emergence of post-military sites is connected to two main factors: (a) the departure of the Soviet army from the territory of the Czech Republic; and (b) the end of compulsory military service connected with the professionalisation of the Czech Army [24].

The Soviet army liberated a significant part of the territory that now comprises the Czech Republic from Nazi occupation, a feat that formed a positive relationship with the newly restored Czechoslovakia. This positive feeling was all but lost when it entered the territory of the Czech Republic, then part of Czechoslovakia, during an anti-liberalisation intervention in 1968 aided by armies of Warsaw Pact countries [25]. An act with which the citizens of Czechoslovakia almost universally disagreed and that had international implications as well as long-standing implications for Czechoslovakia and its successor states [26]. Because not only did the army enter, it stayed, and the stay lasted until 1991 in what was introduced as a temporary deployment. The long term stay resulted in the creation of military and civil amenities for the soldiers and their families that lived with them among ordinary citizens of Czechoslovakia [27]. Tomek [28] speaks of circa 75 thousand soldiers and 30 thousand additional family members that were present but frequently restricted to living inside the Soviet army complexes.

The compulsory military service was a feature of the communist regime in Czechoslovakia and lasted into the days of its successor state, the Czech Republic, which separated from Slovakia in 1993. The compulsory military service was abolished in 2005, marking a shift towards a professional army based on volunteer recruits [29]. This inevitably impacted the number of people involved in the army, as it was then comprised only of professional soldiers and no compulsory recruits who had to be fed, clothed, housed, and trained. Even the number of professional soldiers decreased in the early 1990s [30] together with overall changes in the structure, reduction of equipment, and indeed the military expenses [29]. In this time, military brownfields emerged also in the core areas of Czech cities [31].

Both of these factors intensely influenced the emergence of military brownfields, sites that no army, Soviet or Czechoslovak, needed anymore. These sites, however, were present both inside and outside urban areas. Aside from smaller brownfields, vast areas of unused land remain after being used as military training grounds.

These have been emerging even in more recent years in the Czech Republic, i.e., military training area Brdy closed in 2012 [32] or the training ground of Libavá which was originally established in an area of approximately 327 km^2, was reduced by approximately one fourth in 2016 [33, 34].

Vast amounts of unused land of military origin were transferred free of charge from the state to regional or municipal ownership [35], some of these brownfields have since been redeveloped. Currently, the Czech National database of brownfields lists 266 individual military brownfields on 2679 hectares [36].

16.3.2 Redevelopment of (Military) Brownfields in the Czech Republic

The redevelopment of Czech brownfields and its success are subject to a variety of factors. Frantál et al. [9] concluded that the most important ones are property relations (defined as ownership structure and availability for selling), associated costs, and ecological burden. The redevelopment of individual sites in the territory was examined in detail by Osman et al. [10] who point out that their sample consisted of 101 sites, 21 of which were previously used by the military. Civic amenities and mixed use of retail, service, housing, and administration were the most prevalent types of new use noted in this study. Sport and recreation, specifically, were identified at 11 sites. Frantál et al. [9] also point out that sports, culture, and child-related leisure activities are the most popular types of new use in the Czech Republic, as confirmed by Navratil et al. [20].

The military brownfields have been a point of interest. Mindful of their increasing number after the professionalisation of the armed forces, the public institutions of the Czech Republic launched both research and support programmes in the mid-2000s, ie. "Programme to support the development of areas significantly affected by the reduction or disappearance of military garrisons", which also featured financial support for redevelopment into tourism, leisure, and sports facilities [37]. The special focus on military brownfields has gradually abated, and these days they are covered by general schemes of brownfield regeneration such as the National Brownfield Regeneration Strategy 2019–2024.

Examples of Czech military brownfields that have undergone the process of redevelopment include the larger-scale military training area Brdy, once considered for locating US military equipment, which was converted into a protected landscape [32], smaller scale brownfields in the city of Hodonín redeveloped for mixed use with strong financial benefits for the municipality as in investor [14], a former military airport was successfully converted into an industrial zone in Žatec city [38] (as well as in Pilsen [31].

16.3.3 Transformation of Military Brownfields in Milovice and Ostrava to Emerging Areas for Recreation and Landscape Protection

Military Training Area Milovice—Mladá case study

The brownfield previously served as the largest military base for The Central Group of Forces of the Soviet Army Forces (30,000 soldiers and 5000 civilians—mostly family members of the soldiers) garrisoned in Czechoslovakia after the Warsaw Pact invasion of Czechoslovakia (1968). The Czechoslovak Armed Forces left the military base for the exclusive use of the Soviet Army, although it was still its property, lacking any real decision-making or control. However, it was founded in 1904 for the Austro-Hungarian Army, subsequently used by the Czechoslovak Army and the armed forces of Nazi Germany during the occupation of Czechoslovakia from 1939–1945, and after World War II used by the Czechoslovak Army again. The military area was abolished in 1991 shortly after the departure of Soviet troops. It consisted of a military airport used for training, other training grounds, ammunition dumps, and barracks. Former Czechoslovak Armed Forces installations were also part of this area, which were maintained by Soviet troops in a very limited manner, despite Czechoslovakia's financial contributions.

The territory was left unguarded, which led to equipment theft, vandalism of buildings, and damage to other military structures. In subsequent years, efforts were made to clear the military area of ammunition, and by 2001, soil decontamination from fuel and other petroleum substances had taken place. However, incomplete demolition of barracks and residential buildings for military officers and their families remains a problem in the eastern part of the area. Subsequently, part of the territory was privatized or transferred to the Central Bohemian Region and the Czech Republic's Nature Conservation Agency. On private land, due to easy accessibility from Prague via the D10 highway and the suburbanization of Milovice (from 1200 to 16,000 inhabitants), there was rapid development of recreational areas. These included a children's amusement park called Miraculum and a Tankodrom where visitors could ride Soviet military vehicles or other off-road vehicles. Additionally, both of these areas are connected by an imported original forest railway from Romania. Another use for tourism lies in the space between pastures and a European significant natural reserve, where there is potential for further off-road activities with various vehicles. The original military airfield serves as a training ground for different vehicles and equipment, while also hosting summer music festivals featuring rock, popular, and drum and bass music. The public, after registration, can use a professional golf course with 18 holes. A specific feature is the activity of a non-profit organization within the natural park area. They maintain groups of freely living large European herbivores, including the European bison, wild Exmoor pony, and reintroduced European aurochs (Bison bonasus), as well as the aurochs (Bos primigenius) and the European wild ass (Equus hemionus hydruntinus), using the method of extensive naturalistic grazing.

Military Training Area and Shooting Range Area Ostrava—Bělský Les case study
The area was initially used by the forces of Nazi Germany from 1940 to 1945. Subsequently, it was handed over to the Czechoslovak army, which, like the previous case, abandoned it in favour of the Soviet military in 1968. This territory was also contaminated with oil substances and fuels, including groundwater. Even in this case, after the departure of Soviet troops due to lack of supervision, vandalism occurred. This included theft, looting, destruction, and squatting. Various proposals were made for the use of the area in the 1990s and the first decade of the twenty-first century. However, the primary concern was addressing its decontamination, including removing overgrown shrubs, grass, and trees. This became possible because the land was transferred to the city of Ostrava in 2011, allowing Ostrava to utilise EU funds for revitalising the area. Thanks to the efforts of the Ostrava City Forests, the land was transformed into a forest type that blends with the surrounding environment. The site is located within a large housing estate with 100,000 residents, so a proposal was put forward to convert the area into an educational forest kindergarten for children. The kindergarten includes rural spaces such as sandboxes, climbing structures, outdoor workout areas, and indoor facilities with an education centre. Here, children learn about nature conservation and become familiar with the functions of the forest.

16.3.4 Military Brownfields in Ukraine

The Soviet Union left more than 3000 military bases in Ukraine, where several tens of thousands of people lived. In the early 1990s, the largest group of troops in Europe was stationed on the territory of Ukraine, which after the collapse of the Soviet Union and Ukraine's independence became part of the Armed Forces of Ukraine. During the military reform, the number of military units in the Armed Forces was reduced [44]. Accordingly, some military facilities were abandoned and in most cases are not used to this day. Among the abandoned military facilities can be named the Ukrainian defensive line "Oleksiivska fortress" (Kharkiv region), Pamir radar station (Chernivtsi region), Over-the-horizon radar station "Duga" (OTH RS "Duga") (Kyiv region), Protected Space Communication Node (Ivano-Frankivsk region), Abandoned air defense base in Pashena Balka village (Dnipropetrovsk region), Tarakaniv Fort (Rivne region), Rocket tunnels (Lviv region) etc. [42, 46, 58–61].

There are a large number of military facilities in Ukraine: bases, towns, arsenals, warehouses of ammunition, fuel and lubricants, training grounds, old mine launchers, which pose dangers and risks to the population and the natural environment. They pollute the environment with chemical substances, in particular heavy metals, increase the radiation background, which leads to the degradation of natural complexes [44]. The obsolescence of military materials at the bases led to a series of explosions at military warehouses. In particular, during the years of Ukraine's independence, 16 accidents occurred at military warehouses [45].

Russia's full-scale invasion of Ukraine significantly affected the country's development, negatively affecting the overall quality of life. Thus, during 2023, the Russian occupying forces launched 2.4 thousand missile and 38.1 air and artillery strikes against the positions of the Defense Forces and populated areas of Ukraine [43, 54]. As a result of these strikes, according to the General Prosecutor's Office (OGPU), 11,673 people died, 18,336 Ukrainians were injured from the beginning of the war until the end of 2023 [57], and according to the Kyiv School of Economics (KSE), as of January 2024, the total amount of direct damage caused to the infrastructure of Ukraine increased to 155 billion dollars [55, 62]. The huge amount of ammunition used in war creates significant risks and dangers for the reconstruction of the country and the development of a healthy environment.

Planning for the reconstruction of Ukraine should include a systematic understanding of the long-term consequences and general context of the war. In this aspect, it will be very important to emphasize not only the restoration of infrastructure facilities, but also the development of projects for the reuse and rehabilitation of military brownfields.

16.3.5 *Military Brownfields in Kharkiv*

Kharkiv is the second largest city in Ukraine and the eighth in Eastern Europe in terms of population. This is one of the largest industrial, scientific, educational, trade, tourist, cultural centers of Eastern Ukraine, an important transport and logistic center of North-Eastern Europe [50]. Kharkiv as a post-Soviet city in recent decades has been characterized by processes of depopulation, deindustrialization of the economy, conversion of enterprises of the military-industrial complex, tertiarization and commercialization of urban space etc. These trends are intensified by the processes of globalization, European integration, the transition to the post-industrial stage of development and related processes of economic restructuring [49]. Given the industrial and scientific specifics of the city of Kharkiv, appropriate brownfields are widespread in the city. The following can be named among the military brownfields in Kharkiv.

Buildings of the former Kharkiv Higher Military Aviation School of Pilots named after S. I. Hrytsevts in the Rohan district (Fig. 16.1). This place is also called Voenved. The school operated from 1930 to 1993 and trained fighter pilots for all of Ukraine. Among the graduates can be named Heroes of the Soviet Union, the pilot Ivan Kozhedub and the astronaut Oleksiy Leonov. Currently, the object is in an abandoned state. The MIG-21 monument, several training buildings, hangars, residential, administrative and utility multi-story buildings remain [39].

The abandoned buildings of the Kozhedub Air Force Institute are located at the Holodna Gora district (Fig. 16.2). During 1941–1993 the buildings housed the Kharkiv Higher Military Aviation Engineering School, which became part of the Kharkiv Institute of the Air Force named after Ivan Kozhedub. In 2003 this institute

Fig. 16.1 Buildings of the former Kharkiv Higher Military Aviation School of Pilots named after S. I. Hrytsevts [39]

Fig. 16.2 The abandoned buildings of the Kozhedub Air Force Institute [48]

Fig. 16.3 The abandoned missile base in the Zhukovsky district [41]

was a part of the Kharkiv National University of the Air Force named after Ivan Kozhedub and this building became unnecessary. Currently, the object is in an abandoned state. The MIG-21 monument, several training buildings, hangars, residential, administrative and utility multi-story buildings remain [48].

The abandoned missile base is located in the Zhukovsky district (Fig. 16.3). It was the training base for cadets of the Kharkiv Higher Military Engineering

School of Strategic Missile Forces named after Krylova "Pomerky", where the actions of bringing ballistic missiles into combat readiness were practiced. The large building was a mine layout for the RS-20 (Satan) heavy intercontinental missile layouts. Currently, the object is in the abandoned state with only one big building [41].

The former L.O. Govorov Air Defense Engineering and Radio Technical Academy functioned in 1941–1993 and was located in the city center. In 1993, it was merged with the Kharkiv Higher Military Command and Engineering School of Missile Forces named after M. I. Krylov and the Kharkiv military university was created. In 2003 the university was liquidated and merged with Kharkiv Institute the Air Force named after Ivan Kozhedub forming the Kharkiv National University of the Air Force named after Ivan Kozhedub. The building of the Academy turned out to be unnecessary for the Ministry of Defense of Ukraine, and it was transferred to V.N. Karazin Kharkiv National University [56]. In this case, this object is an example of the transformation of a military object into a civilian one without changing the direction of activity (the educational function of the object remained).

Building of the former L.O. Govorov Air Defense Engineering and Radio Technical Academy [56]

Also, the abandoned military base in Sotrirovka (northern part of the city) can be highlighted among the military brownfields [40].

As we can see, there are not so many abandoned military bases in the city of Kharkiv, and those presented were mostly military educational institutions, which after the independence of Ukraine began to reorganize, premises of institutions became unnecessary, were vacated. Currently they are in a neglected state.

16.4 Conclusions

The exploration of military brownfields in Ukraine, particularly in Kharkiv, highlights significant challenges and opportunities for urban redevelopment and transformation. Ukraine inherited a substantial number of military bases and facilities from the Soviet Union, which played a critical role in the country's military history, but now pose risks to public safety and environmental health due to neglect and aging infrastructure. They contribute to environmental pollution, including chemical contamination and increased radiation levels, posing ongoing risks to local populations and natural ecosystems. Both examples from the Czech Republic demonstrate significant financial demands and long-term removal of environmental burdens from military brownfields left by the Soviet Army, particularly in the form of fuel residues in soil and groundwater. Furthermore, research indicates the necessity of demolishing previous military facilities and buildings due to their decay, vandalism, and squatting if left unattended. Additionally, there appears to be a need for continuous and thorough oversight of organizations conducting territory decontamination. Notably, the decontamination of military brownfields from unexploded ordnance in shooting ranges poses a significant risk.

The war caused Russia's invasion of Ukraine, have further exacerbated the challenges associated with military brownfields, causing extensive damage to infrastructure and resulting in significant loss of life and economic hardship. The reconstruction and redevelopment of Ukraine necessitate a comprehensive understanding of the long-term impacts of conflict, emphasizing not only the restoration of infrastructure but also innovative projects for reusing and rehabilitating military brownfields. The case studies from the Czech Republic demonstrate that international cooperation, particularly within the framework of the European Union (EU), has been effective in providing financial support for the decontamination of military brownfields left by the Soviet Army. This collaboration involved individual ministries and environmental protection agencies, with a focus on preserving the relevant land under state ownership.

Kharkiv, as a major city in Ukraine, faces depopulation and deindustrialization trends, creating opportunities for reimagining and repurposing military brownfields into new urban spaces, including cultural, recreational, educational, and residential facilities. Most of the former military facilities in the city of Kharkiv are mostly in the abandoned state. In our opinion, this is due to the fact that they are buildings of former military schools, and they don't represent special tourist interest. On the other hand, both case studies from the Czech Republic highlight the crucial role of local municipalities or regions and their ability to negotiate with the state regarding the transfer of land ownership. This negotiation allows them to strategically develop military brownfields on the outskirts of cities for landscape protection while promoting sustainable tourism. However, it is essential to consider the intensity of recreational use by private entities and the funding requirements when implementing transformative projects through non-profit organizations. Given the attractiveness of these lands, there is a risk of overusing grant titles if such activities could lead to economic profit for these organizations.

The exploration of military brownfields in Ukraine underscores the complexities of post-Soviet urban development, emphasising the importance of sustainable planning, community engagement, and strategic investment in transforming these sites into valuable assets for the future of Ukrainian cities like Kharkiv. In addition, the military used and dealt with various chemical substances at these facilities. Therefore, it is necessary to conduct comprehensive environmental and sanitary-hygienic studies to determine the suitability of these areas for redevelopment. Conversely, both examples from the Czech Republic demonstrate that developing military brownfields on the cities' peripheries through recreation and landscape protection is feasible and brings attractiveness to these areas for both residents and visitors from outside the respective cities. Additionally, they can serve as best practice examples for other cities in Europe, including Ukraine. The goal is not only to provide an enjoyable experience for the target audience but also to educate them about landscape preservation, cultural identity, and active utilization of their leisure time during weekend tourism.

References

1. Alker S, Joy V, Roberts P, Smith N (2000) The definition of brownfield. J Environ Plan Manag 43:49–69. https://doi.org/10.1080/09640560010766
2. Winkelmann K (2007) Hazards typically associated with different conversion sites and investigation strategies for historic military land-use analysis and risk assessment. In: Spyra W, Katzsch M (eds) Environmental security and public safety. Springer, Dordrecht, pp 109–125. https://doi.org/10.1007/978-1-4020-5644-4_10
3. Morar C, Dulca M, Nagy G (2016) Brownfields regeneration, between sustainable urban development and cultural heritage. The former military sites in Oradea, Romania. J Urban Reg Anal 8:75–84
4. Musa S, Šakić D (2019) Military Brownfields–Potential for tourist revitalization in Bosnia and Herzegovina. GeoJ Tour Geosites 26:1021–1032. https://doi.org/10.30892/gtg.26326-414
5. Turečková K (2021) Specific types and categorizations of brownfields: synthesis of individual approaches. Geographia Technica 16:29–39. https://doi.org/10.21163/GT_2021.162.03
6. Preston PD, Dunk RM, Smith GR, Cavan G (104590, 2023) Not all brownfields are equal: a typological assessment reveals hidden green space in the city. Landsc Urban Plan 229. https://doi.org/10.1016/j.landurbplan.2022.104590
7. Klusáček P, Alexandrescu F, Osman R, Malý J, Kunc J, Dvořák P, Frantál B, Havlíček M, Krejčí T, Martinát S, Skokanová H, Trojan J (2018) Good governance as a strategic choice in brownfield regeneration: regional dynamics from The Czech Republic. Land Use Policy 73:29–39. https://doi.org/10.1016/j.landusepol.2018.01.007
8. Škrabal J, Turečková K, Nevima J (2021) Analysis of brownfield sites in the villages and towns in The Czech Republic. Transylv Rev Adm Sci 17:151–169. https://doi.org/10.24193/tras.64E.8
9. Frantál B, Greer-Wootten B, Klusáček P, Krejčí T, Kunc J, Martinát S (2015) Exploring spatial patterns of urban brownfields regeneration: the case of Brno. Czech Republic Cities 44:9–18. https://doi.org/10.1016/j.cities.2014.12.007

10. Osman R, Frantál B, Klusáček P, Kunc J, Martinát S (2015) Factors affecting brownfield regeneration in post-socialist space: the case of The Czech Republic. Land Use Policy 48:309–316. https://doi.org/10.1016/j.landusepol.2015.06.003
11. Seljamaa E-H, Czarnecka D, Demski D (2017) "Small places, large issues": between military space and post-military place. Folklore Electron J Folklore 70:7–18. https://doi.org/10.7592/FEJF2017.70.introduction
12. Grimski D, Ferber U (2001) Urban brownfields in Europe. Land Contam Reclam 9
13. Bagaeen SG (2006) Brownfield sites as building blocks for sustainable urban environments: a view on international experience in redeveloping former military sites. Urban Des Int 11:117–128. https://doi.org/10.1057/palgrave.udi.9000168
14. Hercik J, Šerý O, Toušek V (2011) Post-military areas in The Czech Republic and their revitalization – examples of the towns of Hodonín and Uherské Hradiště. Geographica 42:107–119
15. Ansenberg U, Marom N (2024) The multiple enactments of contamination: rethinking the remediation and redevelopment of military-industrial brownfields in the Tel Aviv region. Int J Urban Reg Res 48:7–30. https://doi.org/10.1111/1468-2427.13220
16. Franz M, Güles O, Prey G (2008) Place-making and 'green' reuses of brownfields in the Ruhr. Tijd Voor Econ Soc Geog 99:316–328. https://doi.org/10.1111/j.1467-9663.2008.00464.x
17. Liu H-Y (2012) Regeneration of Kinmen military installations through the integration of renewable energy technologies and battlefield resources. Renew Energy 43:165–171. https://doi.org/10.1016/j.renene.2011.11.053
18. Matković I, Jakovčić M (2020) Conversion and sustainable use of abandoned military sites in the Zagreb Urban Agglomeration. Hrvatski geografski glasnik 82:81–106. https://doi.org/10.21861/HGG.2020.82.02.03
19. Alker S, Stone C (2005) Tourism and leisure development on brownfield sites: an opportunity to enhance urban sustainability. Tour Hosp Plann Dev 2:27–38. https://doi.org/10.1080/14790530500072336
20. Navratil J, Krejci T, Martinat S, Pasqualetti MJ, Klusacek P, Frantal B, Tochackova K (2018) Brownfields do not "only live twice": the possibilities for heritage preservation and the enlargement of leisure time activities in Brno, The Czech Republic. Cities 74:52–63. https://doi.org/10.1016/j.cities.2017.11.003
21. De Sousa CA (2006) Unearthing the benefits of brownfield to green space projects: an examination of project use and quality of life impacts. Local Environ 11:577–600. https://doi.org/10.1080/13549830600853510
22. Derek M (2016) Urban recycling: spaces for leisure in polish urban brownfield sites. Geogr Pol 89:555–561
23. Turečková K, Nevima J, Škrabal J, Martinát S (2018) Uncovering patterns of location of brownfields to facilitate their regeneration: some remarks from The Czech Republic. Sustain For 10:1984. https://doi.org/10.3390/su10061984
24. Hercik J, Šimáček P, Szczyrba Z, Smolová I (2014) Military brownfields in The Czech Republic and the potential for their revitalisation, focused on their residential function. Quaestiones Geographicae 33:127–138. https://doi.org/10.2478/quageo-2014-0021
25. Brand M (2021) Intelligence, warning, and policy: the Johnson administration and the 1968 Soviet invasion of Czechoslovakia. Cold War Hist 21:197–214. https://doi.org/10.1080/14682745.2020.1752675
26. Černá M (2015) From "occupation" to "friendly assistance": the "presence" of soviet troops in Czechoslovakia after august 1968. Hung Hist Rev New Ser Acta Historica Academiae Scientiarum Hungaricae 4:114–143
27. Koštialová K (2023) They had beautiful boots and bananas there. (In)visible presence of Soviet Soldiers in a Small Town. Slovenský národopis 71:147–164
28. Tomek P (2017) Life with Soviet Troops in Czechoslovakia and after their withdrawal. Folklore Electron J Folklore 70:97–120
29. Kříž Z (2010) Army and politics in The Czech Republic twenty years after the velvet revolution. Armed Forces Soc 36:627–646. https://doi.org/10.1177/0095327X09358649

30. Ministerstvo obrany: Vývoj skutečných počtů osob v resortu MO ČR v letech 1992–2022. https://mocr.army.cz/scripts/detail.php?id=129653. Last accessed 2024/04/30
31. Šilhánková V (2006) Rekonverze (vojenských) brownfields
32. Klusáček P, Martinát S, Krejčí T, Bartke S (2019) Re-development of a former military training area—the case of Brdy told from a local actors' perspective. Land Use Policy 82:147–157. https://doi.org/10.1016/j.landusepol.2018.11.045
33. Skokanova H, Havlíček M, Klusácek P, Martinat S (2017) Five military training areas-five different trajectories of land cover development? Case studies from The Czech Republic. Geographia Cassoviensis XI:201–213
34. Havlíček M, Skokanová H, Dostál I, Vymazalová M, Pavelková R, Petrovič F (2018) The consequences of establishing military training areas for land use development—A case study of Libavá. Czech Republic Land Use Policy 73:84–94. https://doi.org/10.1016/j.landusepol.2018.01.039
35. Czech Republic: 174/2003 Sb. Zákon o převodu některého nepotřebného vojenského majetku a majetku, s nímž je příslušné hospodařit ..., https://www.zakonyprolidi.cz/cs/2003-174. Last accessed 2024/04/30
36. CzechInvest: O brownfieldech – Brownfieldy, https://www.brownfieldy.cz/o-brownfieldech/. Last accessed 2024/04/30
37. Ministerstvo pro místní rozvoj ČR: Ministerstvo pro místní rozvoj ČR – Program podpory rozvoje území výrazně postižených redukcí nebo zánikem vojenských posádek, https://mmr.gov.cz/cs/temp/archiv-temp-stranek/zaloha-z-narodnich-dotaci/archiv-programu-a-dotaci-regionalni-politiky-(1)/podpora-regionalniho-rozvoje-v-roce-2005/program-podpory-rozvoje-uzemi-vyrazne-postizenych. Last accessed 2024/04/30
38. Hercik J, Szczyrba Z (2012) Post-military areas as space for business opportunities and innovation. Stud Ind Geogr Comm Pol Geogr Soc 19:153–168. https://doi.org/10.24917/20801653.19.12
39. Abandoned Aviation School named after Hrytsevts, Rogan (Voyenved). Findway. Retrieved from: https://find-way.com.ua/oblast/kharkivska/kharkiv/pokynute-aviatsiinomu-uchylyshchi-im-s-i-hrytsevtsa-rohan-voienved. Last accessed 04/05/2024
40. Abandoned military unit, Kharkiv. Findway. Retrieved from: https://find-way.com.ua/oblast/kharkivska/kharkiv/pokynuta-viiskova-chastyna-kharkiv. Last accessed 04/05/2024
41. Abandoned missile base (school), Zhukovsky (Zhuky) Findway. Retrieved from: https://find-way.com.ua/oblast/kharkivska/kharkiv/pokinuta-raketna-baza-uchilishe-pos-gukovskogo-guki-cikavi-miscya-ukraini. Last accessed 04/05/2024
42. Adreeva V (2021) Abandoned military facilities of Ukraine, which were chosen by tourists. 24Travel. Retrieved from: https://travel.24tv.ua/viyskovi-obyekti-ukrayini-yaki-pokinuli-voyaki-oblyubuvali-turisti_n1713952
43. Chernovol K (2024) In 2023, the Russians struck Ukraine with tens of thousands of strikes: infographic https://www.unian.ua/war/u-2023-roci-rosiyani-zavdali-po-ukrajini-desyatki-tisyach-udariv-infografika-12500187.html
44. Havrylenko OP (2008) Ecogeography of Ukraine: education. Manual. Kyiv. Retrieved from: http://pidruchniki.com/15941024/ekologiya/ekologichna_bezpeka_oboronnoyi_diyalnosti
45. Kot G, Matyukhina N (2017) Weapons warehouses in Ukraine: history of explosions, fires and victims. Retrieved from: https://www.bbc.com/ukrainian/features-41414138
46. Lazarev P (2015) Abandoned military facilities of Ukraine: echoes of war. IGotoWorld.Com. Retrieved from: https://ua.igotoworld.com/ua/article/451_zabroshennye-voennye-obyyekty-ukrainy.htm
47. Morar C, Nagy G, Boros L, Gozner M, Niemets L, Sehida K (2021) Heritage, culture and regeneration of the former military areas in the city of Oradea (Romania). Architektúra & urbanizmus 55(1–2):76–87. (Scopus (Q3), Web of Science) (in English). https://doi.org/10.31577/archandurb.2021.55.1-2.6
48. Neglected buildings of the air force institute named Kozheduba, Kharkiv. Findway. Retrieved from: https://find-way.com.ua/oblast/kharkivska/kharkiv/zanedbani-korpusy-instytutu-viiskovo-povitrianykh-syl-kozheduba-kharkiv. Last accessed 04/05/2024

49. Niemets L, Husieva N, Suptelo O, Sehida K, Kravchenko K (2018) Research of Brownfields and Greyfields of the City: Theory and Practice. Sustainable Economic Development and Application of Innovation Management from Regional expansion to Global Growth. Proceedings of the 32nd International Business Information Management Association Conference (IBIMA) (Seville, Spain, 15-16 November 2018). Editor Khalid S. Soliman. Seville, International Business Information Management Association (IBIMA), pp 409–435. https://ibima.org/accepted-paper/research-of-brownfields-and-greyfields-of-the-city-theory-and-practice/
50. Niemets LM, Segida KY (eds) (2017) [Innovation-investment potential as the basis of the region's competitiveness (by example Kharkiv region)]. [Monography]. Kharkiv, Ukraine, V. N. Karazin Kharkiv National University, p 520
51. Niemets K, Sehida K, Niemets L, Kravchenko K, Kobylin P, Telebienieva I, Kliuchko L (2022b) Methodology of human-geographical researches: contemporary approaches and methods. Visnyk of V. N. Karazin Kharkiv National University, Series. Geol Geogr Ecol 56:143–158. https://doi.org/10.26565/2410-7360-2022-56-10
52. Niemets L, Kravchenko K, Sehida K, Telebienieva I, Kliuchko L (2022a) Informational and synergetic component of human-geographical research in the scientific heritage of Kostiantyn Niemets. Visnyk of V. N. Karazin Kharkiv National University, Series. Geol Geogr Ecol 57:153–172. https://doi.org/10.26565/2410-7360-2022-57-12
53. Niemets L, Sehida K, Kobylin P, Şahin TM, Bartosh O, Telebienieva I, Kravchenko K, Kliuchko L (2023) Integrative role of human geography in solving global problems of modernity. Visnyk V. N. Karazin Kharkiv Natl Univer Ser Geol Geogr Ecol 58:171–187. https://periodicals.karazin.ua/geoeco/article/view/22199, https://doi.org/10.26565/2410-7360-2023-58-14
54. Official telegram channel of the Strategic Committee of the Armed Forces of Ukraine (UA War Infographics). Retrieved from: t.me/uawarinfographics. Last accessed 01/01/2024
55. Official web-site of Kyiv School of Economics. Retrieved from: https://kse.ua/ua/
56. Official web-site of the Kharkiv National University of the Air Force named after Ivan Kozhedub. Retrieved from: https://hups.mil.gov.ua/
57. Official web-site of the Office of the Prosecutor General of Ukraine. Retrieved from: https://www.gp.gov.ua/
58. Oleksiivska Fortress (2021) Cycling route through the fortresses of Slobozhanshchyna. https://tourdefort.bike/fortetsi/oleksiivska-fortetsia
59. Ponyatishyn N (2015) Giant secret facilities of the USSR on the territory of Ukraine: "dead" power plants and abandoned underground bases (photo). Firtka. News agency. Retrieved from: https://firtka.if.ua/blog/view/gigantski-sekretni-obekti-srsr-na-teritorii-ukraini-mertvi-elektrostancii-ta-zakinuti-pidzemni-bazi-foto69387
60. Protected Space Communication Node. Guru.Ua. Rertieved from: https://rest.guru.ua/kolomiya/places/631-_zashishennyj_uzel_kosmicheskoj_svyazi/. Last accessed 2024/04/28
61. Veselova – Solovonyuk A (2023). Old bunkers and catacombs: objects of Ukraine that you have not heard of, but they are becoming more and more popular. UA Portal. Retrieved from: https://news.uaportal.com/ukr/section-articles/news-staryie-bunkeryi-i-katakombyi-obektyi-ukrainyi-o-kotoryih-vyi-i-ne-slyishali-a-oni-stanovyatsya-vse-populyarnee-02-10-2023.html (in Urainian)
62. Zanuda A (2024) War in numbers: how Ukraine and Ukrainians have changed over the past year. BBC News Україна https://www.bbc.com/ukrainian/articles/cv28871qdv8o

Chapter 17
European Security and Post-conflict Reconstruction of Ukraine

Marian Dincă and Constantin-Vasile Țoca

Abstract The article presents an analysis of European security developments in the context of the conflict in Ukraine and the prospects for post-conflict reconstruction. European security has always been a challenge and a priority in the international security environment, especially at the beginning of the twenty-first century, when mankind is facing a complex, dynamic and unpredictable security environment, one of profound economic and social imbalances. The article is important for understanding the complexity of the conflict in Ukraine and its effects on European and global security, and provides a well-researched and analytical perspective on the security challenges posed by Russia's actions. The conflict in Ukraine is and will remain a hot topic of the utmost interest due to its political-social and economic-financial implications at the global level. The article emphasizes the importance of an integrated and comprehensive approach to European security in the context of the conflict in Ukraine. The post-conflict reconstruction requires a continuous and coordinated involvement of all relevant actors, both at European and international level, to ensure stability and sustainable security in the region.

Keywords Security · Economic-financial · Military conflict · Reconstruction

17.1 Introduction

Security has always been an essential prerequisite in the overall development process. Conflicts, whatever their nature, not only lead to loss of life, but also damage and destroy infrastructure of all kinds, including social infrastructure, encourage crime, vandalism and public disorder, deteriorates investment and good governance, preventing a normal economic and social life.

M. Dincă · C.-V. Țoca (✉)
University of Oradea, Oradea, Romania
e-mail: ctoca@uoradea.ro

C. Morar et al. (eds.), *Achieving Sustainability in Ukraine through Military Brownfields Redevelopment*, NATO Science for Peace and Security Series C: Environmental Security, https://doi.org/10.1007/978-94-024-2278-8_17

European and Euro-Atlantic security has been and will remain of great topicality and of real interest to all major international players, not just those in Europe and North America. In this context, the Black Sea region and the Western Balkans have a special place, as they have recently emerged as areas of political and military instability and economic and social uncertainty, real sources of conflict.

Due to its proximity and socio-economic influences, the conflict in Ukraine is having an extremely significant impact on European security, generating a series of complex and profound challenges. These challenges affect not only Ukraine as a country directly involved, but also the entire European region, calling into question the principles of collective security and cooperation within the North Atlantic Alliance as well as at the level of the European Union.

In the light of these developments, European security analysis becomes crucial to understanding how European states and international organisations are redefining their security policy and conflict management approaches.

17.2 European Security and the Challenges of the Beginning of the Twenty-First Century

The end of the Cold War, marked by the fall of the Berlin Wall in November 1989, brought remarkable changes in Central and Eastern Europe that would confront the North Atlantic Alliance[1] with a new and very different set of security threats. NATO's extensive enlargement and the challenges posed by the new realities of the international security environment required the Alliance to undergo a rapid and continuous process of transformation and adaptation.

At the beginning of the twenty-first century, we found ourselves faced with a complex and dynamic security environment that was fluid, uncertain and highly unpredictable, characteristics that were to persist and even increase in places, while leaving room for a Euro-Atlantic area of competitive democracies, peace and prosperity. The most significant risks and threats specific to these decades are political, military, economic and social [1, pp. 56–59].

By far, the economic risks are the most significant, because of their scale, speed and scope. They are the implicit result of political decisions and political-military threats, and have immediate and long-lasting implications for our social life.

However, Russia's recent illegal, brutal, unprovoked and unjustified military aggression against sovereign and independent Ukraine has shaken the peace in Europe and fundamentally changed the vision of Euro-Atlantic security, demonstrating once again that the unpredictable never ceases to surprise us.

[1] The North Atlantic Treaty Organization (NATO), also known as the North Atlantic Alliance, is a political-military alliance of 30 countries in Europe and North America, founded on April 4, 1949, in Washington in the United States.

The Black Sea region, whether we refer to the riparian countries or the wider area, has a long and rich history of conflicts and diverging interests, becoming in recent years an area where three major players are vying for influence: NATO, the EU and Russia. In addition to the six littoral states, Russia, Ukraine, Romania, Bulgaria, Turkey and Georgia, the Wider Black Sea Area also includes Armenia and Azerbaijan [1, pp. 95–96].

At present, the Black Sea is a border area between the European Union and NATO on the one hand and the Caucasus region on the other. As a geopolitical area, the Black Sea basin is characterised by frozen conflicts, prolonged by the persistence in the region of Soviet cultural, social and politico-military remnants, the rivalry between Turkey and the Russian Federation for naval supremacy, and attempts by the littoral states and the European Union to develop economic cooperation and strengthen democracy.

The Black Sea region has been and will remain a conflict zone, an area of permanent disputes and periodic tensions between the Russian Federation and the West.

17.3 Economic and Financial Implications of the Conflict in Ukraine

Wars have always brought destruction, sometimes irreversible, and significant costs at all levels, and even more so in modern conflicts such as the one in Ukraine, which is notable for its negative effects in key areas of daily life such as energy, transport and communications, food, education and health. Thus, in addition to the purely military costs and direct destruction in the conflict zones, the economic and social effects are spreading almost in real time, like a tsunami through international social life and all the financial markets, which are increasingly interconnected globally.

The sanctions imposed on the Russian Federation[2] by the Western world and the post-Cold War dependence of European states on its energy resources are striking in all directions, including against those who proposed them. Moreover, Ukraine,[3] a

---

[2] Russia, officially the Russian Federation, is the world's largest country, covering more than one-eighth of the Earth's inhabited area, and the ninth largest in terms of population, with more than 144 million people as of December 2017. The Russian economy is the twelfth largest by nominal GDP and the sixth largest by purchasing power parity in 2015. Russia's rich mineral and energy resources are the largest in the world and the country is one of the world's leading producers of crude oil and natural gas.

[3] Ukraine is an Eastern European country with an economy based on a large industrial and agricultural component, a significant part of which is exported, especially to countries in the European Union and the Commonwealth of Independent States. Ukraine produces almost all types of transport vehicles and spacecraft. Ukraine is also an important producer and exporter of cereals. Last year Ukraine ranked fourth on the list of the world's largest wheat exporters, with an export volume of 33.5 million tonnes. The total amount of maize produced in Ukraine was 42 million tonnes. Ukraine is the world's largest producer of sunflower (17.5 million tonnes in 2021) and sunflower oil (seven million tonnes in 2021). According to the USDA, the country was the largest exporter of

major supplier of raw materials and agricultural products to the EU, suddenly found itself, following the invasion of Russian troops, facing economic and financial blockages and unable to honour its export commitments.

Ukraine is a major global agricultural producer and exporter, but also plays an important role in other industries. The war in the country has disrupted both supply chains and economic activity in the region, with seismic waves felt around the world.

The ongoing military conflict has already affected the last 2 years' agricultural campaign, with Ukrainian farmers cultivating less land and fertilising less, which was bound to lead to lower yields. Nevertheless, farming activities continued in Ukraine, but the main blow was the blockade of Ukrainian ports on the Black Sea and Sea of Azov, the main export routes for grain.

Along with Russia, Ukraine is one of the world's top 10 producers of staple grains, ranking first in sunflower, sixth in maize and seventh in wheat, according to data from the last two agricultural years. Thus Ukraine accounts for 2.7% of world maize production, 3.3% of world wheat production and 28.6% of world sunflower production.

At world level, Ukraine and Russia are responsible for more than 30% of world trade in wheat, 32% in barley, 20% in maize and more than 50% in sunflower oil and oilseeds.

In terms of exports, Ukraine is the world's fifth largest exporter of wheat, accounting on average for over 8% of world exports, the world's fourth largest exporter of maize, accounting for over 12% of total exports, and the largest exporter of sunflower, accounting for over 50% of the value of exports.[4]

Moreover, Ukraine provides 5% of the world's rapeseed production and is the world's second largest producer of barley, the fourth largest barley exporter in the world and the fifth largest rye producer.

In terms of agri-food potential and capacities, Ukraine can provide food for more than 600 million people.

After long and difficult negotiations, with the support of UN representatives, Turkey achieves some progress in late August 2022. But even so, trade is more difficult and costly, given the uncertain situation in the area, the vast minefields in the harbour area, the more or less accidental artillery and rocket attacks on commercial shipments, the control of loading, the need for permanent escort of convoys in the Black Sea and the checking of cargoes as they pass through the Bosphorus Strait.

The situation deteriorated again in the autumn of 2022 following Ukrainian shelling in the Sevastopol area, with Russia suspending the agreement indefinitely. But after only a few days, under pressure from Turkey, commercial activity across

sunflower oil last year, with a global market share of nearly 61%. The six million tonnes of barley exported in 2021 made Ukraine the world's third-largest supplier in this market. In rye, 40% of global trade is conducted by Ukraine, which put it in first place—as in sunflower oil. (Source: https://agroexpert.md/rus/articole/)

[4]Data taken from https://agrobiznes.ro/11768-ucraina-importanta-in-agricultura-ce-exporturi-asigura-la-nivel-mondial/, last accessed 2024/04/20.

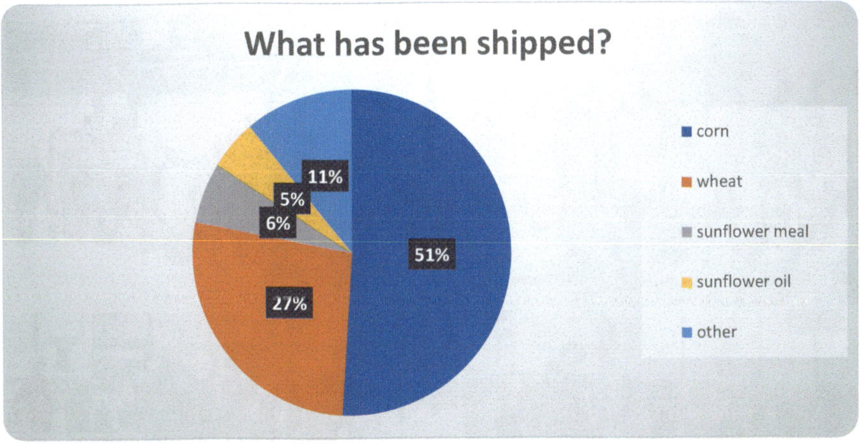

Fig. 17.1 The Black Sea Grains Initiative exports

the Black Sea was resumed and continues today under the same difficult and risky conditions.

What was exported through the Black Sea Grains Initiative?

It is very suggestively illustrated in the following graphic representation (Fig. 17.1). More then half (51%) was given by the corn trade.

The economic-financial dimension of this conflict was doubled by the social implications, in particular for neighbouring countries such as Romania and Poland, which had to manage, at least in the first months of the invasion, a massive and continuous influx of refugees (Fig. 17.2), including many children.

The conflict in Ukraine has significant economic and financial implications, not only for Ukraine but also for the Eastern European region and the wider Black Sea area, and even for the global economy.

In a nutshell, ***the economic and financial implications*** of this conflict mainly affect Ukraine's economy, energy industry, trade, investment, and regional stability as well.

1. ***Impact on Ukraine's Economy:*** The conflict has severely affected Ukraine's economy, with significant declines in GDP and industrial production. Ukraine's economic infrastructure, such as industrial and transport facilities, has suffered considerable damage. Foreign investors have been discouraged from investing in Ukraine due to political uncertainty and the risk associated with the conflict.
2. ***Impact on the Energy Industry:*** Ukraine is a transit route for Russian gas to Europe and the conflict has heightened concerns about energy security in the region. Moreover, there have been instances where Ukraine has had difficulties in securing energy supplies for its population and industry, with major economic and social implications.
3. ***Impact on International Trade:*** The conflict in Ukraine has led to the introduction of trade blockades between Ukraine and separatist regions in the east of the

Fig. 17.2 Ukrainian refugees at the borders of the European Union

country, affecting trade flows. Ukraine's exports, particularly in the metallurgical and agricultural industries, have been hit extremely hard by the conflict.

4. ***Military and Humanitarian Costs:*** Ukraine has had to devote significant resources to defending itself against Russian aggression, which has led to a significant increase in military spending. The conflict generated a wave of internal and external refugees and the government had to allocate resources for humanitarian assistance.
5. ***Risks to Regional Stability and Investment:*** Political instability and the risk of conflict in the region have discouraged many foreign companies from investing in neighbouring countries. Geopolitical risks in the region have led to increased cost of credit for neighbouring countries, which have been perceived as riskier for investors.
6. ***Global Impact:*** The conflict in Ukraine has also affected the prices of some raw materials, such as metals and grains, with global implications, given Ukraine's important role in world production. The conflict in Ukraine has also contributed to heightened geopolitical tensions between Russia and the West, impacting global financial markets and investor confidence.

The conflict in Ukraine therefore has profound economic and financial implications, locally, regionally and globally.

In order to mitigate the negative impact of the conflict on the economy and society, it is crucial to find a lasting political solution and a peace agreement.

17.4 The Reconstruction of Ukraine—A Regional Issue

At the European Economic and Social Committee summit on 24 March 2022, President Christa Schweng remarked: "This invasion has put our security and values at risk, and the EU stands legitimately and firmly by Ukraine, responding with unity and solidarity" and added: "Organised civil society in the EU is one of the building blocks of our democracy: first by addressing the humanitarian, economic and social consequences of the war, but also by demonstrating its determination to help the Ukrainian people uphold European values. We are turning our solidarity into action by giving our unwavering support to Ukraine."[5]

In this context, it is worth noting the unprecedented solidarity shown by civil society organisations, authorities and citizens alike towards people fleeing the war in Ukraine.

Another aspect worthy of note is the information environment, which in the current conflict in Ukraine has proved to be a veritable amalgam of more or less real data and information, with an extremely strong and immediate psycho-social and economic-financial influence, given that the war was experienced almost in real time by the whole of humanity, at least in the first few days, both through TV broadcasts and the most popular social networks.

In this respect, in addition to the humanitarian drama and panic induced in the social environment, most economies and international financial markets reacted accordingly, with fluctuations in line with the actions (destruction of infrastructure, announcement of sanctions, nuclear threat, etc.), as a real interconnection to the situation in the conflict zone.

But by far the biggest economic and financial challenge will be the post-conflict reconstruction of Ukraine. Regardless of when and how the hostilities end, one thing is certain, the destruction is significant both in the industrial and transport infrastructure, in the administrative area, and in housing and many other basic goods and services.

The conflict in Ukraine has caused widespread and devastating destruction across the country, affecting both physical infrastructure and society as a whole. Intense shelling and fighting has left cities such as Mariupol, Severodonetsk and Bahmut almost completely ruined, with residential buildings, hospitals, schools and critical infrastructure in ruins (Fig. 17.3).

In addition to material damage, millions of people were forced to flee their homes, creating a massive humanitarian crisis with waves of refugees and internally displaced people. Agriculture and local industries suffered significant losses, disrupting the economy and causing food insecurity.

Post-conflict reconstruction requires considerable and coordinated international efforts, both to rebuild destroyed infrastructure and to support the social and economic recovery of affected communities.

---

[5] https://www.eesc.europa.eu/ro/news-media/press-releases/razboiul-din-ucraina-si-impactul-sau-economic-social-si-ecologic, official EU website, last accessed 2024/04/2024.

Fig. 17.3 Destruction in Ukraine. (Photo: Profimedia and Twitter)

Commenting on the destruction caused by the Russian invasion and the reconstruction efforts in Ukraine, European Commission President Ursula von der Leyen said: "Russia's unprovoked and unjustified invasion of Ukraine has caused appalling human suffering and massive destruction across the country, forcing millions of innocent Ukrainians to flee their homes. Ukraine can count on the EU's full support. The EU will continue to provide Ukraine with short-term financial support to meet its needs and to enable it to maintain the functioning of basic services.

We are also ready to take a leading role in international reconstruction efforts to help rebuild a democratic and prosperous Ukraine. This means that investments will be linked to reforms that will support Ukraine in pursuing its European path."[6]

The value of the damage has been estimated by experts to be in the hundreds of billions of dollars, with more than $100 billion in the first two months alone, and the reconstruction effort will be a major, long-term effort with global implications.

In 2 years of conflict, the damage caused by Russia's invasion of Ukraine has been estimated at more than $250bn: destruction and losses to agriculture and the economy.

And the bill paid by Russia is exorbitant. By last summer, Moscow had spent more than $167 billion on the invasion of Ukraine, not including the economic losses caused by sanctions, which add several hundred billion dollars.

The reconstruction of Ukraine is "the mission of a generation, which must begin now", German Chancellor Olaf Scholz stressed on Tuesday 25 October 2022, at the opening of a conference in Berlin on the long-term support of this country invaded by Russia on 24 February 2022.[7] It is "no more and no less than a new Marshall Plan of the 21st century", the German leader stressed, in the presence of the President of the European Commission, Ursula von der Leyen.

In the most recent estimates for Ukraine's reconstruction, the World Bank, the UN, the EU and the Kiev authorities estimated a total financial effort of over $400

[6] https://ec.europa.eu/commission/presscorner/detail/ro, last accessed 2023/11/27.

[7] https://www.news.ro/externe, last accessed 2023/11/27.

Fig. 17.4 Ukraine recovery conference, London—June 2023

billion.[8] Destruction in Ukraine has particularly affected Donetsk and Lugansk districts in the east, Kharkov in the north-east and Herson in the south, which are close to the current front line, as well as Kiev. Data show that 10% of Ukraine's real estate was destroyed or damaged.

The destruction of the Kahovka dam and the respective hydroelectric power plant in the south of the country, blamed by Ukrainians on the Russian occupation forces in the area, has significantly affected the environment and agriculture, making life even more difficult for the locals already affected by the horrors of war.

If the conflict were to end now, it would take up to 10 years to fully restore the destroyed infrastructure in Ukraine, according to estimates by government authorities in Kiev. Rebuilding critical infrastructure alone would take about 3 years. The key condition would be that the necessary financial funds are available.

At the 2023 Ukraine Recovery Conference in London (Fig. 17.4),[9] the European Commission, the European Investment Bank (EIB), the European Bank for Reconstruction and Development (EBRD) and the International Finance Corporation (IFC) signed agreements worth over €800 million to mobilise private investment for the recovery and reconstruction of Ukraine's economy. The agreements, which are supported by the European Fund for Sustainable Development Plus, the financing component of the IVCDCI—Global Europe, underline the joint international commitment to mobilise the private sector for Ukraine's recovery and reconstruction.

17.5 Conclusions

The conflict in Ukraine has redefined Europe's security priorities, highlighting the need for a strengthened defense, strong international cooperation and the most comprehensive approach to modern threats. Europe must continue to adapt and innovate in the face of new challenges to ensure long-term stability and security.

---

[8] estimate reported by AFP, published by Agerpres on 15 February 2024.

[9] https://www.urc-international.com/urc-2023, last accessed 2024/02/20.

The economic-financial and social implications of the conflict in Ukraine are wide-ranging, far-reaching, impressive, constantly growing, and rebuilding will be difficult and long-lasting, requiring a major global effort. A comprehensive and coordinated approach is needed to manage these challenges and to find lasting solutions to ensure stability and prosperity in the region.

Therefore, the post-conflict reconstruction of Ukraine will require a coordinated and sustained effort by the Ukrainian government, the private sector and the international community as wel. The success of this process will depend on the ability to implement structural reforms, to secure the necessary financing and to build a solid foundation for sustainable development and long-term peace.

As long as the conflict continues, it cannot be known precisely how much Ukraine's reconstruction needs will total. However, it is important that the main building blocks of this international effort are designed now.

The involvement of the international community is essential to support Ukraine in resolving the conflict and to facilitate reconstruction and economic recovery.

Bibliography

1. Dincă M (2012) The evolution of the North Atlantic Alliance seen geopolitically and geostrategically. University of Oradea Publishing House, Trends and perspectives
2. Rebuilding Ukraine: Principles and Policies – RAND Corporation, 2022
3. Reconstruction in Ukraine: EU's Role and Challenges – BBC News, July 2022
4. Security Challenges in Eastern Europe – NATO Parliamentary Assembly, 2019
5. The impact of Russia's aggression in Ukraine on European and Global Security. J Int Aff, 2016
6. The impact of the Ukraine conflict on European Energy Security – Financial Times, November 2022
7. The Future of European Security Post-Ukraine Conflict – Foreign Policy, January 2023
8. Ukraine's Path to Peace: Analyzing Reconstruction Efforts – The Economist, March 2023
9. Ukraine conflict—background and current developments – US Department of State, 2022
10. https://agrobiznes.ro
11. https://dexonline.ro
12. https://www.eesc.europa.eu
13. https://www.urc-international.com/urc-2023
14. https://www.nato.int
15. https://www.news.ro

Chapter 18
Urban Regeneration of Sensitive Brownfield Sites and Areas for Sustainable Future Scenarios: Models and Strategies

Celestina Fazia

Abstract The regeneration of sensitive disused areas, and in particular military areas, is a topic that has been the subject of several studies due to the topicality and relevance of the issues involved: the presence of abandoned and contaminated areas is widespread and represents a problem due to the risk to which territories and communities are exposed. The objective of the research is to identify the best solutions for the recovery of abandoned military areas; to this end, a careful examination of some case studies (seven Italian and two Spanish) and an analysis of the emerging problems was carried out in order to define the fundamental phases of the regenerative process involving decontamination, environmental rebalancing and re-functionalisation.

According to the proposal, decontaminated sites will have to be part of a single environmental and social infrastructure and be in line with certain qualitative, quantitative and performance parameters set in compliance with EU guidelines. In order to hypothesise a strategic intervention model, nine case studies were analysed, selected to capture the different functional characteristics, size and current conditions of the area. It is important to understand whether the regeneration processes implemented or underway depend on certain geographic, location, size, site contamination, market conveniences or the presence of particular constraints that may inhibit transformation. The strategies defined in this essay can be duplicated in different contexts as the proposed road-map of actions is flexible and adaptable to different context variables.

Keywords Brownfield · Urban regeneration · Site contamination

C. Fazia (✉)
Kore University, Piazza dell'Università, Enna, Italy
e mail: celestina.fazia@unikore.it

C. Morar et al. (eds.), *Achieving Sustainability in Ukraine through Military Brownfields Redevelopment*, NATO Science for Peace and Security Series C: Environmental Security, https://doi.org/10.1007/978-94-024-2278-8_18

18.1 Urban Regeneration of Sensitive Brownfield Sites

18.1.1 Introduction

The subject of regeneration of sensitive brownfield sites, and military areas in particular [1] is the subject of numerous studies as, the presence of abandoned and contaminated areas is widespread, representing a problem because of the risk to which territories and communities are exposed.

Many processes of reorganisation and restructuring of military functions and presences have affected areas and properties in terms of downsizing, reconversion of functions or relocation. These decisions have also led to the demilitarisation and alieanation of a substantial part of military assets, resulting in the emergence of so-called Military Brownfields [1] and the phenomenon of abandoned -and in some cases contaminated- sites, including infrastructure and associated installations.

Military and defence or training facilities such as barracks, fuel and lubricant depots, bases, arsenals, ammunition depots, military infrastructure (ports, heliports, etc.), landfills and hazardous waste disposal areas are ecological bombs that expose inhabitants to serious health risks.

Today, abandoned military areas call for clear and effective responses in terms of recovering the useful functions of contaminated land and restoring environmental quality. There is a need to ensure the reclamation of contaminated soil and groundwater [1], but also to reorganise the land supply by bringing services up to date with new social needs. This is a major challenge, not least because other specific challenges [2] arise for local and other policy makers. Europe and the UN indicate certain environmental parameters to be achieved (Agenda 2030 for example), in terms of limiting soil consumption, de-impermeability, and reducing environmental pollution levels [2, 3]. These environmental 'mandates' are accompanied by others of a social and economic nature.

Through a careful examination of a number of case studies (seven Italian and two Spanish) and an analysis of the emerging problems, three phases of the regenerative process are identified: (1) decontamination; (2) Renaturalisation-environmental rebalancing; (3) refunctionalisation.

According to the proposal, the decontaminated sites will be able to be part of a single environmental and social infrastructure and will have to be in line with certain qualitative/quantitative and performance parameters set in the framework of research and in compliance with EU guidelines, UN indications regarding Agenda 2030 and the Nature Restoration Law (NRL) approved by the European Parliament a few months ago.

18.1.2 Objectives of the Work

The objective of the research is the identification of sensitive military brownfields sites in Italy and Spain decommissioned since the post-war period. The seven case studies were selected in order to capture the different functional characteristics, dimensions and current conditions of the area. It is important to understand whether the regeneration processes implemented or underway depend on certain geographical, location, dimensional characteristics, site contamination or the presence of particular constraints that may inhibit transformation [1, 3].

The aim is to understand what has been done in the different cases; whether the regeneration experiences implemented have been innovative and successful models and what could be done.

Discriminating readings are:

Initial destination
Area/facility characteristics:

- Surface area (range of type: small, medium, large)
- Dating
- Type of ownership, public or in alienation

Relationship to the urbanised context (internal, external in continuity, external in suburban areas)
State of disposal:

- Abandoned
- Underused

Conversion implemented or in progress, typology
Current or planned use:

- Mixing with prevalence of: Services and urban amenities of a social and aggregative nature; Business/office centre
- Monofunctional

Environmental weaknesses and vulnerabilities of the study area and context:

- Contaminated site: level and percentage
- Widespread urban decay [4]: range of incidence
- Lack of services and social deprivation: quantitative data restitution through verification of services and crime indicators

Management complexity:

- Technical
- Economic.

18.1.3 Case Studies Compared

The italian Case Studies are:

- *DIA, in Reggio Calabria,*(see Table 18.1)
- *Mezzacapo Barracks, in Reggio Calabria,* (see Table 18.2)
- *Former USAF Military, Base of Monte Nardello (RC),* (see Table 18.3)

Table 18.1 D.I.A. (Direzione Investigativa Antimafia) in Santa Caterina at Reggio di Calabria

Region:	Calabria (Italy)
City:	**Reggio Calabria**
Name:	D.I.A. (Direzione Investigativa Antimafia) in Santa Caterina at Reggio di Calabria
Initial destination: *Military barracks* *Airport, military infrastructure* *Military base/shooting range, exercises* *Other*	Military barracks
Age of construction:	1943
Property:	–
Size (range): <1000 sqm >1001 e < 10.000 >10,001 sqm	3.675 mq
Relationship to the urbanised context *Interior* *External in continuity* *External in suburban areas*	Interior
State of disuse: *Abandoned* *Underused*	Abandoned
Conversion implemented or in progress:	In progress
Mixed with prevalence of: *Services and urban amenities of a social and aggregative nature* *Business/office centre*	Business centre with offices
Monofunctional:	NO
Environmental weaknesses and vulnerabilities of the study area and context:	1. Presence of an old one-storey building above ground level of approximately 80 square metres that housed the fire-fighting team 2. Presence of the underground tunnel
Contaminated site:	NO
Widespread urban decay:	Limited to certain areas in the St Catherine district
Lack of services and social discomfort:	Critical urban context affected by a high number of organised crime offences
Management complexity:	NO

Table 18.2 Mezzacapo Barracks Reggio Calabria

Region:	Calabria
City:	**Reggio Calabria**
Name:	**Mezzacapo Barracks Reggio Calabria**
Initial destination: *Military barracks* *Airport, military infrastructure* *Military base/shooting range, exercises* *Other*	Military barracks
Age of construction:	1855, rebuilt after the 1908 earthquake
Property:	State Property Agency
Size (range): <1000 sqm >1001 e < 10.000 >10,001 sqm	The building consists of four manufacts.
Relationship to the urbanised context *Interior* *External in continuity* *External in suburban areas*	Central area of the city of Reggio Calabria.
State of disuse: *Abandoned* *Underused*	Underused
Conversion implemented or in progress:	In progress
Mixed with prevalence of: *Services and urban amenities of a social and aggregative nature* *Business/office centre*	Partial recovery
Monofunctional:	Business centre/offices
Environmental weaknesses and vulnerabilities of the study area and context:	No
Contaminated site:	Protected by the Superintendence of Fine Arts and Landscape of Calabria
Widespread urban decay:	No
Lack of services and social discomfort:	Social degradation in the surrounding area
Management complexity:	Critical urban context affected by a high number of crimes attributable to micro-crime

- *Monte Cimone Barracks, in Banne (Trieste),* (see Table 18.4)
- *Former Mardichi Barracks, La Spezia,* (see Table 18.5)
- *Seaplane airport with the* Nervi hangars and *the former ammunition depot in Via Dante Alighieri, Marsala (Trapani),* (see Table 18.6)
- *Former Military Hospital, located at Corso Calatafimi in Palermo,* (see Table 18.7)

The foreign Case Studies are:

- *Antiguo Hospital Militar, en el barrio del Realejo (Granada) y su conversión en E.T.S. de Arquitectura,* (see Table 18.8)
- *Son Busquets (Spain),* (see Table 18.9).

Table 18.3 Military base USA, Roccaforte del greco (RC)

Region:	Calabria
City:	Roccaforte del greco (RC)
Name:	Military base USAF
Initial destination: *Military barracks* *Airport, military infrastructure* *Military base/shooting range, exercises* *Other*	US base for radar control of the Mediterranean skies
Age of construction:	1965
Property:	Ministero of Difesa Italiana
Size (range): <1000 sqm >1001 e < 10.000 >10,001 sqm	>10.001 mq
Relationship to the urbanised context *Interior* *External in continuity* *External in suburban areas*	External in suburban areas
State of disuse: *Abandoned* *Underused*	Abandoned
Conversion implemented or in progress:	NO
Mixed with prevalence of: *Services and urban amenities of a social and aggregative nature* *Business/office centre*	–
Monofunctional:	–
Environmental weaknesses and vulnerabilities of the study area and context:	It is within the Aspromonte National Park, presence of forested areas
Contaminated site:	Yes
Widespread urban decay:	Yes
Lack of services and social discomfort:	No
Management complexity:	The area is distant from urban centres of strategic interest, and it is difficult to give impetus to redevelopment initiatives due to the considerable investments to be foreseen for site clearance and regeneration.

Table 18.4 Former Monte Cimone barracks in Banne in the Trieste area

Region:	Friuli-Venezia Giulia
City:	Trieste
Name:	Former Monte Cimone barracks in Banne in the Trieste area
Initial destination: *Military barracks* *Airport, military infrastructure* *Military base/shooting range, exercises* *Other*	Former Artillery barracks housed the 8th Pasubio artillery group
Age of construction:	Not certain; from the 1930s until the 1990s military use
Property:	–
Size (range): <1000 sqm >1001 e < 10.000 >10,001 sqm	Approx. 170,000 sqm
Relationship to the urbanised context *Interior* *External in continuity* *External in suburban areas*	Outside in suburban areas
State of disuse: *Abandoned* *Underused*	Abandoned
Conversion implemented or in progress:	
Mixed with prevalence of: *Services and urban amenities of a social and aggregative nature* *Business/office Centre*	None. Alienation in progress by the State Property Office.
Monofunctional:	–
Environmental weaknesses and vulnerabilities of the study area and context:	Various ideas and proposals for social and environmental enhancement: Circular development of the territory
Contaminated site:	No
Widespread urban decay:	No, presence of wooded areas; included in breeding sites and areal ecological corridors
Lack of services and social discomfort:	No
Management complexity:	Yes

From the comparison of the nine cases analysed it emerges that the most attractive sites and those that can attract interest to start regeneration/reconstruction processes are those closest to or located within the existing city and in a strategic position for the connection with other services and urban amenities (DIA in Reggio Calabria, for example) [5]; some brownfields are recognised as having historical/architectural

Table 18.5 Former Mar.di.chi. Barracks La Spezia

Region:	Liguria
City:	La Spezia
Name:	Former Mar.di.chi. Barracks La Spezia
Initial destination: *Military barracks* *Airport, military infrastructure* *Military base/shooting range, exercises* *Other*	Former Artillery Barracks
Age of construction:	1969
Property:	–
Size (range): <1000 sqm >1001 e < 10.000 >10,001 sqm	1480 sqm
Relationship to the urbanised context *Interior* *External in continuity* *External in suburban areas*	External in continuity
State of disuse: *Abandoned* *Underused*	Underused
Conversion implemented or in progress:	No
Mixed with prevalence of: *Services and urban amenities of a social and aggregative nature* *Business/office centre*	Headquarters/Internal Revenue Service Offices
Monofunctional:	Yes
Environmental weaknesses and vulnerabilities of the study area and context	Constrained pursuant to Art. 12 of Legislative Decree 42/2004
Contaminated site:	No
Widespread urban decay:	–
Lack of services and social discomfort:	No
Management complexity:	–

value (Former Mardichi Barracks in La Spezia, Mezzacapo Barracks in Reggio Calabria, Former Military Hospital in Palermo) and are partly used; others present vast areas but are abandoned because the investment required for the recovery of large facilities and sites becomes, at times, uneconomic (Monte Cimone Barracks in Banne, Seaplane airport with the Nervi hangars), insufficient attention and policies even though the site may represent an environmental and technological risk (Monte Nardelli Military Base, for example). The selected international cases are hybrid examples (Antiguo Hospital Militar en el barrio del Realejo, in Granada and the Son

Table 18.6 Former seaplane airport with Nervi warehouses, Former ammunition depot in Via Dante Alighieri

Region:	Sicily
City:	Marsala (Trapani)
Name:	Former seaplane airport with Nervi warehouses, Former ammunition depot in Via Dante Alighieri
Initial destination: *Military barracks* *Airport, military infrastructure* *Military base/shooting range, exercises* *Other*	Military Accommodation, Canteen, Hangar, Munitions Depot.
Age of construction:	1935 (Hangar Nervi)
Property:	–
Size (range): <1000 sqm >1001 e < 10.000 >10,001 sqm	~ 120.000 m2
Relationship to the urbanised context *Interior* *External in continuity* *External in suburban areas*	External in continuity
State of disuse: *Abandoned* *Underused*	Abandoned
Conversion implemented or in progress:	Conversion still not implemented
Mixed with prevalence of: *Services and urban amenities of a social and aggregative nature* *Business/office centre*	
Monofunctional:	
Environmental weaknesses and vulnerabilities of the study area and context:	No
Contaminated site:	The intervention area lies in an environmental zone of high landscape-cultural value, being in the coastal area of the Stagnone lagoon, today a protected nature reserve
Widespread urban decay:	The area of the former ammunition depot in via D. Alighieri has been reclaimed and secured
Lack of services and social discomfort:	–
Management complexity:	–

Table 18.7 Former Military Hospital, located at Corso Calatafimi No. 404

Region:	Sicily
City:	Palermo
Name:	Former Military Hospital of the Michele Ferrara Barracks
Initial destination: Military barracks Airport, military infrastructure Military base/shooting range, exercises Other	Military Hospital
Age of construction:	1932
Property:	State property of the Italian Air Force
Size (range): <1000 sqm >1001 e < 10.000 >10,001 sqm	~ 50.000 m^2
Relationship to the urbanised context Interior External in continuity External in suburban areas	Inside the urbanised area
State of disuse: *Abandoned* *Underused*	Abandoned
Conversion implemented or in progress:	Conversion started in November 2020.
Mixed with prevalence of: *Services and urban amenities of a social and aggregative nature* *Business/office Centre*	New destination of use: Carabinieri directional pole.
Monofunctional:	Offices-services and urban facilities
Environmental weaknesses and vulnerabilities of the study area and context:	Directional pole, new departments and warehouses
Contaminated site:	–
Widespread urban decay:	No
Lack of services and social discomfort:	The area of Corso Calatafimi shows widespread urban decay and several abandoned buildings.
Management complexity:	Social unease due to urban decay in the Calatafimi area.

Busquets Barracks in Palma de Majorca) as, while presenting interesting locational factors, they receive different attention: the Hospital Militar en el barrio del Realejo has been transformed into the brand new ETSAG (Escuela Técnica Superiorde Arquitectura), Son Busquets (Spain) abandoned since 2000, is considered the last great urban void in Palma and is still waiting for an architectural and urban renewal to respond to the demand for urban transformation of the surrounding neighbourhoods.

Table 18.8 Antiguo Hospital Militar (actual Escuela Técnica Superiorde Arquitectura) in the Realejo district

Region:	Spain
City:	Granada
Name:	Antiguo Hospital Militar (actual Escuela Técnica Superiorde Arquitectura) in the Realejo district
Initial destination: *Military barracks* *Airport, military infrastructure* *Military base/shooting range, exercises* *Other*	Military Hospital
Age of construction:	Mendoza Palace in the seventeenth century, Admiral Aragón's house in the eighteenth century, Hospital of the Archbishopric in the 19th and the Military Hospital from 1866
Property:	
Size (range): <1000 sqm >1001 e < 10.000 >10,001 sqm	5.996,60 m^2
Relationship to the urbanised context *Interior* *External in continuity* *External in suburban areas*	Inside the city
State of disuse: *Abandoned* *Underused*	Abandoned
Conversion implemented or in progress:	Since 2015, the hospital has been transformed into the brand new ETSAG (Escuela Técnica Superiorde Arquitectura).
Mixed with prevalence of: *Services and urban amenities of a social and aggregative nature* *Business/office centre*	Social and aggregative urban services and facilities
Monofunctional:	
Environmental weaknesses and vulnerabilities of the study area and context:	Problems arising from its location on Mauror Hill related to complex accessibility and urban mobility
Contaminated site:	No
Widespread urban decay:	–
Lack of services and social discomfort:	–
Management complexity:	The project refers to sustainable solutions for a balanced management of resources and their socio-economic viability, overcoming regulatory and administrative barriers through inter-institutional collaboration.

Table 18.9 Barracks—headquarters of the 91st Mixed Artillery Regiment (RAMIX 91), Elaborated by Celestina Fazia

Region:	Spain
City:	Palma (Majorca)
Name:	Old military quarter of Son Busquets
Initial destination: *Military barracks* *Airport, military infrastructure* *Military base/shooting range, exercises* *Other*	Barracks—headquarters of the 91st Mixed Artillery Regiment (RAMIX 91).
Age of construction:	1949 (presumed era)
Property:	Public
Size (range): <1000 sqm >1001 e < 10.000 >10,001 sqm	110.000 m^2 Total built space is 27,743 m^2
Relationship to the urbanised context *Interior* *External in continuity* *External in suburban areas*	***Interior***
State of disuse: *Abandoned* *Underused*	Abandoned The buildings are in a relatively good state of preservation and are considered to have a certain charm because they were built in the tradition of the time.
Conversion implemented or in progress:	Abandoned since 2000, it is regarded as Palma's last great urban void and is still awaiting architectural and urban renewal to respond to the demand for urban transformation of the surrounding neighbourhoods
Mixed with prevalence of: *Services and urban amenities of a social and aggregative nature* *Business/office centre*	There are currently abandoned artefacts and potentially productive greenery. A project is hoped for to create urban services and amenities of a social and aggregative nature
Monofunctional:	
Environmental weaknesses and vulnerabilities of the study area and context:	None
Contaminated site:	–
Widespread urban decay:	Camp Redó, is a very dense peripheral neighbourhood that suffers from serious problems of physical and social degradation, with severe urban discontinuity and poor connection to the historic city centre

(continued)

Table 18.9 (continued)

Lack of services and social discomfort:	The General Urban Development Plan of Palma of 1973, the current plan of 1998 and its imminent successor, contemplate for the urban expansion of Palma the reinforcement of six civic axes (west or east Fàbrica, Ocells, Blanquerna-Cotlliure, Arxiduc, Nuredduna and Foners). Since 1973, only 14% of the 7110 planned 'civic' metres have been implemented.
Management complexity:	The regeneration of the Old Military Quarter of Son Busquets could become an opportunity for the urban planning and development of the Camp Redó district, a densely built-up area with problems of physical and social decay.

18.2 Italian Case Studies

18.2.1 DIA of Reggio Calabria

The site was originally a fuel depot for military use. Used during World War II, it housed three tanks, one of which was demolished to make room for the construction of the Manganelli multifunctional building [5]. Within the area, there was a building of approximately 80 square metres in use by the fire-fighting team, which had been disused for several years. Within the project area with access from Via Enotria level, there was a tunnel used during the Second World War.

For the project of the new DIA (Anti-Mafia Investigative Directorate) headquarters (see Fig. 18.1), several specialised investigations and analyses were carried out: interference analysis with overhead and underground utility networks and design solutions (water network, sewage and stormwater disposal networks, electricity, gas and telecommunication supply network).

The project for DIA headquarters (see Fig. 18.2)-which will also house the 12th Mobile Department and the State Police- was drawn up to meet the needs of the Administration receiving the building and in accordance with the following criteria

- Flexibility of spaces.
- Quality of the working environment (systems to improve acoustic quality), healthy indoor air and respect for the site's microclimatic conditions.
- Energy saving (photovoltaic system on the roof slab).
- Ease of maintenance and durability of materials.

The functional layout, distributional characteristics and orientation of the envelope were designed to improve the building's energy and functional performance by reducing the consumption of resources and the emission of pollutants through:

- controlled mechanical ventilation system with heat recovery;
- primary air diffusion in the environment, domestic hot water production systems;
- draining outdoor paving, which plays an important role in surface water management with the perimeter hedge. Irrigation of the hedge and the portion of lawn following the ramp is done with recycled rainwater.

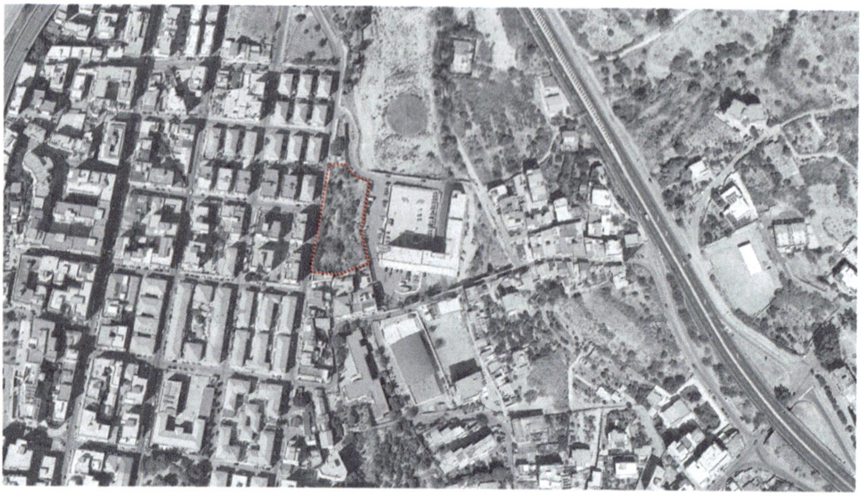

Fig. 18.1 Area of new project of DIA

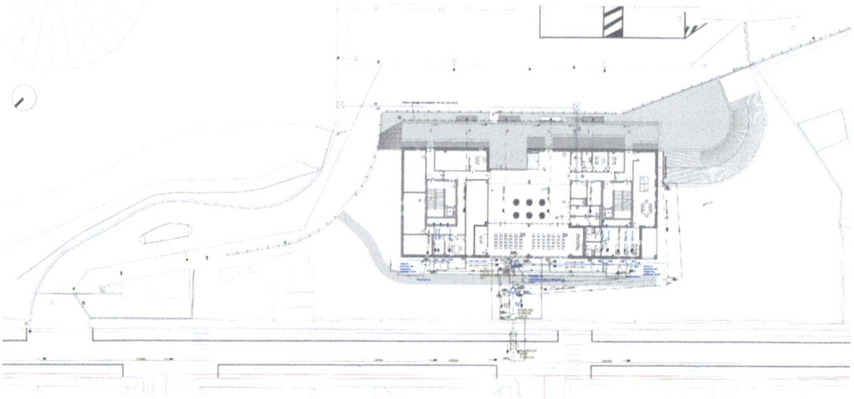

Fig. 18.2 Project of DIA di Reggio Calabria, 2017, PE-RE-01_5 RELAZIONI

18.2.2 Mezzacapo Barracks Reggio Calabria

The barracks are located in the historical centre of the city of Reggio Calabria (see Fig. 18.3) and is protected by the Superintendence of Fine Arts and Landscape of Calabria. It is one of the historical buildings reconstructed after the 1908 earthquake.

The building consists of four bodies. In the centre is a large courtyard, known as the 'Piazza d'Armi' (Square of Arms), where the morning flag-raising ceremony was held. It is in good condition. Completely destroyed by the 1908 earthquake, it was rebuilt. On 30 November 2009, the historic barracks were handed over to the Agenzia del Demanio, Calabrian subsidiary. The latter handed it over to the Prefecture, which, together with the Police Headquarters, opened some offices inside the military structure.

Fig. 18.3 Mezzacapo Barracks Reggio Calabria

18.2.3 Former USAF Military Base of Monte Nardello

The former USAF Military Base (see Fig. 18.4) is located 1800 metres above sea level, on the summit of Mount Nardello and occupies about 3.5 hectares in the municipality of Roccaforte del Greco [6] within the Aspromonte National Park.

It was built in 1965 by the US Army for the radar control of the Mediterranean skies, together with that of Catania. The complex comprised a series of sheds with huge electricity generators inside used to power the radar antennas more than 10 metres high. Provided with accommodation and all the necessary facilities for the military to stay on site, the demarcation of the area was (and still is) entrusted to a massive fence more than 2 metres high.

Following the development of new satellite technologies, the base was decommissioned in 1993 and handed over to the Italian Ministry of Defence [6].

18.2.4 Monte Cimone Barracks in Banne

Owned by the Burgstaller/Bidischini (see Fig. 18.5) family from the early nineteenth century until the 1930s, the complex -known as Mandri- was used as military headquarters until the 1990s. It covers a vast area, over 15 hectares and a total commercial surface of 44,651 m^2. It consists of buildings with architectural and typological-constructive [7] features of cultural interest.

Fig. 18.4 Area of Military base USA, Roccaforte del greco (RC)

Fig. 18.5 Monte Cimone Barracks in Banne, Bidischini-Burgstaller

Today it is decommissioned, owned by the State and subject to alienation, offered for sale to private parties. A former military compendium is located in the historic village of Banne on the Karst plateau, in a hilly area northeast of the centre of Trieste. The area is of environmental value, close to the border with Slovenia, and offers a panoramic view of the entire gulf [1, 5, 7].

18.2.5 Former Mardichi Barracks. La Spezia

The Mardichi area, also known as the former Artillery Barracks, is an architectural complex of great historical importance [8] for the city of La Spezia (see Fig. 18.6). In the past, it housed the headquarters of the Army Cavalry and later the Navy Artillery. This area had a defensive function for the arsenal and represented a true independent citadel capable of withstanding even a siege.

Currently, the complex is subject to constraint under Article 12 of Legislative Decree 42/2004. The Agenzia delle Entrate and the Agenzia del Demanio (State Property Agency) have signed an agreement for the reutilisation of the Mardichi Barracks to house the new headquarters of the Agenzia delle Entrate.

Both the seismic vulnerability of the existing buildings and the presence of polluting materials [9] in the complex were assessed. In the meantime, a fruitful collaboration was initiated with the Superintendence to find a balance between [10] conservation and innovation [11].

Following the practice of the Agenzia del Demanio—D.R. Liguria, the project will be carried out in accordance with sustainability protocols, and the offices will be designed with state-of-the-art facilities adapted to recent smart office innovations.

Fig. 18.6 Seaplane airport and Hangar Nervi, Former ammunition depot in Via Dante Alighieri

18.2.6 The Seaplane Airport with the Nervi Hangars and the Former Ammunition Depot in Via Dante Alighieri, Marsala (Trapani)

Built in the 1930s by the Regia Aeronautica, it was a military seaplane station.

It was built in the Stagnone lagoon, a stretch of coastline from Marsala to Mozia, and the Stagnone islands, classified as a nature reserve. In mid-December 1917, Marsala was home to a French Air Force Biserta CAM detachment that controlled the Strait of Sicily with the function of anti-submarine control. The two large reinforced concrete hangars designed by Pier Luigi Nervi, known as 'Hangar Nervi', were built in 1935 and are of historical and architectural significance. In disuse since the end of World War II, its facilities are still a military area owned by the Italian Air Force and sporadically used for amphibious military aircraft [12] (see Table 18.6).

There are many interesting proposals and solutions, but the complexity of the site, linked to environmental dimensions, urban planning issues and the cost of the intervention, does not encourage the realisation of even partial regeneration projects.

18.2.7 Former Military Hospital, Located at Corso Calatafimi

The former Military Hospital of the buildings identified with the initials 'B', 'C' and 'I' inside the 'Michele Ferrara' Barracks (see Fig. 18.7), formerly known as Villa Stagno, located in Corso Calatafimi no. 404, Palermo, is being rehabilitated [13] for the allocation of the Carabinieri Corps' departments [14].

Fig. 18.7 Former Military Hospital

In November 2020, urban redevelopment works began that will lead to the re-functioning and restoration of the former military hospital, which has been abandoned [15] for about 18 years.

18.3 Foreign Case Studies

18.3.1 Antiguo Hospital Militar en el barrio del Realejo (Granada) y su conversión en E.T.S. de Arquitectura

The former military hospital in the Realejo district was converted into an E.T.S. of Architecture (see Fig. 18.8).

The strategic redevelopment was considered as an opportunity for the socio-economic revitalisation of the degraded neighbourhood and the territorial context.

Fig. 18.8 Antiguo Hospital Militar (actual Escuela Técnica Superiorde Arquitectura) in the Realejo district

Its main characteristics are its relationship with the context, its position and communication with the Campo del Príncipe space, the height of its floors, its internal distribution and the precise vertical hierarchy of the façade openings.

18.3.2 Son Busquets

Son Busquets could assume an important aggregative role in the planning process of the new civic and pedestrian axis -Cotlliure- that would serve to regenerate the neighbourhood and effectively connect it with the historic centre (see Fig. 18.9). In the Camp Redó neighbourhood, the Son Busquets site is also emerging as a valuable solution to the need for new green areas and quality public open spaces. The barracks were closed in 2000, leaving a complex of buildings consisting of 16 ground-floor hangars with an east-west orientation and another 18 smaller hangars with a north-south orientation. These and other artefacts are in a relatively good state of preservation, having been built according to the tradition of the time, with marés load-bearing walls and pitched roofs made of Arabian tiles of historical and cultural value. It is set in a context of interesting vegetation, with productive potential [13–15].

Fig. 18.9 Barracks—headquarters of the 91st Mixed Artillery Regiment

18.4 Discussion

How to intervene? Through the definition of:

- Analysis activities: present and future risk diagnostics, types of contamination and materials found on sites, risk assessment approaches.
- Communication: Risk communication plans to inform/train local communities and listening activities to intercept local demand (participatory construction of rehabilitation actions).
- Redevelopment options. Frameworks for addressing design solutions for reuse respecting the environmental characteristics of military land.
- Risk management prioritisation methodologies.
- Assessment. Strategic environmental assessment scenarios for proposed options and environmental monitoring plans to be used during implementation.

The idea is to regenerate disused military areas, respecting their original conditions, by creating a single 'Social' Infrastructure connected with a system of 'Networks' (see Table 18.10). I would propose a single Social Infrastructure [13–15], composed of several material and immaterial networks that interconnect the different disused military areas. The Social Infrastructure [14, 15] will make a sustainable contribution to the regeneration of the existing city and make the territory resilient. It will contain:

- Physical networks and junction areas for sustainable mobility
- Lifelines, resilient infrastructure for emergencies, and Networks of:
    - *vegetated landscapes and permeable soils for ecosystem quality and Environmental network, ecological network*
    - *urban landscapes and technical infrastructure for dynamic city/water coexistence*
    - *hydraulic regulation and soil defence for resilience and adaptation*
    - *waste infrastructure and contaminated or disused sites to be reclaimed and renaturalized.*

And another:

Networks (frame) of public spaces and infrastructure for services, relational life and sociality

Networks for energy, Networks for information and communication technologies

Networks for culture and identity [13–15].

Social infrastructure is a place of multi-actor and multi-sectoral convergence for social re-appropriation, recycling of resources, the creation of agreements for the management of common heritage [16–19], and the growth of widespread forms of circular economy.

Table 18.10 Area Base Americana Monte Nardello (Aspromonte) in Reggio Calabria (for example)

Type of disused military area (name/location) Relationship to built-up area			**Regeneration options (examples of actions in the military area) Multifunctional area**	**Level and type of risk to be mitigated**	**Project Actions Objectives/priorities (levels)**		**Evaluation/consistency with sustainability principles**
□ **Internal**	□ **External in continuity**	□ **External in suburban areas**			Strategic Impact Area A: Environmental S: Social I/S: Infrastructure and Services	Priority (1–5, decreasing)	
Ownership and availability: □ Public (Ministry of Defence) □ In the process of being decommissioned			1) Technology Park		**A** **S** **I/S**	1	Building a resilient infrastructure and promoting innovation and fair, responsible and sustainable industrialisation Goal 9 Agenda 2030
□ Partially used	□ Totally disused	□ Presence of constraints inhibiting transformations: Aspromonte National Park	2) Cultural and technological start-ups		**A** **S** **I/S**	1	Building a resilient infrastructure and promoting innovation and fair, responsible and sustainable industrialisation Goal 9 Agenda 2030

	3) Rurality "enjoyed" exhibition spaces connected with rural areas for typical local food and wine products. - Creation of an "itinerary" combining visits to special places (e.g. villages, nature parks, remote hamlets)		**A** **S** **I/S**	2	Making cities and human settlements inclusive, safe, resilient and sustainable Goal 11 Agenda 2030
	4) The Aspromonte tourist. Create spaces for organisation and promotion based on a series of interventions that favour local tourist reception.		**S** **I/S**	2	Protect, restore and promote sustainable ecosystem use Terrestrial Goal 15 Agenda 2030

18.5 Conclusions

It is necessary to consider the disused military heritage as a system made up of variable geometry networks of citizenship, knowledge networks, ecological and environmental networks, cultural heritage and tourism offer, and infrastructure.

Various cost/investment issues for the transition contribute to the "redesign" by achieving 4 objectives as a result indicator:

- overcoming the territorial and social divide
- job creation
- use of operational tools for urban, territorial and social regeneration
- proximity policies that aim to change urban development so that functions are accessible to local communities.

In this evolving framework, the role of the Ivb [14] should be understood as:

- a dynamic and resilient system of adaptation to the multiplicity of environmental risks, amplified by climate change on a planetary and urban scale;
- a diffuse, growing and oriented reservoir for the production of ecosystem services and the active and compensatory contrast of the weaknesses produced by soil consumption [13–15];
- primary reference context for the reorganisation of urban metabolism and resource life cycles (primarily water, soil, vegetation cover, waste, energy);
- incremental frame of the new public city characterised by high performative capacities from a landscape and ecological-environmental point of view, associated with processes of social inclusion and economic development.

Finally, one can add a reflection on brownfields regeneration scenarios in Ukraine.

Areas that are more inland or close to the city centre could be regenerated as landing areas for priority social needs, identifying mixed and changing functions in relation to social demand.

Underlying the regeneration of areas must be the recovery of environmental values and green and blue infrastructure/connections [16–19].

Each military area will be specialised by reorganising the offer of services (see Table 18.11) according to the characteristics of the territory and the evolution of needs, guaranteeing levels of provision of basic services (e.g. 30% of the area), restoration of ecosystem services and biodiversity (e.g. 25%), flexible offer of spaces according to "needs" (e.g. 45%). All regenerated areas will be part of a single infrastructure.

Table 18.11 Regeneration scenarios in Ukraine

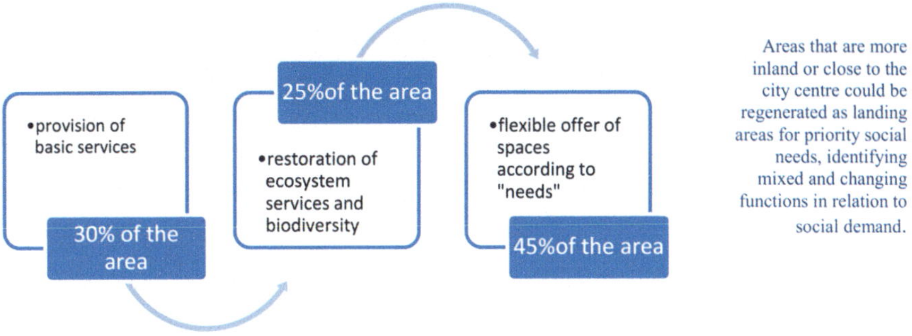

References

1. Morar C, Dulca M, Nagy G (2016) Brownfields regeneration, between sustainable urban development and cultural heritage. The former military sites in Oradea, Romania. J Urban Reg Anal III(5, 1):75–84
2. Bermann L, Orr D, Forrester T (2009) Improving community health: brownfields and health monitoring. Environ Pract 11(3):190–195
3. Sarni W (2009) Greening brownfields: remediation through sustainable development. McGraw-Hill, New York
4. Morar C, Berman L, Unkart S, Erdal S: Sustainable brownfields redevelopment in the European Union: an overview of policy and funding frameworks. https://www.ncbi.nlm.nih.gov/pmc/articles/PMC8959022/ 17/09/23, 00:59
5. Stephen H (2014) Winston Churchill Fellowship Trust – Military Base consolidations US experience: interim fellowship report July. Winston Churchill Memorial Trust, Londra
6. Lopez-Pineiro S (2020) A glossary of urban voids. Jovis, Berlino
7. Lozano Muñoz A (2008) La gestión del suelo militar desafectado. Ministerio de Defensa, Madrid
8. Ponzini D, Vani M (2014) Planning for military real estate conversion: collaborative practices and urban redevelopment projects in two Italian cities. Urban Res Pract 7(1):56–73
9. AAVV, In città il Demanio cede 6 immobili ma tre caserme restano invendute – Il Piccolo (gelocal.it)
10. Camerin F, Gastaldi F, (eds): Rigenerare le aree militari dismesse. Prospettive, dibattiti e riconversioni in Italia, Spagna e in contesti internazionali, Maggioli editore, Santarcangelo di Romagna (RN) (2021)
11. SiciliaNews24 (2020) Ex ospedale militare di Palermo sarà sede di uffici dei carabinieri, 26 Novembre Palermo, 2020
12. Artioli F (2012) Public real estate between administrative reforms and financial constraints. A comparative analysis of the re-use of military assets in Italy and France working papers du Programme Villes & territoires, 2012–04. Science Po, Paris
13. European Commission (2009) Mainstreaming sustainable development into EU policies: 2009 review of the European Union Strategy for Sustainable Development. European Commission, Brussels
14. Gasparrini C (2023) *Significante e significati". Infrastrutture verdi e blu, INU ed.* Infrastrutture verdi e blu – Urbanistica informazioni
15. Munafò M (2022) (a cura di): Consumo di suolo, dinamiche territoriali e servizi ecosistemici. Edizione Report SNPA 32/22
16. AAVV, http://www.naturalmentebrancaleone.org. Last accessed 20.1.2014

17. Relazione generale del progetto esecutivo, PE-RE-01_5 RELAZIONI E DOCUMENTI, Agenzia del Demanio, Direzione Regionale Calabria, 2018 in: Nuova sede della D.I.A. – Direzione Investigativa Antimafia a Reggio Calabria (RC) (mateng.it). Last accessed 20.1.2014
18. AAVV, Nardello, gli USA e l'Area 51 della Calabria: Aspromont Horizon (icalabresi.it). Last accessed 20.1.2014
19. Ministero della Difesa (2020) Relazione sulla performance Anno 2019 (art. 10, c. 1, lettera b del D. Lgs. n. 150 del 2009). https://performance.gov.it/performance/relazioniperformance/documento/1171. Last accessed 14.1.2014

Chapter 19
Beyond the Limits of the City: Strategies to Regenerate Fragile Territories

Chiara Garau, Paolo Nesi, and Paola Zamperlin

Abstract Implementing sustainable urban policies in fragile contexts that have recently emerged from wars may be challenging due to the complexity of emerging priorities and the fragmentation of policymaking within government structures and hierarchies. This paper starts by examining these issues, focusing not only on brownfield sites but also on the whole city, extending beyond its boundaries. The underlying premise is that planning for a vulnerable and fragile city may initially appear more difficult but may ultimately be converted into a targeted strategic planning approach. The authors will first introduce the concepts of which are the best regeneration strategies for a devastated by-conflict context. Following that, a concise overview is provided on fragility, robustness, and antifragility in relation to complex social systems, with a specific focus on cities. Finally, the authors will analyse the elements that contribute to the vulnerability of cities and propose the use of artificial intelligence, namely digital twins, as a potential solution for reconstructing cities devastated by warfare. The objective is to hypothesise on the feasibility of reconstructing an enhanced urban area using the existing digital city plan as a foundation.

This paper is the result of the joint work of the authors. In particular, "Abstract". "Strategic Antifragile Planning and Moving from Theory to Practice" and "Discussion: The Value of a Possible Model and Applications" were written by all authors. C.G. wrote "Introduction" and "Fragility and Conflict: Starting from What?", P.Z. wrote "The Ukraine (anti)Fragile Context". P.N. wrote "Conclusions". P.Z. and P.N. wrote "and "Snap4City for (Anti)fragile Contexts".

C. Garau (✉)
Department of Civil and Environmental Engineering and Architecture (DICAAR), University of Cagliari, Cagliari, Italy
e-mail: cgarau@unica.it

P. Nesi
DISIT Lab, Department of Information Engineering, University of Florence, Florence, Italy

P. Zamperlin
Department of SAGAS, University of Florence, Florence, Italy

C. Morar et al. (eds.), *Achieving Sustainability in Ukraine through Military Brownfields Redevelopment*, NATO Science for Peace and Security Series C: Environmental Security, https://doi.org/10.1007/978-94-024-2278-8_19

Keywords Fragile territories · Regeneration strategies · Snap4City · Digital twins · Devastated by-conflict context · Indicators

19.1 Introduction

The course of human history has been plagued by an almost uninterrupted series of various armed wars—ranging from small-scale to large-scale, from regional to global—resulting in devastating societal harm and the destruction of urban infrastructures.

The ongoing tragedy continues as countless individuals remain deprived of fundamental services due to the conflicts. The process of reconstructing nations and communities that have been devastated by war may be seen as a sequence of disconnected tasks undertaken (and often enforced) by international organisations and governments, to advance political and other interests. The consequence is that the disasters of conflicts are often joined by the disasters of rebuilding, disregarding sustainable development. Despite the absence of a solid and unified theory/methodology on post-war reconstruction, researchers have made efforts over time to examine strategies for mitigating the impacts of war emergencies. International institutions and national governments have started to establish specific goals and obligations for disaster and vulnerability proneness reduction. The academics agree that nations that have undergone a conflict have two primary challenges: The two essential objectives are: (1) the need to restore the economy under the physical, social, economic, and cultural aspects, and (2) the imperative to avoid the reoccurrence of armed conflicts.

The requirement to rebuild places devastated by traumatic and violent events inevitably brings forward the issue of mitigation methodologies, which mostly depend on political and strategic decisions. Currently, there is a lack of operational protocols that allow for a successful rebuilding process systematically and cohesively, including all aspects from initial emergency management to the finalisation of reconstruction, including physical, social, economic, and cultural aspects.

Based on these premises, the authors first introduce the concepts of which are the best regeneration strategies for a devastated by-conflict context, after having framed the Ukraine context under the (anti)fragile paradigm. Secondly, they thoroughly examine the characteristics of a fragile territory subject to post-war reconstruction, by considering the terms of fragility, robustness, and antifragility about complex social systems. They subsequently propose an innovative approach to constructing more sustainable settings by proposing digital twin-based tools as immersive and potential solutions for reconstructing devastated cities. The authors acknowledge the inherent difficulties in providing universally applicable solutions and emphasising the importance of finding methods for shortening the post-war recovery period, which currently stands at 25 years [1]. Equally significant is the objective of reducing the rate of nations returning to civil war within 10 years to 40% [2], (44% in 5 years [3]) to enhance reconstruction planning in a more dynamic, proactive, and sustainable manner. The authors conclude the paper by providing final reflections and offering recommendations for future research.

19.2 The Ukraine (Anti)Fragile Context

Recent studies has identified the Russian invasion of Ukraine as further dramatic evidence of the urban nature of contemporary warfare [4].Cities have become increasingly unavoidable battlegrounds in contemporary times, surpassing the days when military strategists pushed for open battle in extra-urban contexts. As stated by A. King, the longest and most intense battles of the last quarter of a century have occurred in cities, not in the field, and "*the wars in Iraq, Syria, the Donbas, Libya, and Yemen have all been predominantly fought not only for but, often, actually inside cities*" [5]. The reasons for these choices, according to the authors, are to be explained by the combination of crucial factors such as urbanisation, demographics, asymmetry, and consequently for the extent of the socio-economic damage that is intended to cause. Regarding the Ukrainian war, it is evident that Russia employs a brutal combination of siege warfare, involving extensive and random bombing of cities. In response, Ukraine adopts an urbanised defence approach, utilising the urban terrain to its advantage and the build cities as fortresses and operational bases.

In line with this trend towards the urbanisation of warfare, it is evident that the invasion in Ukraine has mostly focused on urban areas. These areas have been severely affected by the destructive consequences of war, so much so that some scholars have referred to it as a true *urbicide* (Ljungkvist 2022, *passim*).

Before the 2022 invasion, Ukraine had an urbanisation rate of over 70%, and the majority of its 44 million inhabitants lived in cities with more than 100,000 residents. It is readily comprehensive that in order to have authority over the whole nation, it is essential to possess control over all of its cities.

During the course of the conflict, a number of Ukrainian cities, including Kiev, Kharkiv, Mykolaiv, Odessa, and Kherson, have seen significant levels of shelling and bombing. Chernihiv, located in the northern region, has been estimated to have suffered around 70% destruction, while Mariupol has been almost completely razed after prolonged indiscriminate bombardments.

However, the rural areas are also severely affected, and essential services, such as those ensuring access to drinking water, are compromised. According to Strokal et al. [6], the war has significantly impacted Ukraine's environment, particularly its soil and water resources. About 35% of Ukrainian territories are affected by soil destruction, with 130,000 km^2 of land mined or damaged, rendering some agricultural areas, especially in regions like Kharkiv, Mykolaiv, Kherson, Zaporizhzhia, Kyiv, and Chernihiv, unsuitable for crop cultivation as of May 2023. The Kakhovka Hydropower Dam's damage resulted in crop yield losses of 100 thousand tons in the Kherson region and disrupted water access for 31 irrigation systems in southern Ukraine, exacerbating soil degradation and pollution. Water resources have suffered extensively: 724 hydrotechnical systems, 160 water treatment and sewage systems, and 22 dams have been destroyed, causing widespread water pollution. Up to 90% of the irrigation system in southern Ukraine is lost, and fishery has decreased by 67% due to the damaged hydraulic systems. The Kakhovka Dam's disruption led to the flooding of areas with polluted water, affecting 31 water supply and drainage facilities, leaving 13 villages without centralised water, and flooding 4 solid household waste landfills.

The war has led to significant losses in Ukraine's agricultural sector, with over 1.5 million agricultural fields damaged by December 4, 2023, accounting for about 5.65% of the total sown area. The most affected crops include wheat (5.78% of its total cultivated area), sunflower (1.56%), maize (1.2%), and rapeseed (2.65%) [7].

Furthermore, the impact of environmental hazards on eastern Ukraine, specifically the industrialised Donbas region, highlights the need for proactive and long-term management of conflicts in urban areas that have been significantly influenced by human activities, such as coal mining and nuclear tests [8].

This represents the first stage of a complex reconstruction process that also has to include the aimf of sustainable development [9].

19.2.1 Fragility and Conflict: Starting from What?

Conflicts and fragility are closely interconnected, and although not all fragile contexts experience a war, many have an underlying conflict and a significant risk of potential conflict. When these terminologies are paired with the conceptual framework of territory, also complexity becomes a prominent consideration. It is widely agreed that conflict is the most extreme expression of fragility. Therefore, the examination of the primary causes of fragility is closely connected with the evaluation of the primary causes of conflict. The OECD report defined five dimensions of fragility: political, societal, economic, environmental, and security [10]. All of these dimensions contributing to fragility are interlinked, making it difficult to determine the primary factors driving fragility. Some of these dimensions may serve as causes or effects (in most cases via cascading effects) of other dimensions and attempting to examine the reasons for fragility could be a difficult undertaking.

By considering a territory, several definitions of fragility or fragile state pertaining to a limited ability to effectively administer its people and territory, as well as to a lack of capability to establish positive and mutually beneficial relationships with society [11].

Fragility refers to the situation of being unable to fulfil two fundamental functions: ensuring security and providing economic opportunities. In what way starting from a context that has recently emerged from a conflict?

Before responding to this question, it helps to understand the precise meaning of the terms fragility, robustness, and antifragility in relation to complex social systems, with a specific focus on cities.

Cecchini and Blecic [12] effectively depicted these principles. Following Taleb's definition [13], they consider items or systems as fragile if they are susceptible to damage, ruin, or destruction when faced with unexpected events of disturbances. Robust items or systems are the opposite of fragile ones. Resilient systems are also distinct from fragile ones because they are not affected by events of disturbances, being more capable to react and recover. Robustness allows items or systems to cope with perturbations without being affected, while resilience enables them to absorb or recover from disturbances. Antifragile items or systems are not only susceptible to unexpected events of disturbances, but they may also benefit from them and even enhance their performance.

Cecchini and Blecic effectively underlined that the terms fragility and antifragility may be seen as representing a dynamic nature, while robustness and resilience are essentially static [12]. The same authors underline as Taleb states also that some entities flourish and expand as a result of unexpected impacts or disturbances, lack of order, and factors that induce stress, and they get pleasure from undertaking daring endeavours, taking risks, and embracing unpredictability. Although fragility is common, there is currently no specific term to describe its opposite. The concept of antifragility could be what is just described. Antifragility surpasses the concepts of "elastic resilience" and robustness. Resilient entities have the ability to withstand and recover from shocks, whereas antifragile entities not only withstand shocks but also benefit and develop from them. Namely, the principle of antifragility not only leads to improvement based on experience, but also enables more effective reorganisation in response to the shock, or to the impact of the event.

Based on these principles and according to the authors of this paper, post-war environments as well as post-war cities should behave as antifragile systems that act on the political, societal, economic, environmental, and security aspects. They underline that although it is unthinkable to benefit from conflict, seeing these contexts in this manner could potentially reduce the time it takes for a post-war city to recover and return to "normal" everyday life by 25 years. In the next paragraph, the authors will explain how the notion of antifragility applies to a strategic antifragile planning. They will then explore how it is possible to move from theory to practice.

19.2.2 Strategic Antifragile Planning and Moving from Theory to Practice

Within the literature, several studies have been presented for examining the city as a complex organism and also as an antifragile system, with a specific emphasis on the uncertainty surrounding public policy [14–18]. Based on the required synthesis in this paper, the authors only provide the key components of antifragility focusing on the main objective of this research and beginning with the analyses from the aforementioned publications. Therefore, the key aspects of antifragility are: (1) The natural relationship that the antifragile with uncertainty, complexity, and self-organisation; (2) The necessary of a Smart Decision Support System for Anti-fragile Territories (SDSSfAT); (3) the necessity to have data, big data for the realisation of a digital twin for antifragile territories and cities.

Consequently, and considering the points just listed, antifragile strategic planning enhances the antifragility of the city and could be articulated by considering: (1) the framework of Lefebvre's '*right to the city*' and '*to urban life*' [19] in which everyone has the freedom to live, to move, to health, to learn, to work, to choose one's life plan. (2) The SDSSfAT can increase the opportunities and possibilities of understanding the city and, in this way, it could be possible to have shared visions, to create and choose the best-shared scenarios; to acquire skills and put them into action. (3) The data are necessary to create predictive conditions and anomaly detection solutions, to better express the freedoms of point (1). Rebuilding a city in a concerted manner starting from its digital models exploiting processes of simulation and analysis (the so-called digital twin), can help to better address previous critical issues and to consider all the aspects and interrelationships among the several entities and services in a simulated and thus cheap conditions.

The process of antifragile reconstruction should be able to:

1. Construct a shared, in-depth, and detailed knowledge base for each sector through an integrated analysis (political, economic, social, urban, environmental, cultural, functional, and operational, etc.).
2. Facilitate understanding and fully consider the local context, and the several relationships among subsystems.
3. Prepare and share objectives with local stakeholders.
   - divide objectives into reachable sub-objectives and define detailed programs/steps and projects to reach them.
   - have a continuous development, follow-up, and improvement process (periodic and scheduled inspections).

In a nutshell, the creation of a digital twin for antifragile territories and cities can help to best answer the question: in terms of design possibilities for redevelopment and reconstruction, which regeneration strategies for thinking about a better future?

The next paragraphs show the hypotheses of how we can move from theory to practice, by considering the Snap4City (paragraph 3) as the most adaptive tool for Smart City Digital Twin applied in (Anti)fragile territories (paragraph 4).

19.3 Snap4City for (Anti)Fragile Contexts

Snap4City is an entirely open-source platform, operating across numerous cities and regions, developed under the leadership of the DISIT Lab at the University of Florence, Italy. This platform offers numerous strengths that suggest how to perform monitoring and control on operations and performing tactic and strategic planning. It can be applied to the implementation in a reconstruction scenario or at the initiation of reconstruction, at meso and macro scales.

Snap4City is GDPR compliant, has passed the PENTest, and is an EOSC platform, an official Node-RED library, an official E015 API, and more. One as the option to either install the platform or provide it "as a Service" or have it installed at its location (by personally or via accredited Snap4City companies) [20]. By its nature, it facilitates the realisation of a wide range of services in smart cities and integrated IoT/IoE (Internet of Things/Internet of Everything) domains. In general, it enables the definition of strategies, implementation of control rooms, creation of applications based on artificial intelligence, explainable artificial intelligence, simulations, digital twins, and thus of decision support systems driven by the value of key performance indicators (KPI), essential in the daily management of both cities and industrial plants, offering cross-sectional or vertical solutions [21].

Snap4City's robust structure and the wide range of tools permit to provide an in-depth understanding of the context and formulate reliable forecasts. It effectively offers timely alerts upon detecting abnormalities and/or performing simulations. The aforementioned data is used to strategize and implement real-time interventions with the objective of improving services for citizens in multiple areas, such as mobility, transport, energy, government, tourism, environment, and civil engineering. The digital twin includes global and local aspects of the city in an integrated manner [22]. At the global level, it includes structural, functional, and data information about buildings, roads, services, traffic flow, people flow, environmental aspects, and more.

Figure 19.1 shows the general overview of the Scenario tool in Snap4city.

On this wide knowledge base of information, it is possible to identify scenarios on which to perform simulations, predictions, anomaly detection, causality analysis, and also what-if analysis. The what-if analysis techniques are going to answer questions such as: "How is going to work service X or KPI Y if the city would be changed/designed according to scenario Z?" [23]. To this end, the editing of scenarios is playing a relevant role [24].

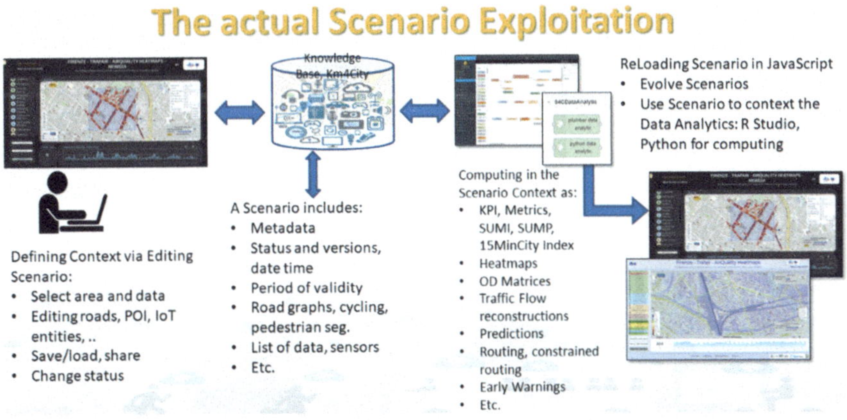

Fig. 19.1 General overview of the Scenario tool in Snap4city

19.4 Discussion: The Value of a Possible Model and Applications

The post-conflict reconstruction process should be treated as a proactive and multidisciplinary approach toward the dynamics of rebuilding social, political, economic, and spatial systems more holistically and systemically.

An effective antifragile planning process requires not just focusing on design decisions but also on constructing cognitive frameworks, the identification of qualitative and quantitative indicators, and the availability of tools capable of developing analysis scenarios adhering to reality and in real-time with high reliability in the prediction result. In this study, the authors opted to envision a particular context by utilising Snap4city as a digital twin representation for the city of Oradea in Romania. This choice was motivated because the dissemination of this research was accomplished within this city, and the authors aimed to develop a practical application for the intended audience (Fig. 19.2). The applicability of this case study extends to all cities inside Ukraine.

At this stage, the authors considered important to reply to these questions: how to guarantee virtual post-conflict depiction for antifragile settings? How can urban areas be shown using intervention priority criteria that are beneficial for municipal governing bodies? A plausible response would include an analysis of indicators that possess the ability to distinguish between the tangible and intangible aspects of urban reality. Like any other standard, antifragility must be quantifiable to be used as a measurement of a territory's capacity to learn from shocks and adjust to its risks and fragility. Having a comprehensive understanding of a specific context, including its advantages and disadvantages, together with the awareness of the priority

Fig. 19.2 Snap4city dashboard of digital twin reconstruction of Oradea (Romania)

levels of activity, enables a more efficient restoration of the same territory. Generally, indicators are employed to effectively manage the fragility of a city or territory during the process of urban transformation. They can accurately describe and assess various parameters and variables in an unbiased way, including different geographical scales and the timing of phases associated with the event (emergency-reconstruction). Furthermore, a set of indicators can provide and assess the extent of spatial recovery/ reconstruction by considering several levels of analysis, such as global and local, individual, and social/community.

Based on these premises, the authors consider the following categories of indicators—also identifiable as Key Performance Indicators (KPIs)—but considering the specific Ukrainian context, burdened by widespread and profound destruction:

1. Indicator of reconstruction (Ir):
   - *Material Categories* (private and public architectural heritage; progress of expenditure; energy consumption; waste management; mobility and traffic; ...)
   - *Immaterial Categories* (demographic attractiveness; economic dynamism; social welfare, intangible cultural heritage, ...).
2. strategic objectives of the reconstruction (SOr)
   - *Features of actions* (responsiveness; transparency; timeliness; Resilience antifragility; Reliability, ...).
   - *Priorities and importance of actions* (i.e.: strong; moderate; weak).

The authors believe that Snap4City is the ideal vehicle for generating this post-war scenario. in fact, Snap4City has a strong and flexible structure that can be easily adaptable to diverse contexts, thus enabling the ability to create custom KPIs relevant to the spatial context and the available data.

Once a scenario is created, it can be modified from the current state, saved, exported/imported, and shared using a smart data model for immediate use in KPI and metric calculations.

Decision-makers face the challenge of balancing data accuracy, time, and costs to achieve the necessary level of information for decision-making. Tools often lean towards either speed or accuracy. In a context requiring resilience, such as Ukraine's (anti)fragile territory, adopting a method that is cost-effective, timely, and precise is crucial. In contrast to conventional functional-based or indices-based approaches, the Q-FRAM method presents a more efficient and informative solution by establishing key performance indicators (KPIs) and systemically computes synthetic indicators [25]. From this viewpoint, evaluating resilience to activate recovery processes in an (anti)fragile logic involves measuring the capacity to maintain necessary operations under both anticipated and unforeseen circumstances. The myriad of human, organizational, and technical factors, coupled with their rapidly evolving nature, presents significant challenges in fully comprehending and overseeing system functions.

To manage operational control effectively, it's crucial to acknowledge the high variability and uncertainty as constant factors. Q-FRAM aims to quantify system resilience at any given moment through a System Resilience Index (SRI), serving as a measure of resilience. The design principles for this indicator include ease of use and understanding, data availability, cost-effectiveness, relevance, and consensus-building among stakeholders through a co-design approach. Q-FRAM offers a rapid, cost-efficient way to assess socio-technical system resilience, enhancing strategic, tactical, and operational decision-making. It guides in defining KPIs based on system models and formulating indicators through a theoretical approach. Specific evaluation criteria allowed the assessment of this approach, including the non-regression property, the rebound effect, computational complexity, accuracy in measuring the framework's ability to handle both tangible and intangible assets, and value of information, emphasizing the methodology's practical effectiveness without delving into impractical details.

The respect for all these properties is strategic in (anti)fragile contexts.

19.5 Conclusions

The idea of fragility is multifaceted and often includes five dimensions: political, societal, economic, environmental, and security. The adoption of the Antifragile approach has significant potential in the context of post-conflict rebuilding scenarios.

In fact, for a comprehensive exploration of Ukraine's post-war reconstruction and development economics necessitates contributions from various domains, including: national security enhancement, sustainable economic development strategies, tourism and hospitality sector advancement, marketing for territorial development, and urban and infrastructure transformation, among others. This is particularly crucial in light of the full-scale Russian invasion in February 2022.

Topics like governance, ICT systems for environmental protection, business management standardisation, adaptive systems for manufacturing, insurance sector roles, technological progress in processing industries, corporate sustainability, tourism development, marketing strategies, urban waste management, and infrastructure modernisation assume significant importance [26].

Snap4City is the most adaptive tool for Smart City Digital Twin applied in (Anti) fragile territories, particularly for the post-conflict rebuilding contexts.

The research's future implications are focused on accurately defining the indicators described in the previous section and incorporating them into a context visualisation approach using the digital twin concept.

The authors acknowledge the inherent difficulty in generating universally applicable replies. Nevertheless, it is necessary to identify shared elements across diverse post-conflict rebuilding situations to adopt a more dynamic, proactive, and sustainable approach to reconstruction planning.

Acknowledgements This study was supported by the MUR through two projects: (1) "WEAKI TRANSIT: WEAK-demand areas Innovative TRANsport Shared services for Italian Towns (Project protocol: 20174ARRHT_004; CUP Code: F74I19001290001), financed with the PRIN 2017 (Research Projects of National Relevance) programme; (2) "SMART3R-FLITS: SMART Transport for Travellers and Freight Logistics Integration Towards Sustainability" (Project protocol: 2022J38SR9; CUP Code: F53D23005630006), financed with the PRIN 2022 (Research Projects of National Relevance) program. We authorise the MIUR to reproduce and distribute reprints for Governmental purposes, notwithstanding any copyright notations thereon. Any opinions, findings, and conclusions, or recommendations expressed in this material are those of the authors and do not necessarily reflect the views of the MUR.

References

1. TRANSITION REPORT 2022–23—BUSINESS UNUSUAL
2. Collier P, Hoeffler A, Söderbom M (2008) Post-conflict risks. J Peace Res 45(4):461–478. Available at: https://doi.org/10.1177/0022343308091356
3. Schwartz RA (2023) Undermining the State from Within. Cambridge University Press
4. Ljungkvist K (2022) A New Horizon in Urban Warfare in Ukraine? Scandinavian Journal of Military Studies 5(1):91–98. https://doi.org/10.31374/sjms.165; Hook K, Marcantonio R (2023) Environmental dimensions of conflict and paralyzed responses: the ongoing case of Ukraine and future implications for urban warfare. Small Wars Insurg 34(8):1400–1428. https://doi.org/10.1080/09592318.2022.2035098
5. King A (2022a) Will inter-state war take place in cities? J Strateg Stud 45(1):69–95. https://doi.org/10.1080/01402390.2021.1991797
6. Strokal V, Berezhniak Y, Naumovska O, Palamarchuk S (2024) The impact of the Russian-Ukrainian war on the soil-surface water interactions, EGU General Assembly 2024, Vienna, Austria, 14–19 Apr 2024, EGU24-1115. https://doi.org/10.5194/egusphere-egu24-1115
7. Drozd S, Kussul N, Yailymova H (2024) Evaluating the impact of armed conflict on agricultural sector in Ukraine through Remote Sensing and Machine Learning, EGU General Assembly 2024, Vienna, Austria, 14–19 Apr 2024, EGU24-10494. https://doi.org/10.5194/egusphere-egu24-10494
8. Hook K, Marcantonio R (2023) Environmental dimensions of conflict and paralyzed responses: the ongoing case of Ukraine and future implications for urban warfare. Small Wars Insurg 34(8):1400–1428. https://doi.org/10.1080/09592318.2022.2035098
9. Sydorova E, Sydorov O, Kakovkina O (2022) Anti-crisis strategy of state governance in the wartime: an example of Ukraine. Sustain Debate Brasília 13(2):175–189. https://doi.org/10.18472/SustDeb.v13n2.2022.44169
10. OECD (2016) States of Fragility 2016: understanding violence. OECD Publishing
11. OECD (2011) Supporting statebuilding in situations of conflict and fragility: policy guidance. OECD, Paris
12. Blečić I, Cecchini A (2024) Urban policy design for antifragility. In: Fragility and antifragility in cities and regions. Edward Elgar Publishing, pp 71–90
13. Taleb NN (2012) Antifragile: things that gain from disorder, 1st edn. Random House, New York
14. Carta M (2020) Progettare città aumentate antifragili. In: Panglossismo: l'architetto postpandemico. Pacini, pp 24–27
15. Garau C, Desogus G, Zamperlin P (2020) Governing technology-based urbanism. Technocratic governance or progressive planning? In: Willis KS, Aurigi A (eds) The routledge companion to smart cities. Routledge, New York, pp. 157–174
16. Blečić I, Cecchini A (2020) Antifragile planning. Planning Theory 19(2):172–192

17. Sartorio FS, Aelbrecht P, Kamalipour H, Frank A (2021) Towards an antifragile urban form: a research agenda for advancing resilience in the built environment. Urban Des Int *26*:135–158
18. Shearer AW et al (2021) Conceptualizing a model of antifragility for dense urban areas. J Digit Landsc Archit 6:75–84
19. Lefebvre H (1968) Le Droit à la ville. Anthropos, Paris
20. Bellini P, Fanfani M, Nesi P, Pantaleo G (2024) Snap4City Dashboard Manager: a tool for creating and distributing complex and interactive dashboards with no or low coding. Submitted on Software X. Elsevier
21. KPIs, such as Sustainable Urban Mobility Indicators (SUMI) and Sustainable Urban Mobility Plans (SUMP), are required to assess city mobility and transport management conditions/facilities. See also: Wefering F., et al. (2014). Guidelines. Developing and implementing a sustainable urban mobility plan. For SUMI consider: https://transport.ec.europa.eu/system/files/2020-09/sumi_wp1_harmonisation_guidelines.pdf
22. Adreani L et al (2023) Implementing integrated digital twin modelling and representation into the Snap4City platform for smart city solutions. Multimed Tools Appl 83:37121. Springer
23. Bellini P, Bilotta S, Ipsaro Palesi LA, Nesi P, Pantaleo G (2022) Vehicular traffic flow reconstruction analysis to mitigate scenarios with large city changes. IEEE Access, ISSN: 2169-3536. https://ieeexplore.ieee.org/document/9984661
24. Adreani L, et al (2024) Smart city scenario editor for general what-if analysis. submitted on Sensors MDPI 2024.
25. Bellini E, Coconea L, Nesi P (2020) A functional resonance analysis method driven resilience quantification for socio-technical systems. IEEE Syst J 14(1):1234–1244. https://doi.org/10.1109/JSYST.2019.2905713
26. Cirella GT (ed) (2024) Handbook on post-war reconstruction and development economics of Ukraine, Contributions to Economics. Springer, Cham

Chapter 20
Spatial-Functional Transformation of Military Brownfields on the Example of Sites in Sarajevo and East Sarajevo, Bosnia and Herzegovina

Jelena Golijanin, Saša Golijanin, Cezar Morar, Tin Lukić, and Aleksandar Valjarević

Abstract It is fact that almost all urban areas in Bosnia and Herzegovina had barracks and other military facilities or training grounds that over time acquired new functions or were still not valorized. The spatial analysis in the paper focused on the military brownfields of the capital and largest city of this country, which is divided by the entity border. That is precisely why were analyzed four examples of spatial-functional transformation of military brownfields in the area of Sarajevo in the Federation of Bosnia and Herzegovina and East Sarajevo in the Republic of Srpska. With a combination of field research, GIS, satellite and cartographic data sources, we obtained the results of spatial changes, the current state and functions of abandoned military facilities, as well as land. It was found that the largest number of buildings and especially land received a residential function, but also became an area for the construction and organization of university campuses or other public institutions. During the research were also found problems related to the reduction of green areas and the expansion of settlements, the unresolved issue of property-legal relations, the problem of housing refugees who are still using individual build-

---

J. Golijanin (✉)
Faculty of Philosophy, Department of Geography, University of East Sarajevo, Pale, Bosnia and Herzegovina
e-mail: jelena.golijanin@ff.ues.rs.ba

S. Golijanin
SG Engineering, Istočna Ilidža, Bosnia and Herzegovina

C. Morar
Department of Geography, Tourism and Territorial Planning, University of Oradea, Oradea, Bihor, Romania

T. Lukić
Faculty of Sciences, Department of Geography, Tourism and Hotel Management, University of Novi Sad, Novi Sad, Serbia

A. Valjarević
Faculty of Geography, University of Belgrade, Belgrade, Serbia

C. Morar et al. (eds.), *Achieving Sustainability in Ukraine through Military Brownfields Redevelopment*, NATO Science for Peace and Security Series C: Environmental Security, https://doi.org/10.1007/978-94-024-2278-8_20

ings for housing, etc. Although military bases are located in urbanized zones, the presence of unused (ruins) facilities was also established. Nevertheless, there are significantly more positive examples of the new purpose of these buildings and land, where we particularly highlight the sports and recreational role that improved and enriched life in both cities.

Keywords Military brownfields · Spatial-functional transformation · GIS · East Sarajevo · Sarajevo

20.1 Introduction

Military brownfields are the oldest form of brownfield. Throughout history, numerous military structures have lost their importance and purpose [1]. The specificity of military brownfields is that their abandonment is not caused by economic, but political reasons [2]. The transition processes in the post-socialist countries of Central and Eastern Europe included the process of demilitarization, and military brownfields appeared on the site of a large number of former military facilities [3]. Of course, military brownfields are not a specific challenge of spatial development only in the post-socialist countries of Europe, but the speed of changes conditioned by transition processes and the dynamics of the creation of brownfields were more pronounced in these countries compared to other regions [4].

The existence and origin of abandoned military facilities has been present for a long period of time in numerous countries around the world, so Bosnia and Herzegovina are no exception. These facilities were abandoned either due to changes in the doctrine of war, changes in technology, or the drawing of new boundaries. Their regeneration and reuse are primarily related to the interest of local authorities, whereby in some cases abandoned military facilities are used for the purpose of tourism, where they serve as accommodation facilities, and sometimes they themselves represent potential tourist attractions. Examples of military brownfields with a possible tourist function in Bosnia and Herzegovina were analyzed in the paper [5]. Modern research on military brownfields tells the world that they are used as places for new tourist destinations, or for functional purposes such as administrative and residential buildings, sports and recreational fields, airports, warehouses, garages, etc. [3–8].

Bosnia and Herzegovina were one of the republics within Yugoslavia. Considering its position (the central republic of the SFRY), it has the largest number of military brownfields compared to other republics of the former Yugoslavia. Many of these objects suffered destruction, and an even greater number were repurposed. The process of secession from Yugoslavia in the early 90s was followed by armed conflicts and a war lasting less than 4 years. After the end of the war, new borders were established that divided the capital into two parts, Sarajevo and East Sarajevo (originally called Srpsko Sarajevo) [9, 10]. The paper analyzed isolated examples of military brownfields on the territory of both Sarajevo (Fig. 20.1).

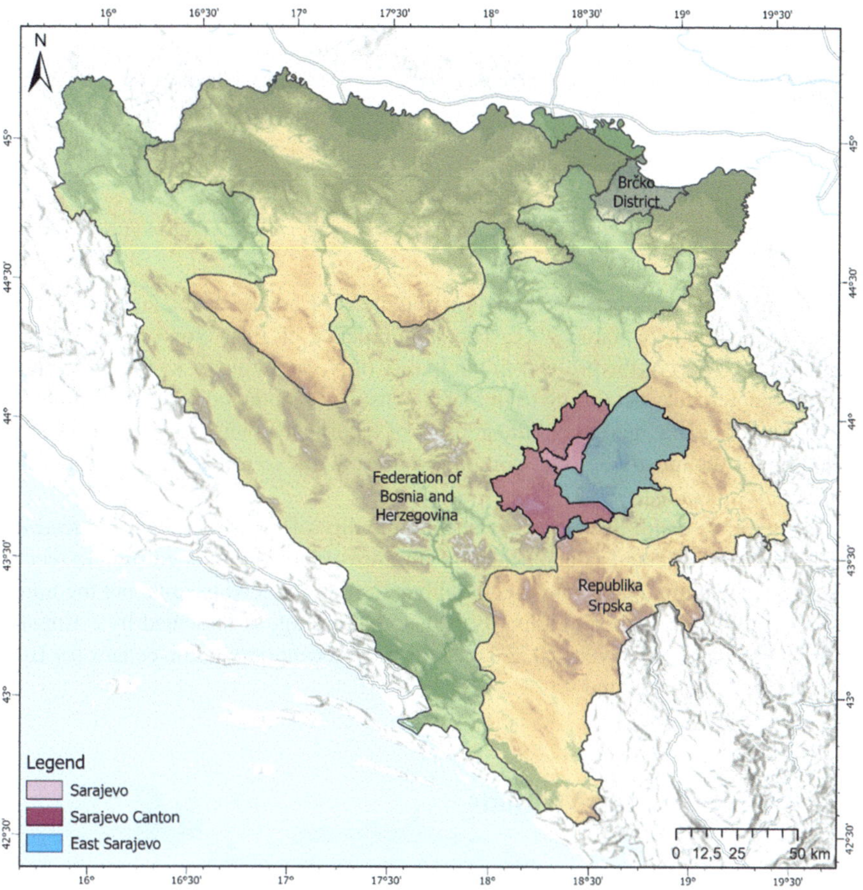

Fig. 20.1 Bosnia and Herzegovina divided in two entities and one district, with the location of Sarajevo and East Sarajevo

20.1.1 General Data on the Researched Area

We analyzed four military brownfields in the area of three border municipalities in both entities. Municipalities are Istočno Novo Sarajevo in Republica Srpska and Novo Sarajevo and Novi Grad in Federation of BiH. Significant changes were observed in the researched area—changes in area and population by municipalities (1991 and assessments for 2022) [11–13].

The entity's boundary did not respect the boundaries of the settlements system and did not respect the boundaries of the settlements itself as the network of settlements, which led to the division of settlements between the two entities. Major changes and disintegration settlements occurred in the area of the capital city of Bosnia and Herzegovina—Sarajevo, where, in 1992, the city of East Sarajevo was founded [14].

Table 20.1 Changes in area and population by municipalities before and after the civil war in BiH

Municipality	Area before war (km²)	Population (census 1991)	Area after war (km²)	Population (assessments for 2022)
Istočno Novo Sarajevo (RS)	–	4112 (assessment) [15]	37,9	12552 [13]
Novo Sarajevo (FBiH)	47,1	95089 [11]	9,2	62813 [12]
Novi Grad (FBiH)	48	136616 [11]	47,3	122614 [12]

Changes in the number of population and areas of newly established municipal borders are shown in Table 20.1. The municipality of East Novo Sarajevo was formed from a part of the pre-war municipality of Novo Sarajevo, so the population data for that territory according to the 1991 census was obtained by estimation [15]. There is a noticeable increase in the number of inhabitants, which is the result of the relocation of the mostly Serbian population from Sarajevo to East Sarajevo and the Republika Srpska entity.

On the other hand, in Sarajevo, the municipality of Novo Sarajevo reduced its area, which also affected a decrease in the number of inhabitants according to recent data. Also, the municipality of Novi Grad has slightly reduced its area, but the number of inhabitants in this municipality is also decreasing, as indicated by estimates based on the data for the year 2022, since the most recent population census for BiH has not yet been conducted.

20.2 Materials and Methods

An insight into the literature reveals the lack of records on military brownfields and a relatively small number of papers and research dealing with their conversion. In most countries, the question of military areas is riddled with security restrictions, due to which data remains inaccessible and scientific research is difficult. Authors from post-socialist countries point out another possible reason for the poor research of military brownfields, which stems from the fact that industrial brownfields, the largest in size, have long been the focus of research, which has put other types of brownfields in the background [2, 4].

Military brownfields, as well as brownfields in general, are the subject of research in various scientific disciplines—geography, economics, architecture and urbanism, ecology, sociology, history, political science, etc. Considering the actual or assumed soil and water pollution, a significant part of the research is devoted to the aspect of environmental rehabilitation of military brownfields. Equally represented are research on economic models of management and management of brownfield conversion and regeneration processes, as well as models of participation of the public and the business sector in these processes. Part of the literature deals with the aspect of cultural and historical value of military brownfields and the potential of future use in the function of tourist infrastructure. There is a series of manuscripts that investigate the genesis, distribution and potential of repurposing military brownfields, among which the previously mentioned research in the post-socialist countries of Central and Eastern Europe stand out.

In the research of spatial-functional transformation of military brownfields in Sarajevo, we used many available written, satellite and cartographic data sources. Also, field work was inevitable, and it provided precise information about the current purpose of all individual buildings within the analyzed military bases. The brownfields included in the analysis differ in size: from individual buildings or building plots to large military terrains (land). Surface-wise, the largest military base included in the analysis are in the territory of the municipality of Istočno Novo Sarajevo (especially the military base of Slobodan Princip Seljo), while the smallest base, Jusuf Džonlić, are in the territory of the municipality of Novi Grad. We could not use the cadastral data on the facilities because they are not available, mainly due to the protection and confidentiality of the information, as they were former military facilities. Therefore, we focused on the analysis of available satellite images from 2002 to 2022 (with the lack of data in period from 2004 to 2006 and for 2010, because there were no available satellite images) [16].

In the research, we used satellites and cartographic data sources, such as Google Earth Historical imagery, OpenStreetMap, ArcGIS, etc. [16, 17]. We have singled out military brownfields with areas that occupy:

- Municipality Istočno Novo Sarajevo—military base Slobodan Princip Seljo 1,272 km^2 (3,35%) and Slaviša Vajner Čiča 0,240 km^2 (0,63%).
- Municipality Novo Sarajevo—military base Maršal Tito 0,192 km^2 (2,08%).
- Municipality Novi Grad—military base Jusuf Džonlić 0,109 km^2 (0,23%) (Fig. 20.2).

20.3 Results and Discussion

Through the analysis of satellite images in the period 2002–2022, marked changes were observed in terms of spatial-functional transformation of the analyzed military brownfields. For more detailed spatial analysis will be used, Google Earth satellite images, urban plans of areas of interest, as well as field research. These analyzes will help to determine what is the current purpose of certain buildings, how much the number of green areas decreased or increased in certain examples, what are the dominant functions of these buildings today (Table 20.2 and Fig. 6). Based on these data, detailed maps were drawn with a precise spatial analysis of the current use of buildings and land for the four analyzed military bases. The total area of the military brownfields included in the analysis is 1,8159 km^2, i.e. 181,59 ha (Table 20.2).

The models of conversion of military brownfields differ primarily by functional conversion, i.e. new way of use, with variations associated with different placement in space and different ways of financing conversion activities, as well as with different forms of spatial intervention. The first model of conversion in urban areas refers

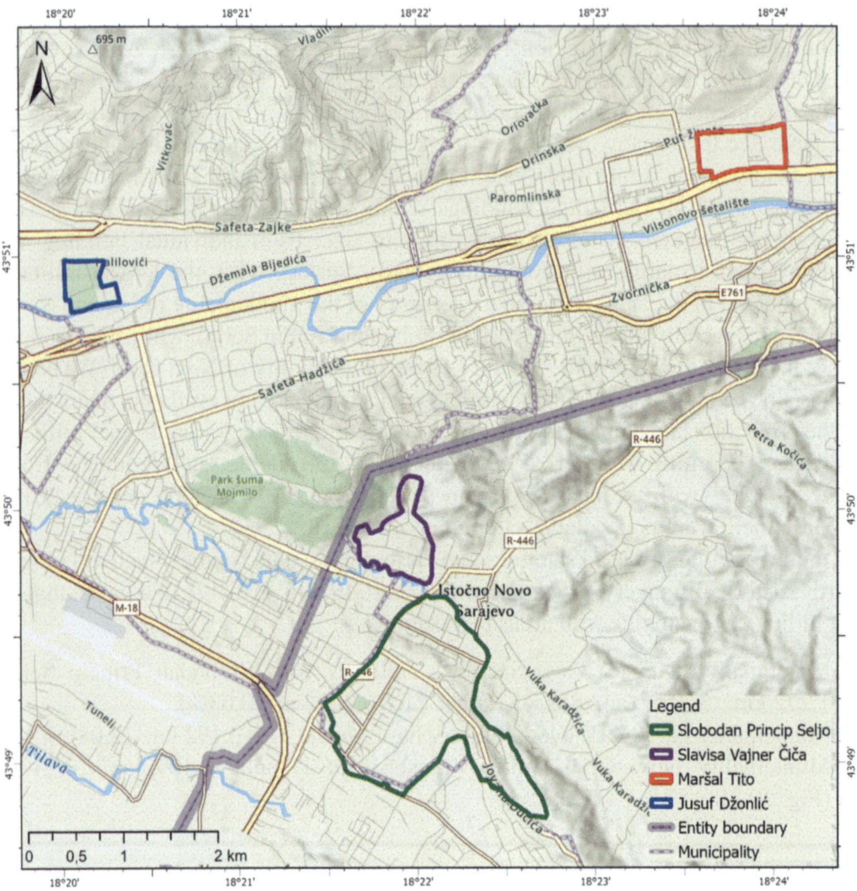

Fig. 20.2 Location and size of analyzed military brownfields in East Sarajevo and Sarajevo

to the basic function of the city—residence. Characteristic of the residential model of military brownfields that, as a rule, is preceded by the removal of existing built structures and planning of the new settlement, whereby quality green areas are sought to be preserved as new settlement parks. The obvious conversion of military brownfields into residential use can be seen in the example of the largest military base Slobodan Princip Seljo, which today is the main urban area of the municipality Istočno Novo Sarajevo with the functions of living, working, and also part of the Campus of the University of East Sarajevo. Slobodan Princip Seljo, underwent the greatest spatial-functional transformation (Figs. 20.3 and 20.4). On the example of this military base, the greatest reduction of green areas, which previously formed the basis of the space of this base, is noticeable, while the conversion of land for the function of residence has increased significantly. Certain objects were given a commercial role, and others an educational and government role. In addition to the

Table 20.2 Functional transformation of the use of military brownfield buildings and land

Military base		Slobod. P.Seljo		Slaviša V. Čiča		Maršal Tito		Jusuf Džonlić	
Area (km²)		1,2740		0,2399		0,1911		0,1109	
Usage		km²	%	km²	%	km²	%	km²	%
Commercial	Planned building	0,0516	4,05	0,0092	3,82	–	–	–	–
	Reconstructed	0,0019	0,15	0,0032	1,31	0,0029	1,51	0,0151	13,60
	Total	**0,0536**	**4,20**	**0,0124**	**5,13**	**0,0029**	**1,51**	**0,0151**	**13,60**
Educational	Planned building	0,0057	0,45	–	–	–	–	–	–
	Reconstructed	0,0057	0,45	0,0031	1,29	0,0158	8,28	–	–
	Total	**0,0114**	**0,90**	**0,0031**	**1,29**	**0,0158**	**8,28**	–	–
Government	Planned building	0,0076	0,60	0,0015	0,62	0,0070	3,68	–	–
	Reconstructed	0,004	0,31	–	–	–	–	0,0003	0,27
	Total	**0,0116**	**0,91**	**0,0015**	**0,62**	**0,0070**	**3,68**	**0,0003**	**0,27**
Residential	Planned building	0,1487	11,67	0,0036	1,50	–	–	–	–
	Reconstructed	–	–	–	–	–	–	–	–
	Total	**0,1487**	**11,67**	**0,0036**	**1,50**	–	–	–	–
Sport/ recreation	Planned building	0,0424	3,33	0,0141	5,86	0,0023	1,23	0,0266	23,98
	Reconstructed	–	–	0,0119	4,97	0,0180	9,44	–	–
	Total	**0,0424**	**3,33**	**0,0260**	**10,83**	**0,0203**	**10,67**	**0,0266**	**23,98**
Green area	**Total**	**0,2355**	**18,49**	**0,1280**	**53,34**	**0,0763**	**39,91**	**0,0329**	**29,65**
Unused, ruins	**Total**	**–**	**–**	**0,0028**	**1,17**	–	–	–	–
Infrastructure	**Total**	**0,7708**	**60,50**	**0,0581**	**26,12**	**0,0687**	**35,96**	**0,0360**	**32,51**

Fig. 20.3 Area of Lukavica, a settlement built on the land of the Slobodan Princip Seljo military base: (**a**) Construction of the settlement; (**b**) Current layout

located and mapped facilities of various uses, green areas, etc., infrastructure land zones representing roads, parking spaces, spaces around buildings, etc. have also been defined. Infrastructural zones are the most represented in the military base Slobodan Princip Seljo (60,5%), because they follow the current construction of the settlement.

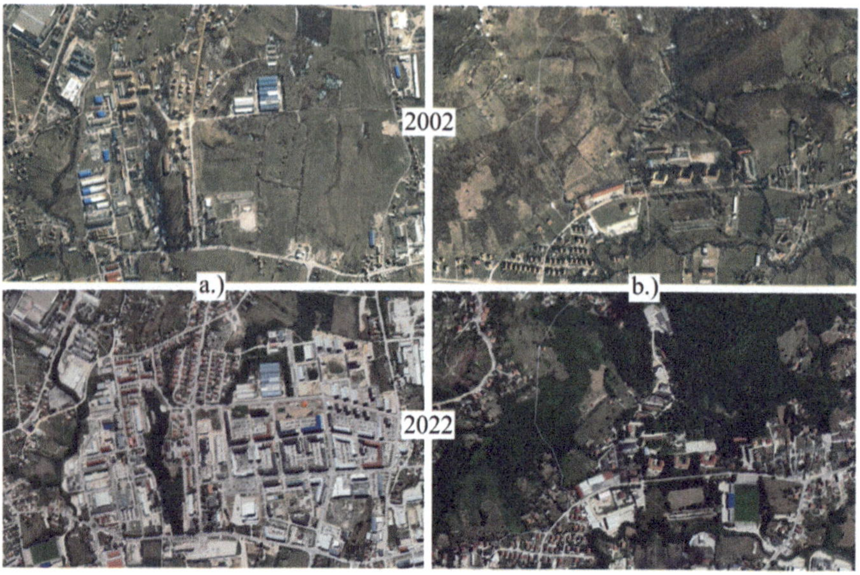

Fig. 20.4 The former military areas of the city of Istočno Sarajevo through time: (**a**) Area of military base Slobodan Princip Seljo (today the basic urban zone of the municipality of INS), (**b**) Area of military base Slaviša Vajner Čiča (today part of the UES Campus and sports and recreation area)

Relatively large areas and rich open spaces of the former military base in urban areas enable their conversion into a sports and recreational function. Usually these are spaces located in urban areas, but a little more peripheral, as well as about public sector investments. The sports-recreational use model most often completely transforms the military brownfield due to specific program requirements, spatial and architectural standards for the construction of sports buildings, with the possibility of retaining and reusing former military training grounds. We have registered the dominant form of this transformation on the example of the military bases Slaviša Vajner Čiča and Jusuf Džonlić.

Military base Slaviša Vajner Čiča it has sports-recreational function, less residential and work functions and partly the Campus of the University of East Sarajevo. This base has the highest level of green and sports-recreational areas (over 64%), compared to all analyzed bases. The conversion of other buildings and spaces in general is aimed at commercial, educational and residential use. In the area of this military base, we registered a certain percentage of unused buildings through field research. In one part of the old barracks, refugees are housed who, almost 30 years after the end of the war, still have not resolved their housing issue. The future urban plan in this area envisages the construction of an aqua park complex with open and closed pools and other accompanying facilities.

Also, the military base Jusuf Džonlić today is the biggest park in Sarajevo. The sports-recreation center "Safet Zajko" is now located here, which together with the associated green areas covers about 54% of the area of the former military base

(Table 20.2). The complex of the former Jusuf Džonlić military base today has a sports-recreational and educational function (Center for Education, Sports and Recreation "Safet Zajko"). This area has been transformed into the largest park in Sarajevo, with a variety of sports and recreational facilities with lots of green areas, an athletics track, a football stadium, basketball and children's playgrounds and a botanical garden with three artificial lakes and a waterfall, as well as numerous other facilities for children and adults. Today, the "Safet Zajko" center is a favorite place for rest and recreation.

An example of a model of conversion from military to public and social use, i.e. business and innovation infrastructure, is highlighted. Military, educational and business-innovation purposes require spatial distribution organized by departments, with common closed and open spaces. It is precisely for this reason that it is easiest to place the new educational and business-innovation purpose in the pavilion-designed complexes of the former barracks. Military base Maršal Tito, in addition to fairly preserved areas of green areas, directs the basic conversion of former military facilities to an educational role (Campus of the University of Sarajevo), as well as a sports and recreational function. Also, the US Embassy is now located there, which adds to its role as a government function. This military base is in one of the most urban parts of the city, which makes it an extremely attractive location for various purposes. However, the urban plan foresees the construction of a modern university complex (Fig. 20.5). Through field observations, we established that the construction of new buildings for the needs of the campus is in progress.

The spatial distribution of military brownfields where conversion was carried out, that is, certain processes of urban regeneration were initiated, points to the role of the geographical factor of the specific location in the local spatial-functional structure in the success of the conversion. All the analyzed bases are in an urban area (Fig. 20.2), which is also confirmed by individual studies according to which the majority of successfully regenerated locations are located near the center of large cities [18], i.e. the importance of the location of brownfields in the urban area for the implementation of conversion and urban regeneration (Fig. 20.6).

Some problems that were established during the research are also highlighted. Main problem with military property has been active since 2005, and it became actualized again with the request of the NATO Summit in Tallinn held in 2010, by which BiH had to register military property as a condition for activating the Action Plan for NATO Membership (MAP). A big problem that slows down the implementation of certain plans for these former military facilities is lack of legal and political will. Also, there are series of unresolved issues, from ownership to infrastructure and development challenges. After the war, certain facilities served as collective centers for refugees and displaced persons, and some of the facilities still have that role today, which slows down the implementation of plans. Development plans and their implementation are ongoing, but the deterioration of some buildings due to lack of use is also noticeable.

Fig. 20.5 Military braunfield Maršal Tito today servs as Campus of the University of Sarajevo: (**a**) Current situation; (**b**) Urbanistic plan for the construction of a modern university complex

20.4 Conclusion

Military brownfields require a specially designed approach to transformation, urban planning, management, and implementation of the urban regeneration process. All military property, the property of the former JNA in Bosnia and Herzegovina, is divided into prospective and nonprospective. These are military bases, warehouses, training grounds, ammunition depots, military resorts, military catering facilities, radio relay centers, aviation repair institutes, training facilities and numerous others. Prospective property remains for use by the Armed Forces of Bosnia and Herzegovina, while non-prospective property should be handed over for use by civilian structures. Given the lack of legal regulation in state property, here we come to the problem of how prospective and non-prospective property should be registered.

Some of these facilities have long been in use by civilian structures, but the ownership of these facilities is still being questioned. An example of an open approach to development plans, regarding the conversion of military brownfields, is clearly visible in the example of the municipality of Istočno Novo Sarajevo. On that area,

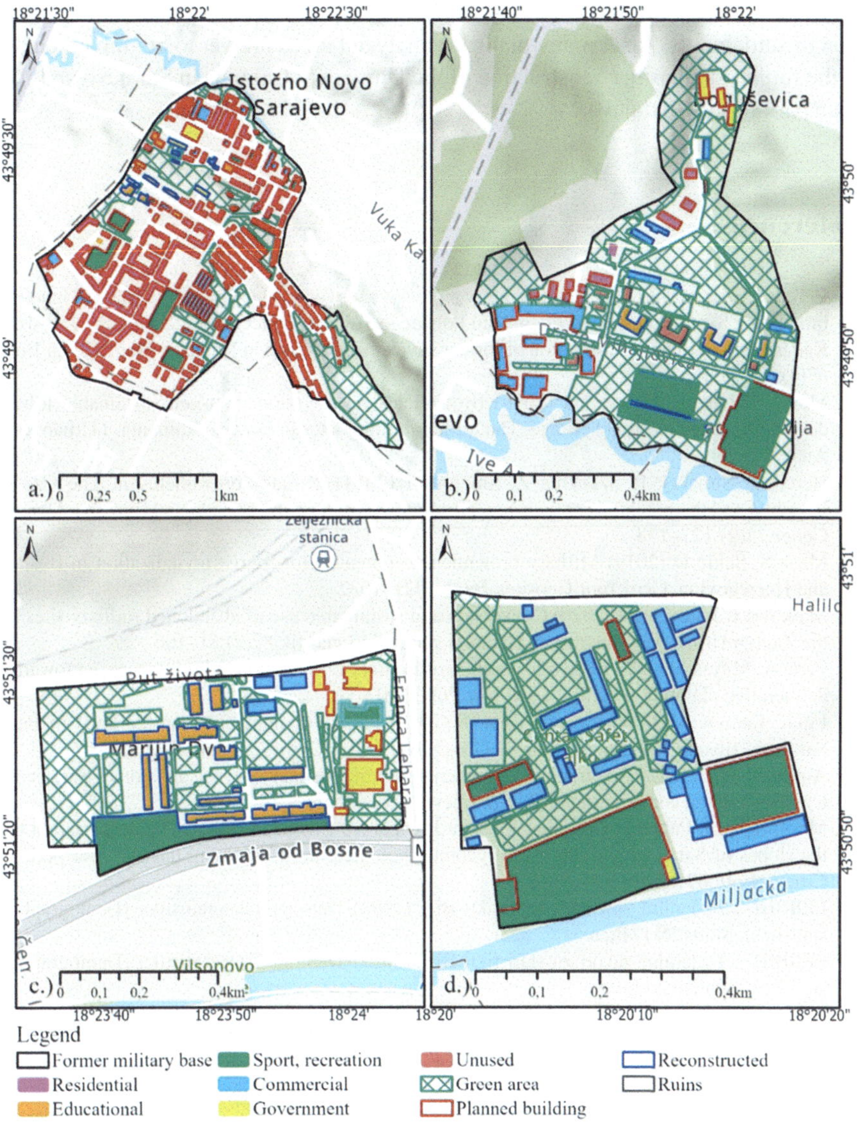

Fig. 20.6 A detailed analysis of the use of objects on an example: (**a**) Military base Slobodan Princip Seljo; (**b**) Military base Slaviša Vajner Čiča; (**c**) Military base Maršal Tito and (**d**) Military base Jusuf Džonlić

which was once a military training ground and base, there are now schools (primary and secondary), university facilities, government institutions, sports and recreational grounds and many residential buildings.

The functions have transformed from the original military, through certain temporary functions (collective accommodation of refugees), to modern functions that are in the service of work, recreation, housing, education and, in general, the expansion of settlements. Also, many plans for analyzed areas are yet to be implemented in the future, such as the construction of the University Campus in Sarajevo and the aqua park in East Sarajevo.

References

1. Hercik J, Szczyrba Z (2012) Post-military areas as space for business opportunities and innovation. Prace Komisji Geografii Przemysłu Polskiego Towarzystwa Geograficznego 19:153–168
2. Kádár K (2014) The rehabilitation of former soviet military sites in Hungary. Hung Geogr Bull 63(4):437–456
3. Morar C, Dulca M, Nagy G (2016) Brownfields regeneration, between sustainable urban development and cultural heritage. The former military sites in Oradea, Romania. J Urban Reg Anal 8(1):75–84
4. Hercik J, Šimaček P, Szczyrba Z, Smolova I (2014) Military brownfields in The Czech Republic and the potential for their revitalisation, focused on their residential function. Quaest Geogr 33(2):127–138
5. Musa S, Šakić D (2019) Military brownfields—potential for tourist revitalization in Bosnia and Herzegovina. Geoj Tour Geosites 26(3):1021–1032
6. Matković I, Jakovčić M (2020) Conversion and sustainable use of abandoned military sites in the Zagreb Urban Agglomeration. Hrvatski geografski glasnik 82(2):81–106
7. Perić A, Miljuš M (2021) The regeneration of military brownfields in Serbia: moving towards deliberative planning practice? Land Use Policy 102
8. Lukić Tanović M, Golijanin J, Šušnjar S (2019) Impact of population on the karst of East Sarajevo (Bosnia and Herzegovina). J Geogr Inst "Jovan Cvijić" SASA 69(2):95–107
9. Aquilué I, Roca E (2016) Urban development after the Bosnian War: the division of Sarajevo's territory and the construction of East Sarajevo. Cities 58:152–163
10. Martín-Díaz J, Palma P, Golijanin J, Nofre J, Oliva M, Čengić N (2018) The urbanisation on the slopes of Sarajevo and the rise of geomorphological hazards during the post-war period. Cities 72:60–69
11. FZSBiH – Federalni zavod za statistiku BiH (1998) Demografska statistika [Demographic statistics]. Statistički bilten
12. FZSBiH – Federalni zavod za statistiku BiH (2023) Demografska statistika [Demographic statistics]. Statistički bilten
13. RZSRS – Republički zavod za statistiku Republike Srpske (2023) Demografska statistika [Demographic statistics]. Statistički bilten
14. Lukić Tanović M, Marinković D (2017) The disintegration of settlements in Bosnia and Herzegovina – the example of Sarajevo/East Sarajevo. Geobalcanica:183–189
15. Lukić Tanović M, Marinković D, Majić A (2019) Promena broja stanovnika Grada Istočno Sarajevo u međupopisnom periodu 1991–2013 [Population changes in the City of East Sarajevo in the intercensal period from1991–2013]. Stanovništvo 1–17
16. Google Earth historical imagery. https://earth.google.com/, last accessed 2023/11/25
17. OpenStreetMap contributors. https://planet.openstreetmap.org/, last accessed 2023/11/15
18. Frantál B, Kunc J, Klusáček P, Martinát S (2015) Assessing success factors of brownfields regeneration: international and inter-stakeholder perspective. Transylv Rev Adm Sci 11(44):91–107

Chapter 21
An Overview of Environmental Impact of the Long-Term Mining, Ambient Dose Equivalent Rates in Urban Air and Remediation Proposals

Ljiljana Gulan, Aleksandar Valjarević, Dušica Spasić, and Uroš Durlević

Abstract The long-term operation of formerly the most important mining area in Europe, the "Trepča" facilities (Pb-Zn mine, Pb flotation and smelter, chemical industry, battery factory, Zn metallurgy), led to significant pollution of the urban environment of the nearest town Kosovska Mitrovica. Besides known chemical pollution due to mining processes, there is a radiological impact due to radionuclides present in air, soil and water. Preliminary, it can be established by the measurements of ambient gamma dose rates in air in urban, suburb and industrial parts of study area. About 39 locations were subjected to measurements of ambient dose equivalent rates by Geiger counter Radex model RD1503^{+}, and spatial distribution maps were made based on GIS (Geographic Information System) methodology. The wide range of results (0.07–0.33 μSv/h) was discussed regarding a few aspects including influences of geology, mining and urban features. A long-term mining left serious consequences on environment and negative effects on health. Some solutions for remediation and reuse of abandoned landfills were proposed for a healthier future.

Keywords Environmental pollution · "Trepča" mine · Ambient dose equivalent rate · Remediation

21.1 Introduction

The mining industry is one of the main sources in the economic development of any country. However, this progress is often followed by side effects. Anthropogenic activities such as mining and smelting of metal ore lead to environmental pollution

L. Gulan · D. Spasić
Faculty of Sciences and Mathematics, University of Priština, Kosovska Mitrovica, Serbia

A. Valjarević (✉) · U. Durlević
Faculty of Geography, University of Belgrade, Belgrade, Serbia
e-mail: aleksandar.valjarevic@gef.bg.ac.rs

C. Morar et al. (eds.), *Achieving Sustainability in Ukraine through Military Brownfields Redevelopment*, NATO Science for Peace and Security Series C: Environmental Security, https://doi.org/10.1007/978-94-024-2278-8_21

and are accompanied by an enormous amount of production waste in the form of tailings, considering that all material is deposited of as waste after the extraction the mineral or after the process of exploitation. In general, tailings ponds resulting from the mining production of metallic mineral ores contain low content of nutrients and organic matter, and an increased concentration of heavy metals [13]. Consequently, the toxicity, mobility and bioavailability of heavy metals in soils depend on their concentrations in waste material, chemical form, binding state, physicochemical properties and other environmental factors [26]. The soil has an important role in the uptake of these elements by vegetation and animals; these elements can be transferred to the hydrosphere and biosphere. Acidic drainage and wind transport of dust are the main effects causing the dispersion of pollution [28]. Therefore, due to the spread of heavy metals from the tailings, contamination of soil, air and water of mining area is inevitable. Since tailings are usually formed near the water flows, urban, suburban or agricultural areas, they often pose a major problem for human health and the environment.

On the other hand, metal ore mining leads to an increase in normally occurring radioactive materials (NORMs). By the nature and origin, the geological formations contain a certain amount of natural radionuclides (uranium, thorium, potassium). As a result of changes in the geological-geochemical systems (ores, rocks, underground waters) that are accompanied by hydrothermal processes, primary rocks are enriched with minerals containing radionuclides. In this way, the content of natural radionuclides in the soil increases naturally. Technologically Enhanced Naturally Occurring Radioactive Material (TENORM) occurring if technological procedures affect the increase concentration of natural radionuclides. When an increased level of natural radioactivity occurs as a result of human (anthropogenic) activities (such as the exploitation and processing of ores and minerals, digging tunnels and working in mines, the production and use of phosphate ores in the production of fertilizers, the disposal of slag and ash and the emission of gases into the atmosphere from thermal power plant) natural radioactivity is technologically modified. Such technological activities can change and significantly increase the level of natural radioactivity, lead to greater population exposure to the natural radiation and affect the quality of the environment. In such conditions, continuous monitoring and control is necessary in order to reduce radiation exposure.

The area of Kosovska Mitrovica, in the northern part of Kosovo and Metohija, represents a zone of great interest for environmental pollution studies. There is severe environmental contamination caused by the activities of the mining industry, i.e. due to the both, the presence of heavy metals and radionuclides in the area. Radio-eco-toxicological elements represent a health risk to inhabitants in the area, which can lead to significant exposure and harmful implications, due to their proximity to settlements. In the recent two decades, there are numerous studies which dealt with contamination of urban, suburb and agricultural soils in Kosovska Mitrovica and its surroundings. Most of them reported the diffuse levels of contamination by heavy metals and radionuclides, assessing the environmental risk [2, 4, 6, 8, 11, 12, 20, 23, 29].

In this regard, environmental protection is needed for this area. This study aimed to consider the main sources of pollution, determine levels of radioactivity in the air and provide potential suggestions for remediation of the demaged environment. It is necessary to implement some technical measures for the rehabilitation of degraded area, with an emphasis on hot spots. As a first step, soft recovery technologies such as phytoremediation would be feasible.

21.2 Materials and Methods

21.2.1 Study Area

The main interest of this study is an area of Mining Metallurgy Chemical Combine "Trepča" which belongs to an administrative unit of the Kosovska Mitrovica area. The city of Kosovska Mitrovica (N 42°53′, E20°52′, an altitude of 510 m) is located in the northern part of Kosovo and Metohija at the foot of the Zvečan extinct volcano cone (799 m). Kosovska Mitrovica and its immediate surroundings (Fig. 21.1) include a territory of about 154 km^2 (a broader area is about 316 km^2). The topographical position is partly in the Kosovo basin on the alluvial field of the Ibar River and Sitnica River. It is surrounded by the steep slopes of Mts. Kopaonik, Rogozna and Mokra Gora (with different morphological and geological composition) from the east, north and west. The landscape is wooded; there are oak, beech, pine and spruce forests on the surrounding mountains. At lower altitudes, where alluvial soil prevails, poplars and willows thrive.

Geology and Climatic Morphologically area of Kosovska Mitrovica is mainly made up of an alluvial plain. Earlier tectonic movements determined the direction of valley formation by the erosive power of the Ibar River and its tributaries. From a geological and geomorphological point of view, the territory around Kosovska Mitrovica belongs to the Dinaric system. Large tectonic movements and intense

Fig. 21.1 City of Kosovska Mitrovica

Tertiary volcanism caused the transformation of the structure. The oldest forms of folds are Cretaceous, while the first ones are assumed to belong to the Hercynian orogeny, together with the formation of the initial magmatism of the old Paleozoik [10]. Later tectonic movements as well as external factors created green and black serpentine metamorphic rocks. Serpentines are usually carriers of the asbestos mineral, which is formed from hydrothermal intrusions in the aforementioned rocks. Apart from asbestos, chromites are also formed from magma in serpentines. However, most of the soil around Kosovska Mitrovica consists of flysch which belongs to the older Cretaceous period. During this period, thick layers of gravel, clay, marl, limestone are located above the serpentine. Gravel appears in the form of tiles (flakes) of green marl and yellow clay. Limestone layers occur more on the western side of Kosovska Mitrovica. Eruptive rocks associated with great volcanic activity were created during the Oligocene orogeny. Intrusions of these eruptions exist in several places. Quaternary sediments in the vicinity of Kosovska Mitrovica are found in alluvial and diluvial horizons, in layered silt, sand, gravel and humus clays, covering an area of nearly 40 km^2 of the wider Kosovska Mitrovica area [7].

The differences in the geological composition of the rocks influence different levels of underground water. Serpentine flysch and shale zones are poor in groundwater; due to the multitude of cracks water is lost by sinking into the depths. In limestone zones, limestone waters descend to the rock horizons that retain water (they are usually not deep), and emerge on the surface as contact springs. "Trepča" deposit with its wider surroundings belongs to the Vardar zone with the direction of extension north-northwest and south-southeast [7]. The deposit belongs to the hydrothermal-metasomatic deposit type. Hydrothermal minerals are represented by: sulfides, carbonates and oxides. Among sulfides, the most important are: galena, sphalerite, pyrrhotite, pyrite, arsenopyrite, plumosite, argentite, etc. As followers, there are: quartz, calcite, dolomite, rhodochrosite, etc. The most economically important minerals are: galena, which, in addition to lead, is the main carrier of silver and bismuth, then sphalerite (a zinc mineral with 0.2% cadmium) [21].

Higher values of natural radionuclides in the soil around Kosovska Mitrovica are partly a consequence of the existence of a mining zone (several basic metal mineralizations). High concentrations of radionuclides in the soil can be linked to the existence of a deep fault zone and a seismogenic fault zone nearby [10]. Volcanic activity causes faults, which consist of fractured and fractured metamorphic rocks, formed at high pressures and temperatures. For this reason, faults are extremely permeable environments suitable for the transfer of radionuclides to the surface by groundwater [3, 31]. In addition, the relative mobility of radionuclides and their decay products during metasomatic activity (the process of metamorphosis of the primary rock by changing the chemical composition due to the action of aqueous solutions) affect the mobility towards the surface of the fault, increasing the concentrations of uranium, thorium and potassium.

The surroundings of Kosovska Mitrovica have a moderate continental climate with frontal contact of Mediterranean air currents. There is a great range of extreme temperatures and non-uniform distribution of rainfall [9]. Temperature directly affects the intensity of evaporation, air humidity, cloudiness and precipitation. The temperature in winter is around 0 °C, in summer around 20 °C (an average annual air temperature is 10.8 °C). The intensity of insolation in the area of Kosovska Mitrovica amounts to 1681 h during the year. The average annual value of air humidity is 77%. The amount of precipitation varies from 600 mm for places of lower altitude to 1100 mm for mountainous ones; during the year there are 50–60 windy days with the predominant direction being the northwest [1].

Overview of Mine Production The opening of the lead and zinc mine in Stari trg was crucial for the economic development of Kosovska Mitrovica and the wider region. The mining activities in this region are known from ancient and medieval times. However, an intensive production started in the past century, when an English company "Trepča Mines Limited" showed interest in this region.

The part of Trepča complex (in study area) consists of the largest and most important mine Stari Trg, flotation plants in Prvi Tunel, lead metallurgy and energy plants (in Zvečan), zinc metallurgy plant, chemical industry complex and industrial processing (in the southern part of Kosovska Mitrovica). The mine in Stari trg and the flotation in Zvečan began operating in 1930; a thermal power plant was built for the needs of the mine and flotation. Melting and refinery of lead in Zvečan started in 1939; lead concentrates obtained in flotation contained about 80% lead. Until 1984, about 24,000,000 t of ore was processed in Zvečan flotation, with an average content of Pb + Zn of 11.9% [33]. During the Second World War, the Stari Trg mine produced 40% of lead used in the German military industry.

After the Second World War, the economic development has been intensified, especially in the field of mining, metallurgy and energetic. From 1947 to 1953, eight gas generators were manufactured. Silver and gold electrolysis was put into operation in 1955. Within the Trepča complex, in the southern part of the city of Kosovska Mitrovica, the processing facilities: a super-phosphate and sulfuric acid factories were founded in 1959, zinc metallurgy began production in 1961, zinc electrolysis (operated from 1966 to 1986) and factory for the production of lead batteries started operating in 1967. The annual capacity of plant for production waste lead accumulators was 20,000 t, and production of mineral fertilizers was 170,000 t. Among the industrial giants, Trepča complex was one of the biggest European mine and main producer of lead, zinc, silver, gold, cadmium, bismuth and products from these metals. The production was expanding outside the area. At the time, the industrial/mining complex Trepča included 22 mines, 12 flotation, 2 metallurgical industries, chemical industry and industrial processing. In the period of full production Trepča employed up to 22,000 workers and the daily production in mines was around 10,000 t of mined ores [24]. Around 1985, the flotation in Zvečan was dismantled, thus freeing up space for the expansion of lead metallurgy. After 1990, there is a decline in production for several reasons.

The facilities in Zvečan operated continuously from 1939 to 1999. In that period, Trepča complex produced 3.284.568 t of raw lead, 4.110 t of silver and 3.296 t of bismuth and other commercial products [25]. After the arrival of UNMIK (United Nations Interim Administration Mission in Kosovo) and KFOR (Kosovo Force) in

1999, the work of the lead smelter in Zvečan was suspended, under the pretext of non-compliance with environmental standards. UNMIK took over the management. Air pollution due to lead metallurgy was noted, although the purified metallurgical gases are released into the atmosphere through a chimney of 305 m height. Production was stopped in 2000, except some small battery recycling (extraction of lead from used batteries) took place in the refinery in Zvečan. In the last 20 years, only about 20,000 t of lead were refined. There is a dilemma about the future work of lead metallurgy in Zvečan. The zinc metallurgy was also closed down in 2000. Chemical industry now produces a small amount of aluminium sulphate. Most of the other Trepča facilities are either inactive or working at very low capacities.

Flotation Tailings and Landfills During the 60 years of operation (1939–1999), Trepča produced enormous amount of flotation tailings, metallurgical slag, intermediate products, sludge, sediment, ash, etc. By mining and flotation processing of metallic minerals, Trepča established tailing ponds. In the process of creation and expansion of these tailings, waste material was transported and deposited near the banks of Ibar and Sitnica rivers. The tailings are polymetallic; their mineral compositions are mostly heavy metals, as they were extracted from ore. A landscape where the tailings are situated is presented in Fig. 21.2. There are a few industrial landfills in the industrial and urban zone of Kosovska Mitrovica and Zvečan (Table 21.1). They have large masses and occupy a large area of mostly fertile agricultural land next to rivers. Flotation tailings are very similar in their physical and chemical characteristics, while the landfills of lead metallurgy, zinc metallurgy and chemical industry are different in chemical composition.

There are four types of deposited tailings, waste materials and intermediate products of various technological processes:

Flotation tailings arose during beneficiation of lead-zinc ores in Zvečan and Prvi Tunel, and have been transported in the form of pulp to the prepared landfill. Flotation tailing Žitkovac (Fig. 21.3) is located on the left bank of the Ibar River, about 1.5 km from the Zvečan flotation facility and around 4 km northeast from Kosovska Mitrovica. It was active from 1962 to 1974. The tailing contains: 14–25% Fe, 8–12% S, 18–30% SiO2, 0.3–0.4% Pb, 0.3% Zn, etc. It was partially re-cultivated 15 years ago.

Next to the smelter and refinery in Zvečan, there is a flotation tailing Gornje Polje (Fig. 21.4), located on the right bank to the Ibar River between the city zones of Kosovska Mitrovica and Zvečan. It was active in the period from 1932 to 1962. The flotation tailings were deposited at two locations next to the Ibar River. The flotation tailings Gornje polje were only partially rehabilitated in the 1980s. Besides these tailings, there is a flotation tailing pond Žarkov Potok (Fig. 21.5), an active landfill site only 2 km northeast from Kosovska Mitrovica, which contains tailings from flotation in Zvečan and Prvi Tunel. It was constructed in 1975 by building a dam 87 m high and with a capacity of $8.2 \cdot 10^6$ m^3 of tailings disposal. The tailing pond is in the central part of the landfill. Reclamation of the landfill was never undertaken. There are two channels for drainage of atmospheric waters from the

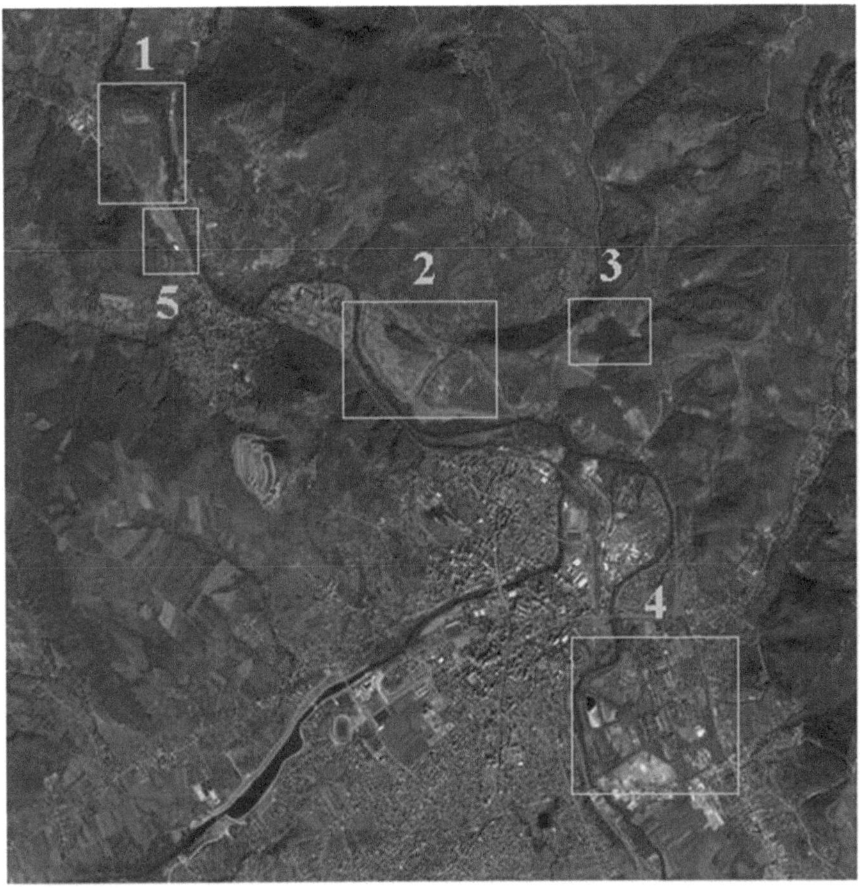

Fig. 21.2 Landfills of Trepča complex in study area: 1—Žitkovac; 2—Gornje Polje; 3—Žarkov Potok; 4—Šupkovac; 5—Rudare

Table 21.1 Landfills in Zvečan and Kosovska Mitrovica

Location	Landfill		Situation	Area (ha)	Waste amount (10^6 tons)
	Name	Type			
Zvečan	Žitkovac	Flotation tailing	Passive	26	8.5
	Gornje Polje	Landfill	Passive	50	12
	Gornje Polje	Flotation tailing	Passive-lead slug	6	2.5
	Žarkov Potok	Tailing pond	Active	20	9
	Rudare	Landfill	Passive-slag and ash	10	0.1
K. Mitrovica	Šupkovac	Landfill	Passive	35	4.2

Fig. 21.3 Žitkovac tailings, partially re-cultivated

landfill, which stop the breakthrough of water towards the dam and its erosion. The composition of the waste is changeable: depending on weather, there is a separation of pyrite and pyrrhotite for sulphuric acid production. During the construction of the landfill, cleaned water from the tailings of the landfill was discharged into the Ibar River through the collector.

Landfills of lead metallurgy were created by storing materials. On the northern part of the landfill Gornje Polje (Table 21.1), waste from the Pb smelter in Zvečan have been disposed of for a long time, so the landfill has taken on the form of a cone. Deep cuts have been created as a result of erosion on both sides of the dam, through which comes discharge of atmospheric waters from the landfill directly into the Ibar. In addition to production of lead-zinc concentrates, flotation slag from shaft furnaces is also obtained. Further processing of slag is needed due to the toxic components and negative effects on the environment. A stock of slag also contains useful metals: 30–35% Fe, about 12% ZnO, 20–22% SiO_2, about 1.5% Pb, and is treated as a secondary raw material, interesting for recycling. It mainly consists of andesitic rocks, siderite and ancherite, followed by alkaline carbonates, quartz, pyrite, pyrrhotite, clay mass, etc.

Landfills of zinc metallurgy were created by storage of materials from zinc concentrate processing and mineral fertilizer production. The residue from leaching and pyrite slag was disposed in Šupkovac (Fig. 21.6) next to the right bank of Sitnica

Fig. 21.4 The Zvečan smelter and Gornje Polje tailings (above); tailings from different positions (below): images made in the winter (left) and in the spring (right)

in the southern part of Kosovska Mitrovica. There is about 450,000 tons of this material with zinc content above 20%. This slag also contains extremely toxic metals: lead and cadmium. Chemical industry landfills contain about 600,000 tons of pyrite (an area of 14 ha) for sulfuric acid production, also deposited on the right bank of the Sitnica. In the process of phosphoric acid production, phospho-gypsum is separated as a waste material. About 400,000 tons of this material (an area of 17 ha) was deposited next to Sitnica River. Both of these chemical industry landfills are acidic in nature and endanger the underground waters.

Landfills of energy plants were created in the industrial circle in Zvečan, when raw and dry lignite was processed and used as fuel in the power plant and gas generators. The waste material of burned lignite is slag and ash. They are stored in Rudare, on the right bank of the Ibar River (Fig. 21.2, marked as 5). There is 100,000 tons of this material at the landfill (an area of 10 ha). This landfill is partially re-cultivated and dislocated.

Environmental Pollution and Consequences Environmental problems in the territory of study area arose at the time of the beginning of its industrialization (in

Fig. 21.5 Žarkov Potok tailing pond (above); a dam (below)

1930). Intensive production in the field of mining, non-ferrous metallurgy and chemical industry and other following industries, took place with the use of outdated technologies that led to a series of problems in terms of environmental protection. All industrial tailings are in the environment with urban, suburb, agricultural, recreational and cultural potential. All of them are in close proximity to water streams or on the banks. Over the years, the material from the deposits have scattered in the environment through the Aeolian process, gravity, precipitation and water flows. They seriously threaten watercourses, the alluvium of the Ibar River, roads and the overall ecosystem. Mining-metallurgical-chemical landfills and tailings are exposed to the influence of atmospheric precipitation (rain, snow, etc.) and wind, resulting in various types of air, soil and river pollution. Due to the fact that these industrial landfills are located directly on the banks of rivers, they are a constant threat to the quality of these river courses. Atmospheric water and waters from tailing ponds are released into the Ibar River through drainage systems in the tailings. Bad water management around the tailings/tailing dam is, most frequently, the main reason for tailing dam collapse during heavy rains, surface runoff entering the dam area and

Fig. 21.6 Šupkovac tailing pond

erosion of dam via groundwater. Then leaching contamination highly endangered the water bodies. The flotation tailing deposits also left consequences on the agricultural land. The location of the industrial complex in the direction of the prevailing winds towards the city and settlements resulted in high air pollution. Some tailing deposits have not ever been rehabilitated. However, technical and biological reclamation is needed for most of landfills; it should be successfully carried out to cover a larger area.

The lack of attention to the environmental consequences and costs during the operation of facilities resulted in widespread damage to water and soil resources around Trepča flotation, chemical and metallurgical units. Even today, 25 years later, UNMIK and the Kosovo Government have done little to improve the bad situation; they have not taken significant action to mitigate air, water and soil contamination, probably due to a lack of financial support.

21.2.2 Measurements of Ambient Dose Equivalent Rate

In order to get preliminary information about the radioactivity levels in an area or to assess the radiological risk of exposure, the measurements of terrestrial gamma radiation should be carried out. Due to different geological characteristics of an area, the variation of terrestrial gamma radiation is usually greater than the cosmic ones. Elevated levels of radiation are associated with igneous rocks, while sedimentary rocks have lower levels of radioactivity. In that regard, it is important and desirable to measure the gamma absorbed dose rate or ambient dose equivalent rate everywhere in human environment, where it is possible.

Instruments for gamma dose rate measurements are calibrated and based on an ambient dose equivalent from which the dose estimation is possible. For penetrating gamma radiation, ambient dose equivalent H*(10) at a point of the radiation field is the dose equivalent that would be produced by the corresponding expanded and aligned field at a depth of 10 mm in the ICRU sphere, on the radius opposing the direction of the aligned field [34].

Ambient dose equivalent rates were measured using a Radex monitor model RD1503$^+$ (Fig. 21.6) at 39 points of study area (Fig. 21.7). The points were selected near the flotation tailings and other landfills, as well as further away from them in the surrounding area.

Measurements were taken at the height of 1 m above the ground in the middle of the day. On the day of the measurement, and on the previous two days, there was no rainfall. Measurement for each point was recorded when the reading on the detector was stable (at least four measurements for a 40 s cycle). The range of the detector is from 0.00 to 9.99 μSv/h. The instrument has a linear response to gamma radiation between 0.1 and 1.25 MeV, which covers the most of gamma radiation emitted from terrestrial sources. The measurement uncertainty for gamma rays is ±15%; factory-calibration was done with 137Cs at 5 μSv/h. The detection of high-energy particles from cosmic radiation is negligible due to the low detector's response.

21.2.3 GIS Analysis and AHP Methods

Isodose mapping or radiological mapping of the study area was plotted using ARCGIS software. The coordinates of each sampling point was converted into the degree decimal unit. The main methods within this research are interpolation, zonal statistics and two types of advanced. The zonal statistics methods together with AHP and AHP modified methods gave satisfactory scientific results, especially in land analyses. These results can be divided on statistical and numerical. The statistical methods are implemented to provide better analyses of fields with higher ambient dose equivalent rates.

The AHP methods are good in determining the process of selecting process of remediation and process of the brown field adaptation, poisoned land adoption. The

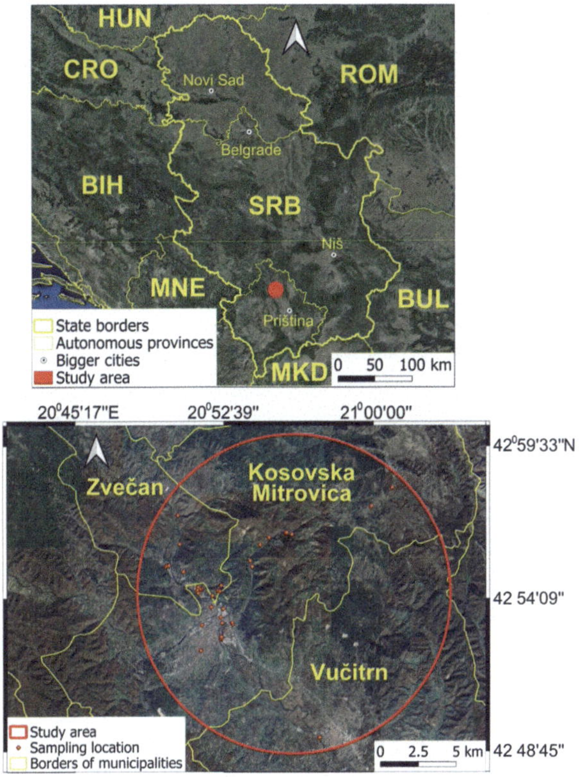

Fig. 21.7 Map of study area

GIS-MCDA procedure implemented in ArcGIS 10.5 software in this research is based on six steps: setting the goal of multi-criteria analysis/problem definition; determination of criteria; standardization of criteria values; determining the weight of the criteria; the results of the analysis (unification of standardized values and weights of criteria); and on the validation of results. Weighting coefficients for the criteria concerning the significance are determined through the application of the Analytic Hierarchy Process (AHP) [36]. The soil type used for this research is from the analogue pedology map in ratio of 1:200,000. This map is digitized and vectorized with the special view on the parts of the Kosovska Mitrovica. Zoning is approved with the help of spatial plane of the Republic of Serbia in ration of 1:10,000 divided on shifts. The slope used and analyzed from the Digital Elevation Model data in resolution of 30 m as well 10 m. Distance from forest in the municipality estimated by tools approved in ArcGIS 10.5 environment. Distance from the main rivers that flows in the investigated area as the Ibar, Ljusa and Sitnica rivers analyzed by digital cartographic tool in the ArcGIS environment [5].

The AHP methods are very important in classification of land and soil degradation. Another important distance matrix tool can be labeled in GIS software. The distance matrix software is valuable in the combination with buffer zonallity. The buffer zonal statistics is good in creating real distance from the brown field zones. The distance (buffer zones) in this research divided according to the related problems of the municipality [35].

Satellite recordings within this research downloaded from the two types of satellite missions, Landsat 9, and Copernicus 2. The resolution of Landsat 9 images was 30 m [16] and resolution of Copernicus data was 10 m [14]. The all satellite images covered visual specter and Digital elevation model extension.

21.3 Results and Discussion

The results of ambient dose equivalent rates are shown in the Table 21.2, and the isodose mapping is presented in Fig. 21.8. The results of ambient dose equivalent rates ranged from 0.07–0.33 μSv/h.

According to the obtained results and presented map, a low radiological impact associated with the location near landfills and tailings is noticeable compared to other locations in the surrounding area, especially in the northwestern part of study area. Even more, obtained gamma dose rates in area of landfills/tailings are mainly lower than usual daily value of ambient dose equivalent rate reported for city of Kosovska Mitrovica; an average annual value of 126 μSv/h [30]. Therefore, lower values of ambient dose equivalent rates correspond to artificial activities, while higher values correspond to the natural occurrence of radionuclides. This means a lower radiological risk from radionuclides in area of industrial landfills than in the surrounding environment of Kosovska Mitrovica. As already mentioned, the increased radioactivity in the area mainly stems from geological structures, existence of faults and ore occurrences.

The main results included spatial analyses of industrial area of Trepča complex - landfills in study area: 1 - Žitkovac; 2—Gornje Polje; 3 - Žarkov Potok; 4—Šupkovac; 5 - Rudare. The Industrial Trepča complex has high possibility of environmental accident and in that way the remediation process requires long time. The Weighting criteria divided in seven classes (C1-C7) and selected on the next spatial attributes are: C1-Soil type, C2-Zoning according Spatial plan, C3-Distance from surface water bodies, C4-Distance from protected areas, C5-Slope, C6-Distance from forests, C7-Distance from the main rivers (Table 21.3).

The soil type in this site is semi-permeable and permeable. According to the spatial plane and zonal statistics the land is industrial. The average distance from the surface water is 100 m. The average distance from the protected area is 600 m. The slope is higher of 20°. The average distance to the forest is 500 m. The average distance from the closest river is 300 m. A multi-criteria analysis of tailings and landfills are presented in Table 21.4.

The Žitkovac has Impermeable land and according spatial plane and approved zonal statistics the land is residential. The distance from the surface water is longer of 600 m. The average distance from the protected area is 600 m. The slope is about

Table 21.2 Coordinates of 39 sampling locations, elevations and values of ambient dose equivalent rates (ADER) in the area of Kosovska Mitrovica

Sampling location	N	E	Elevation (m)	ADER (μSv/h)
1	42°53.493′	20°52.498′	518	0.13
2	42°53.289′	20°52.557′	511	0.10
3	42°53.782′	20°52.618′	509	0.11
4	42°52.718′	20°52.528′	508	0.13
5	42°52.737′	20°52.546′	514	0.09
6	42°52,796′	20°52,530′	508	0.08
7	42°53.246′	20°53.094′	529	0.13
8	42°53.302′	20°53.064′	527	0.16
9	42°54.248′	20°52.096′	506	0.10
10	42°54.423′	20°52.350′	582	0.16
11	42°54,413′	20°52,442′	571	0.10
12	42°55.283′	20°54.057′	600	0.12
13	42°56,450′	20°55,693′	893	0.11
14	42°56.483′	20°55.739′	882	0.16
15	42°58.097′	21°00.945′	1478	0.10
16	42°57.432′	20°59.918′	1255	0.12
17	42°56.415′	20°56.008′	995	0.15
18	42°56.053′	20°54.436′	827	0.11
19	42°56.307′	20°54.881′	776	0.12
20	42°55.486′	20°53.962′	630	0.11
21	42°54.481′	20°51.565′	520	0.14
22	42°54.483′	20°51.550′	523	0.11
23	42°54.485′	20°51.502′	544	0.07
24	42°54.534′	20°51.381′	561	0.10
25	42°54.430′	20°51.351′	524	0.09
26	42°54.329′	20°51.387′	516	0.08
27	42°55.275′	20°49.804′	500	0.16
28	42°55.308′	20°49.872′	524	0.17
29	42°55.331′	20°49.941′	520	0.13
30	42°54.610′	20°52.515′	599	0.12
31	42°49.251′	20°57.352′	548	0.15
32	42°49.239′	20°57.369′	521	0.14
33	42°52.314′	20°51.522′	512	0.15
34	42°53.211′	20°51.584′	521	0.16
35	42°53.849′	20°52.033′	543	0.20
36	42°55.092′	20°50.677′	590	0.22
37	42°57.098′	20°50.386′	524	0.33
38	42°57.094′	20°50.399′	519	0.26
39	42°56.066′	20°50.486′	525	0.25

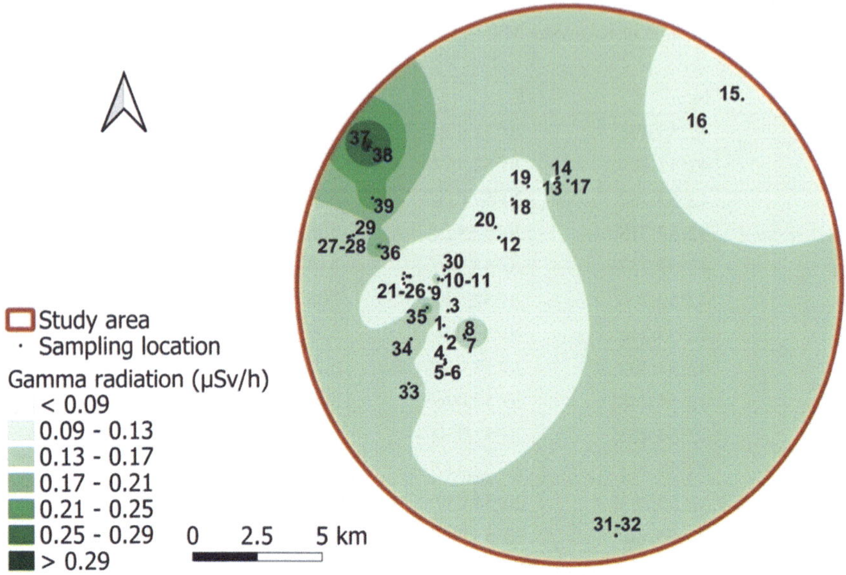

Fig. 21.8 Spatial distribution of ambient dose equivalent rates in study area

Table 21.3 Weighting coefficients for the criteria

Criteria	C1	C2	C3	C4	C5	C6	Weight
C1	1.00	2.00	3.00	3.00	3.00	3.00	0.328
C2	0.50	1.00	2.00	3.00	3.00	2.00	0.226
C3	0.33	0.50	1.00	2.00	3.00	2.00	0.159
C4	0.33	0.33	0.50	1.00	3.00	2.00	0.125
C5	0.33	0.33	0.33	0.33	1.00	2.00	0.085
C6	0.33	0.50	0.50	0.50	0.50	1.00	0.077
C7	0.29	0.40	0.40	0.40	0.40	1.00	0.055

3°. The average distance to the forest is 500 m. The average distance from the closest river is longer of 1000 m.

The Žarkov Potok is have Impermeable and Semi-permeable soil types. According spatial plane and approved zonal statistics the land is residential. The distance from the surface water is 100 m. The average distance from the protected area is 300 m. The slope is 2°. The average distance to the forest is 300 m. The average distance from the closest river is longer of 600 m.

The Šupkovac have permeable and Semi-permeable soil types. According to the spatial plane and approved zonal statistics the land is residential. The distance from the surface water is 300 m. The average distance from the protected area is 600 m. The slope is 6°. The average distance to the forest is 600 m. The average distance from the closest river is longer of 300 m.

Table 21.4 Multi-criteria analysis of flotation tailings/landfills

Criteria	Low	Medium	High
C1	Impermeable	Semi-permeable	Permeable
C2	Industrial	Commercial	Residential
C3	<50 m	50-200 m	>600 m
C4	<300 m	300–600 m	>600 m
C5	>20	3–10	<3
C6	<500 m	500–1000 m	>1000 m
C7	<30 m	30–100 m	>100 m

The Rudare have permeable and Semi-permeable soil types. According spatial plane and approved zonal statistics the land is residential. The distance from the surface water is 100 m. The average distance from the protected area is 1000 m. The slope is 2°. The average distance to the forest is 300 m. The average distance from the closest river is longer of 100 m.

21.4 Proposal of Remediation (Rehabilitation, Re-cultivation) Measures for Landfills

21.4.1 Phytoremediation

Greenery is an important factor for living and working in an urban environment, because it promotes aeration, mitigates climatic extremes, regulates air humidity, binds harmful gases, reduces noise, fixes the soil, prevents dust from rising, and has a number of other useful functions necessary for life and health. Kosovska Mitrovica is in a very critical state in terms of greenery; the parks were mostly destroyed by the construction of new buildings. Due to decades of high level of environmental contamination of the area due to industry, the flora and fauna has been degraded a lot. The most urgent interventions are tailings of lead metallurgy Gornje Polje and tailings in Šupkovac.

The best practices for rehabilitation/re-cultivation of the tailing ponds of the mining and metallurgical industry Trepča should be phytoremediation with native plants. It is also most cost-effective technique for reclaiming heavy metal contaminated soil [22]. More than 400 plant species are identified as natural hyper accumulators of high quantities of more various heavy metals [18]. It is known that plants have a natural potential to remove various types of pollution from nature. Up to now, the largest number of soil remediation research has been done with herbaceous plants, but more recently the possibilities of using woody plants in phytoremediation of polluted soil are being explored. Herbaceous plants are better remediators, they absorb pollution much more, but the advantage of woody species is that they do not enter the food chain [32].

The selection of woody species that can be successfully applied in phytoremediation is limited. The best results were obtained with poplars and willows, due to their high biomass production, developed root system that purifies the soil and intensive use of groundwater to prevent further spread of pollution [27]. They have the possibility of several cycles of decontamination without the obligatory harvesting of the whole plant and re-sowing every year, and after planting the maintenance costs are relatively low [19].

21.4.2 Possibilities of Reuse the Landfills

Agricultural areas in urban and peri-urban areas are particularly susceptible to retention of a large proportion of pollutants. Depending on the ability of a species to uptake heavy metals, plants growing in soils with elevated heavy metal concentrations accumulate pollutants in various concentrations in their edible parts. Entering the human food chain via crops or through groundwater leaching, these pollutants may then accumulate in poisonous concentrations which can seriously endanger human health. The arable land in Kosovska Mitrovica area spreads heavy metal poisons into subsistence foods. Rehabilitation of residential public spaces and agricultural land consisting of removal the first top layer of soil and replacing it with fresh uncontaminated soil is a way to prevent uptake of heavy metals by human being. It will also be desirable to cover the contaminated soil by a layer of 40 cm of clay (this was shown as a good practice) to avoid further flow into deeper layers of soil and groundwater. Seeding grass areas resistant to gases from the air and with a resistant root system would be suitable.

Many interesting possibilities can be implemented to reuse the abandoned landfills/tailings such as construction of recreational facilities, parks, sports halls, business or shopping centers, even a new residential complex. One of the most desirable, but still very expensive investments can be the green energy transition. A solar energy plant is the best way to reuse brown field zones, and to obtain safe clean energy for the future generation and healthier environment. For example, for the Šupkovac tailing pond (N 42.884, E 20.882) on an area of about 12 ha, the annual energy production of a 10,000 kWh solar power plant will be about $12.7 \bullet 10^6$ kWh; calculated parameters are shown in Fig. 21.9 [15].

21.5 Conclusion

The Trepča complex represents one of the major sources of air, soil and underground water pollution in the area of Kosovska Mitrovica. Besides mining, Pb-Zn metallurgy and chemical industry, both of branches are known to be major polluters of the Kosovska Mitrovica area: in the southern part zinc metallurgy, the production

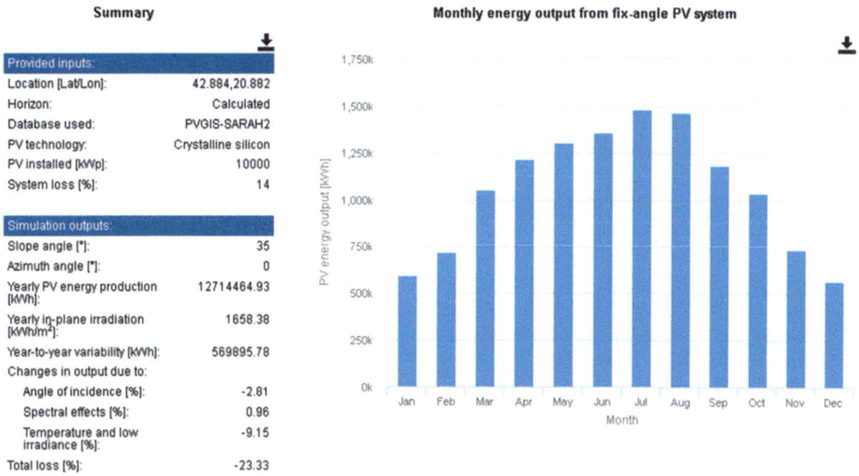

Fig. 21.9 Projected yearly production of a solar energy plant for the Šupkovac tailings

of sulfurous and phosphoric acid, artificial fertilizers and the production of lead batteries, and in the northern part metallurgy (smelling and refining) of lead.

The problems of environmental preservation and improvement are numerous and complex. There are still enormous amount of flotation tailing, metallurgical slag, intermediate products, sludge and ash are deposited near the banks of Ibar and Sitnica rivers. They permanently polluted water and soil by heavy metals. The priority of intervention remediation measures is very high for the most of landfills.

A good choice in the tailing remediation would be the use of woody species in phytoremediation, such as poplars and willows, which have shown high efficiency in reducing air pollution. Future development should be aimed at rehabilitating the Trepča complex in accordance with the EU legislation, remediation of most landfills based on good practice examples, and reusing these sites for green energy transition (the best solution) and recreational, business or residential facilities.

This study showed a lower radiological risk from radionuclides in the area of industrial landfills corresponding to artificial activities, but a higher risk corresponding to the natural ore occurrence and geological characteristics of the surrounding environment of Kosovska Mitrovica.

References

1. Barać M, Stanišić S, Parlić M, Bukumirić Z, Vitas N, Manojlović P (2010) Lokalni ekološki akcioni plan Severne Kosovske Mitrovice – monografija. Opština Severna Kosovska Mitrovica, Fakultet tehničkih nauka, Kosovska Mitrovica i Medicinski fakultet, Kosovska Mitrovica. (In Serbian)
2. Barać N, Škrivanj S, Bukumirić Z, Živojinović D, Manojlović D, Barać M, Petrović R, Ćorac A (2016) Distribution and mobility of heavy elements in floodplain agricultural soils along the

Ibar River (Southern Serbia and Northern Kosovo). Chemometric investigation of pollutant sources and ecological risk assessment. Environ Sci Pollut Res 23:9000–9011
3. Baykara O, Dogru M (2006) Measurements of radon and uranium concentration in water and soil samples from East Anatolian Active Fault Systems (Turkey). Radiat Meas 41:362–367
4. Borgna L, Di Lella LA, Nannoni F, Pisani A, Pizzetti E, Protano G, Riccobono F, Rossi S (2009) The high contents of lead in soils of northern Kosovo. J Geochem Explor 101:137–146
5. Bozdağ A, Yavuz F, Günay AS (2016) AHP and GIS based land suitability analysis for Cihanbeyli (Turkey) county. Environ Earth Sci 75:1–15
6. Di Lella LA, Frati L, Loppi S, Protano G, Riccobono F (2004) Environmental distribution of uranium and other trace elements at selected Kosovo sites. Chemosphere 56:861–865
7. Dimitrijevic MD (1997) Geology of Yugoslavia. Geol. Inst. GEMINI, Belgrade
8. Galjak J, Đokić J, Dervišević I, Milentijević G, Mojsić M, Živković B (2022) Assessment of pollution and distribution of heavy metals in the soil near the flotation tailings GornjePolje. Pol J Environ Stud 31(5):4097–4106
9. Gavrilov MB, Marković SB, Janc N, Nikolić M, Valjarević A, Komac B, Zorn M, Punišić M, Bačević N (2018) Assessing average annual air temperature trends using the Mann–Kendall test in Kosovo. Acta Geogr Slov 58(1):7–25
10. Geological Atlas of Serbia (2002). Ministry of environment and spatial Planning Republic of Serbia, Bareks, Belgrade, Serbia
11. Gulan L, Jakšić T, Milenkovic B, Stajic JM, Vasić P, Simić Z, Zlatić N (2020) Mosses as bioindicators of radionuclide and metal pollution in northern Kosovo and Metohija mountain region. J Radioanal Nucl Chem 326(1):315–327
12. Gulan L, Milenkovic B, Stajic JM, Vuckovic B, Krstic D, Zeremski T, Ninkov J (2013) Correlation between radioactivity levels and heavy metal content in the soils of the North Kosovska Mitrovica environment. Environ Sci: Processes Impacts 15:1735–1742
13. He ZL, Yanga XE, Stoffellab PJ (2005) Trace elements in agroecosystems and impacts on the environment. Review. J Trace Elements Med Biol 19:125–140
14. https://data.marine.copernicus.eu/register. Accessed on 10 May 2024
15. https://re.jrc.ec.europa.eu/pvg_tools/en/. Accessed on 17 May 2024
16. https://search.earthdata.nasa.gov/search?fst0=land%20surface. Accessed on 9 May 2024
17. https://www.quarta-rad.ru/en/catalog/dozimetr-radiometr-radon/dozimetr-radex-rd1503/. Accessed on 10 Apr 2024
18. Freeman JL, Persans MW, Nieman K, Albrecht C, Peer W, Pickering IJ, Salt DE (2004) Increased glutathione biosynthesis play a role in nickel tolerance in Thlaspi nickel hyperaccomulator. Plant Cell 16:2176–2191
19. Kebert M (2014) Biohemijska i fiziološka karakterizacija klonova topole (Populus spp.) u procesu fitoekstrakcije bakra, nikla i kadmijuma. Doctoral dissertation. University of Novi Sad, Faculty of sciences, Novi Sad. (In Serbian)
20. Milentijević G, Nedeljković B, Lekić M, Nikić Z, Ristović I, Djokić J (2016) Application of a method for intelligent multi-criteria analysis of the environmental impact of tailing ponds in northern Kosovo and Metohija. Energies 9(11):935
21. Monthel J, Vadala P, Leistel JM, Cottard F, Ilic M, Strumberger A, Tosovic R, Stepanovic A (2002) Mineral deposits and mining districts of Serbia. Compilation map and GIS databases. BRGM/RC-51448-FR
22. Muhammad S, Shah MT, Khan S (2011) Heavy metal concentrations in soil and wild plants growing around Pb–Zn sulfide terrain in the Kohistan region, northern Pakistan. Microchem J 99:67–75
23. Nannoni F, Protano G, Riccobono F (2011) Fractionation and geochemical mobility of heavy elements in soils of a mining area in northern Kosovo. Geoderma 161:63–73
24. Nedeljković B, Milentijević G, Lazić M (2007) Zaštita životne sredine u neaktivnim industrijskim područjima. In: Proceedings of the Zbornik radova “Zaštita životne sredine u industrijskim područjima”, Kosovska Mitrovica, pp 8–25

25. Nikolić B (1999) Sixty years of lead metallurgy "Trepča" in Zvečan. Metallurgy 5(2):141–156. (In Serbian)
26. Nyamangara J (1998) Use of sequential extraction to evaluate zinc and copper in a soil amended with sewage sludge and inorganic metal salts. Agri Ecosyst Environ 69:135–141
27. Pilipović A, Orlović S, Trudić B, Katanić M, Vasić V, Kebert M (2016) Testiranje klonova topola (Populus spp.) i vrba(Salix spp.) za fitoremedijaciju herbicida kroz ispitivanje uticaja na njihove fiziološke procese. Topola 197(198):35–49
28. Rodríguez L, Ruiz E, Alonso-Azcárate J, Rincón J (2009) Heavy metal distribution and chemical speciation in tailings and soils around a Pb–Zn mine in Spain. J Environ Manag 90(2):1106–1116
29. Šajn R, Aliu M, Stafilov T, Alijagić J (2013) Heavy metal contamination of topsoil around a lead and zinc smelter in Kosovska Mitrovica/Mitrovice, Kosovo/Kosovë. J Geochem Explor 134:1–16
30. Serbian Radiation and Nuclear Safety and Security Directorate (2020). Report on the level of exposure of the population to ionizing radiation from the environment in the Republic of Serbia in 2019, Belgrade
31. Singh S, Kumar M, Kumar MR (2005) The study of indoor radon in dwellings of Bathinda district, Punjab, India and its correlation with uranium and radon exhalation rate in soil. Radiat Meas 39:535–542
32. Štrbac M, Manojlović M, Čabilovski R, Petković K, Kovačević D, Pilipović A (2022) Woody plants in phytoremediation of pollution of agricultural land with nitrates and pesticides. Topola/Poplar 210:73–87
33. Trajković S, Nikolić B, Kamberović Ž, Bajić S (2022) Development of the lead and zinc mining and metallurgy in the "Trepča" mine – Stari Trg. Bakar 47(2):11–30. (In Serbian)
34. UNSCEAR-United Nations scientific committee on the effects of atomic Radiation. Sources and effects of ionizing radiation. Report to general assembly with scientific annexes. United Nations, New York (1993)
35. Valjarević A, Djekić T, Stevanović V, Ivanović R, Jandziković B (2016) GIS numerical and remote sensing analyses of forest changes in the Toplica region for the period of 1953–2013. Appl Geogr 92:131–139
36. Valjarević A (2024) Long-term remote sensing-based methods for monitoring air pollution and cloud cover in the Balkan countries. Environ Sci Pollut Res 31:27155–27171

Chapter 22
Delineation of the Hostilities' Impact on Urban Environment by LiDAR Data Processing (a Case Study of Kharkiv)

Sergiy Kostrikov, Liudmyla Niemets, David Robinson, Kostiantyn Mezentsev, Kateryna Kravchenko, and Denys Serohin

Abstract Our paper presents an approach for extracting buildings (BE-technique) and detecting changes (CD-technique) in urban morphology, using multitemporal LiDAR (Light Detection and Ranging) point clouds. This approach can be especially useful for monitoring urban areas that have been affected by military impact. The paper demonstrates the relevant application for the city of Kharkiv, Ukraine, which has suffered from the hostilities caused by the Russian aggression. Our research uses an application of a novel stand-alone package, that reconstructs the urban environment from Airborne LiDAR data. The software employs calibrated models that have been tested on Kharkiv, with a LiDAR survey in 2019 as the primary dataset, and on two districts of Tallinn, Estonia, with LiDAR surveys in 2017 and 2022 as primary and secondary datasets, correspondingly. Our research also refers to a site survey completed in Fall 2022 as to a secondary dataset of Kharkiv, so that to identify the features of the military impact on residential buildings. The paper visualizes the spatial patterns of urban changes in these three selected sites then. Key geometric values of the featured changes (area of footprint, volume of polyhedron, weighted height) were analyzed by the descriptive statistics and statistical data analysis in *Wolfram Mathematica*. This research effort may be highly valuable for the post-war recovery of the city of Kharkiv.

Keywords LiDAR survey · Desktop software · Building extraction. Changes detected · Statistical analysis

S. Kostrikov (✉) · L. Niemets · K. Kravchenko · D. Serohin
V.N. Karazin Kharkiv National University, Kharkiv, Ukraine
e-mail: sergiy.kostrikov@karazin.ua

D. Robinson
Rutgers University, Piscataway, NJ, USA

K. Mezentsev
Taras Shevchenko National University of Kyiv, Kyiv, Ukraine

C. Morar et al. (eds.), *Achieving Sustainability in Ukraine through Military Brownfields Redevelopment*, NATO Science for Peace and Security Series C: Environmental Security, https://doi.org/10.1007/978-94-024-2278-8_22

22.1 Introduction to Research Issue

Contemporary urban growth has caused a number of challenges in recent years, that require the introduction of innovative technology to urban studies. This includes among all others the Urban Remote Sensing (URS) methodology with its LiDAR (Light Detection and Ranging, lidar) technique. Airborne and Terrestrial LiDAR, both remote sensing mapping technologies, have seen significant advancement over the past few decades [1–6]. Airborne LiDAR platforms (ALS—airborne laser scanning, *aircraft / helicopter/drone-UAV* based) typically generates dense, georeferenced point clouds that map the Earth's surface, while Mobile LiDAR systems (MLS—mobile laser scanning, *terrestrial* based) from ground vehicles—the detailed building exteriors. The procedures of *Automated Feature Extraction* (AFE) and Change Detection, the byproducts of ALS and MLS, are increasingly sought after in urban planning for its ability to create detailed three-dimensional city models. These models are versatile, with potential applications spanning over a hundred uses across two dozen industries [7–11].

Resulted models are particularly valuable in municipal governance, emergency response, and urban infrastructure management, where the demand for high-precision models of urban morphology derived from LiDAR data is on the rise. Achieving such precision often involves combining LiDAR with aerial imagery or enhancing it with MLS data to accurately capture building facades. The importance of a three-dimensional urban model as of a research tool is becoming increasingly recognized, especially as the field of Urban Studies expands in complexity [12–15]. Hence, the process of automatically extracting buildings and man-made structures from LiDAR point clouds, along with generating corresponding topographical data, stands as a significant challenge in the field of research and development. This is crucial for optimization of the urban surveillance systems and for enhancing urban settings through the application of informatics tools and digital infrastructures [16–21].

The number of challenges mentioned above, that accompanies the contemporary urban growth, becomes exceedingly extreme when an urban environment has suffered from the hostilities. The city of Kharkiv, Ukraine, has faced an excruciating experience, especially—for the first several months of Russian aggression. *Our paper aims* to contribute to recognizing the destruction of civilian features, mainly—residential housing blocks, in Kharkiv, Ukraine within some selected sites in the city district that suffered worst from the hostilities.

22.2 Approach and Software

22.2.1 LiDAR Data for Building Extraction Fundamentals

We acknowledge that LiDAR remote sensing is essentially a method for collecting data, which is then processed to gather details about the earth surface, plant life, and human-made structures like buildings, bridges, roads, and power lines, from a specific distance away from a point of measurement. The standard data resolution for our processing technique ranges from **5 to 120 lidar points per square meter.** The foundational AFE-principles shape our core algorithmic architecture, known as High Polyhedral Modeling (HPM), which is developed within the context of an integrated, multifaceted approach [22]. This encompasses three distinct processing and modeling methodologies: *Building Extraction (BE)*, *Change Detection (CD)*, and *Digital Elevation Model Generation (DEM-G)*. Collectively, these research schemes form the comprehensive process for ALS and MLS data analysis. The HPM workflow incorporates the use of external software for the initial general preprocessing of point clouds.

As we already emphasized above, almost all LiDAR platforms can be categorized for either ALS, or MLS types. It can be posited that the prevalent use of MLS techniques is not primarily for the extraction of building facades, but rather for the identification of urban road features [23], or for the delineation of infrastructure networks using a hybrid ALS / MLS approach [24]. However, within the scope of our approach, we assert that a fundamental aspect of automated feature extraction is the establishment of a connection between mobile and airborne LiDAR systems. This leads to the seemingly straightforward conclusion that a combined model of features derived from both ALS, and MLS data represents the most sophisticated, but also—the most efficient approach (Fig. 22.1) [4].

Since the technique of building extraction is a premise for all other following implementations, some of its blocks should be briefly presented more in details according to our algorithmic consequence:

- Importing **.LAS* file to our desktop application internal spatial data format;
- Providing point classification, delineating the ground / non-ground points, and extracting ground points for further *DEM-G*;
- Accomplishing non-ground point classification;
- Providing *DEM-G*: generating two ground (topographic) grids: filled and unfilled; an unfilled grid is completely based on ground points, so it may have *NoData* gaps within its spatial extent; gaps spatially correspond to the non-ground points; the gaps are eliminated by extrapolation of values of surrounding cells in a filled grid;
- Providing *BE*: extracting discrete features by either using generated grids and delineated building feature footprints, or apply to the third-party (external) footprints imported;
- Object filtering: discarding cars, trees, power lines, other man-made features;

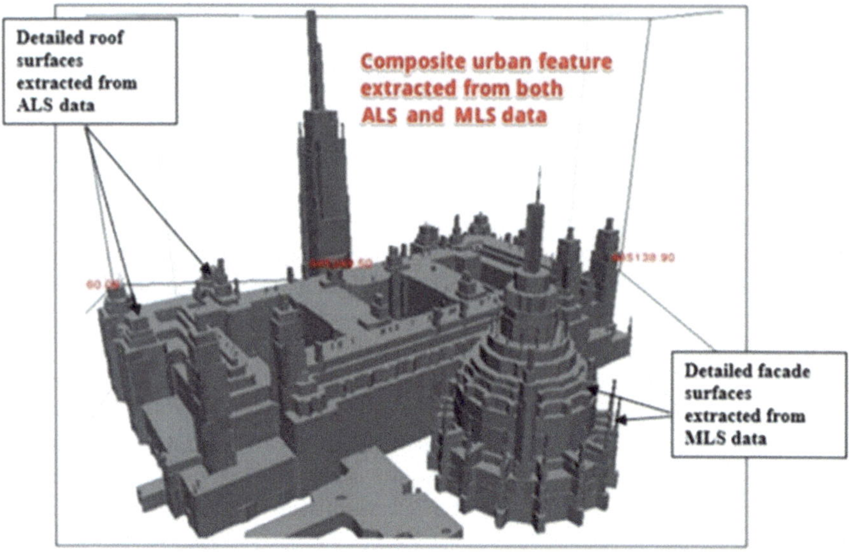

Fig. 22.1 A composite model of the Canadian parliament building extracted from an ALS / MLS point cloud of the downtown of Ottawa by applying *BE*-technique with the original author's desktop software

- Generating simplified pyramidal building model based on ALS data;
- Either arising building facades on the base of MLS datasets, or missing this step, if MLS data is unavailable;
- If MLS is available, the facades can be risen by covering points within footprint boundaries with the triangulated mesh;
- Converting 3D-models either into *.OBJ* format, or into *GeoTIFF* (*a .TIFF* 6.0 compliant raster file) for further desktop visualizing, as well as into *.glTF* for visualizing in the web. Each file has represented one separate building extracted. All these files have been referenced to separate image files. The latter contain textures to be draped on the *3D* models of buildings.

22.2.2 Change Detection with the City Change Management *Software*

The *Change Detection* method relies entirely on the HPM *BE*-functionality to track alterations in urban architecture over time through the automated juxtaposition and scrutiny of dual LiDAR Point Clouds. This instrument pinpoints the modifications in location and configuration of edifices as three-dimensional models [25]. Our *CD*-technique encompasses the trio of universally recognized steps essential for authentic 3D change detection: (1) aligning coordinate systems; (2) conducting spatial and spectral comparisons; (3) presenting and analyzing the changes detected. While considering the established classification of urban modifications, it is necessary to emphasize that it would typically include three principal categories—positive, negative, and no change along with several subcategories. Then we should define two key types of structural alterations as *Added* and *Removed* ones, which align with the partial CD scenarios of "new (partly)" and "demolished (partly)" changes within this classification [26].

Comparing two multi-temporal sets of building models (for changes be obtained as a difference between *primary*, and *secondary* sets of models) by employing *CD*-functionality includes several consequent steps: (1) to build 2D index (so that to enhance the performance on huge datasets): (2) to find intersecting building footprints; (3) to build a correspondence table of the buildings in two datasets, which would find the links between buildings in one dataset with buildings in another dataset; (4) to connect those components in a correspondence table, which identify the features extracted. These components form the list of changes.

The finalized *CD*-results should contain the following: (1) a collection of geometry changes in *.ThreeDFM* data format: one file for one geometric change; (2) an *SQLite* format file with the DB structure that supports the *Report.db* file. Once again, we mandatory need two repeated surveys of different temporal benchmarks to provide change detection technique with output of either *Added,* or *Removed* resulted classes, as well as both.

We applied that versatile technique, which offers two out of three possible key solutions in "an object-based change type determination approach" [27]: (1) either point cloud level's *CD*-parameters are obtained, using the local geometric variations between two multitemporal point clouds—primary and secondary, or (2) *CD*-parameters are derived for feature-based change type determination from either building models, or from urban green areas, for example [28].

This paper's first author had spearheaded and primarily contributed to the development of the *City Change Management* (*CCM*) stand-alone application (for OS *Windows*) from 2016 to 2019, that was accomplished as a non-commercial project. The development employed *C++ 11.0* | *Python 3.7* tools for the algorithmic core and *C# MS .NET* for the interface solutions. The preliminary version of this software firstly was presented within the outcomes of urban studies based on Airborne LiDAR data processing in Kharkiv region [29].

By selecting menu options such as *Workbench= > Building Extraction Technology= > Find Building Changes*, a user activates *CD*-tool, which adds to *BE*-tool functionality of urban studies with *CCM*, enabling a fast and fully automatic detection of changes in the size and the shape correspondingly to two multitemporal benchmarks (see Fig. 22.2).

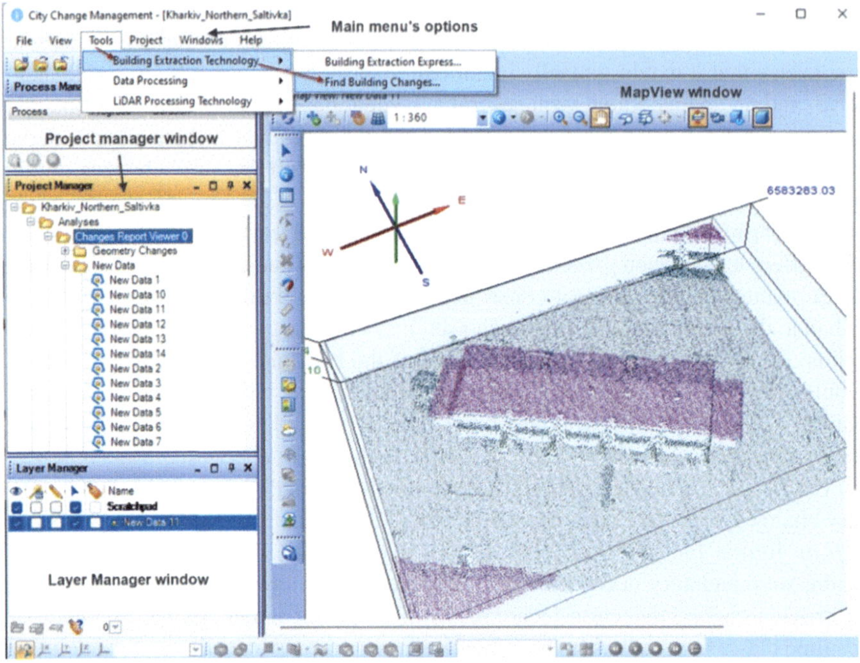

Fig. 22.2 Key interface elements and *Workbench* menu in the *CCM* interface that initiates *BE* and *CD* procedures

22.3 Regional Aspect of the Study

22.3.1 Urban Areas of Research

The key site of this study is some parcel of an urban area of Kharkiv—one of those mostly suffered from hostilities (a corresponding set of ALS data collected was labeled as *KHNS* dataset—the Northern Saltivka city district of Kharkiv). In order to understand the urban morphology caused by military impact, it is necessary to compare this Kharkiv site with other areas from alternative locations, which were not suffered from hostilities. The relevant LiDAR datasets were affiliated with two areas in Tallinn, Estonia: *Lasnamäe* city district (labeled as *EST-1*) and *Mustamäe* city district (labeled as *EST-2*). Kharkiv is located in the North East of Ukraine. It is the second biggest urban area in this country. Its total urban territory is about 350 km2. The city grew steadily before the war, since there were no natural obstacles to its expansion.

The first author of this paper has been working for years with the *EOS Data Analytics* Company (www.eos.com) as a lidar expert in the field of urban remote sensing, especially, overseeing several Airborne LiDAR surveys already mentioned above [29] of various Kharkiv districts in 2017–2019, initiated by the Company.

We need to find an alternative location or locations to this Kharkiv site, which would have a comparable urban structure such as similar types of residential, commercial and other buildings with a dominant share of the residential ones. For example, the Northen Saltivka city district was mainly built in the seventies-eighties of the last century, during the Soviet era, and was supplemented by new, more European high-rise residential buildings at the beginning of the millennium. We seek for a similar urban area, that would been covered by available ALS datasets. This is not an easy task, because almost none of the post-Soviet countries have been scanned by any LiDAR survey even at a small industrial scale. Except for the Baltic countries, and especially—for Estonia.

22.3.2 ALS Datasets of Study Sites

The initial (Primary) KHNS-dataset was acquired in June 2019 using a Riegl VQ-1560 II-S sensor mounted on a helicopter. This resulted in a dataset for a selected urban area (68 sq. km) that covered our current AOI. The latter was defined as the origin for the *Primary KHNS*-dataset, which is a test site of 4 (2*2) sq. km. The number of points defined for this test site was around 122 million. The raw streaming ALS data was further processed with the *Riegl RiProcess* software, and the resulted colorized point clouds were exported as the *.LAS* 1.4 files.

Unlike our previous experience with several North American urban sites, where we received preprocessed and reliable combined ALS/MLS datasets from the local authorities, we faced the challenge of raw LiDAR streaming in this research. As we already stated, a helicopter was the hardware platform for a sensor to collect ALS datasets, while a land vehicle was used to attempt to gather some minimal MLS ones to rise building facades, where is possible. Regrettably, even using some advanced methods for creating the combined ALS/MLS point clouds [30], we failed to obtain a sustainable composite data output. Thus, we couldn't achieve the required MLS input from these data for model calibration within our main High Polyhedral Modeling workflow.

The necessity for URS-monitoring of Kharkiv area to support the post-war recovery efforts became a pressing matter, that was debated by the local government, the research community, and NGOs, after the violent conflicts ceased in the outskirts of Kharkiv in September, 2022. The authors proposed their concept of automated change detection in urban morphology, based on several primary datasets from 2017–2019. With the permission of the local authorities, our group of volunteers conducted the second ALS survey on a small area of 4 sq. km, which already was part of a deserted urban zone with no social or other activities. The survey was carried out in mid-November, 2022 using the same surveying equipment and preprocessing software that the Company had donated. Some from the authors of this paper were directly involved in this event as the surveying specialists.

The relevant free data for two alternatives to Kharkiv locations in Tallinn, Estonia were obtained from the national geoportal of spatial data. The data is annually updated in Dec-Jan (Geoportal, 2022). This data is in *.LAZ* (archived) format and its general characteristics can be found on the Spatial Data page of the Geoportal. The initial, *EST-1 / EST-2* surveys (*Primary*) took place in December 2017. It produced about 80 million points in each dataset merged from four files with an average point density of 20 points / sq. m. The subsequent surveys (*Secondary*) were completed in December 2022. It yielded about 120 million points in each dataset combined from four files (four 1 km^2 squares piece each site) with an average point density of 30 points / m^2. We realize that a challenging accuracy issue may be caused by the following circumstances: while the change detection time gap for the *KHNS* dataset is only 3 years, for *EST-1* and *EST-2* it is up to 5 years.

22.4 Processing / Modeling Results

22.4.1 Accuracy Estimation and Visualization of CD-*Features*

The output 11 columns' attribute tables of *CD*-features exported from *CCM* to *Excel*= > the *Wolfram Mathematica 13.2 (WM)* analytic & modeling software and edited in the latter kept only five columns finally. They describe the geometric properties of the changes detected: *Change ID* (which can be labeled on a map); *ChangeType* (either *Added* or *Removed*); *Area*, **A** (*m^2*)—the footprint size of the *CD*-feature; *Volume*, **V** (*m^3*)—the polyhedral volume of the *CD*-feature; *SizeZ* (*m*)—the weighted *Height*, ***H*** of the feature with the change. The height was calculated both as a spatially weighted height [31], and as a smoothed value, using some of the morphometric methods reported for estimating building heights [32].

As a result, the *EST1* dataset featured 48 *Added* (*EST1* data) and 67 *Removed* (*EST1R* data) *CD*-features, while *EST2* had 175 *Added* (*EST2* data) and 174 *Removed* (*EST2R* data) ones. These changes were likely due to the planned urban reconstructions carried out between 2017–2022 in these two districts of Tallinn. In contrast, the *KHNS* dataset revealed 215 *Added* (*KHNS* data) and 51 *Removed* (*KHNSR* data) *CD*-features. Field measurement-based accuracy for *CD*-feature delineation was understandably feasible for us to define only within the *KHNS* dataset. Technical documents from the municipality, detailing planned changes in urban structure, accounted for just 85 of the 215 detected *CD*-features. The *Removed CD*-objects are presumed to be solely related to military impact, as urban development (UD) in 2019–2021 did not involve any "negative changes". Field measurements were possible for 37 out of the 51 *Removed CD*-features.

A summary of the accuracy findings is as follows. There is a measurable geometric deviation of the lidar-modeled *CD*-features from their actual field measurements. For the Added class, the deviations are: *Area*: from 0.8 to 1.3%; *Volume*: from 1.5 to 2.8%; *Height*: up to 11%. For the *Removed* class, the deviations are: *Area*: from 0.4 to 0.8%; *Volume*: from 1.0 to 2.0%; *Height*: up to 7%. These deviations significantly surpass the declared precision of the LiDAR monitoring equipment (15–30 cm), which could be attributed to various factors. Due to the lack of alternatives, these precision assessments had to be extrapolated to the Estonian datasets.

Thus, by utilizing ALS data collected in 2017 and 2022 for the multi-temporal *EST1* and *EST2* datasets, and those ones obtained in 2019 and 2022 for the *KHNS* dataset, we discerned variations in the *BE*-models created from the primary and secondary segments of each dataset, in accordance with the *BE* and *CD* techniques previously briefly outlined in this paper. In line with the primary objective of this study, we conducted a comparative analysis of the urban morphology changes in the *KHNS* site with *EST1*, *EST2* sites. This included a direct visual comparison of the *CD*-features' spatial distribution across the entire area of the three main study sites (see Fig. 22.3, 22.4, and 22.5).

The alterations in urban morphology identified within three key sites of our study have been rendered alongside the models of the unaffected buildings as *3D* Scenes in the illustrations provided. This approach arguably offers a more comprehensive representation, than merely presenting the findings of analogous thematic studies through basic 2D schemes, especially when the research aims to delineate up to four distinct categories of changes [33].

Additionally, both building model footprints, and centroids of *CD*-feature footprints, as well the topographic grid produced, were all transferred to the fully-functional GIS—*QGIS*. They were exported as a raster layer along with sets of point and polygonal features. Consequently, users have an option to identify buildings with alterations and access a distinct *CD*-layer containing all essential attributes. All this should be done within the familiar environment of a standard GIS interface, rather than specialized software like *CCM*.

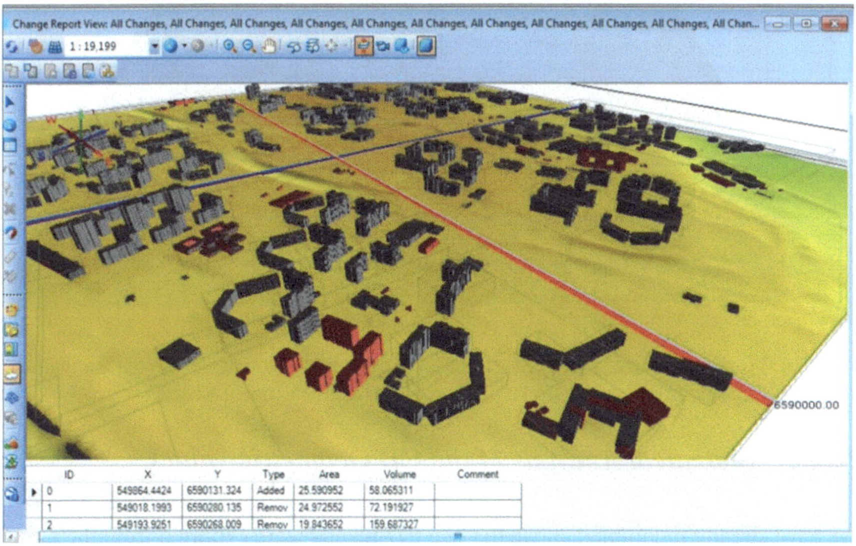

Fig. 22.3 Visualization of *CD*-features (colored in red) in the *CCM* interface for the multitemporal (2017–2022) LiDAR dataset *EST1*: *Lasnamäe* city district of Tallinn

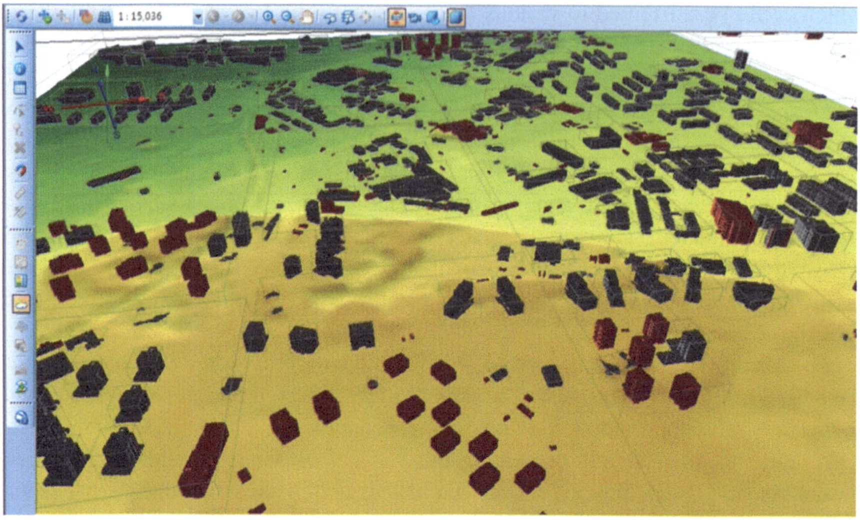

Fig. 22.4 Visualization of *CD*-features (colored in red) in the *CCM* interface for the multitemporal (2017–2022) LiDAR dataset *EST2*: *Mustamäe* city district of Tallinn

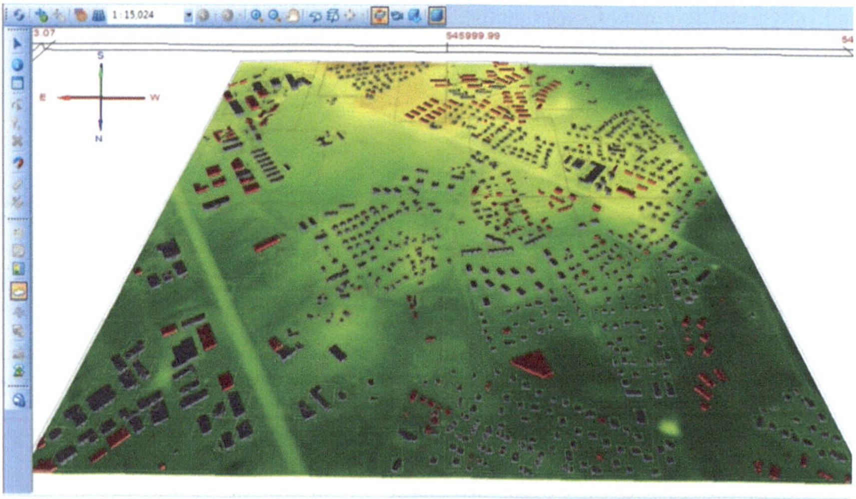

Fig. 22.5 Visualization of *CD*-features (colored in red) in the *CCM* interface for the multitemporal (2019–2022) LiDAR dataset *KHNS*: *Northern Saltivka* city district of Kharkiv

22.4.2 *Statistical Analysis of the* CD *Geometric Values*

We have already mentioned above that the *CD*-results were exported from the *CCM* software tables to *MS Excel*. Then the *Excel* tables were reexported to *WM* software for further editing and derivative processing. Three basic parameters of a *CD*-feature

A, ***V***, and ***H*** have been recognized as the most representative ones for formalization of geometric changes at each site and for both types of these changes. Such metrics are essential for identifying standardized attributes of data consistency and for the subsequent categorization and pattern recognition within two distinct urban configurations — one of typical development and another—affected by the hostilities. Had the frameworks of this text allowed to visualize huge tables of corresponding derivative data, these descriptive values would have been seen as the basic ones for broadly accepted numerical comparison of datasets [34]. Also, we specifically selected the *WM* software for our statistical data and model analysis due to its flexibility in uncovering patterns in any chosen dataset without necessarily enforcing a predefined data distribution.

Descriptive statistics summarized quantitative characteristics of all detected changes through three study sites, while their numbers of *CD*-features were presented above already. Thus, the total volume of the *Added* CD-class, considered as newly constructed features, is approximately 2.6 times greater for the *EST2* site compared to *EST1*. This could indicate a higher rate of this period development activity in the *Mustamäe* district relatively to t*he Lasnamäe*. The ratio (*EST2/EST1*) for the *Removed* CD-class is even higher, its approximate value = 6.5, highlighting a significant volume of negative changes in *Mustamäe*. For the *KHNS* site, the *Added* CD-class shows slightly fewer changes during 2 years (2020–2021), if compared to *EST2* (ratio *KHNS/EST2* = *0.84*), but it demonstrates more than double increase, if compared with the five-years (2017–2022) changes in *EST1 (KHNS/EST1* = 2.18).

The highest value of the *Removed* CD-class (negative changes) for *EST2* can be attributed to the characteristics of routine urban development in *Mustamäe*, which involved a considerable number of demolitions. The *KHNS* site, affected by military hostilities (drastic impact in March–May 2022), formally presents the smallest volume of negative detected changes. However, it is only 1000 m^3 fewer, than the value for *EST1*, which was resulted from 5 years of routine UD. What is more, the UD-information from the Kharkiv municipality indicates no demolished buildings in *KHNS* during 2019–2021.

Descriptive statistics also show a significant average *CD*-feature size for positive CD across the three datasets: the largest one in *EST1*, and about 1.5 times smaller changes in *KHNS*, what can be explained by the scale of local UD. The largest individual demolition changes occurred in *EST2*, as expected due to the UD in the area. Surprisingly, at least few single demolished changes in *KHNS* are definitely larger, than those ones in *EST2*, suggesting an external influence not related to UD, potentially indicative of those military hostilities permanently mentioned in our study.

Due to three fundamental geometric *CD*-parameters—*Area (**A**), Volume (**V**)*, and *Height (**H**)*—it's important to note that the area of a *CD*-feature's footprint and its volume are calculated using standard polyhedral formulas, whereas its representative height is determined as a spatially weighted figure. However, the three pairs of geometric values (*A*↔*V*, *V*↔*H*, *A*↔*H*) can form a linear correlation matrix of several relationships, where higher values in the matrix suggest a closer resemblance of

Table 22.1 Population correlation matrix of geometric *CD*-parameters

Dataset, *CD*-values	*EST1* *A V,H*	*EST2* *A, V,H*	*KHNS* *A V,H*	*EST1R* *A, V,H*	*EST2R* *A, V,H*	*KHNSR* *A V,H*
A	* 0.89*	* 0.85*	* 0.97*	* 0.99*	* 0.98*	* 0.94*
V	** 0.68	**0.55**	** 0.50**	** 0.26**	** 0.34**	** 0.56**
H	0.41** **	0.33** **	0.48** **	0.18** **	0.30** **	0.33** **

*Significance on the level of 0,01
**Significance on the level of 0,05

a *CD*-feature to a regular geometric shape. The relationships can be expressed as follows:

$$\text{For}\, \boldsymbol{A}\ \text{and}\, \boldsymbol{V} : A \leftrightarrow V : Volume = Area^{*}Height, \tag{22.1}$$

$$\text{For}\, \boldsymbol{V}\ \text{and}\, \boldsymbol{H} : V \leftrightarrow H : Height = Volume\, /\, Area, \tag{22.2}$$

$$\text{For}\, \boldsymbol{A}\ \text{and}\, \boldsymbol{H} : A \leftrightarrow H : Area = Volume\, /\, Height. \tag{22.3}$$

These relationships mirrored by Eqs. (22.1–22.3) are foundational in understanding the spatial characteristics of changes detected in an urban environment. The population linear correlation coefficient, (**p**), calculated for these pairs (22.1–22.3), understandably—within a one given dataset, indicates a strong linear relationship between ***A*** and ***V*** across all datasets (Table 22.1, so that to make a table compacter only one correlation value is prescribed in the matrix instead of possible two in the same pair, e.g., $\mathbf{p}_{A,V}$ and $\mathbf{p}_{V,A}$). Such strong relationship was expected, because the volume of a feature is directly related to its area, when a height is considered. However, this correlation is not as strong as expected for the *Added CD*-class, suggesting that the newly constructed features may have more variability in their geometric proportions, which could be due to a range of factors such as architectural design, purpose, or constraints during urban construction.

Thus, across the entire set of parameters (*A*↔*V, V*↔*H, A*↔*H*), the **p**-value relationships are roughly equivalent within the corresponding matrix cells for a given *CD*-class, except for the *A*↔*H* relationship, which remains the weakest link. The *Removed* CD-class exhibits slightly stronger **p**-values for the *A*↔*V* pair and weaker values for other two pairs compared to the positive changes. Furthermore, the demolished alterations within the *KHNSR* dataset show significantly stronger **p**-values for *V*↔*H* and *A*↔*H* relationships, if compared to the other two datasets, where negative changes are solely attributed to routine UD and not to any hostilities.

22.5 Conclusion

The three methods of derivative processing — feature extraction / change detection, *CD*-feature visualizing, and descriptive statistics of changes — demonstrate the efficacy of possible pattern recognition and classification on this base: sites undergoing standard urban development (without military impact) show noticeable distinctions in geometric change patterns compared to the data from the Northern Saltivka city district in Kharkiv. This distinction enables the subsequent statistical model analyses to identify formalized characteristics that differentiate between urban areas affected by hostilities and those ones experiencing normal development. Due to this, the value of this research, particularly for post-war monitoring and reconstruction efforts in Ukrainian cities, cannot be overstated.

Disclosure of Interests The authors have no competing interests to declare that are relevant to the content of this article.

References

1. Baltsavias E (1999) A comparison between photogrammetry and laser scanning. ISPRS J Photogramm Remote Sens 54:83–94
2. Sampath A, Shan J (2007) Building boundary tracing and regularization from airborne LiDAR point clouds. Photogramm Eng Remote Sens 73(7):805–812
3. Shan J, Sampath A (2009) Urban dem generation from raw LiDAR data: a labeling algorithm and its performance. Photogramm Eng Remote Sens 75:427–442
4. Stal C, Tack F, De Maeyer P et al (2013) Airborne photogrammetry and LiDAR for DSM extraction and 3D change detection over an urban area – a comparative study. Int J Remote Sens 34(4):1087–1110
5. Dong P, Chen Q (2018) LiDAR remote sensing and applications. CRC Press, Boca Raton
6. Chio S-H, Hou K-W (2021) Application of a hand-held LiDAR scanner for the urban cadastral detail survey in digitized cadastral area of Taiwan Urban City. Remote Sens 13(24):4981
7. Wang R (2013) 3D building modeling using images and LiDAR: a review. Int J Image Data Fus 4(4):273–292
8. Biljecki F, Stoter J, Ledoux H et al (2015) Applications of 3D city models: state of the art review. ISPRS Int J Geo Inf 4:2842–2889
9. Kostrikov S, Pudlo R, Kostrikova A, Bubnov D (2019) Studying of urban features by the multifunctional approach to LiDAR data processing. In: 2019 Joint Urban Remote Sensing Event (JURSE), Vannes, France, pp 1–4. URL: https://ieeexplore.ieee.org/document/8809063. Accessed 8 March 2024
10. Kada M (2022) 3D reconstruction of simple buildings from point clouds using neural networks with continuous convolutions (convpoint). Int Arch Photogramm Remote Sens Spat Inf Sci XLVIII-4/W4-2022:61–66
11. Župan R, Vinković A, Nikçi R et al (2023) Automatic 3D building model generation from airborne LiDAR data and OpenStreetMap using procedural modeling. Information 14(7):394
12. da Silveira LG, Musse SR (2006) Real-Time Generation of Populated Virtual Cities. In Proceedings of the ACM Symposium on Virtual Reality Software and Technology, pp 155–164. Limassol, Cyprus, 1–3 November 2006; Association for Computing Machinery, New York

13. Rottensteiner F (2014) Results of the ISPRS benchmark on urban object detection and 3D building reconstruction. ISPRS J Photogram Remote Sens 93:256–271
14. Buyuksalih G, Baskaraca P, Bayburt S et al (2019) 3D city modelling of Istanbul based on Lidar data and panoramic images—issues and challenges. Int Arch Photogramm Remote Sens Spat Inf Sci XLII-4/W12:51–60
15. Esch T, Zeidler J, Palacios-Lopez D et al (2020) Towards a large-scale 3D modeling of the built environment—joint analysis of TanDEM-X, Sentinel-2 and open street map data. Remote Sens 12:2391
16. Fruh C, Zakhor A (2003) Constructing 3D city models by merging aerial and ground views. IEEE CG&A 23:52–61
17. Caoa R, Zhanga Y, Zhao Z (2017) 3D building roof reconstruction from airborne LiDAR point-clouds: a framework based on a spatial database. Int J Geog Inf Sci 31:1359–1380
18. Kostrikov S (2019) Urban remote sensing with LiDAR for the Smart City concept implementation. Visnyk of V N Karazin Kharkiv National University, series. Geol Geograp Ecol 50:101–124
19. Choei N-Y, Kim H, Kim S (2020) Improving infrastructure installation planning processes using procedural modeling. Land 9:48
20. Kostrikov S, Seryogin D (2022) Urbogeosystemic approach to agglomeration study within the urban remote sensing frameworks. In: Battisti A, Baiani S (eds) Sustainable development dimensions and urban agglomeration. IntechOpen, London, pp 251–273
21. Kostrikov S, Kravchenko K, Serohin D et al (2023) The performance of the digital city projects in urban studies of the megalopolises (the case studies of Kharkiv and Dnipro cities). Visnyk of V N Karazin Kharkiv National University, Series. Geol Geograp Ecol 59:140–165
22. Kostrikov S, Pudlo R, Bubnov D et al (2020) Automated extraction of heavyweight and lightweight models of urban features from LiDAR point clouds by specialized web-software. Adv Sci Technol Eng Syst J 5:72–95
23. Wang J, Lindenbergh R, Menenti M (2017) SigVox—a 3D feature matching algorithm for automatic street object recognition in mobile laser scanning point clouds. ISPRS J Photogram Remote Sens 128:111–129
24. Wang Y, Chen Q, Liu L et al (2018) Systematic comparison of power line classification methods from ALS and MLS point cloud data. Remote Sens 10:1222
25. Stilla U, Xu Y (2023) Change detection of urban objects using 3D point clouds: a review. ISPRS J Photogram Remote Sens 197:228–255
26. Tian J, Chaabouni-Chouayakh H, Reinartz P (2011) 3D building change detection from high resolution spaceborne stereo imagery. In: In: 2011 IEEE international workshop on multi-platform/multi-sensor remote sensing and mapping, Xiamen, China, pp 1–7
27. Zhang L, Zhang Z, Zhang J et al (2023) Object-based 3D building change detection using point-level change indicators. Int J Appl Earth Obs Geoinf 118:103293
28. Morar C, Lukic T, Valjarevic A et al (2022) Spatiotemporal analysis of urban green areas using change detection: a case study of Kharkiv. Ukraine Front Environ Sci 10:823129
29. Kostrikov S, Niemets L, Sehida K et al (2018) Geoinformation approach to the urban geographic system research (cases studies of Kharkiv region). Visnyk of V N Karazin Kharkiv National University, series. Geol Geograp Ecol 49:107–124
30. Cheng L, Tong L, Wu Y et al (2015) Shiftable leading point method for high accuracy registration of airborne and terrestrial LiDAR data. Remote Sens 7:1915–1936
31. Getis A, Ord JK (1992) The analysis of spatial association by distance statistics. Geogr Anal 24:189–206
32. Godlowska J, Kazowski W (2019) Testing various morphometric methods for determining the vertical profile of wind speed above Krakow, Poland. Bound – Layer Meteorol 172:107–132
33. Yadav R, Nascetti A, Ban Y (2022) Building change detection using multi-temporal airborne LiDAR data. Int Arch Photogramm Remote Sens Spatial Inf Sci XLIII-B3-2022:1377–1383
34. Niroumand H, Zain M, Jamil M (2013) Statistical methods for comparison of data sets of construction methods and building evaluation. Procedia Soc 89:218–221

Chapter 23
Transformation of Brownfields in Western Ukraine

Kuzyshyn Andrii, Liudmyla Kliuchko, Kateryna Kravchenko, and Daria Venhryn

Abstract The article examines the concept of brownfields, which gained popularity in the 1990s, and explores successful foreign experience in brownfield redevelopment. Brownfields became widespread in Europe after the collapse of socialism due to deindustrialization and outdated industries, especially in Central and Eastern European countries like Ukraine. Despite the challenges, recent geopolitical events have brought changes to brownfield utilization policies in Ukraine. The authors analyze the factors influencing their formation, highlighting the role of historical, temporal, and internal geographical factors in reconstruction models. The article investigates the potential of brownfields for revitalizing economically developed territories through functional transformation, emphasizing the effectiveness of creating industrial parks on abandoned lands. Special attention is given to studying the experience of using neglected territories that were previously industrialized within regions in western Ukraine. The Russian-Ukrainian war has had a significant impact on this situation, leading to the relocation of enterprises from hazardous regions. The research indicates a correlation between the location of post-industrial developed projects and their subsequent use. Despite their significance, brownfields in the Western Ukrainian region primarily former enterprises from the planned economy era, often remain overlooked. Their attractiveness for reconstruction depends on location, available infrastructure, logistics, and the investment climate. The use of brownfields in the region involves creating industrial parks to rejuvenate economic activity or relocating enterprises from conflict-affected areas.

Keywords Brownfields · Reindustrialization of territories · Western Ukraine

K. Andrii (✉)
Ternopil Volodymyr Hnatiuk National Pedagogical University, Ternopil, Ukraine
e-mail: kuzyshyn_a@tnpu.edu.ua

L. Kliuchko · K. Kravchenko · D. Venhryn
V.N. Karazin Kharkiv National University, Kharkiv, Ukraine
e-mail: ludmila.klychko@karazin.ua; kateryna.kravchenko@karazin.ua; daria.venhryn@student.karazin.ua

C. Morar et al. (eds.), *Achieving Sustainability in Ukraine through Military Brownfields Redevelopment*, NATO Science for Peace and Security Series C: Environmental Security, https://doi.org/10.1007/978-94-024-2278-8_23

23.1 Introduction and Review of Previous Scientific Developments in This Area

23.1.1 Introduction

Traditionally, brownfields are associated with industrial spaces. The term emerged in the early 1990s when practitioners and researchers observed how new regulatory frameworks, developed to protect the environment (as a side effect), hindered the reuse (cleanup and redevelopment) of former industrial and commercial sites. The first brownfields appeared in the United States because many industrial facilities were moved abroad. The vacated territories, which were recognized as polluted as a result of previous industrial activity, needed to be revitalized. In 2003, the United States adopted a brownfields redevelopment program. One of the most prominent examples of a brownfield is the Highline Industrial Park in New York City, which was created on the site of an unusable railroad in 2004–2014. The Highline Industrial Park in New York City, created on the site of a disused railroad The Pearl District (Portland, Oregon) was once an abandoned industrial area dominated by railroad stations, warehouses, and factories. In the 1990s, a group of developers and urban planners began to revitalize the area, turning it into one of Portland's most vibrant neighborhoods. Today, the Pearl District is a mixture of residential buildings and shops.

Many European countries (Germany, France, Czech Republic, Romania) have experienced issues with the use of derelict areas since the 1970s. Brownfields emerged in large numbers immediately after the collapse of socialism and the transition from a centralized planned economy to a market economy in Central and Eastern European countries. Brownfields are associated with deindustrialization and the overvalued and non-competitive industry in Ukraine. The enterprises simply lacked the capability to adapt to market conditions while their owners had enough financial resources and flexibility to continue their industrial use. In many regions of Western Ukraine, redevelopment proved to be less effective than in other areas of the country. Potential investors are afraid of the risks and uncertainties associated with the regeneration of brownfields and prefer projects in new areas. The situation changed in the last decade when Russian armed aggression forced a reconsideration of the policy regarding the use of such industrial sites specifically in Western Ukraine.

23.1.2 *Theoretical and Methodological Features of Brownfields Research*

This work has the following objectives. Firstly, to identify the key factors influencing why only certain areas of former industrial development have been successfully regenerated and repurposed within the Western Ukrainian region, while others have not. Secondly, to determine the justification of industrial parks as a key tool for regenerating former industrial sites. Thirdly, to investigate why the private sector invests in some regeneration zones but not others, and to understand the local and specific factors influencing the decisions of investors and developers. This study can be applied by central and regional government bodies, regional development agencies, urban planners, and other decision-makers involved in effectively utilizing territories [1, 2].

The formation of brownfields is influenced by temporal and historical factors. However, the role of internal geographical factors is also important, as they influence the actual situation and redevelopment models. M. Soldak believes that the strategic goal of initiating revitalization processes in brownfield areas is the modernization of the economy in old industrial districts by creating a new industrial base through the preservation and development of medium- and high-tech production [9].

Frantál B., Josef Kunc J., Nováková E., Klusáček P., Martinát S., Osman R synthesized most of the contemporary approaches to defining brownfields in their research [4].

In a broad sense, brownfields are areas that were previously industrial or commercial activities and need to be reconstructed or rehabilitated. This phenomenon is quite common both in global practice and in Ukraine. Research by Ray Perez and Rodrigo Eugenio [7] indicates that 2.8 million such sites recorded within the territory of the EU [8]. They was a source of ecological threats, inconvenience, and discomfort for the local population.

It is believed that they can be used to renew the economy within the territory that has been industrially developed. This problem can be overcome through revitalization, which involves repurposing abandoned industrial facilities to meet the emerging socio-cultural needs of the city and its residents. Revitalization provides an opportunity for urban development by densifying urban areas, revitalizing depressed areas, reducing costs associated with urban sprawl (such as environmental degradation, infrastructure expansion, increased transportation costs) [3].

Brownfields have two options for implementation. The first is within the territory of an enterprise that has ceased to operate and produce for various reasons. The second is the territory associated with former mining operations. In both cases, this issue has a spatial dimension and is related to the regeneration of the territory. The global organization CABERNET (Concerted Action on Brownfield and Economic Regeneration Network, EU) has developed an A-B-C classification, aimed at analyzing the potential for the effective utilization of brownfields.

According to the classification established in global practice, three forms of organizing the reuse of obsolete industrial zones have emerged.Form A is applied to minimally contaminated sites located in commercially attractive areas, which can be rev italized by private investors to generate high profits. The role of the government in these projects is limited to necessary approvals and permits within the city's development planning framework. Form B is applied to sites with moderate or severe contamination. Projects are implemented through public-private partnerships, which are becoming increasingly common. Various stages of the process are regulated through relevant agreements and contracts. Sometimes, direct financial assistance from the government sector is allocated for the reconstruction of production, subject to the completion of planned socio-economic tasks. Form C involves socially oriented projects, primarily executed in situations involving significant contamination of industrial zones, uncertainty about the brownfield's specialization, and prospects for commercial land use. However, there are several modern studies that critically evaluate the functioning of this model (Franz et al., Vojvodíková et al.) and even propose supplementing it with Form D, where the site poses a danger due to pollution or degradation of buildings.

Pollution Risk Assessment
When we work on brownfields, we always need to work with risk assessment and analyze whether the risk is acceptable or not.

Advantages of Brownfield Projects:
- Reducing construction on new, non-polluted sites;
- Reducing the threat to public health;
- Improved air and water quality;
- Short- and long-term job creation;
- Local economic growth and investment;
- Increase in the value of surrounding real estate;
- Recovery of tax revenues;
- Increase in the tax base;
- Efficient use of existing infrastructure;
- Renewed use of existing commercial facilities;
- Revitalization of the district;
- Slowing down urban sprawl.

There are many positive practices of implementing brownfields in the planning of territories in different countries.

23.2 Positive Aspects of Brownfields Practice in Western Ukraine

23.2.1 Spatial Aspects of Industrial Parks Organization as a Tool for Brownfields Regeneration

Usually, the initiative to introduce brownfields is associated with the initiatives of the communities themselves, who are interested in redeveloping their own territories. However, the possible results of such activities can be of two kinds. On the one hand, the successful functioning of the industrial or mining areas themselves. On the other hand, it can result in the pollution of the territory (environmental problems) or the existence of outdated or inappropriate housing and social infrastructure. The development of industrial parks within brownfields is an effective process of qualitatively utilizing land that already has a certain level of developed production infrastructure. Global experience suggests that establishing industrial parks within brownfield areas is an effective organizational and economic tool for accelerating the socio-economic development of these territories and creating "points of regional growth" by applying a system of stimulating organizational and economic conditions for their participants.

Industrial parks of the "greenfield" type, created on previously undeveloped industrial land, will be primarily attractive to large companies seeking space for constructing industrial facilities tailored to their specific needs. Industrial parks of the "brownfield" type, established on the basis of industrial sites with infrastructure reconstruction or capital works according to the specialization of the industrial park and its participants, are likely to attract interest from small companies whose production complexes can be accommodated in standard premises. The "time to market" (the time from investment initiation to product release and sales) for participants in brownfield-type industrial parks is expected to reduce from two to three years to six to nine months [7].

Most examples of brownfields in Western Ukraine (includes the territory of seven regions of the country) are industrial parks. These are the territories planned for organizing the management of certain land use.

Industrial parks were included in the strategies of socio-economic development of most regions of Western Ukraine: in 2006—Rivne region, in 2007—Ivano-Frankivsk and Lviv regions, in 2008—Ternopil and Zakarpattia regions, and in 2012—Volyn region. This was the first step in introducing brownfields in these areas.

According to part 3, paragraph 1, article 1 of the Law of Ukraine "On Industrial Parks", an industrial park is a territory determined by the initiator of the industrial park creation in accordance with the urban planning documentation and equipped with the appropriate infrastructure, within which the participants of the industrial park may carry out economic activities in the field of industrial production, as well as research activities in the field of information and telecommunications, under the conditions specified by this Law and the agreement on economic activity within the industrial park [6].

23.2.2 Brownfields Within the Western Ukrainian Region: Challenges of War and Spatial Organization

Today, local community websites offer information and proposals for the use of abandoned territories that have already been industrialized. However, such proposals have not yet been systematized and are mostly fragmentary (Table 23.1).

The construction of five processing plants is being completed on the territory of a former brick factory in Rivne: for the packaging of cereals, the production of teas, spices, pasta, and gluten-free sweets. The total area of the industrial complex is 6.5 hectares, according to the Rivne Regional State Administration. The work began in

Table 23.1 Brownfields for socio-economic development in the plans for the territories of the Western Ukrainian regions in 2020–2023 (proposed by the authors according to [5, 6, 7, 10])

Administrative territory (territorial community)	Number of plots bleached for brownfields	Area, hectares	Planned profile of activities	Number of newly created jobs, people
Volyn region				
Vyshniv	1	41	Feed mill	***
Kovel	3	***	***	***
Transcarpathian region				
Perechin	1	16,5		745
Solomonovo	1	7,9		
Solotvyno	1			
Ivano-Frankivsk region				
Dolyna	1	27,1	Mechanical engineering, Forestry and woodworking industry	***
Hryplyn (Ivano-Frankivsk)	1	42	Solar power plant, Construction industry enterprises	***
Lviv region				
Brody	1	***	***	***
Vinnyky	3	***	***	***
Lviv	4	98	Mechanical engineering, Light industry	7500
Stryj	1	13,95	Food production, Electrotechnical engineering, Furniture industry	700
Solonycya	1	7,9	***	***
Rivne region				
Rivne	2		***	***
Demydivka	2		Light industry Food industry	1400
Ternopil region				
Ternopil	1	15	Food industry	***
Chernivtsi region				
Rukshyn	1		Food industry	***

2021, the dismantling of the old structures took eight months, and the new premises appeared a year later. Today, interior finishing is ongoing, the improvement of the territory is in progress, and commissioning has already begun. Investments in construction materials exceed UAH 75 million for today. Most of the equipment is made in Ukraine. Several dozen workers, mostly people who were internally displaced, are currently employed here. In total, more than 800 workplaces will be created. In the future, the industrial complex also plans to start processing fish and milk. This example is typical for most brownfields in Western Ukraine that have undergone redevelopment. Summarized information looks as follows.

According to the available information from open sources, within this region, 21 post-industrial brownfield regeneration projects identified, mainly located in the central and most urbanized parts of the regions. The total value of these projects by the end of the third quarter of 2022 amounted to 1.1 billion euros, primarily attributed to land issues and infrastructure aspects.

Tthe largest group consisted of obsolete objects of various purposes. Almost all of these projects were implemented by state institutions, mainly local government bodies. The second largest group of enterprises consisted of commercial projects carried out exclusively by private entities, mainly without the support of state funds. However, it is worth noting that the average cost of projects in this category is traditionally higher. Post-industrial brownfields were also often used to create various types of investment zones aimed at supporting local economic development and, in some cases, repurposed for new types of industrial activities. These projects represent a field of interaction between state structures, which are active in the stage of reclamation and infrastructure improvement, and private investors, who redevelop the site for economically beneficial use.

The war in Ukraine has become another reason for the revitalization of brownfields. Currently, relocation is possible to one of the sixteen regions. As of October 2023, 840 Ukrainian enterprises have been successfully relocated from dangerous regions with the support of the state. 667 of these are already successfully operating at their new locations, while 239 companies are still searching for a suitable location or mode of transportation. More than 650 companies refused to relocate due to the de-occupation of the territories where they are located. Some companies are also returning to their previous locations due to improved security in the Kharkiv, Chernihiv and Sumy regions. 44 companies have already returned.

A total of 1816 companies have registered applications for relocation on the corresponding online platform, ProZoro. However, 601 companies refused to relocate due to the de-occupation of the territories of Ukraine. Most companies were relocated to Lviv (30% of relocated companies), Zakarpattia (17%), Chernivtsi (11%), Ivano-Frankivsk (8%), Khmelnytskyi (7%), and Ternopil (7%) regions.

Among the relocated enterprises that have already resumed their operations at the new location, the largest share is made up of: wholesale and retail trade, repair of motor vehicles and motorcycles (40.24% of the total number of relocated companies); processing industry (31.71%); information and telecommunications (6.34%); professional, scientific and technical activities (5.85%); and construction (4.15%). This associate the impact of location with a local development potential or the area competitiveness, being the resultant of environmental conditions, economic potential, and social capital.

Relocation has significantly changed the spatial features of the territories where it is carried out. In particular, some regions are gradually getting rid of the term "agrarian-industrial" due to such processes.

Another point related to revitalization is changes in the housing market in terms of quantity and quality. The increase in demand has generated a kind of boom in the quantitative indicators of construction and housing prices.

Additional activation of the use of spatial opportunities is associated with interest in buildings that have ceased to be used for their intended purpose and have become a kind of ghosts on the territories of Ukrainian cities. Let's consider the case of Ternopil city. Several factories have been preserved here, built in the late 1980s or 1990s. It was a period of gigantomania in Ukraine, a Soviet disease of self-assertion. There are two examples of radio equipment factories—Saturn and Orion—that created monumental buildings and refused to maintain them (Fig. 23.1). Their fate was different. A certain part of the Saturn plant was given new life by the construction of the largest shopping and entertainment center in the region (Fig. 23.2). The territory was developed. This has significantly "revitalized" the entire surrounding area.

Fig. 23.1 Brownfields of industrial facilities in Ternopil (Orion and Saturn)

Fig. 23.2 Brownfields of industrial facilities in Ternopil (Orion and Saturn)

But this example is rather an exception. We do not emphasize that the industrial facility has changed its functionality into a shopping and entertainment centre.

The central location of brownfields and their good transport accessibility positively correlate with their conversion into commercial objects and offices, and negatively correlate with their conversion into residential and public services.

23.3 Conclusion

Brownfields are an important tool for planning the territory within urban settlements. Their implementation should be well thought out and take into account all components of spatial change. Institutional measures for the restoration of neglected territories should be aimed at cultivating special institutions to promote the revitalization of brownfields in order to modernize the economy of the regions.

As a result of this research, it was found that there is a relationship between the location of a post-industrial brownfield regeneration project and the new use of the site.

The number of brownfields is significant and underreported in the Western regions of Ukraine. Traditionally, these are former enterprises that operated under a planned economy and were unable to adapt to market conditions. Although these regions are not predominantly highly urbanized, brownfields are traditionally associated with urban settlements. The attractiveness for further use of such territories is influenced by their location, the presence of functioning infrastructure, logistics, and the ability to optimally utilize available space. Another important factor driving investor interest in such spaces is the creation of a favorable investment climate.

The use of brownfields in Western regions of Ukraine is associated with two main aspects. The first is the establishment of industrial parks, which are designed to "revive" the economic life of old industrial districts (here, it is possible to draw on the experience of European countries in addressing this issue). The second option is the relocation of enterprises from those territories of Ukraine that have been occupied or were located in the front-line zone as a result of the Russian-Ukrainian war (a forced option, but it allows for the intensification of the use of old industrial sites for further use).

Several factors hinder the regeneration of brownfields. These include the excessive length of time between the cessation of initial industrial activity and the decision to redevelop the renewable sites, as well as the information asymmetry that is unfavorable to potential investors. In addition, the generally negative image of brownfields is cited as a deterrent to redevelopment. Even the terrain, irregular plot shape, and water bodies can act as deterrents to land development.

References

1. Černík J, Kunc J, Martinát S (2016) Territorial-Technical and Socio-Eeconomic aspects of successful brownfield regeneration: a case study of the liberec region (Czech Republic). Geographia Technica 11(2):22–38
2. Cotič B, Momirski LA (2020) Inventory of brownfield sites in Slovenia: towards a new methodology. PROSTOR 1(59):28166–28179
3. Jarczewski W, Koj J (2023) Spatial factors affecting the functional diversity of regenerated brownfields: the case of Silesian Voivodeship (Poland). Moravian Geograph Rep 31(2):84–94
4. Frantál B, Josef KJ, Nováková E, Klusáček P, Martinát S, Osman R (2013) Location matters! Exploring brownfields regeneration in a spatial context (a case study of the south Moravian Region, Czech Republic). Moravian Geograph Rep 21(2):5–19
5. Franz M, Güles O, Prey G (2008) Place-making and 'green' reuses of brownfields in the Ruhr. Tijdschr Econ Soc Geogr 99:316–328
6. Industrial parks in Ukraine: Tax incentives., https://tax.gov.ua/mass-media/news/638981.html; Marchyshynec OV, Marchyshynec SM (2017) Industrial parks as a tool for attracting investment in the real sector of the region's economy. Econ Soc 9:16–22
7. Perez AP, Eugenio NR (2018) Status of local soil contamination in Europe: Revision of the indicator «Progress in the management contaminated sites in Europe», EUR 29124 EN. Publications Office of the European Union. URL: https://doi.org/10.2760/093804
8. Soldak M (2021) Institutional aspect of brownfields revitalization: the case of Ukraine. J Europ Econom 20(2):315–340
9. Sytnyk N, Humeniuk V, Sych O, Hlevatska N (2022) Revitalization of rural areas of the Carpathian region in the context of EU macro-regional strategy. J Settl Spat Plan 13(1):33–44
10. LNCS Homepage., http://www.springer.com/lncs. Last accessed 2023/10/25

Chapter 24
Transforming Brownfields: Urban Renewal in Ukrainian Cities

Kostyantyn Mezentsev, Liudmyla Niemets, and Kateryna Sehida

Abstract In Ukrainian research, the term "brownfield" emerged relatively recently, about 10–15 years ago. However, brownfields in the urban spaces of Ukrainian cities began to appear massively from the late 1980s, stemming from the economic problems of the Soviet Union and its subsequent dissolution. Today, several generations of brownfield transformation have evolved, starting in different sequential periods. Yet, even today, in certain cities, brownfields located adjacent to each other may (not) regenerate using different models, correlating with different generations of their transformation. Thus, it's not a linear process of changing approaches but rather a gradual diversification. The Russian war in Ukraine significantly increased the number and spatial scale of brownfields, as well as their concentration. In addition to all existing problems, there are war-driven challenges of destruction of industrial urban spaces and relocation of industrial facilities. The aim of this study is to attempt to identify the key characteristics of four generations of brownfield transformation in Ukrainian cities. To achieve this, various brownfield regeneration projects across the country were analyzed. The authors conducted both thematic analysis and observations of such spaces in cities.

Keywords Brownfield regeneration · Transformation · Post-socialist · Ukraine

K. Mezentsev (✉)
Taras Shevchenko National University of Kyiv, Kyiv, Ukraine

V.N. Karazin Kharkiv National University University, Kharkiv, Ukraine
e-mail: mezentsev@knu.ua

L. Niemets · K. Sehida
V.N. Karazin Kharkiv National University University, Kharkiv, Ukraine
e-mail: ludmila.nemets@karazin.ua; kateryna.sehida@karazin.ua

C. Morar et al. (eds.), *Achieving Sustainability in Ukraine through Military Brownfields Redevelopment*, NATO Science for Peace and Security Series C: Environmental Security, https://doi.org/10.1007/978-94-024-2278-8_24

24.1 Introduction

Brownfield regeneration is apparently one of the youngest (yet highly important) research fields of urban studies [2]. In Ukrainian research, the term "brownfield" emerged relatively recently, about 10–15 years ago. However, brownfields in the urban spaces of Ukrainian cities began to appear massively from the late 1980s, coinciding with the economic problems of the Soviet Union and its subsequent dissolution. To some extent, the excessive industrialization of Soviet cities was artificially maintained by the Soviet authorities more from ideological considerations than economic expediency, but such necessity diminished after the dissolution. This was compounded by the economic crisis caused by the prolonged transition to market relations and deindustrialization. As Soldak's research [11] showed, almost 80% of surveyed stakeholders in over 30 cities consider deindustrialization to be the main cause of brownfield emergence. Nearly 30% placed processes of property transformation second among the factors contributing to the emergence of brownfields. Therefore, practically all cities in Ukraine have been grappling with the problem of declining and neglected industrial spaces since the 1990s.

Subsequently, the transformation of brownfields was not clearly regulated in legislation, both at the national and municipal levels. Therefore, one can speak of a certain chaos, lack of regulation, and fragmentation in brownfield revitalization efforts. At first glance, it may be assumed that over time there was a gradual transition from urban "strategies" of doing nothing with brownfields to the emergence of conceptual approaches of a new generation, implemented under slogans such as "city for people," "city comfortable for all," and so on. However, the analysis of various brownfield regeneration projects in Ukrainian cities suggests that such changes are non-linear. In fact, one can speak of several generations of brownfield transformation, which began in different consecutive periods. Yet even today, in certain cities, brownfields located adjacent to each other may (not) regenerate using different models, which can be associated with different generations of their transformation. Thus, it is not a linear process of changing approaches but rather a gradual diversification of them.

Therefore, the aim of this research is to attempt to identify the key characteristics of four generations of brownfield transformations in Ukrainian cities. To achieve this, various brownfield regeneration projects across the country were analyzed. The authors conducted both thematic analysis and observations of such spaces in cities.

24.2 Brownfield Regeneration: Theoretical Frameworks

Brownfields in cities can be seen as both an expected and unexpected outcome of industrial development. The expectation is linked to corresponding social and economic transformations that have led, one way or another, to the closure of industrial enterprises. At the same time, the unexpectedness is associated with the fact that in

Ukrainian cities, these areas have remained unchanged for a long time, continuing to be used for industrial purposes. However, over a relatively short period, the appearance of brownfields has become quite common and has been accompanied by the deprivation of nearby residential areas.

Brownfields are mainly located in fully or partly developed urban areas that cannot be suitably and effectively utilized without regeneration, requiring intervention to bring them back to beneficial use. This is complicated by the problem of real or perceived contamination on one hand and the desire to preserve industrial buildings and areas as cultural heritage sites on the other. Brownfields significantly vary in size, location, condition, duration of dereliction, market value, and extent of contamination [5]. Accordingly, their regeneration may require the development of projects of different scales and content. Moreover, the local context matters. Understanding how residents perceive brownfields is important. Numerous communications by the authors with residents of traditionally industrial Ukrainian cities indicate that some perceive brownfields as a given that does not require intervention ("let it be"), while others insist on their revival solely as industrial territories (reindustrialization, not necessarily advanced or high-tech, but the same as they were).

Special interest is attributed to military brownfields, characterized by specific features [6, 7]:

- the functional diversity (heterogeneity)—administrative and residential buildings, warehouses, garages, etc.;
- a highly individual matter for each municipality;
- a non-existence of functional communities and a lack of interest in public issues.

The study of brownfield transformations can be approached from various perspectives. Some emphasize the brownfield as a part of urban space, the use of which should yield maximum economic benefits. Accordingly, the focus is on the efficiency of investment and changes or restoration of functions. However, when considering brownfields from the perspective of accommodating the interests of all stakeholders, the emphasis shifts to the effectiveness of negotiations between them, seeking compromises, and community involvement. Similarly, the focus varies when viewing brownfield regeneration through the lens of social relations between local residents of adjacent areas and newcomers in terms of newly built gentrification, or through the lens of land contamination, assessing relevant risks, and effective remediation methods.

Particular emphasis is placed on sustainable brownfield regeneration, the so-called "marrying of the brownfield and sustainability concepts" [3], with a shift in focus from increasing land value to the protection of public health and safety. However, as noted by Bagaeen [1], there is no general agreement on how sustainable development as a concept can be translated into the practice of brownfield regeneration.

Doak and Karadimitriou [4] identified 5 themes of the brownfield discourses:

- *Market-generated brownfield development: the neoliberal discourse* (economic utility, private sector involvement, housing, etc.);

- *Negotiated (or equitable) brownfield development: the participatory discourse* (community involvement, empowerment, active citizenship, etc.);
- *Organisationally efficient brownfield development: the managerial discourse* (good practice, efficiency, effectiveness, technical guidance, etc.);
- *Technically efficient brownfield development: the scientific discourse* (risk assessment, remediation techniques, land use, etc.);
- *Sustainable brownfield development: the sustainability discourse* (environmental efficiency, quality of life, social equity, etc.).

Another important aspect of studying brownfield transformations is their differences in capitalist and post-socialist cities. Sometimes these transformations are even juxtaposed. However, the most significant difference between capitalist and post-socialist countries regarding brownfield regeneration is a certain time delay [10]. While in capitalist countries, brownfield regeneration began to receive significant attention as early as the 1970s, in other cities, they emerged after the collapse of socialism and were maintained until the early 1990s. Similarly, while brownfield regeneration became a relevant social topic in capitalist cities in the 1990s, in post-socialist cities, it became a publicly discussed topic after 2010. This implies a two-decade delay, allowing for the consideration of paths already taken and successful experiences, but also implying a phase of "maturation." Although the experiences of many cities show the effectiveness of public-private partnership models and the implementation of not purely commercial projects, in many post-socialist cities, commercial interests prevail in regeneration projects.

The factors that determine the specificity of brownfield regeneration are multilevel. Hercik et al. [6] discuss three levels: general factors (national or regional development policies; spatial planning strategies; economic instruments; regeneration management), location factors (location, transport links, socio-demographic structure, community involvement, etc.), and site-specific factors (size, previous use, soil quality and extent of contamination, available infrastructure, ownership relations, etc.).

Finally, we can observe a variety of approaches to brownfield regeneration, regardless of the type of city and the timeframe of regeneration—including do-nothing, demolishing, conservation, remediation, reuse, reconstruction, renovation, urban healing, and so on.

Thus, to identify the generations of brownfield transformation in Ukrainian cities, which will be discussed in the next section, we took into account the principles of considering the diversity of brownfields, the diversity of discourses, the diversity of approaches, and the specificities of Ukrainian cities as post-socialist with a two-decade delay, sometimes compressed catching up with cities that have already gone through this path, and sometimes attempts to find their own "recipes".

24.3 Results: The Four Generations of Brownfield Regeneration in Ukraine

With the emergence of the first brownfields in Ukrainian cities, the main approach was "do-nothing". A significant number of other, primarily social problems, overshadowed the issue of urban space regeneration. Industrial enterprises closed down, significant industrial and warehouse premises and lands were abandoned. This effectively transformed former successful industrial territories into "rusty fields". The question of regeneration was not on the agenda of either the municipal authorities or in the focus of national strategies.

Despite the fact that over time, initially in individual cities, and then with much broader coverage, strategies, visions, and concrete revitalization programs for brownfields began to be developed, until now many former industrial areas remain abandoned, unused, and neglected. This is especially true for large industrial cities, where development is driven by the most competitive industries, but there is no need for new spaces for development. This is primarily due to the fact that population decline does not generate demand for new housing construction, or there are available plots in cities that require significantly lower capital investment for the development of new functions compared to brownfields (especially with a high level of soil and water pollution, etc.). Therefore, such a (non)strategy of brownfield transformation has proven to be quite persistent (Fig. 24.1).

a b

Fig. 24.1 Do-nothing approach: Kryvyi Rih (**a**) and Stebnyk (**b**). (Photo: K. Mezentsev, 2017 (**a**) and 2015 (**b**))

In the late 1990s to the early 2000s, cities began to undergo changes associated with the need for the development of new retail, entertainment, and office infrastructure. Brownfields, predominantly located in central and midtown areas of cities, became the most attractive for such developments. However, these were not urban strategies for brownfield regeneration but rather spontaneous and fragmented "take-overs" of individual industrial premises, or plots of industrial land by various new owners. They either placed new establishments and office centers in former modernized (to varying degrees) manufacture premises or demolished them and built new ones. Sometimes, it was a mix of demolition, reconstruction, and new construction.

A vivid example is the territory of the former "Bilshovyk" plant in Kyiv, founded in the late nineteenth century. Certain plots of the plant territory were transferred to other owners. Retail and entertainment complex was opened on them, as well as various small industries, offices, and food establishments. The rest of the territory continued to partially function as a factory (Fig. 24.2).

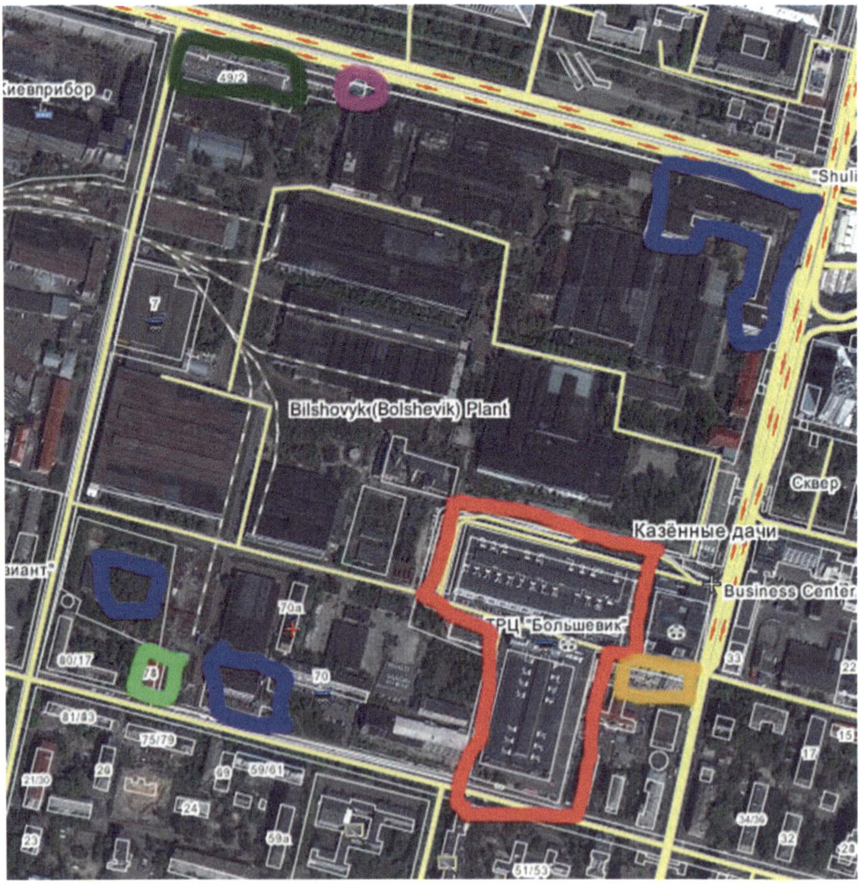

Fig. 24.2 Spontaneous fragmented commercial (non-transparent) brownfield conversion—From Bilshovyk plant to Cosmo Multimall, Kyiv

Another function within this approach to brownfield regeneration was residential. There was also a fairly chaotic and fragmented construction of new residential complexes or individual buildings on the territory of former industrial enterprises.

From 2010 onwards, spontaneous and fragmented brownfield regeneration was replaced by conceptual approaches. Its essence lies in promoting new concepts for former industrial spaces. Typically, these concepts are attractive and vibrant, promoting the idea of renewal and revitalization of neglected urban areas, relying on slogans of urban renewal, transforming it into a comfortable, modern, socially-oriented space.

For instance, on the territory of the former Darnytsky Silk Factory in Kyiv, the concept of creating the Platform Art Factory (Art-zavod Platforma) was proposed. Another example is the ongoing revitalization of the territory of the former crane factory in Lviv by transforming it into a "creative space" (offices, sports and food areas, public spaces, etc.) (Figs. 24.3 and 24.4).

An alternative to creating creative, art, and office spaces is the implementation of concepts for entirely new residential estates. One of the most vivid examples, both literally and figuratively, is the Comfort Town residential complex built on the territory of the former Vulcan chemical plant in Kyiv. However, this new conceptual space has essentially turned into a colorful "island," surrounded by old gray

Fig. 24.3 Conceptual brownfield regeneration: Art-zavod Platforma as creative cluster in Kyiv. (Photo: https://artzavodplatforma.com (**a**) and K.Mezentsev, 2018 (**b**))

Fig. 24.4 Conceptual brownfield regeneration: KIVSH as creative space in Lviv. (Photo: K. Mezentsev, 2023)

a b

Fig. 24.5 Conceptual brownfield regeneration: Comfort Town housing estate in Kyiv. (Photo: https://comforttown.com.ua (**a**) and V. Dudin, 2018 (**b**))

Soviet-era buildings, with a striking lack of contact and understanding between the new-build gentrifiers and their socialist city neighbors, and raising the open question of "brownfield blues" regarding the inherited potential environmental hazards [9] (Fig. 24.5).

The third generation of brownfield transformation is associated with the conversion of former military territories. A significant portion of these areas remained abandoned after the closure of military bases, undergoing minimal changes and adhering to a "do-nothing" approach. However, since the late 2010s, many cities have gradually demolished these areas and replaced them with extensive residential developments. This phenomenon is not unique to post-socialist countries. Several studies in Central European cities have shown a similar trend, where the conversion of military bases has facilitated significant new development, particularly in housing. One of the most common uses for revitalized military brownfields is their residential function, which is often positively evaluated by the public (Fig. 24.6).

The Russian war in Ukraine has resulted in significant levels of direct and indirect urbicide [8]. In particular, cities have seen the emergence of vast areas of new "brownfields," which have become "brown" not due to long-term neglect and underinvestment, but as a result of bombardments and missile strikes. Additionally, due to negligent management and a lack of interest in urban development, the occupying authorities after 2014 have been creating new brownfields in the occupied cities.

Such territories will require the development of new approaches to (post)-war strategic mass (instantly emerged) brownfield regeneration (Fig. 24.7).

Thus, four generations of brownfield regeneration can be identified (see Table 24.1), the change of which is not a linear process. They can coexist in the same cities or appear in reverse or random sequence.

Fig. 24.6 Military brownfield residentialisation: 'Great Britain' housing estate in Lviv. (Photo: K. Mezentsev, 2023)

Fig. 24.7 (Post)-war strategical mass (instantly emerged) brownfield regeneration: War-caused brownfield in Kramatorsk. (Photo: D. Rashevska, 2024)

Table 24.1 The four generations of brownfield regeneration in Ukraine

N	Type	Key approach	New functions	Period
	All	Do-nothing	–	From the early 1990s
I	Industrial brownfields	Spontaneous fragmented commercial (non-transparent) brownfield conversion	Malls, offices	From the late 1990s—Early 2000s
II	Industrial brownfields	Conceptual brownfield regeneration	Art-, creative, coworking, and food spaces (IIa) Housing projects (IIb)	From the 2010s
III	Military brownfields	Brownfield residentialisation	New housing estates	From the late 2010s
IV	War-caused brownfields	(Post)-war strategical mass (instantly emerged) brownfield regeneration	?	After Feb. 2022

24.4 Conclusions. War and Brownfields—With the Future in Mind

The overlap of different generations of brownfield transformations is not a unique phenomenon in Ukrainian cities. However, the Russian war in Ukraine has significantly increased the number and spatial extent of brownfields, as well as their concentration. In addition to all existing problems, there are challenges posed by the mass destruction of industrial areas in cities and the forced relocation of industrial facilities. Therefore, Ukrainian cities are faced with the task of searching for effective post-war strategies for all types of old and new brownfields—pre-war brownfields, destroyed (in directly urbicided cities), and temporarily occupied (in indirectly urbicided cities) industrial areas.

We can reasonably expect a "window of opportunity" associated with the influx of post-war investments and intentions to implement flagship projects. However, new challenges will arise in finding ways to prevent non-military (new-build) urbicide (due to the need for rapid reconstruction not only of housing but also of industrial facilities), to prevent excessive post-war touristification of ruined industrial sites that have become symbols of Ukrainian resistance to Russian aggression (such as Azovstal or Avdiivka coke-chemical plant), and to address the issue of numerous 'places of pain and shame' and pride at the same time.

References

1. Bagaeen SG (2006) Brownfield sites as building blocks for sustainable urban environments: a view on international experience in redeveloping former military sites. Urban Des Int 11:117–128
2. Berki M (2012) Post-1990 urban brownfield regeneration in Central and Eastern Europe: a theoretical concept. In: Csapó T, Balogh A (eds) Development of the Settlement Network in the Central European Countriesz, pp 137–144
3. Dixon T, Raco M, Catney P, Lerner DN (2007) Sustainable brownfield regeneration. Blackwell Publishing, p 378
4. Doak J, Karadimitriou N (2007) (Re)development, complexity and networks: a framework for research. Urban Stud 44(2):209–229
5. Frantál B, Kunc J, Nováková E, Klusáček P, Martinát S, Osman R (2013) Location matters! Exploring brownfields regeneration in a spatial context (case study of the South Moravian Region, Czech Republic). Moravian Geograph Rep 21(2):5–19
6. Hercik J, Šimáček P, Szczyrba Z, Smolová I (2014) Military brownfields in The Czech Republic and the potential for their revitalisation, focused on their residential function. Quaestiones Geographicae 33(2):127–138
7. Klusáček P, Krejčí T, Martinát S, Kunc J, Osman R, Frantál B (2013) Regeneration of agricultural brownfields in The Czech Republic—case study of the south Moravian region. Acta Universitatis Agriculturae et Silviculturae Mendelianae Brunensis 61(2):549–561
8. Mezentsev K, Mezentsev O (2022) War and the city: lessons from urbicide in Ukraine. Czasopismo Geograficzne 93(3):495–521
9. Mezentsev K, Gentile M, Mezentseva N, Stebletska I (2019) An Island of civilization in a sea of delay? Indifference and fragmentation along the rugged shorelines of Kiev's newbuild archipelago. J Urban Aff 41(5):654–678
10. Osman R, Frantal B, Klusáček P, Kunc J, Martinat S (2015) Factors affecting brownfield regeneration in post-socialist space: the case of The Czech Republic. Land Use Policy 48:309–316
11. Soldak M (2021) The institutional aspect of the revitalization of brownfields: the case of Ukraine. J Eur Econ 20(2(77)):315–340. (In Ukrainian)

Chapter 25
Involvement of Military Brownfields for Tourism: The Case of Ukraine and Turkey

M. Tahsin Sahin, Ievgeniia Telebienieva, Liudmyla Kliuchko, Alexandru Tiba, Jelena Golijanin, and Pavlo Kobylin

Abstract In recent years, the reuse and restoration of brownfield sites and their contribution to the economy of the region/country have become topical issues. Especially former industrial and war sites are among the most important representatives of brownfields. Especially old battlefields remain idle and abandoned to their fate. Renovating these old battlefields and bringing them into tourism is valuable in terms of reusing brownfields. This reconstruction contributes to the economy of the region and the country and is also important in terms of environmental sustainability. The aim of this study is to create a detailed land use map using Geographic Information Systems in the process of changing brownfields in Turkey and Ukraine into historical national parks, to introduce the use of historical national parks and to contribute to the process of protecting and managing historical heritage. For this purpose, potential former battlefields were identified and their potential for tourism was evaluated. As a result, Turkey has declared 3 important old battlefield regions as national parks in the past and brought them to tourism. In this context, Ukraine not only has brought certain regions to tourism, but also has potential regions to be brought to tourism.

Keywords Ukraine · Turkey · Military brownfields · Tourism

M. T. Sahin (✉)
Akdeniz Üniversitesi, Antalya, Turkey
e-mail: tahsinsahin@akdeniz.edu.tr

I. Telebienieva · L. Kliuchko · P. Kobylin
V.N.Karazin Kharkiv National University, Kharkiv, Ukraine
e-mail: eugene.telebeneva@karazin.ua; ludmila.klychko@karazin.ua; pavlo.kobylin@karazin.ua

A. Tiba
University of Oradea, Oradea, Romania
e-mail: alexandrutiba@gmail.com

J. Golijanin (✉)
University of East Sarajevo, Republika Srpska, Bosnia and Herzegovina
e-mail: jelena.golijanin@ff.ues.rs.ba

C. Morar et al. (eds.), *Achieving Sustainability in Ukraine through Military Brownfields Redevelopment*, NATO Science for Peace and Security Series C: Environmental Security, https://doi.org/10.1007/978-94-024-2278-8_25

25.1 Introduction

The concept of brownfields represents a global issue related to the reuse of abandoned and contaminated sites [1]. Such sites are often created by the cessation of industrial activities and can cause environmental contamination [2]. This is associated with economic, social and environmental risks, and cleaning up, redeveloping and putting these sites to beneficial use is a major challenge for policy makers [3]. Across Europe, the presence of derelict and contaminated land is a regional concern [4] and these sites often remain idle, despite their potential to improve environmental quality and stimulate the local economy.

There are many definitions and interpretations of "brownfields" in the European context. One of the most common definitions is as follows: "Brownfields" are areas affected by previous uses of the site and the surrounding land, derelict and underutilized, which may have real or perceived contamination problems, mostly developed in urban areas, and require intervention to "return them to beneficial use". For example, areas left behind after war. In particular, military brownfields deserve special attention in terms of the environmental and economic impacts of these sites. The presence of derelict and contaminated land across Europe represents a regional concern [5]. Brownfield sites present particular challenges to policymakers in terms of returning land to beneficial use and environmental quality, requiring cleanup of contaminated soil and groundwater [6]. The redevelopment of such sites is an important part of a sustainable urban regeneration and can provide several advantages [7]. Military brownfields represent a specific category of brownfields and are common in many countries across Europe. The socio-economic changes in Central and Eastern Europe after the collapse of the Soviet Union have left many military sites abandoned and derelict [8]. While this situation brings economic and environmental challenges, it also offers opportunities for redevelopment by unlocking the potential of these sites [9]. In particular, former military sites in Oradea, Romania, are an example of this transformation and have the potential to contribute to the local economy, environmental quality and social well-being [10, 11]. Following the collapse of the USSR, Central and Eastern Europe experienced large-scale socio-economic changes [12]. This process resulted in a reduction of military activities and the abandonment of military installations [13]. This led to the emergence of military brownfields. Military brownfields include military bases, arsenals, ammunition depots, fuel depots and training areas [14]. While these sites cause problems such as environmental pollution and economic waste, they also present a potential redevelopment opportunity [15]. The former military sites in Oradea, Romania, provide an example of this transformation [16]. Affected by the disarmament process in the late twentieth and early twenty-first century, these sites can be transformed into important resources with potential contributions to the local economy and environment. Therefore, the sustainable redevelopment of military brownfields offers an important opportunity for regional development and environmental improvements [17]. In this transformation process, the redevelopment of former military sites can contribute to regional development by improving environmental quality and increasing social benefits, while providing new opportunities for the local economy [18]. The former military sites in Oradea, Romania, represent an example of this transformation process. A mix of late nineteenth and early twentieth century military facilities, they are strategically located in or around the city center [19]. However, with the decline of military activity, these areas were vacated and fell into disuse. Through redevelopment, they can be transformed into new economic activities and urban green spaces, thus contributing to the sustainable development of the city [20]. The transformation of former military sites in Oradea is helping the region to take a step towards a sustainable future [21]. This transformation project has brought Oradea not only local attention, but also national and international attention. Successful redevelopment in the region can serve as a model for similar regeneration projects in other cities and emphasize the importance of such projects [17]. The experience in Oradea can serve as an inspiration for other regions in similar situations, showing that sustainable redevelopment of former military sites is possible [22]. This project serves as a model for future urban regeneration projects not only in Oradea, but also in other cities in the region [18]. Separate issues military brownfields and tourism were revealed during the research of theory and practice of brownfields and greyfields of the city [23], the study of the heritage, culture and regeneration of the former military areas in a city [24], integral assessment of ethnic tourism in Ukraine [25], tourism development in Ukraine due to the coronavirus Covid-19 pandemic [26], features of pilgrimage tourism in Ukraine [27], rural tourism in Ukraine [28].

The association of brownfields with tourism and the revitalization and tourism development of these areas has been an important research topic. Abandoned and contaminated brownfields are associated with economic, social and environmental problems, but have great potential for the tourism industry [29]. These areas consist of a variety of sources such as contaminated factory sites, abandoned mining sites, and military installations that have emerged after the end of industrial activities [30]. Brownfields, especially in urban areas, can be brought into tourism as part of urban regeneration projects and contribute to sustainable urban tourism development [31]. However, the development of policies and strategies for the tourism development of such sites faces several challenges. These challenges include cleaning up contaminated areas, making them suitable for tourism use, and increasing public trust in these areas [16]. Therefore, it is crucial to identify appropriate policies and strategies for sustainable urban development in relation to the use of brownfield sites for tourism. In addition, measures should be taken to increase public trust in brownfield sites and improve environmental quality. In this way, brownfield sites can be reused for economic, social and environmental benefits. Bringing brownfields into tourism can bring significant benefits to the country in many ways [32]. First of all, opening these areas to tourism can create new job opportunities and revitalize the local economy. In addition, tourism activities can also help preserve and protect the cultural and historical heritage of the region [33]. In addition, by opening brownfield sites to tourism, environmental quality can be improved and green areas can be increased. In this way, the quality of life of the inhabitants of the region will improve, while contributing to the protection of the natural environment and sustainable development goals [13].

As can be seen, many projects related to brownfields are being implemented. In this context, especially bringing brownfields to tourism is one of the most important of these projects. In this context, this research aims to analyze the situation of bringing historical military sites in Turkey and Ukraine into tourism.

The aim of this study is to create a detailed land use map using Geographic Information Systems in the process of changing brownfields in Turkey into historical national parks, to introduce the use of historical national parks and to contribute to the process of protecting and managing historical heritage. Next, the findings will be discussed and finally the evaluation will be made.

25.2 Methods

The research was designed as a case and identification study. Firstly, historically important places in Turkey and Ukraine where brownfields were changed into historical national parks were identified and demarcated on Google Earth and these areas were digitized as polygons and integrated into ArcGIS 10.8.1, a Geographic Information Systems software. Using ArcGIS, these polygons were integrated into the GIS database and a detailed digital map of the park was created, including the general view of the historical national park. Then, using the Classification tool on this map, three different categories were identified as Brown Areas, Forested Areas and Transition Areas and pixel values of these areas were determined by taking samples for each category. This is an important step to identify the different ecosystems and uses of the historical national park. In the next step, Interactive Supervised Classification analysis was applied to obtain the land use map of the historical national park. In addition, the areas covered by the classes in the category within the historical national park are shown on the map in ha. In this analysis, the locations of the brown areas within the park were determined by using the identified examples and classification algorithms, and which historical elements are located in which parts of the park were revealed in detail. Finally, the land use map and the locations of the historical elements were integrated into the location map and the locations of the monuments, landmarks, recreation areas, etc. within the park were visually marked. These markings were used to indicate important points within the park.

25.3 Military Brownfields in Turkey

First of all, 3 military historical national parks in Turkey were identified. These areas have an important value in the history of Turkey. 3 historical areas are important for the establishment of the Republic of Turkey and the struggle for independence. On the other hand, these areas are important for current and future generations to remember and remember the past periods. Map 25.1 shows the analyzed regions.

The national park area is the place where the Great Offensive and the Commander-in-Chief battles took place (Map 25.2). This area, which consists of two parts, Afyonkarahisar and Dumlupınar, is the center of the Great Offensive launched by the Turkish army against the Greeks and the Afyon Front, the most important front of the War of Independence. The Great Offensive, in which the Turkish army attacked for the first time, marked the beginning of a great victory for the Turkish nation. It is known that the Greek army left the battlefield on August 31 after the clashes between August 26 and August 30. This historical site has been protected as a national park in order not to be forgotten and passed on to future generations. The park includes a total of 10 monuments and martyrdoms in Afyonkarahisar and Dumlupınar, 2 museums and 1 promotional park.

Opened to visitors in 993, the Commander-in-Chief Historical National Park stands out as one of Turkey's most important historical and natural sites. The park is located in the area where the Battle of the Commander-in-Chief, one of the turning points of the Turkish War of Independence, was fought under the leadership of Mustafa Kemal Atatürk, the founder of the Republic of Turkey (August 26–30, 1922). It contains Atatürk's headquarters, battle positions, trenches and other historical remains. In addition to this historical richness, the park is also rich in flora and fauna. The Bashkomutan Historical National Park offers both a rich historical

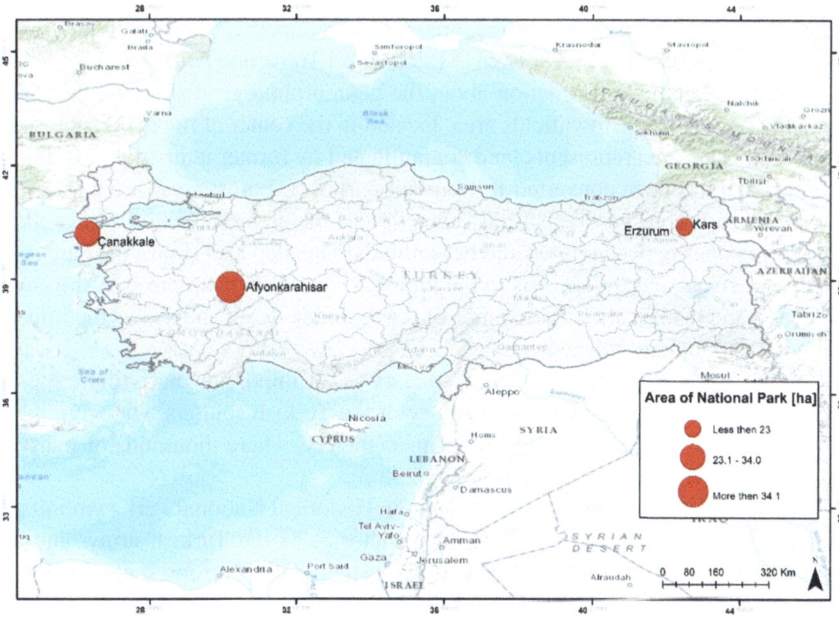

Map 25.1 Turkey Military brownfields. **Brownfield of Başkomutan History National Park**

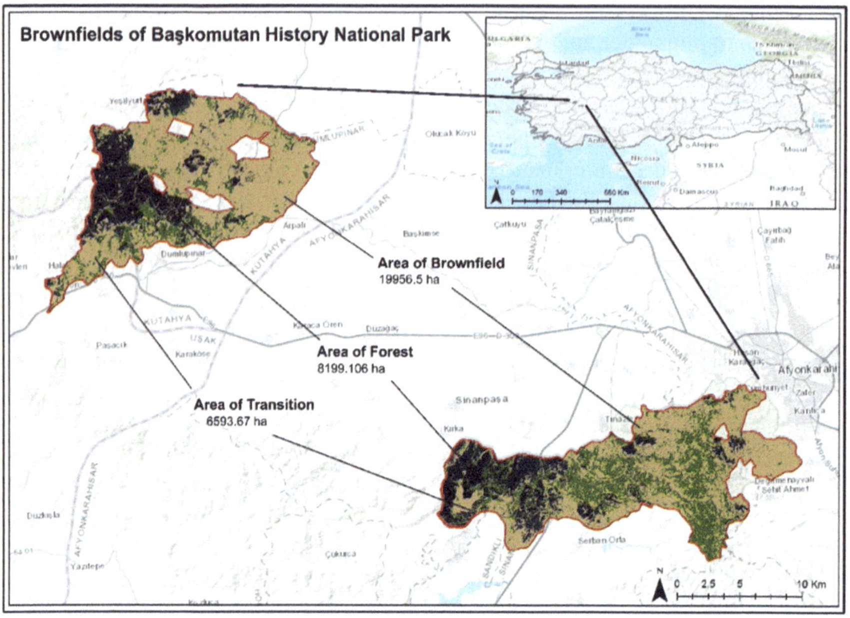

Map 25.2 Brownfield of Başkomutan History National Park

and natural experience, providing visitors with a unique perspective of Turkey's national history and natural beauty. Map 25.1 shows the Bashkomutan Historical National Park and the surrounding areas. The mapped areas are divided into Brownfields (19,956 ha), Forest Area (8199 ha) and Transition Area (6593 ha). Map 25.2 provides detailed information about the geomorphological structure and land use in the region. The Brownfields area, located in the center of the park, is marked in dark red. This area represents land contaminated by former industrial or military activities and has been converted for tourism purposes. Forest areas are marked in dark green and represent the natural vegetation of the area. Transition areas represent buffer zones between forest and brownfields. The map also shows the names of mountains, rivers and settlements in the region. Overall, the map presents the environmental and historical characteristics of the region in detail and shows the impact of natural and human-induced changes (Fig. 25.1).

Colonel Reşat Çiğiltepe Martyrs Cemetery is an important martyrdom site in Sarıkamış. This place was built in memory of the Turkish soldiers who were martyred during the Sarıkamış Operation. This cemetery, where thousands of martyrs lie, is flooded with visitors every year (Fig. 25.2).

The Victory Monument in the Başkomutan Historical National Park symbolizes the Great Offensive, one of the historical victories of the Turkish army, and an important period of the War of Independence. The monument represents the determination and heroism of the Turkish nation in the struggle for independence.

Fig. 25.1 Colonel Resat Cigiltepe Martyrs Cemetery

Fig. 25.2 Victory monument in Bashkomutan Historical National Park

Thousands of visitors visit this monument every year to honor the memory of the heroes of the national struggle.

The Martyr Sancaktar Mehmetçik Monument in the Başkomutan Historical National Park was built to commemorate the heroic Mehmetçiks who were martyred in the Sarıkamış Operation during the Turkish nation's struggle for independence. This monument is open to visitors in order to keep alive the cherished memory of the soldiers who lost their lives in the Sarıkamış Operation and to convey their sacrifices to future generations. The monument is visited by thousands of people every year and a moment of silence is observed in memory of the fallen martyrs (Fig. 25.3).

Tinaztepe Recreation Area in Başkomutan Historical National Park is an important recreation area that offers visitors the opportunity to spend time in touch with nature. This area offers visitors the opportunity to rest and picnic by combining historical and natural beauties. Surrounded by lush pine forests, Tinaztepe Recreation Area allows visitors to spend time in a peaceful atmosphere.

The Kocatepe Monument in the Başkomutan Historical National Park stands out as an important monument symbolizing the determination and perseverance of the Turkish nation in the struggle for independence. This monument represents the Great Offensive, one of the critical periods of the War of Independence, and the victory. This monument, dedicated to the ideals of independence and freedom of the Turkish nation, is visited by thousands of visitors every year.

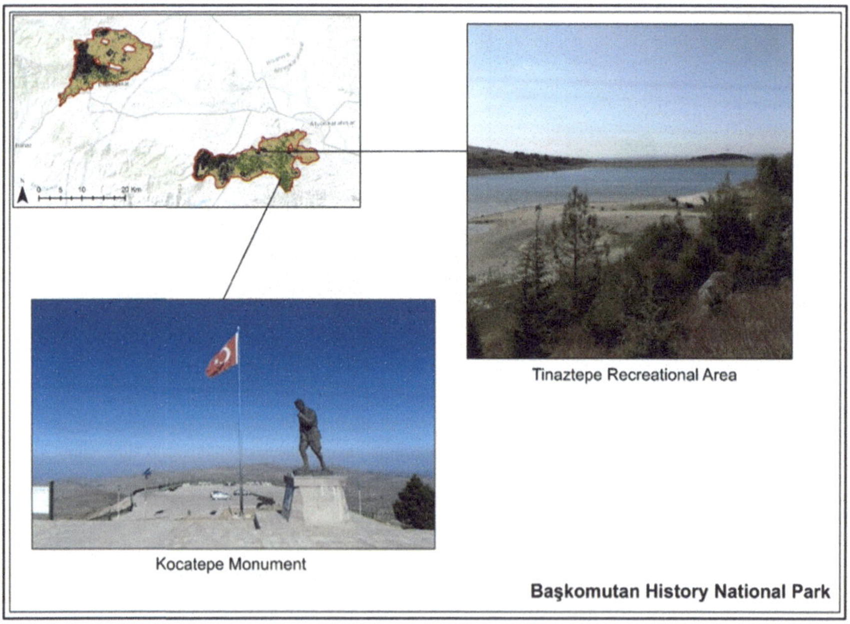

Fig. 25.3 Tinaztepe Rest Area

25.3.1 *Brownfield of Gelibolu Peninsula History National Park*

The Gallipoli Peninsula is one of the most visited battlefields in Turkey (Map 25.3). This national park is of great importance not only for Turks but also for many countries such as Australia and New Zealand. Because the Battle of Gallipoli symbolizes the Turkish nation's great struggle against other powerful states in difficult times. The number of visitors who come here is a sign of gratitude to the heroes who protected their nation. Declared a national park in 1973, the area includes 56 local monuments and martyrdoms and 35 foreign monuments and cemeteries.

The Gallipoli Peninsula Historical National Park is one of Turkey's important historical and natural heritages. The Gallipoli Peninsula was the center of the Gallipoli Front in 1915 during World War I and has a great historical significance. The National Park includes battlefields, martyrdoms, monuments, tunnels and other historical remains in this historical area. The park is also remarkable for its natural beauty, providing an important ecosystem with its rich biodiversity, endemic plant and animal species. The Gallipoli Peninsula Historical National Park has an important role in the preservation of historical heritage and sustainability of wildlife, offering visitors both a historical and natural experience. Map 25.2 shows the Gallipoli Peninsula Historical National Park and its surroundings. The mapped areas are classified as Brownfield (9789 ha), Transition Area (11,734 ha) and Forest Area (12,418 ha). The Brownfield area is marked in dark red and is located in the center of the peninsula. This area represents land affected by past conflicts and

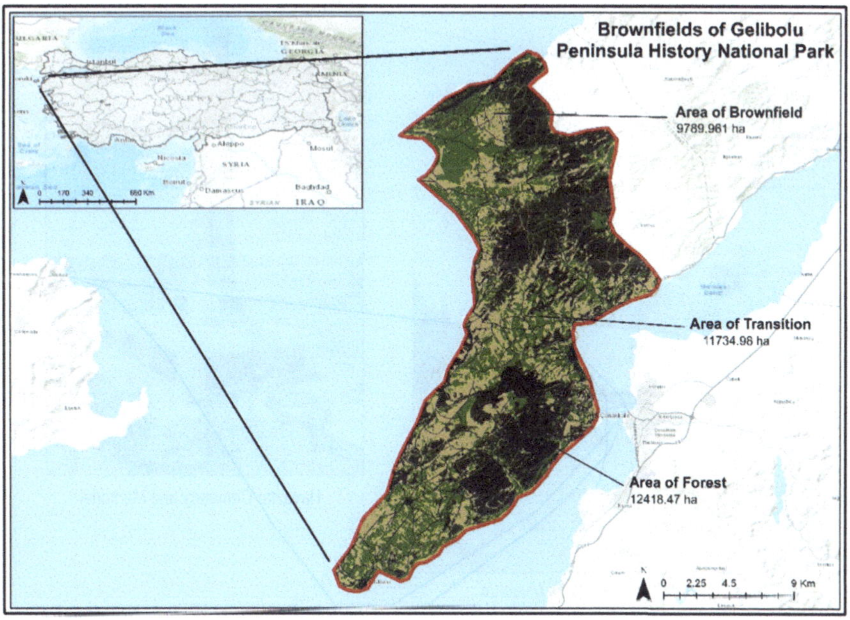

Map 25.3 Brownfield of Gelibolu Peninsula History National Park

military activities. These areas have been restored and restored for tourism in order to preserve the historical heritage of the region and increase its tourism potential. The Transition Area serves as a buffer zone between the Brownfield area and the Forest Area. The Forest Area is marked in dark green and constitutes the natural vegetation of the peninsula. The map also shows the names of the bays, coastlines and settlements in the region. Overall, the map provides detailed information about the geomorphological structure, land use and historical significance of the region. Map 25.3 contributes to the protection of the natural and cultural heritage of the region (Fig. 25.4).

The Gallipoli Peninsula is one of the most visited battlefields in Turkey (Map 25.3). This national park is of great importance not only for Turks but also for many countries such as Australia and New Zealand. Because the Battle of Gallipoli symbolizes the Turkish nation's great struggle against other powerful states in difficult times.

The number of visitors who come here is a sign of gratitude to the heroes who protected their nation. Declared a national park in 1973, the area includes 56 local monuments and martyrdoms and 35 foreign monuments and cemeteries.

Çanakkale Martyrs Hospital Museum is a historical hospital building where wounded soldiers were treated and martyred soldiers were sent to their last journey

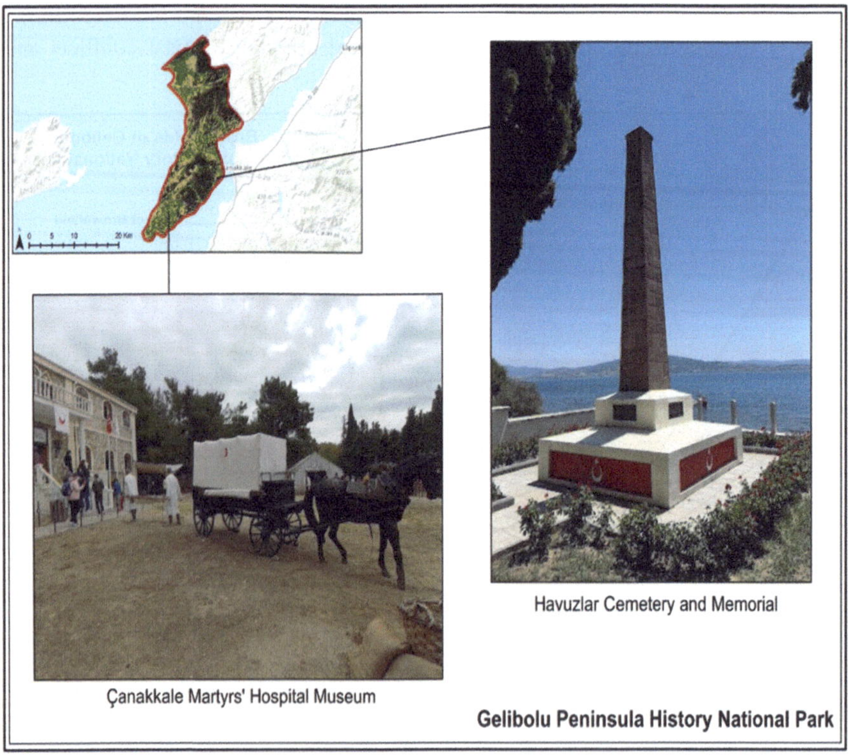

Fig. 25.4 Çanakkale Martyrs Hospital Museum

during the Çanakkale War. The museum exhibits medical equipment, photographs and documents from the war period. Visitors have the opportunity to get to know the medical and human dimension of the Çanakkale War more closely by visiting here.

Havuzlar Martyrdom and Memorial is an important memorial and martyrdom site built in memory of the soldiers who lost their lives during the Battle of Çanakkale. It is located in the Havuzlar region, which was one of the most violent points of the war. By visiting here, visitors can feel more closely what happened in the Gallipoli War and come together with respect to the memory of the martyrs. gelebilirler (Fig. 25.5).

Akbaş Martyrdom is an important martyrdom site where Turkish soldiers who lost their lives during the Gallipoli War were buried. This martyrdom is located in one of the most intense areas of the war and is home to the graves of thousands of Turkish soldiers. By visiting here, visitors can vividly feel the painful memories of the Gallipoli War and come together in respect for the cherished souls of the martyrs.

The Gallipoli Martyrs Memorial is an important memorial and martyrdom site commemorating the Turkish and Anzac soldiers who lost their lives in the Battle of Gallipoli. This memorial is located on the Gallipoli Front, one of the most difficult points of the war and reminds visitors of the bitter memories of the war. Thousands of people visit the site every year to honor the memory of the fallen (Fig. 25.6).

Seddülbahir Martyrdom is an important martyrdom area where Turkish and Anzac soldiers who lost their lives in the Gallipoli War were buried. This martyrdom is located in the Seddülbahir region and is one of the most intense areas of the

Fig. 25.5 Akbas Martyrdom

Fig. 25.6 Seddülbahir Martyrdom

war. By visiting here, visitors can experience the painful memories of the Gallipoli War and come together with respect for the sainted souls of the martyrs.

Yahya Çavuş Martyrdom and Monument is an important martyrdom and memorial site built in memory of the Turkish soldiers who lost their lives in the Gallipoli War. This martyrdom and memorial is located at one of the most intense points of the war and reminds visitors of the painful memories of the war. Thousands of people visit the site every year to pay their respects to the cherished memory of the martyrs.

25.3.2 Brownfield of Sarıkamış National Park

Sarıkamış National Park is located in Sarıkamış district, one of Turkey's natural and historical riches. The National Park is located in the region where the Sarıkamış Operation took place and this operation during World War I is an important historical event. The park was established in memory of the Turkish soldiers who lost their lives in the Sarıkamış Operation.

The park is also notable for its rich natural beauty; the forested areas in the region harbor an important biodiversity with endemic plant and animal species. Sarıkamış National Park has an important role in the preservation of historical heritage and sustainability of wildlife and offers visitors both a historical and natural experience. Map 25.3 provides a detailed view of Sarıkamış National Park and its surrounding areas. The mapped areas are classified under three different categories: Brownfield (408,705 ha), Forest Area (1032.02 ha) and Transition Area (172.94 ha). The area marked as Brownfield is highlighted in dark red and located in the center of the park. This area represents land that has been affected and destroyed as a result of past conflicts and military activities. In order to preserve the historical heritage of the area and to increase its tourism potential, the Brownfields have been restored and opened to tourism. The Forest Area is represented by complex patterns of shades of green and covers an area of 1032.02 hectares. This area constitutes the natural vegetation of Sarıkamış National Park. The Transition Area covers 172.94 hectares and serves as a buffer zone between the Brownfield and Forest Areas. This zone provides a transition between two different land use types. The map also shows the names and altitudes of the mountains in the region. Overall, this map provides detailed information about the geomorphological structure, land use and historical significance of Sarıkamış National Park and aims to contribute to the conservation of the natural and cultural heritage of the region.

It is a destination located in the Eastern Anatolia Region, between Erzurum and Kars provinces (Map 25.4). It is known as the place where 60 thousand soldiers froze to death according to some sources and 90 thousand soldiers according to others. It is claimed that Enver Pasha was solely responsible for the martyrdom of 90 thousand soldiers by directing the Turkish army against Russian troops in Sarıkamış. In Sarıkamış, the "Martyrdom Monument" was built in memory of the Turkish

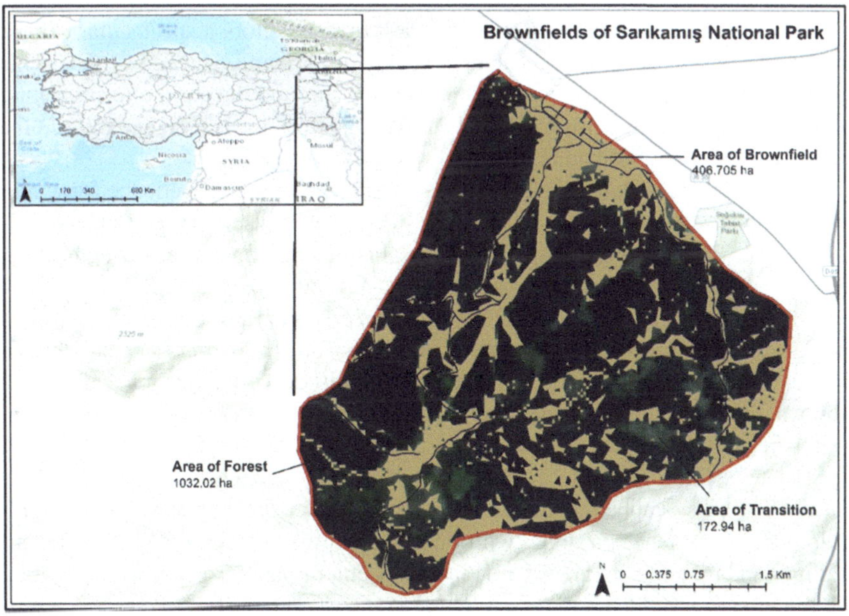

Map 25.4 Brownfield of Sarıkamış National Park

Fig. 25.7 Cable Car in Sarıkamış National Park

soldiers who fought without stopping for a moment, despite the -40 degree cold and hunger. Thousands of people visit this monument every year.

The Cable Car in Sarıkamış National Park offers visitors a spectacular experience with a bird's eye view of the natural beauty of Sarıkamış (Fig. 25.7). The cable car provides easy access to the mountainous areas around the park, offering nature walks and scenic views. Visitors can take the cable car to the heights and see the magnificent view of Sarıkamış.

The Ski Center in Sarıkamış National Park offers a unique experience for winter sports enthusiasts. It appeals to skiers of all levels with its slopes of various difficulty levels. The white snow-capped mountains of Sarıkamış offer ski enthusiasts the opportunity to experience an unforgettable adventure.sunmaktadır (Fig. 25.8).

Soğuksu Picnic Area in Sarıkamış National Park is an important recreation and picnic area that attracts attention with its natural beauties and offers a peaceful environment to visitors. Surrounded by pine forests, this area is an ideal place for those who want to spend time in touch with nature. Visitors can have a picnic and take a walk here.

Sarıkamış Ski Center is one of the most important winter tourism centers in Turkey. The snow-capped mountains of Sarıkamış offer ski enthusiasts an unforgettable skiing experience on slopes with varying degrees of difficulty. With its modern facilities and magnificent view, Sarıkamış Ski Center is an indispensable destination for winter sports enthusiasts.

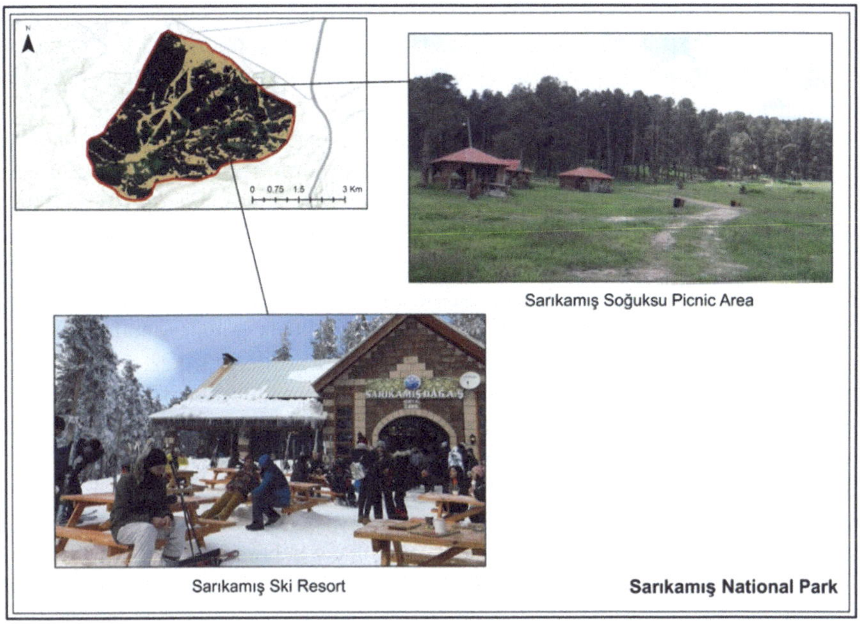

Fig. 25.8 Soguksu Picnic Area in Sarikamis National Park

25.3.3 Military Brownfield in Ukraine

During the period of the Soviet Union, there were more than 3000 military bases of various sizes and degrees of secrecy in Ukraine. There were military bases here, where tens of thousands of people lived. In the early 1990s, the largest group of troops in Europe was located on the territory of Ukraine. All these formations became part of the Armed Forces of Ukraine after receiving the status of independence. During the next military reform, the number of military units in the Armed Forces was reduced, the number of additional technical units was accordingly reduced, and the weapons ranges were increased. Decay has covered the territory of former military facilities. As a result, the economic difficulties in this area have caused a number of different social and environmental problems.

In Ukraine, there are a large number of military facilities: bases, towns, arsenals, ammunition depots, fuel and lubricant depots, fuel, training grounds, tank training grounds, landfills and landfills for hazardous waste, and old mine launchers, which are particularly radioactive. Most of these facilities pose real and potential hazards and risks to the population and the natural environment. They pollute the environment with chemicals, including heavy metals, and increase the radiation background. All this leads to the degradation of natural complexes and requires a specialized approach to cleaning and restoring the soil and its possible further use [41].

The use of military brownfields for civilian purposes (for example, for housing construction, tourism, agricultural projects, etc.) can be important for the socio-economic development of regions. However, previous military use and its possible consequences must be taken into account.

Ukraine has a significant tourist and recreational potential, but there is a constant decrease in the volume of tourist flows. Thus, according to the World Tourism Organization, in 2010, Ukraine was visited by 21.2 million foreign tourists, and this volume amounted to 2.17 million people in 2022 [45]. Russia's full-scale invasion of Ukraine also makes the development of this sphere in Ukraine even more vulnerable. In particular, foreign diplomatic institutions do not recommend visiting Ukraine without an urgent need for security reasons. This is understandable, since Russian aggression takes the form of missile, air and artillery strikes. Thus, during 2023, the Russian occupying forces inflicted 2.4 thousand missile strikes and 38.1 thousand airstrikes and artillery shelling on the positions of the Defense Forces and populated areas in Ukraine [34, 44]. According to the Office of the Prosecutor General (OGPU), in general, 11,673 people were killed, and 18,336 Ukrainians were injured from the beginning of the Russian invasion until the end of 2023 [47]. In addition, according to the calculations of the Kyiv School of Economics (KSE), as of January 2024, the total amount of direct damage caused to the infrastructure of Ukraine has increased to 155 billion dollars. We are talking about the loss of housing stock, infrastructure and industrial enterprises [35, 46].

At the same time, after the victory, Ukraine will be able to become one of the largest tourist centers in the world, and therefore it is necessary to increase the number of objects that can be involved in tourist activities. Among them may be objects that have suffered damage, as well as military brownfields, military museums.

Analysis of military brownfields in Ukraine allowed to highlight three types of military brownfields in Ukraine:

- *The first type* is related to those military brownfields which are not used actively in tourist activity and can become potentially tourist objects.
- *The second type* is related to those military brownfields which are not used in tourist activity and can't be tourist objects.
- *The third type* is related to the former military objects which used in tourist activity (currently they are museums, Historic and Cultural Reserves or tourist objects).

25.3.3.1 Military Brownfield Not Used in Tourist Activity in Ukraine (The First Type)

Let's analyze some military brownfield not used in tourist activity in Ukraine.

Tarakaniv Fort or Duben Fort is located in Tarakaniv village of the Dubno district, Rivne region (Fig. 25.9). This architectural monument was built at the end of

Fig. 25.9 Tarakaniv Fort (Rivne region) and its location [40, 52]

Fig. 25.10 Abandoned air defense base in Pashena Balka village (Dnipropetrovsk region) and its location [36, 40]

the XIX century on the initiative of the tsarist government of Russia to strengthen the western border with Austria and protect the Lviv-Kyiv railway line. The shape of the fort resembles a rhombus with sides of 240 m. In the middle of it there were two-story barracks and everything for the life of the garrison, from a bakery to medical facilities, a church. The fortress was surrounded by two rows of massive earth ramparts, between which a 14-m ditch with casemates was dug. For its intended purpose, the fort was used only during the First World War, when it was occupied by Russian, Austrian, and Polish troops. Now this fort is in a neglected state [36, 37].

Abandoned air defense base is located in Pashena Balka village of Dnipropetrovsk region (Fig. 25.10). It defended Dnipropetrovsk region and Dnipro city from the air attacks, since this city had a lot of military plants and it has strategic value. The base resembled an anthill: a large number of underground bunkers, tunnels and utility rooms. Now the base is abandoned, robed by looters. The rest of the ruins of the air defense base have grown over and now resemble the ghost town of Pripyat [36].

Protected Space Communication Node is located in the city of Horodenka, Ivano-Frankivsk region (Fig. 25.11). A high-powered bunker was a secret military

Fig. 25.11 Protected space communication node (Horodenka village, Ivano-Frankivsk region) and its location [40, 51] (Protected space communication node; GoogleMaps)

Fig. 25.12 Over-the-horizon radar station "Duga" (Kyiv region) and its location [36, 40]

facility and built in the 1980s. Its main purpose was to support space communication for the USSR troops. Four shafts are symmetrically located in the corners of the bunker, in which retractable antennas used to function. There are many different secret rooms, training grounds and other service premises throughout the territory. The place is in the abandoned state, but it is especially popular among history buffs and fans of military events [51].

25.3.3.2 Military Brownfields Not Used in Tourist Activity and Can't Be Tourist Objects in Ukraine (The Second Type)

Let's analyze some military brownfields not used in tourist activity and can't be tourist objects in Ukraine. The bright example can be the **Over-the-horizon radar station "Duga" (OTH RS "Duga")**. It is located in the city of Pripyat near the Chornobyl NPP (Kyiv region) (Fig. 25.12). This facility was launched in 1985, it is also called Chornobyl 2. The Duga radar system was used to detect intercontinental ballistic missile launches. It is 150 m high and 800 m long. It was damaged by radiation as a result of the accident at the Chernobyl nuclear power plant in 1986. High radiation has been recorded at the station, so it cannot be used for tourist purposes [36, 37].

25.3.3.3 Former Military Objects Used in Tourist Activity in Ukraine (The Third Type)

Odessa catacombs are located both in the city of Odesa and in the village of Nerubaiske (Fig. 25.13). The total length of the Odesa catacombs is estimated to be approximately 2.5 thousand km. During the Second World War, a partisan unit under the command of V. Molodtsov-Badayev hid there. Guerrillas conducted quite successful subversive activities, using dungeons as a permanent base. Nowadays excursions are held in the Odesa catacombs, and the Museum of Partisan Glory is located in the village of Nerubaiske [36, 37].

Pamir radar station (the village of Verkhniy Yalovets, Chernivtsi region) (Fig. 25.14). The station is located in the highlands (mountain of Tomnatyk (1565 m). It was launched in 1960 to detect the enemy in the vast surrounding area. Taking into account the weather conditions and the nuances of the landscape, the working spaces were made in the shape of spheres and were completely autonomous. In 1995, the station ceased to exist and was abandoned. Now the station is popular among tourists. The tourist festival "Pamir FEST" is held here, rafting and

Fig. 25.13 Odessa catacombs and its location [author of the photo—Yurii Perebaev, journalist of IGotoWorld.com [36, 40]

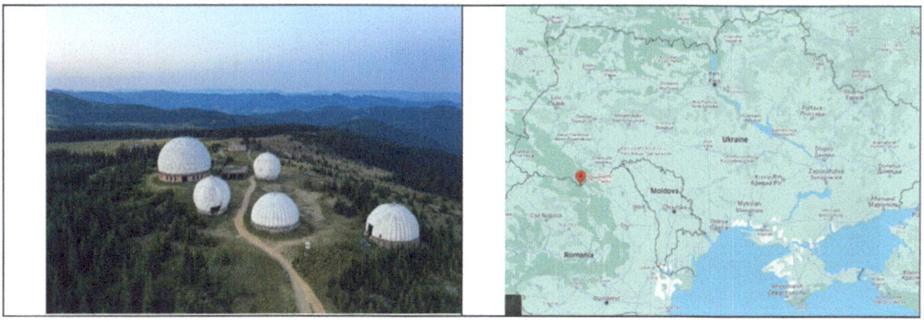

Fig. 25.14 Pamir radar station (Chernivtsi region) and its location [40, 42]

jeep tours are available nearby. The station can be seen in Jamala's music video for the song "1944", the winner of the Eurovision Song Contest 2016 [38].

Military-historical complex "Skelya" is located in the city of Korosten (Zhytomyr region) in the dungeons of the former command post of the Korosten fortified district (Fig. 25.15). It was built in 1934–1938. During the World War II, units of the fifth Army together with machine gun battalions delayed the enemy's offensive on Kyiv on this section of the front until the beginning of August 1941. After the war, local residents lived in the empty bunker for some time, then it housed a spare control point of the Korosten city administration, and in 2005 a museum complex was created in the defense structure. The basis of the museum is a complex of the underground command post. The bunker had working rooms, dining rooms, recreation rooms, 30 rooms in total [43].

National Historical and Architectural Museum "Kyiv Fortress" is a museum complex, which was created on the basis of the Hospital Fortification of the Kyiv Fortress (Fig. 25.16). It includes three exhibitions: "Fortification on the territory of Ukraine", "History of the use of the structures of the Kyiv Fortress", an open-air exhibition with defensive ramparts, a fortress wall and caponiers, as well, the stone and earth Lysohirsky fort which is the largest in Europe with an area of about 120

Fig. 25.15 Military-historical complex "Skelya" (Zhytomyr region) and its location [40, 43]

Fig. 25.16 State historical and cultural reserve "Field of the Poltava Battle" and its location [40, 50]

hectares. The fortress buildings, which are part of the Kyiv fortress, were built for almost 200 years from the 70-th to the XIX centuries. Today, the Kyiv Fortress is one of the largest surviving fortifications in the world [Kyiv Fortress].

State historical and cultural reserve "Poltava battlefield" is located in the city of Poltava (Fig. 25.17). It is a complex of monuments of history and monumental art and a museum dedicated to the events of the Great Northern War and the Battle of Poltava (1709). The "Poltava battlefield" complex has a total of 23 attractions. There are archeological monuments on the territory of the reserve, namely 30 ancient burial mounds and 4 hillforts. The reserve is an important scientific and methodical center and an interesting tourist location of the Poltava region. It is visited by tourists from all continents of the planet.

The Ukrainian defensive line "Oleksiivska fortress" is located near the village of Oleksiivka, Lozivsky district, Kharkiv region (Fig. 25.18). This is one of the fortifications of the Ukrainian Defense Line, which was built in 1731 to protect against enemy attacks. The fortress is earthen, almost square in plan, with four bastions. The area of the serf yard is more than 1.5 hectares. The length of the fortress along the outer side of the base of the rampart is more than 1030 m. Before the full-scale invasion of Ukraine, the "Family Fest Oleksiivska fortress" Festival was held

Fig. 25.17 State historical and cultural reserve "Poltava battlefield" and its location [40, 49]

Fig. 25.18 State historical and cultural reserve "Field of the Poltava Battle" and its location [39, 40]

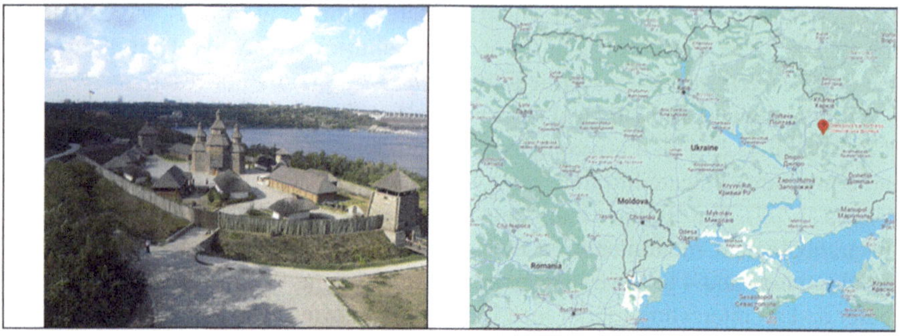

Fig. 25.19 Museum of the History of the Zaporizhzhya Cossacks of the National Reserve "Khortytsia" and its location [40, 48]

on the territory of the fortress, and it was also included in the route of the annual Tour De Fort bicycle tour through the fortresses of Slobozhanshchyna [39].

The Museum of the History of the Zaporizhzhya Cossacks of the National Reserve "Khortytsia" is located on Khortytsia, Zaporizhzhya (Fig. 25.19). Zaporizhzhya Sich is a center of the struggle of the Ukrainian people for their social liberation. The Zaporizhzhya Sich was founded around 1554 by the Volyn prince Dmytro Vyshnevetskyi on the island of Mala Khortytsia to defend against the Tatar invasion. It was destroyed in 1709 by Petro I's order, because he considered it as one of the centers of the liberation struggle of the Ukrainian people. The idea of creating an architectural ensemble in Khortytsia is a generalized image of the Cossack capitals of the XVI–XVIII centuries, appeared as early as 1960. In 1965 the island was granted the status of a State Reserve. In 1983 the Museum of the History of the Zaporizhzhya Cossacks of the National Reserve "Khortytsia" was opened [Khortytsia].

25.4 Conclusion

There are three types of military brownfields in Ukraine related to the level of the use of the object in tourism in Ukraine. Some object are abandoned and not used in tourist activity, but some of them are used in tourist activity. Today, when the Ukrainian-Russian war is going on, the military museums within the former military objects or other objects which were used in the tourist activity before the war, don't work. We hope they will open after the war, after the Victory of Ukraine. It is reasonably to develop military brownfields which are not used in the tourist activity now after the war. It will develop the tourist attractiveness of Ukraine in general and wide tourist potential of Ukraine in particular.

The areas where important battles have been filed for Türkiye from the past are now used as a national park. In this context, tourism was brought. These areas under the direction of the Ministry of Culture and Tourism are supervised by the state and sustainability is ensured.

It provides an important added value as tourism. The institutional procedure of identifying, selecting and declaring the military heritage can be an important part of the upgradingprocess, but it is not the only basis for value creation.

In this context, for the sustainability of the martyrdom areas, the Ministry of Culture, the Ministry of Urban Development and Tourism, the Ministry of Defense, Local Authorities and Civil Society should take part.

After the Ukraine-Russian war, the models in Turkey can be examined and applied to the appropriate areas in Ukraine.

References

1. Oliver L, Ferber U, Grimski D, Millar K, Nathanail P (2005) The scale and nature of European brownfields. In CABERNET (2005). Proceedings from the CABERNET 2005 international conference on managing urban land LQM Ltd, Nottingham
2. Wernstedt K, Hersh R (2006) Brownfields regulatory reform and policy innovation in practice. Prog Plan 65(1):7–74. https://doi.org/10.1016/j.progress.2005.10.004
3. Thomas MR (2003) Brownfield redevelopment: information issues and the regeneration workforce. Local Econ 18(1):80–89
4. Frantál B, Greer-Wootten B, Klusáček P, Krejčí T, Kunc J, Martinát S (2015) Exploring spatial patterns of urban brownfields regeneration: the case of Brno, Czech Republic. Cities 44:9–18
5. Bardos R, Jones S, Stephenson I, Menger P, Beumer V, Neonato F, Maring L, Ferber U, Track T, Wendler K (2016) Optimising the value from the soft re-use of brownfield sites. Sci Total Environ 563–564:769–782
6. Vanheusden B (2009) Research article: recent developments in European policy regarding brownfield remediation. Environ Pract: J Nat Assoc Environ Prof 11(4):256–262. https://doi.org/10.1017/s1466046609990202
7. Dixon T, Raco M, Catney P, Lerner DN (eds) (2007) Sustainable brownfield regeneration: liveable places from problem spaces. Wiley
8. Osman R, Frantál B, Klusáček P, Kunc J, Martinát S (2015) Factors affecting brownfield regeneration in post-communist space: the case of The Czech Republic. Land Use Policy 48:309–316
9. Seelarbokus CB (2021) The military and the environment: the neglected core barrier. In: International environmental cooperation and the global sustainability capital framework. Elsevier, pp 349–434
10. Morar C, Dulca M, Nagy G (2020) Brownfields regeneration, between sustainable urban development and cultural heritage. The former military sites in Oradea, Romania. J Urban Reg Anal 8(1). https://doi.org/10.37043/jura.2016.8.1.6
11. Rey E, Laprise M, Lufkin S (2022) Urban brownfield regeneration projects: complexities and issues. In: Neighbourhoods in transition. The urban book series. Springer, Cham. https://doi.org/10.1007/978-3-030-82208-8_4
12. Rainnie A, Herod A, McGrath-Champ S (2011) Review and positions: global production networks and labour. Compet Chang 15(2):155–169
13. Franz M, Pahlen G, Nathanail P, Okuniek N, Koj A (2006) Sustainable development and brownfield regeneration. What defines the quality of derelict land recycling? Environ Sci 3(2):135–151. https://doi.org/10.1080/15693430600800873

14. Hercik J, Šimáček P, Szczyrba Z, Smolová I (2014) Military brownfields in The Czech Republic and the potential for their revitalisation, focused on their residential function. Quaest Geogr 33(2):127–138. https://doi.org/10.2478/quageo-2014-0021
15. Wan Y, Chen S, Liu J, Jin L (2024) Brownfield-related studies in the context of climate change: a comprehensive review and future prospects. Heliyon 10(4):e25784. https://doi.org/10.1016/j.heliyon.2024.e25784
16. De Sousa CA (2003) Turning brownfields into green space in the City of Toronto. Landsc Urban Plan 62(4):181–198. https://doi.org/10.1016/s0169-2046(02)00149-4
17. Dixon T (2007) The property development industry and sustainable urban brownfield regeneration in England: an analysis of case studies in Thames gateway and greater Manchester. Urban Stud 44(12):2379–2400
18. Grimski D, Ferber U (2001) Urban brownfields in Europe. Land Contam Reclamation 9(1):143–148
19. Far MS (2011) An evaluation on sustainable brownfield regeneration opportunities in periphery area of Nicosia buffer zone. Unpublished. https://doi.org/10.13140/RG.2.2.10931.76324
20. Loures L (2015) Post-industrial landscapes as drivers for urban redevelopment: public versus expert perspectives towards the benefits and barriers of the reuse of post-industrial sites in urban areas. Habitat Int 45:72–81. https://doi.org/10.1016/j.habitatint.2014.06.028
21. Loures L, Panagopoulos T (2007) From derelict industrial areas towards multifunctional landscapes and urban renaissance. WSEAS Trans Environ Dev 3(10):181–188
22. Ferber U, Grimski D, Millar K, Nathanail P (2006) Sustainable brownfield regeneration: CABERNET network report. University of Nottingham, Nottingham
23. Niemets L, Husieva N, Suptelo O, Sehida K, Kravchenko K (2018a) Research of Brownfields and Greyfields of the City: Theory and practice. Sustainable economic development and application of innovation management from regional expansion to global growth. Proceedings of the 32nd International Business Information Management Association Conference (IBIMA) (Seville, Spain, 15–16 November 2018). Editor Khalid S. Soliman. Seville, International Business Information Management Association (IBIMA), p 409–435. https://ibima.org/accepted-paper/research-of-brownfields-and-greyfields-of-the-city-theory-and-practice/
24. Morar C, Nagy G, Boros L, Gozner M, Niemets L, Sehida K (2021) Heritage, culture and regeneration of the former military areas in the city of Oradea (Romania). Architektúra & urbanizmus 55(1–2):76–87. (Scopus (Q3), Web of science) (in English). https://doi.org/10.31577/archandurb.2021.55.1-2.6
25. Niemets L, Kandyba Y, Kobylin P, Kostrikov S, Dobrovolskaya N, Telebienieva I (2021) Integral assessment of ethnic tourism in Ukraine: resource provision and regional features. Innovation management and information technology impact on global economy in the era of pandemic. Proceedings of the 37th International Business Information Management Association Conference (IBIMA) (Cordoba, Spain, May 30–31, 2021). Editor Khalid S. Soliman. Seville, International Business Information Management Association (IBIMA), pp. 2077–2082 (Scopus, web of science) (in English). https://ibima.org/accepted-paper/integral-assessment-of-ethnic-tourism-in-ukraine-resource-provision-and-regional-features/
26. Niemets L, Kostrikov S, Kandyba Y, Dobrovolskaya N, Kobylin P, Kucheriava H (2020) Features and problematic issues of the tourism development in Ukraine due to the coronavirus Covid-19 pandemic. Proceedings of the 36th International Business Information Management Association Conference (IBIMA) (Granada, Spain 4–5 November, 2020). Editor Khalid S. Soliman. Seville, International Business Information Management Association (IBIMA), p 1799–1807 (Scopus, Web of Science) (in English). https://ibima.org/accepted-paper/features-and-problematic-issues-of-the-tourism-development-in-ukraine-due-to-the-coronavirus-covid-19-pandemic/
27. Niemets L, Telebienieva I, Sehida K, Krainiukov O, Kucheriava H, Pohrebskyi T (2019) Features of pilgrimage tourism in Ukraine (a case study of Uman). Vision 2025: education excellence and management of innovations through sustainable economic competitive advantage. Proceedings of the 34nd International Business Information Management

Association Conference (IBIMA) (Madrid, Spain, 13–14 November, 2019). Editor Khalid S. Soliman. Madrid, International Business Information Management Association (IBIMA), p 509–515 (Scopus, Web of Science) (in English). https://ibima.org/accepted-paper/features-of-pilgrimage-tourism-in-ukraine-a-case-study-of-uman/
28. Niemets L, Sehida K, Lohvinova M, Kraynukov O, Kliuchko L (2018b) Rural tourism in Ukraine: Peculiarities and trends of development. Sustainable economic development and application of innovation management from regional expansion to global growth. Proceedings of the 32nd International Business Information Management Association Conference (IBIMA) (Seville, Spain, 15–16 November 2018), Editor Khalid S. Soliman. Seville, International Business Information Management Association (IBIMA), p 290–301 (Scopus, Web of Science) (in English). https://ibima.org/accepted-paper/rural-tourism-in-ukraine-peculiarities-and-trends-of-development/
29. Garb Y, Jackson J (2010) Brownfields in The Czech Republic, 1989–2009: the long path to integrated land management. J Urban Regen Renew (3):3, 263–276. https://doi.org/10.13140/2.1.2099.7767
30. Kempa T, Marschalko M, Yilmaz I, Lacková E, Kubečka K, Stalmachová B, Bouchal T, Bednárik M, Drusa M, Bendová M (2013) In-situ remediation of the contaminated soils in Ostrava city (Czech Republic) by steam curing/vapor. Eng Geol 154:42–55. https://doi.org/10.1016/j.enggeo.2012.12.008
31. Ahmad N, Zhu Y, Ibrahim M, Waqas M, Waheed A (2018) Development of a standard brownfield definition, guidelines, and evaluation index system for brownfield redevelopment in developing countries: the case of Pakistan. Sustain For 10(12):4347. https://doi.org/10.3390/su10124347
32. Ibrahim H, Al-Hagla K, Nassar D (2020) Towards a Sustainable City: brownfields as a potential for urban development. J Eng Appl Sci 67(8):1903–1922
33. Vujičić DM, Vasiljević DA, Hose TA, Tasić N, Morar C, Đurić A, Marković SB (2018) A multi-criteria decision analysis with special reference to loess and archaeological sites in Serbia (could geosciences and archaeology cohabitate?). Open Geosci 10(1):333–343. https://doi.org/10.1515/geo-2018-0026
34. Chernovol K (2024) In 2023, the Russians struck Ukraine with tens of thousands of strikes: infographic https://www.unian.ua/war/u-2023-roci-rosiyani-zavdali-po-ukrajini-desyatki-tisyach-udariv-infografika-12500187.html
35. Zanuda A (2024) War in numbers: how Ukraine and Ukrainians have changed over the past year. BBC News Україна. https://www.bbc.com/ukrainian/articles/cv28871qdv8o
36. Lazarev P (2015) Abandoned military facilities of Ukraine: echoes of war. IGotoWorld.Com. Retrieved from: https://ua.igotoworld.com/ua/article/451_zabroshennye-voennye-obyyekty-ukrainy.htm
37. Veselova-Solovonyuk A (2023) Old bunkers and catacombs: objects of Ukraine that you have not heard of, but they are becoming more and more popular. UA portal. Retrieved from: https://news.uaportal.com/ukr/section-articles/news-staryie-bunkeryi-i-katakombyi-obektyi-ukrainyi-o-kotoryih-vyi-i-ne-slyishali-a-oni-stanovyatsya-vse-populyarnee-02-10-2023.html (in Urainian)
38. Adreeva V (2021) Abandoned military facilities of Ukraine, which were chosen by tourists. 24Travel. Rertieved from: https://travel.24tv.ua/viyskovi-obyekti-ukrayini-yaki-pokinuli-voyaki-oblyubuvali-turisti_n1713952
39. Oleksiivska Fortress (2021) Cycling route through the fortresses of Slobozhanshchyna. https://tourdefort.bike/fortetsi/oleksiivska-fortetsia
40. GoogleMaps. Rertieved from: https://www.google.com.ua/maps/, last Accessed 28 Apr 2024
41. Havrylenko OP (2008) Ecogeography of Ukraine: education. Manual, Kyiv. Retrieved from: http://pidruchniki.com/15941024/ekologiya/ekologichna_bezpeka_oboronnoyi_diyalnosti
42. It became known how much Ukrainian tourism lost due to the war (2023) Finance.ua. Rertieved from: https://news.finance.ua/ru/stalo-izvestno-skol-ko-turizm-ukrainy-nedopoluchil-iz-za-voyny

43. Military-historical museum "Skelya" in the city of Korosten (Zhytomyr Region) (2018). Travel-al.com.ua. Retrieved from: http://travel-al.com.ua/2018/10/07/skelya/
44. Official telegram channel of the Strategic Committee of the Armed Forces of Ukraine. (UA War Infographics). Retrieved from: t.me/uawarinfographics, last accessed 01/01/2024
45. Official website of the World Tourism Organization (WTO). Retrieved from: https://www.unwto.org/tourism-data/global-and-regional-tourism-performance, last accessed 01/01/2024
46. Official web-site of Kyiv School of Economics. Retrieved from: https://kse.ua/ua/
47. Official web-site of the Office of the Prosecutor General of Ukraine. Retrieved from: https://www.gp.gov.ua/
48. Official web-site of the Historical Reverse "Khortytsia". Retrieved from: https://hortica.zp.ua/ua/guide/north/91-museum/, last accessed 2024/04/28
49. Official site of the National Historical and Cultural Reserve "Poltava battlefield". Retrieved from: https://battle-poltava.org.ua/uk/, last accessed 2024/04/28
50. Official site of the National Historical and Architectural Museum of the Kyiv Fortress. Retrieved from: https://www.kyiv-fortress.com.ua/, last accessed 2024/04/28
51. Protected Space Communication Node. Guru.Ua. Rertieved from: https://rest.guru.ua/kolomiya/places/631-_zashishennyj_uzel_kosmicheskoj_svyazi/, last Accessed 28 Apr 2024
52. Tarakaniv Fort. Castles and fortresses of Ukraine. Retrieved from: https://ua-castles.at.ua/index/tarakanovskaja_krepost/0-11, last Accessed 28 Apr 2024

Chapter 26
Historical Context and Political Conditions of Revitalization of Degraded Military Areas: Case Study "Garnizon" in Gdańsk

Jan A. Wendt and Agnieszka Bógdał-Brzezińska

Abstract This article was created as a result of research on historical and political factors influencing the revitalization process of "Garnizon" in Gdańsk. The authors assumed that historical and political processes, leading to changes in the political system or changes in state borders, also influence changes in spatial development, including changes in the functions of objects originally intended for military purposes. This article presents the history of the creation and use of the "Garrison" for the needs of the German army before World War I, and then analyzes the historical changes in the uses of the discussed spatial complex in the twentieth and twenty-first centuries. Political processes become the background for historical analysis: reconstruction policy, historical policy (politics of memory) and revitalization policy inscribed in the context of Poland's integration policy with the European Communities. In a broad assessment of the revitalization processes, it was emphasized that thanks to the use of EU funds, it was possible to make the "Garrison" a multi-functional and, in the broad meaning of this statement, a self-sufficient part of Wrzeszcz with a prestigious and attractive character. The case of the "Garrison" in Gdańsk-Wrzeszcz is part of the list of successful initiatives to transform post-military areas of a historical nature, and the very idea of revitalizing the garrison increased the attractiveness of the district as a place of residence and life.

Keywords "Garrison" · Gdańsk · Historical factors · Political factors · Revitalization · Wrzeszcz

J. A. Wendt (✉)
Gdańsk University, Gdańsk, Poland
e-mail: jan.wendt@ug.edu.pl

A. Bógdał-Brzezińska
Warsaw University, Warsaw, Poland
e-mail: bogdal@uw.edu.pl

C. Morar et al. (eds.), *Achieving Sustainability in Ukraine through Military Brownfields Redevelopment*, NATO Science for Peace and Security Series C: Environmental Security, https://doi.org/10.1007/978-94-024-2278-8_26

26.1 Introduction

Historical and political processes, leading to changes in political systems or changes in borders, also influence changes in spatial development. This applies both to the rank assigned to it and to the functions it performs. It is relatively easy, with small investments, to introduce new functions in residential districts, but the development and revitalization of areas used by the army poses much more problems [1–4]. An example of a well-conducted revitalization of degraded military areas is the "Garrison" area, which for over a hundred years was an area of military barracks and was used for military purposes [5]. Currently, the former area of the historic "Garrison" serves residential, service and commercial functions and is a popular place for recreation and meetings not only for local residents. The subject of this study is the analysis of the historical context and assessment of the impact of political factors on the use of the "Garrison" military complex in the Wrzeszcz district of Gdańsk, which was subject to revitalization processes after Poland's accession to the EU.

The revitalization of this military-degraded area raises the question about the most important factors influencing the successful transformation of space, which is the research question of this work. Thanks to preliminary research, analysis of the literature on the subject and field research, the hypothesis that political factors had the greatest influence on the success of revitalization seems justified. Both on a macro and local scale.

The garrison (former barracks) in Wrzeszcz, a district of Gdańsk, covers an area of approximately 25 ha, historically associated with the military formations stationed here and performing barracks and defense functions. The area occupied by the barracks is located on the border of the districts of Wrzeszcz (German: Langfuhr) and Strzyża Górna (German: Hochstrieß) (see Fig. 26.1).

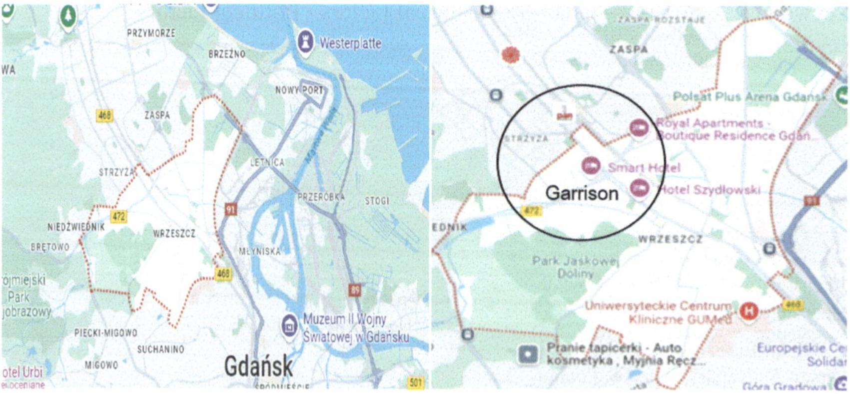

Fig. 26.1 Gdańsk—Wrzeszcz and the location of the revitalized "Garnizon" area. (Source: own elaboration based on Google maps)

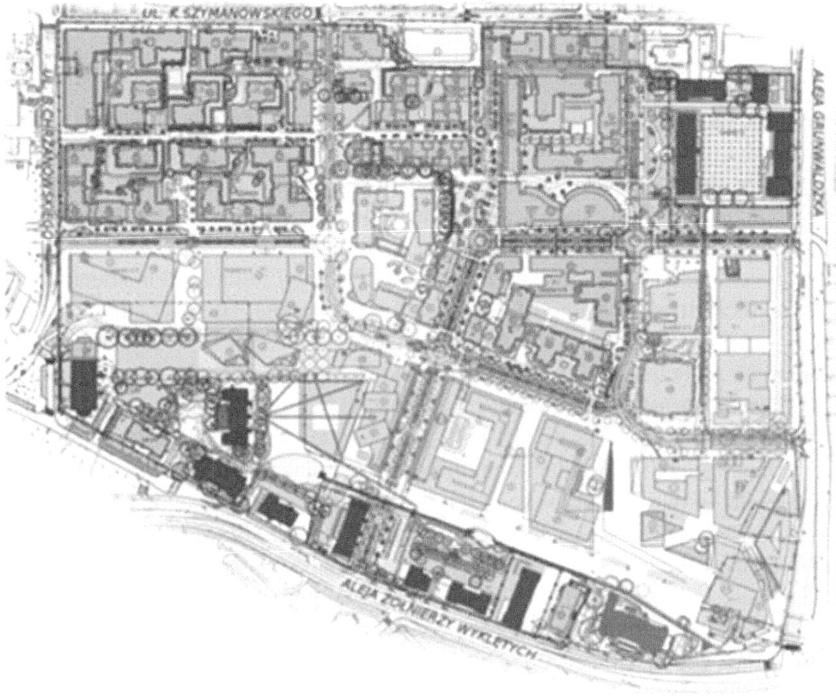

Fig. 26.2 Layout of the Garrison in Gdańsk—Wrzeszcz. (Source: [6, p. 237])

The exact location covers a quarter of the city separated from the rest of the city by the current Aleja Grunwaldzka and Karola Szymanowskiego, Bernarda Chrzanowskiego and Juliusz Słowacki streets and Żołnierzy Wyklętych (see Fig. 26.2). It should also be emphasized that some of the garrison facilities have been entered into the Register of Monuments of the Pomeranian Voivodeship since December 29, 2004 under the registration number A-1753.

This spatially defined area of the former barracks, currently located in the center of Wrzeszcz, a commercial district of Gdańsk, was located on the outskirts of the city at the time when the first military buildings were built. At the end of the nineteenth century, the area served a residential function for the local inhabitants of the former villages of Strzyża Dolna and Wrzeszcz, but the expanding garrison changed it to a military function. When the name "Garrizon" is used in this article, it means historical military facilities, and when the name "Garrizon" is used, it means a modern revitalized area.

26.2 Historical Conditions for the Creation of the Garrison and the Changes of its Hosts

Gdańsk, the largest port of the Polish-Lithuanian Commonwealth, was occupied by the Kingdom of Prussia after the partitions of Poland. From the main commercial centre of the country, it became one of many regional centers of the Prussian state. And it remained such a city after the transformation of Prussia into the Second German Reich in 1871. Like in every large city, military units were stationed here, but the development of the city in the nineteenth century influenced the decision to move out their districts of the Main Town and create one, compact center, a new military garrison [7]. Gdańsk was the headquarters of the elite first Adjutant Hussar Regiment No. 1 belonging to the 36th Division of the German Empire, part of the XVII German Army Corps. The First Hussar Regiment was also known as Totenkopfhusaren, i.e. Hussars of Death, basically Death's Head Hussars after the symbol they wore on the Heiduckenmütze or Flugelmütze, i.e. on the hussar cap [8]. For the needs of the first and second horse regiments and their support, an area on the outskirts of the city was selected (see Fig. 26.3). The villages of Wrzeszcz and Strzyża Dolna were incorporated into the administrative boundaries of Gdańsk only in 1814, were sparsely populated (Wrzeszcz had only 908 inhabitants in 1814) and offered a relatively large area necessary for horse units.

The choice was also determined by its convenient location in terms of transport, as in 1870 the Gdańsk-Wrzeszcz railway station was built next to the railway line, as well as access to water, as the area was crossed by the Strzyża stream. The choice

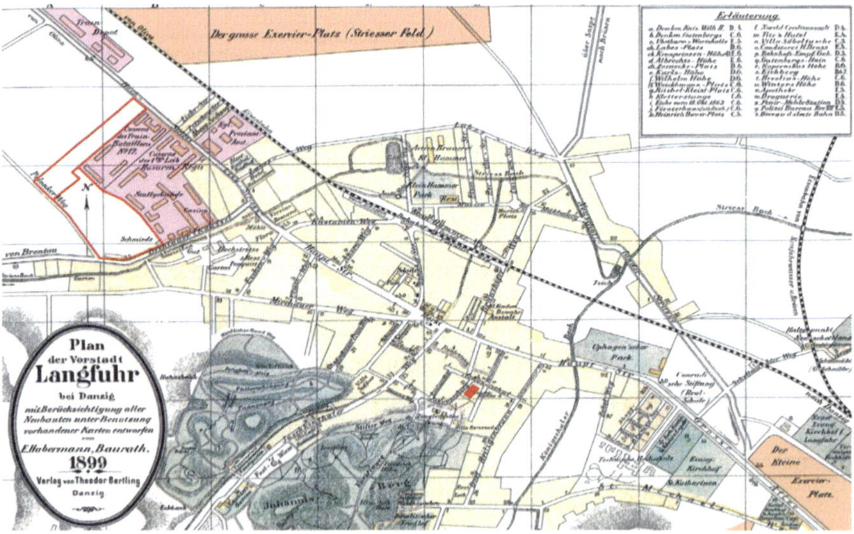

Fig. 26.3 Plan of the Wrzeszcz district in Gdańsk (Danzig-Langfuhr) with the garrison area in 1899. (Source: https://commons.wikimedia.org/wiki/File:Plan_der_Vorstadt_Langfuhr_bei_Danzig.JPG)

of this location allowed the creation of barracks from scratch for the first Hussar Regiment, the second Adjutant Hussar Regiment named after Queen Victoria of Prussia and the 17th Tabor Battalion [9] (see Fig. 26.4).

An important element when planning the constructed buildings was the preservation of large free spaces between them [11]. They facilitated the manoeuvring and movement of horse units and allowed for the density of buildings as the unit developed. The unit was officially handed over to the army on September 14, 1901, when

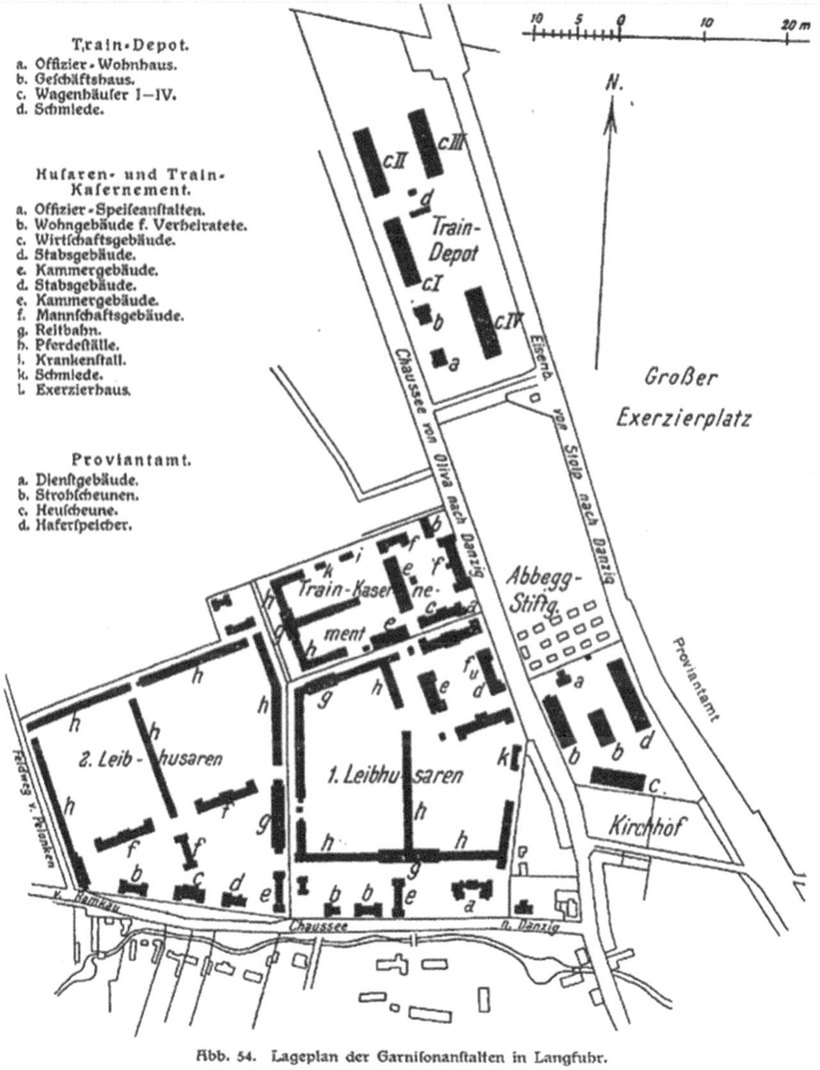

Fig. 26.4 Map of garisson in Langfuhr (Polish: Wrzeszcz) from the beginning of twentieth Century. (Source: [10, p. 121])

a ceremonial march to the garrison of the first and second Hussar Regiments of the Bodyguard took place in the presence of Emperor Wilhelm II (see Figs. 26.5, 26.6, and 26.7).

The barracks of the Gdańsk hussars were separated from the city (Wrzeszcz district) by a wall, thus creating a compact, separate development. However, this did not mean complete isolation. Balls were often organized in the officers' mess, to which representatives of the city authorities and local elites were also invited. The permanent stay of the army influenced the development of Wrzeszcz. The number of service points working for the army, hairdressers and tailors increased, and large stores from the city center opened their branches in Wrzeszcz, counting on wealthy customers, who included numerous officers in both regiments [12]. The importance of the garrison increased enormously after the appointment of the heir to the throne, Kronprinz Wilhelm, as the commander of the first Regiment in 1911, and his sister, Princess Louise Victoria, as the honorary commander of the second Regiment, while leaving the unit under the command of August von Mackensen [8]. The garrison not only became a place for stationing troops, but also gained political importance. Although Kronprinz Wilhelm and his wife did not live in the garrison, but in a villa located at today's Grunwaldzka Street (formerly Hauptstrasse 98), still, while acting as commander of the first Regiment, he often stayed in the barracks. And the

Fig. 26.5 Kaiser Wilhelm II in the new barracks (14.08.1901). (Source: https://pl.m.wikipedia.org/wiki/Plik:Cesarz_Wilhelm_II_w_Garnizonie_we_Wrzeszczu.jpg)

Fig. 26.6 Fragment of the buildings of the barracks of the first Leibhusaren Regiment. (Source: https://www.danzig-online.pl/page43.html)

Fig. 26.7 Fragment of the buildings of the barracks of the second Leibhusaren Regiment. (Source: https://www.danzig-online.pl/page43.html)

presence of the heir to the throne favoured new investments, the development of the garrison, and therefore also the development of Wrzeszcz.

In 1914, units from Gdańsk were sent to the Western Front, where they suffered significant losses. Only a small part of them returned to the garrison in January 1919, and in the spring of the same year both regiments were disbanded.

In the period between the First and Second World Wars, as a result of peace provisions, Gdańsk became a quasi-independent state called the Free City of Gdańsk [13], which was established after peace talks on November 25, 1920. The city did not have its own army, but a new Gdańsk police was created to maintain order. Most of the Gdańsk police units were stationed in the former German hussar barracks in Wrzeszcz. The headquarters of the Command Staff was also located on the premises of the former hussar barracks. It was also the seat of the seventh Police Headquarters in Gdańsk (Husarenkaserne, Langfuhr, Hauptstraße 71 (currently Wrzeszcz, ul. Grunwaldzka). After the outbreak of World War II (1939), the Free City of Gdańsk was incorporated into the Third Reich, only to return to it after the end of World War II Polish.

26.3 Political Factors: From Reconstruction Policy to Revitalization Policy

The experiences of war-affected countries are a valuable point of reference in the diagnosis and assessment of common areas pursued by the authorities' reconstruction and revitalization policies. In the case of Poland, areas that were subject to the policy of infrastructural and socio-demographic adaptation after 1945 deserve attention. Following World War II, Poland's borders moved from east to west [14], which resulted in the resettlement of the population (estimated at over 1.8 million people) to new areas previously belonging to Germany, including Pomerania. The war campaign of 1945 was conducted by the Soviet army in the spirit of total war, i.e. without respect for the rights of the civilian population and with the intention of total destruction of urban infrastructure. Therefore, Gdańsk, as a key city in the region, experienced almost complete destruction of the urban fabric, and the Polish communist authorities were faced with the need not so much for revitalization, but for the complete reconstruction of urban space.

The policy of rebuilding urban infrastructure in Poland after World War II includes several spectacular cases. In the post-war history of Poland, the experience of rebuilding entire urban complexes was associated with political propaganda strengthening social trust in the new "people's government". The plan to rebuild the capital city of Warsaw, completely destroyed in 1944 by the German army, was implemented as a large national undertaking [15], although alternative scenarios for the city reconstruction process appeared [16]. Propaganda slogans such as "The whole nation is building its capital" found in the press and posters resonated in the memory of several generations of Poles [17]. In addition to administrative bodies, social institutions, such as the Social Fund for the Reconstruction of the Capital (SFOS), were also involved in the reconstruction. Moreover, the policy of rebuilding the urban fabric of Warsaw, Wrocław and Gdańsk was based on the best reflexes of the population, good faith and civic generosity, starting from free clean-up works (so-called social deeds) to collecting private funds dedicated to reconstruction.

As for the discussed case of the "Garrison" post-military complex, it should be noted that as a result of the occupation of Gdańsk by the Soviet army in 1945, the Wrzeszcz district was ruined on a relatively smaller scale than the oldest part of the city, where the losses amounted to over 90% of buildings [18, 19]. The reconstruction policy applied to the city by the Polish communist authorities was not uniform. A decision was made to rebuild the so-called The Main Town, but also Wrzeszcz, was intended to be built in the spirit of socialist realism, implementing the Gdańsk Residential District project to facilitate the settlement of new residents and the location of offices [20]. Ultimately, however, socialist realist architecture appeared in Wrzeszcz on a smaller scale than originally planned.

The fate of the "Garrison" during communism was part of the post-war policy of strengthening Poland's influence in its new borders on the so-called Western and Northern Territories, which also included Pomerania. From this perspective, the "Garrison" was not perceived as a historic building, but a complex of military buildings taken over for the defence purposes of the new regime. In the first period after the occupation of Gdańsk by the Soviet army, its troops were stationed in the "Garrison", and the Polish communist authorities maintained the military character of the facility. In the years 1945–1949, the 16th Kashubian Infantry Division was stationed. After it, the seventh Lusatian Landing Division, called "Blue Berets" after its characteristic headgear [21]. It is worth noting, however, that in the case of the Garrison in Gdańsk-Wrzeszcz, "as in other places in Poland, although historic buildings sometimes symbolized the manifesto of the invaders' power and were associated with a turbulent period in the history of Poland, they were not interpreted pejoratively" [22, p. 332]. Moreover, the usefulness of the objects in question during the communist period influenced the continuation of the policy towards historic urban complexes during the systemic transformation after 1989. The policy of revitalization then began to be combined in Poland with historical policy (politics of memory), defined as reviving the past and including it in current social life. Polish sociological research has noticed both positive and negative effects of this relationship. "The relationship between revitalization and memory has another dimension – political. Revitalization depends on the directions of policy pursued, both at the local, national and international level. Memory involved in the revitalization process is often politicized—it is enough to trace the variability of street names, the fate of monuments and commemorative plaques. The danger of this phenomenon lies in simultaneous uniformity and selectivity, leading to the creation of a vision of history and interpretation of events in a one-sided, yet unique, necessary and binding way. Therefore, it seems that the distant past and its memory are closer to revitalization projects due to their relative neutrality than the recent past" [23, p. 76–77].

The reconstruction of the district where the Garrison complex was located required a decision of not only urban or economic importance, but above all—political. Historical policy was present from the first moments of the Polish side taking power in post-war Gdańsk. An element of this policy was emphasizing the Polish roots of the city itself, its historical connections with Poland and the active presence of the Polish minority in pre-war Gdańsk. The political propaganda shaped in this way was intended to facilitate the assimilation of the resettled population from the former eastern borderlands of the Second Polish Republic in the city [24]. It should be noted that they inhabited a space whose urban dimensions (the shape of urban infrastructure) and those relating to lifestyle and culture reminded us of the forced migration of the German population [25]. This led to fear of the shortness of life in Gdańsk, fuelled by Cold War agitation. Until 1989, the German cultural heritage of both the Main Town rebuilt from the foundations and the Wrzeszcz district was talked about reluctantly or falsely. Only Poland's aspirations to join the EU and the slow generational change brought about the emergence of positive pro-German cultural codes in Wrzeszcz, the most popular of which was associated with the figure of the Nobel Prize-winning writer Gunter Grass, born in Dolny Wrzeszcz [26].

26.4 Revitalization and Current Development

The political, economic and social transformation after the collapse of the former Soviet Union changed Poland's defense policy and in the mid-1990s, decisions were made to liquidate military units and the garrison in Wrzeszcz. Only a few military institutions have survived in the area of the former garrison (e.g. the headquarters of the Military Property Agency, Polish: Agencja Mienia Wojskowego = AMW). The destruction of buildings and the former area occupied by the army continued. The Military Property Agency had no funds either for investments or even for full security of the area it managed. However, an attempt was made to sell part of the former barracks area by offering 8 hectares of the former garrison at a tender, but no investors were found. The policy of revitalizing post-industrial areas and areas abandoned by the army was just taking shape in Poland. More than fifty towns and former Soviet garrisons in Poland were waiting for investments and revitalization of often very degraded areas. To this number we can add the areas after the liquidation of over six hundred Polish units, which were liquidated as part of the reorganization of the Polish army and its adaptation to the challenges faced by the army at the beginning of the twenty-first century. There was a lack of funds, so it is not surprising that investors chose only locations in the largest cities, attractive in terms of potential housing investments, such as the garrison in Gdańsk.

An element of the revitalization policy after regaining independence was incorporating historic buildings with previous military use into the civilian urban fabric. Therefore, the functions of objects were changed. It was also connected with the process of transformation of the economic system, as former military facilities belonging to the state treasury were transformed into local government property

(1990s) or were subject to privatization. Historic buildings located in attractive city districts were perceived as potential attractive monuments or future seats of cultural institutions. The 1990s were the period of the beginnings of cooperation between the nascent private sector and local authorities to save the historical objects in question as part of a private-public partnership [27]. The attitude towards historical post-military facilities in the former Eastern Bloc countries was also influenced by the political will to join international integration structures. This is also the case in Poland, where in the 2000s new revitalization trends appeared related to integration with the European Union. They are based on European funds [28, p. 11]. Both in the pre-accession period and after 2004, revitalization plans and revitalization policies took into account trends specific to highly developed Western countries. One of them involved the adaptation of post-industrial complexes into multifunctional modern residential, commercial and service buildings. This trend was used to revitalize the Garrison facilities.

The revitalization policy in Poland after 1989 was conducted by the Office of Housing and Urban Development, which in 1995–2003 prepared the Act on the revitalization of Polish cities. After Poland joined the European Union (2004), revitalization took on a new political meaning. The entities responsible for its implementation became local government authorities, and at the central level—subsequent ministries responsible for regional policy and infrastructure. After 2004, the revitalization policy developed in EU countries based on the values of social participation and projects financed from citizens' budgets became an attractive tool for the modernization of urban space in Poland. Revitalization was defined as a public service, and development activities in the revitalized spaces were interpreted as the implementation of a public-private partnership. In these conditions, in the changed political and economic situation, the attractiveness of the area of the former garrison in Wrzeszcz increased. In the next tender, AMW offered most of the area of the former barracks, which was purchased by the Hossa Investment Group (November 2005) for housing construction with service, cultural and recreational functions. The revitalization process began with the demolition of devastated buildings and tidying up the area. Four historic buildings that served as headquarters, barracks and offices in the former garrison were renovated. The remaining buildings on the side of Szymanowskiego Street were demolished and replaced with residential buildings in accordance with the area revitalization plan presented by the Hoss group. The former barracks were transformed into a prestigious residential district with service, recreational and cultural functions.

The value and attractiveness of the area increased when in 2007, in close proximity to the degraded military areas of the garrison, a shopping center was opened—Baltic Gallery (Polish: Galeria Bałtycka), one of the largest in Gdańsk. Its construction had started only a year earlier. A modern center with over 200 shops and numerous restaurants has increased the value of neighbouring areas, and the former military barracks have become a potentially interesting location.

Already 2 years after accession to the EU, the authorities of Gdańsk made a decision, both economic and political, about the future use of the post-military areas of the Garrison in question. The decision-making body was the City Council,

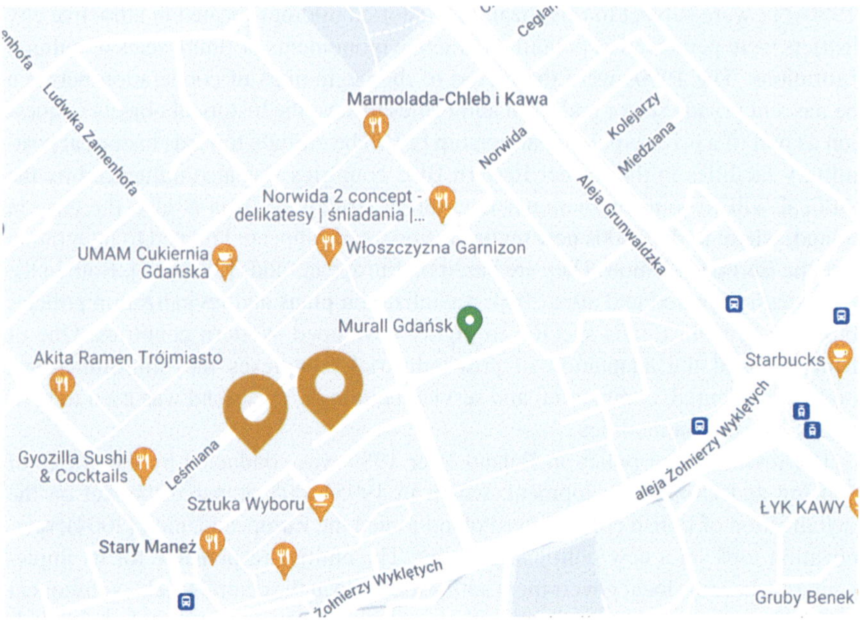

Fig. 26.8 New "Garnizon"—housing investments. Source Own elaboration

which authorized solutions as part of the preliminary spatial development plan. Its important element was conservation supervision over nineteenth-century buildings, including the objects in the register of monuments. The background related to the Western European trend of striving for sustainable development of urban infrastructure is the project of increasing the proportion of green areas in connection with recreational facilities (including walking). An inventory of old trees was recommended.

A system of service and commercial points was also included as an element of the project. In the latter case, it was recommended not to use the space for large-scale development (shopping mall), which was a revitalization trend in other large Polish cities, e.g. Poznań (Stary Browar), Łódź (Manufaktura).

A new spatial development plan was adopted [29], according to which a residential and service development area was marked out, with a minimum service area of 8000 m^2; the possibility of building commercial facilities was limited to 2000 m^2, pedestrian routes, the maximum height of buildings, and communication accessibility from the streets limiting the garrison area were established. The mapping of the frontage at the junction of Grunwaldzka, Chrzanowskiego and Szymanowskiego streets was placed under conservation protection, and a decision was made to recreate the historical spatial structure, taking into account the location of given military units (see Fig. 26.8).

The planned residential infrastructure of an apartment type includes office buildings, shops, craft service points, parking lots and cultural institutions. The new "Garnizon" has become a multifunctional district with a resident-friendly space in which residential, office and service functions interact harmoniously (Figs. 26.9, and 26.10).

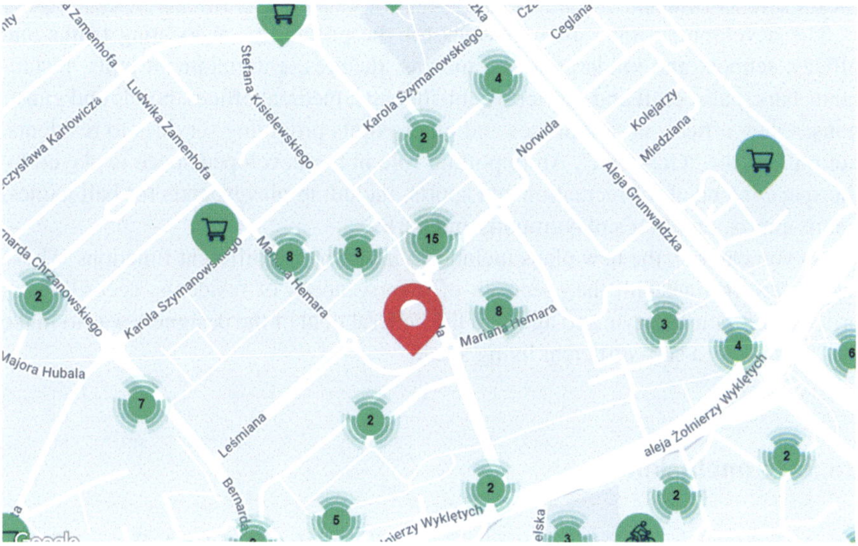

Fig. 26.9 Shops and services. (Source: Own elaboration)

Fig. 26.10 Shops and services (zoom view Norwida str.). (Source: Own elaboration)

It is worth emphasizing that already at the planning stage and, more importantly, also during the implementation of the military area revitalization project, it was possible to give the space completely new functions. Commercial and service development closely related to communication routes that constitute the main axis of the city's car transport (Aleja Grunwaldzka) and the nearby railway line with the Szybka Kolej Miejska (SKM, the Gdańsk equivalent of the metro line), which is important for the city's residents.

The development plan also provides for the possibility of locating banks and offices, schools and kindergartens, cinemas, theatres, entertainment halls, restaurants, bars, pubs, craft and service establishments, medical offices, hotels and guesthouses, law offices, notary offices and other points providing services to residents, not only of the "Garnizon". An important role in the developed space is played by the space intended for recreation and sports, including playgrounds for ball games, tennis courts, children's playgrounds and parks.

As you can see, the new plans included a multitude of different functions. These tasks were in line with the spectrum of diverse needs of residents, according to: criterion of their lifestyle and age, and the general goal of the designers was to make the "Garrison" a self-sufficient living space.

26.5 Conclusion

The historical infrastructure of the "Garnizon" complex in Gdańsk-Wrzeszcz was created in the nineteenth century as a specific military facility. Its location at one of the outskirts of the district allowed at the same time to maintain a certain distance from the everyday rhythm of life in the civilian part of the city, and on the other hand to increase the sense of security and prestige of the people neighbouring the barracks. Due to the size and different functions of individual objects, the "Garnizon" was an example of the urban dimension of political propaganda, focused on the power of the German Empire.

The well-thought-out spatial structure of the facility and the lack of damage to the barracks fabric during both world wars meant that the "Garrison" became the natural headquarters of police or military units both in the Free City of Gdańsk (1920–1939) and during the period of People's Poland (1945–1989).

Over time, the historically designated functions of the facility were integrated into the context of political changes, which was most visible in two periods of systemic transformation in Poland. In the first case, the defensive functions of the "Garrison" were used by the people's authorities to strengthen the propaganda of the Polishness of Gdańsk, in the second case—after Poland's accession to the European Union, there was a complete change in function and a comprehensive reconstruction of the facility, in accordance with the policy of sustainable urban development and post-industrial adaptation of space. It seems that the case of the "Garnizon" in Gdańsk-Wrzeszcz fits into the list of successful initiatives to transform post-military areas of a historical nature, and the very idea of revitalizing the "Garnizon" increased the attractiveness of the district as a place of residence and life.

Acknowledgments This study was funded by NATO programme: Emerging Security Challenges Division; Science for Peace and Security Programme "Achieving Sustainability in Ukraine through Military Brownfields Redevelopment".

Disclosure of Interests The authors have no competing interests to declare that are relevant to the content of this article.

References

1. Ćwiertniak R (2010) Znaczenie procesu rewitalizacji obiektów i obszarów powojskowych w rozwoju lokalnym i regionalnym miast i gmin. In: Noworól A (ed) Obszarowe ujęcie rewitalizacji w Polsce. Monografie i studia Instytutu Spraw Publicznych Uniwersytetu Jagiellońskiego. Instytut Spraw Publicznych Uniwersytetu Jagiellońskiego w Krakowie, Kraków, pp 29–46
2. Jarczewski W, Kuryło M (2010) Regeneration of post-military areas in Poland. Europa XXI 21:117–133. https://doi.org/10.7163/Eu21.2010.21.9
3. Jarczewski W, Kułaczkowska A (2019) (eds) Rewitalizacja. Raport o stanie polskich miast. Instytut Rozwoju Miast i Regionów, Warszawa–Kraków
4. Rynio D (2019) The revitalisation of post-military areas as an element of consistent city structure building. Acta Scientiarum Polonorum Administratio Locorum 18(4):411–424. https://doi.org/10.31648/aspal.4409
5. Kowalski KM (2018) (red.): Od Czarnych Huzarów do Niebieskich Beretów. Kosycarz Foto Press, Gdańsk
6. Woyciechowski M (2018) Wybrane aspekty tworzenia struktur mieszkaniowych o charakterze miejskim na przykładzie założeń: Wiszące Ogrody i Garnizon w Gdańsk. Studia Urbanistyczne 187:228–244
7. Daniluk J (2011) Wrzeszcz skoszarowany. In: Szczepańska K, Szczepański J (eds) Wędrówki po Wrzeszczu. Wydawnictwo słowo/obraz terytoria, Gdańsk, pp 110–130
8. Gruszczyński W (2019) Leibhufaren. Historia Czarnych Huzarów. Wydawnictwo Region i Gdański Kantor Wydawniczy, Gdańsk
9. Duchnowski TM (2016) Dyslokacja i skład niemieckiego XVII Korpusu Armijnego w latach 1890–1900. Europa Orientalis Studia z Dziejów Europy Wschodniej i Państw Bałtyckich 6:35–57. https://doi.org/10.12775/EO.2015.002
10. Danzig und seine Bauten (1908) Architekten und Ingenieurverein zu Danzig. Wilhelm Ernst & Sohn, Berlin
11. Piwek A (2018) Urbanistyka i architektura dawnego garnizonu w Gdańsku Strzyży. In: Kowalski KM (ed) Od Czarnych Huzarów do Niebieskich Beretów. Kosycarz Foto Press, Gdańsk, pp 521–615
12. Hirsch R (2011) Uwagi o architekturze koszar huzarów we Wrzeszczu. In: Szczepańska K, Szczepański J (eds) Wędrówki po Wrzeszczu. Wydawnictwo słowo/obraz terytoria, Gdańsk, pp 131–149
13. Bógdał-Brzezińska A, Wendt JA (2020) Status międzynarodowy Wolnego Miasta Gdańska w świetle refleksji Ludwika Ehrlicha. In: Grzebyk P, Tarnogórski R (eds) Siła prawa zamiast prawa siły. Ludwik Ehrlich i jego wkład w rozwój nauki prawa międzynarodowego oraz nauki o stosunkach międzynarodowych. PISM, Warszawa, pp 307–322

14. Wendt J, Raczyński M (1999) Granice Polski w XX wieku. Kwartalnik Geograficzny 3(11):14–25
15. Kurcewicz U (2021) Ideologiczne oblicze odbudowy śródmieścia Warszawy ze zniszczeń po II wojnie światowej. In: Grabarczyk T, Pogońska-Pol M (eds) Oblicza wojny, t.4, Miasto i wojna. Wydawnictwa Uniwersytetu Łódzkiego, Łódź, pp 369–392
16. Getka-Kenig M (2021) Plan londyński. Niezrealizowana wizja odbudowy Warszawy (1945–1946). Instytut Pileckiego, Warszawa
17. Odbudowa Warszawy po drugiej wojnie światowej na plakatach propagandowych. https://szczecin.ap.gov.pl/pl/Odbudowa+Warszawy+po+drugiej+wojnie+%C5%9Bwiatowej+na+plakatach+propagandowych. Last accessed 2023/11/21
18. Greczanik-Filipp I (ed) (1997) Wspomnienia z odbudowy Głównego Miasta, t.2. Wydawnictwo Marpress, Gdańsk
19. Gawlicki M (2016) Zabytkowa architektura Gdańska w latach 1945–1951. Kształtowanie koncepcji konserwacji i odbudowy. Wydawnictwo słowo/obraz terytoria, Gdańsk
20. Koprowski K (2023) Jak Wrzeszcz z ruin odbudowano. O Grunwaldzkiej Dzielnicy Mieszkaniowej. https://szczecin.ap.gov.pl/pl/Odbudowa+Warszawy+po+drugiej+wojnie+%C5%9Bwiatowej+na+plakatach+propagandowych. Last accessed 2023/11/20
21. Pacek B, Polak A, Mazurek W (2014) Łużycka Dywizja Desantowa 1963–1986. AON, Bellona, Warszawa
22. Kwasek M, Grębowski K, Przewłócki J (2018) Garnizon w Gdańsku wystąpienia kształtowania atrakcyjnych przestrzeni miejskich na terenach powojskowych. Studia komitetu przestrzennego zagospodarowania kraju PAN 192:325–333
23. Smolarkiewicz E (2012) Rewitalizacja jako forma przywoływania przeszłości. In: Derejski K, Kubera J, Lisiecki S, Macyra R (eds) Deklinacja odnowy miast. Z dyskusji nad rewitalizacją w Polsce. Wydawnictwo Naukowe Wydziału Nauk Społecznych Uniwersytetu im. Adama Mickiewicza w Poznaniu, Poznań, pp 69–78
24. Bykowska S (2005) Gdańsk – miasto (szybko) odzyskane. Biuletyn Instytutu Pamięci Narodowej 09-10:35–44
25. Bykowska S (2015–2016) Gdańsk 1945 – koniec i początek miejskiej wspólnoty. Rocznik Gdański, t LXXV–LXXVI, 97–109
26. Dymnicka M, Szczepański J (2017) Polityki pamięci i tożsamości wobec (nie)chcianego dziedzictwa. Od Gdańska do Gdańzigu. In: Bierwiaczonek K, Dymnicka M, Kajdanek K, Nawrocki T (eds) Miasto, przestrzeń, tożsamość: studium trzech miast: Gdańsk, Gliwice, Wrocław. Wydawnictwo Naukowe Scholar, Warszawa, pp 81–100
27. Kołomański T (2019) Public-private partnership as a chance for revitalising historic objects in Central Europe. Wiadomosci Konserwatorskie – J Herit Conserv 59:90–96
28. Ciesiółka P (2017) Rewitalizacja w polityce rozwoju kraju. Rozwój Regionalny i Polityka Regionalna 39:9–28
29. Uchwała nr LIV/1825/06 Rady Miasta Gdańska z dnia 31 sierpnia 2006 roku w sprawie uchwalenia miejscowego planu zagospodarowania przestrzennego Wrzeszcz- byłe koszary przy ul. Słowackiego w mieście Gdańsku, (2006)

Part III
Our Future: Young Researchers and Brownfields

Chapter 27
Migration and Resilience of Ukraine Communities: Challenges in the War Conditions

Oleksandr Bodnariuk and Iryna Gavrikova

Abstract The issue of involuntary internal migration is increasingly worsening within society. The military aggression by Russia towards Ukraine, including the annexation of the Autonomous Republic of Crimea and the ongoing military conflict in Donbas, has led to an unparalleled displacement of both internal and external Ukrainian populations. The compelled migration of populations has presented Ukraine with significant challenges across various domains: socioeconomic, political, legal, and socio-psychological. Drawing upon extensive analysis conducted by both foreign and Ukrainian scholars on forced migrations, the primary issues faced by Ukrainian internally displaced persons (IDPs) were identified. In the article, the temporal, structural, dynamic, and regional characteristics of forced displacements during both the initial (2014) and subsequent (2022) waves of migration were established. The impact of warfare and associated internal population displacement on Ukrainian communities was examined, aiming to uncover both the potential advantages and risks faced by these communities amidst conflict, towards achieving stability. Territorial features of the displacement of IDPs have been revealed. During the first wave of displacement in 2014–2018, the main target regions for displacement were the eastern regions (Donetsk, Luhansk, Kharkiv, Zaporizhia). This was connected with the presence of relatives, friends, and socio-cultural closeness. During the second wave of displacement in 2022, the feature of a larger-scale population movement was the expansion of geography to the western regions. People were no longer oriented towards similar mentalities, religions, or linguistic features but were in search of a safe place. Based on the analytics of Ukrainian migrations, a SWOT analysis of forced migrations for the sustainability of host communities was created.

Keywords Forced migration · Internally displaced persons (IDPs) · Population displacement · Resilience of communities · Sustainable communities · Ukrainian communities

O. Bodnariuk (✉) · I. Gavrikova
V.N. Karazin Kharkiv National University, Kharkiv, Ukraine
e-mail: oleksandr.bodnariuk@student.karazin.ua; iryna.gavrikova@student.karazin.ua

C. Morar et al. (eds.), *Achieving Sustainability in Ukraine through Military Brownfields Redevelopment*, NATO Science for Peace and Security Series C: Environmental Security, https://doi.org/10.1007/978-94-024-2278-8_27

27.1 Relevance of Research

The stability of communities in each country characterizes the reproduction of the population in spatial and temporal certainty. Providing favorable conditions for the development of Ukraine's population is considered one of the most important national interests underlying state policy and national security. The demographic factor is key to ensuring stable and safe development of the country, and the problems of optimal demographic development should be recognized as the most important for the state, as they affect the result of its functioning. The development of labor potential and aggregate national income directly depends on the demographic indicators of the working population and the dynamics of demographic development. Analysis of the current demographic situation, as well as its dynamics in recent years, indicates the existence of a deep demographic crisis in Ukraine, alongside socio-economic problems. One of the main factors influencing the sustainable development of communities in Ukraine is the occupation of the territory of the Autonomous Republic of Crimea, the invasion of the territory of Donetsk and Luhansk regions in 2014, as well as the full-scale war with Russia since 2022.

The unprecedented scale of population displacement presents a difficult challenge for internally displaced persons (IDPs) and also constitutes a significant problem for host communities. These challenges encompass a wide range of issues, including humanitarian, social, economic, and infrastructural concerns. Internal displacement is a complex process that significantly affects communities where displaced persons find temporary shelter. These displacements transform communities, altering their demographic characteristics, economic landscape, and social structure. Communities are faced with the need to adapt to new realities, including a sharp increase in population, changes in age and gender structure, and an increased load on social infrastructure. In challenging conditions, communities must provide IDPs with everything they need for a comfortable life, while ensuring stability and the ability to develop their region. Understanding the dynamics of these processes and identifying the features and patterns accompanying population displacement, especially in the context of war and mass displacement, becomes crucial. Analyzing current data allows for a better understanding of the current situation, predicting future trends, and developing effective strategies for restoring and maintaining community resilience. Therefore, studying the impact of war and related internal population displacement on Ukrainian communities remains timely and relevant. This study will focus on areas with the highest number of IDPs, the changes occurring in these communities, and the challenges they face in adapting to the new demographic and socio-economic reality.

27.2 Analysis of Previous Research

Jarosław Wątróbski, Aleksandra Bączkiewicz, Ewa Ziemba, Wojciech Sałabun in the research "Sustainable cities and communities assessment using the DARIA-TOPSIS method" tried to give the answer to the question: "How can sustainable cities and communities (society) be measured". Efficient evaluation of urban and community sustainability holds paramount importance in fostering sustainable development. This study proposes an innovative sustainability assessment methodology that integrates the Multi-Criteria Decision Analysis (MCDA) framework with the variability of alternative performance measurement. This approach was employed to evaluate sustainable cities and communities across 26 European countries we can apply it to Ukrainian communities [13].

Niemets, L., Husieva, N., Sehida, K., Lohvynova, M., Suptelo, O. (2019) «Forced internal displacement in Ukraine: Realities, Socio-Economic Problems, Perspectives» based on the analysis of numerous studies of forced migrations by foreign and Ukrainian scholars figured out that the primary challenges faced by internally displaced persons (IDPs) in Ukraine can be categorized into three main groups: socio-economic, institutional and legal, and social and psychological [6]. Within the socio-economic realm, issues such as housing shortages, unemployment or underemployment among migrants, limited incomes, restricted access to social services, interruption of social welfare payments and pensions, and challenges encountered by individuals with disabilities are prominently emphasized. The paper "Migration Challenges of the tewenty-first Century: Experience of Ukraine" investigates the structural dynamics and spatial characteristics of internal population displacement in Ukraine, discerning the factors influencing IDPs' decisions regarding their choice of residence region [11].

Analysis of the features of the IDPs of Georgia and Ukraine in work "Forced migrations of Georgia and Ukraine: reasons for the emergence, process features, implications for countries". The authors Elizbarashvili, N., Niemets, L. M., Meladze, G., Sehida, K. Y., Lohvynova, M. O., Meladze, M., & Sidamonidze, D. identified that IDPs pose a significant challenge for both countries (Georgia and Ukraine), overwhelming governmental capacities. Housing remains a persisting issue for IDPs in both nations. In Ukraine, apart from housing challenges, IDPs encounter difficulties in defining their legal status, accessing social benefits, and securing employment. Neglecting these issues exacerbates societal tensions, fuels local conflicts between migrants and residents, and ultimately results in recurrent internal displacement. Addressing the needs of forced migrants should be the foremost priority in policymaking concerning IDPs [2].

The analysis of the main social rights of internally displaced persons and their realization after Russia's full-scale invasion in Ukraine is presented in the work by Fedorova, Alla; Hamuľák, Ondrej. Social rights of internally displaced persons in Ukraine: new challenges in the context of Russian full scale military aggression. The special attention was also drawn to the the importance of sufficient international support and assistance for IDPs' social protection in Ukraine.

Changes and alterations in Ukraine's demographic stability from 1989 to 2021 prior to the ongoing extensive conflict, and evaluating the numerical, compositional, and geographical demographic shifts resulting from it. Roman Tesliuk, in his work "Demographic Sustainability of Ukraine and Its Changes Caused by the Russian-Ukrainian War", used index analysis tools to study changes in the reproductive and age-sex balances in Ukraine and its regions. It was discovered that from the mid-1990s onward, Ukraine has encountered a profound demographic reproduction crisis characterized by significant population reduction and worsening age and gender equilibrium [12].

The purpose of this study is to establish the temporal, structural, dynamic, and regional features of forced displacements during the first wave (2014) and the second wave (2022) of migration, to investigate and analyze how war and related internal population displacement have affected Ukrainian communities, and to identify opportunities and threats for Ukrainian communities in the context of war on the path to stability.

27.3 Presentation of the Main Material

Forced population displacement and changes in communities after the start of the war in 2014. In 2014, Ukraine faced the problem of internally displaced persons (IDPs). This happened for the second time in its recent history. The first case of mass population displacement occurred as a result of the accident at the Chernobyl nuclear power plant in 1986, when more than 116,000 people were evacuated from radioactively contaminated areas [14].

A significant number of people were forced to leave their places of permanent residence due to Russian aggression and the occupation of Crimea and certain areas of Donetsk and Luhansk regions. According to the Ministry of Social Policy, as of June 2016, 1,785,740 IDPs were registered in Ukraine, of which almost 1.1 million were women and 700,000 were men. By July 2018, the official number of IDPs in Ukraine exceeded 1.5 million citizens, which accounts for 3.6% of the total population. The Ministry of Social Policy reports that the largest number of IDPs are registered in Donetsk, Luhansk, Kharkiv, Zaporizhzhia, Dnipropetrovsk, and Kyiv regions, while the least number of IDPs are located in Ternopil, Chernivtsi, Rivne, Zakarpattia, Ivano-Frankivsk, and Volyn regions [10].

The population from the temporarily occupied territories chose the urbanized regions of Ukraine to live, as the main part of the IDPs lived in cities before the forced resettlement. Consequently, the main reason for choosing eastern regions for living is their proximity in religious, cultural, and linguistic aspects [9]. Regarding employment, a significant portion of the population was employed in the metallurgical, chemical industry, and machine building sectors, which is why IDPs also sought employment opportunities at similar enterprises. All these facts indicate the uneven regional distribution of IDPs in Ukraine, leading to an excessive social and administrative burden on host communities, their local labor markets, and the social infrastructure of settlement regions. Simultaneously, the choice of regions close to the places of previous residence by a significant portion of the settlers indicates the intention to return to the abandoned settlements. However, a significant number of IDPs only register outside the occupied territories of Donetsk and Luhansk regions to receive social benefits and then return home. Sometimes, hundreds and even thousands of migrants are registered at one address, as reported multiple times in the media [1].

According to the Unified Information Database on Internally Displaced Persons, approximately 62% of IDPs are women. The predominance of women can be attributed to various reasons, ranging from men staying at home to take care of the house or relatives, to reluctance to register as an IDP to avoid possible mobilization into the Ukrainian army. This gender structure among IDPs is mainly formed by immigrants from eastern Ukraine [10]. There is parity between the number of men and women among people from the Autonomous Republic of Crimea. Women predominate among the working population in all regions. For example, in Zaporizhzhia and Vinnytsia regions, women make up 69%, while in Ivano-Frankivsk region, it's 67%. Among the able-bodied IDPs, the proportion of youth is significant, with more than 40% of immigrants being under the age of 35. This age structure reflects a significant potential of labor resources, as young people are more mobile and creative, have modern education, and are capable of engaging in new types of activities [10].

Internally displaced persons of working age have a high educational and qualification level, as most of them have higher education. According to the Ukrainian State Employment Service [15], the distribution of IDPs by education differs significantly from the educational level of the total number of unemployed. People with higher education make up more than 70%, with vocational and technical education at 19%, and primary and secondary education at 11%. At the same time, among all the unemployed, the share of persons with higher education is less than 45%, with professional and technical education at 35%, and secondary education at 20%. It is noteworthy that the share of immigrants with higher education predominates in all regions except Zhytomyr region (47.1%) and Ternopil region (47.5%). The high educational and qualification level of IDPs indicates significant accumulated human capital, which can become an independent factor in the development of high-tech industries and services with intensive use of knowledge in areas where internally displaced persons are concentrated [7].

The mass displacement of the population has reduced the overall quality of life in Ukraine and has affected the socio-economic situation in the country and its individual regions. Constant challenges related to the adaptation and integration of IDPs into new host communities, as well as the legal status and the need of IDPs for legal support, create risks to the social and political stability of the country.

The reduction in the number of IDPs during 2019–2020 may be related to the general demographic decline in Ukraine, which is observed by significant emigration, a high share of elderly people, and a significant share of unemployed among IDPs. The main reasons include the return to the place of permanent residence in the Russian-occupied territories of Ukraine or Crimea, emigration abroad in 2018 (5% of IDPs reported that they have already found work abroad and are leaving soon or intend to emigrate in the near future to find work), successful integration of IDPs into local communities, as a result of which they cease to confirm their status as IDPs, and reluctance to confirm IDP status due to bureaucratic barriers and periodic checks [12].

Population displacement and community changes after the start of a full-scale invasion in 2022. After the full-scale invasion of the Russian Federation into the territory of Ukraine, which began on February 24, 2022, our state faced mass forced migrations of the population. As a result of active hostilities, millions of civilians were forced to leave their places of permanent residence. Some of these people migrated within the country, becoming internally displaced persons (IDPs). As of the beginning of 2024, three million 413 thousand Ukrainians remain internally displaced persons within Ukraine [3, 4]. A significant share of citizens was forced to leave the country; according to UNHCR estimates, five million 974 thousand displaced persons are abroad [4]. It is worth noting that according to calculations by the IOM, four million 445 thousand people returned to their permanent place of residence in Ukraine after relocation, with 26% returning from abroad [5].

As of the beginning of 2024, three million 413 thousand internally displaced persons (IDPs) were registered in Ukraine, of which 61% are women and 39% are men [4]. Displacement has affected all regions of Ukraine, but the largest influx of forced migrants is observed in the regions adjacent to the war zone. It's worth noting that the vast majority of people forced to move choose to relocate within their own region (where there are no active hostilities) or to a neighboring region that borders their region of permanent residence [26]. Thus, the regions adjacent to the combat zone, such as Dnipropetrovsk, Poltava, Kharkiv, and Odesa, observe a high number of registrations of internally displaced persons [4].

Another notable feature is the attraction of internally displaced persons to the safest regions, located in the West of Ukraine. Therefore, many displaced persons are accepted in Lviv, Ivano-Frankivsk, Zakarpattia, and Chernivtsi regions. It should be noted that the majority of IDPs migrating to the west of Ukraine have minor children and undertake longer migrations compared to those who relocate within their own or neighboring regions [19–21, 24, 25]. Additionally, a large number of IDPs are registered in Kyiv and the Kyiv region [4]. This can be explained by the significant socio-economic potential of the region and the country's best air defense system [22, 23].

Population displacement has had a significant impact on host communities, especially in regions that received the largest number of IDPs, such as Kharkiv, Dnipropetrovsk, Lviv, and Kyiv. Population growth in these regions has placed additional pressure on local infrastructure, including housing, healthcare, education, and social services. Communities have faced challenges related to the integration of IDPs, provision of their needs, and involvement in local social and economic life.

Addressing these challenges has required concerted efforts from the government, local authorities, international organizations, and civil society. Various assistance programs for IDPs have been developed and implemented, including the provision of temporary housing, access to education and medical care, psychological support, and assistance in finding employment. One of the most urgent challenges for host communities has been the need to provide internally displaced persons with housing. Many IDPs left their homes with minimal belongings, hoping for a quick return. However, the protracted nature of the conflict has forced them to seek long-term shelter. This has placed additional pressure on the real estate market in host areas, where there was already high competition for affordable housing [8].

The increase in the number of residents in communities has created an additional burden on the social services system, particularly on healthcare, education, and social protection. Medical facilities have experienced an increase in demand for services, especially given the psychological trauma associated with war. Schools and preschools have had to adapt to receiving a large number of migrant children, requiring additional resources and staff training. The integration of IDPs into the local economy is a key aspect of their inclusion in the community and ensuring their sustainability. This involves not only providing access to the labor market but also supporting entrepreneurship and professional development. However, many IDPs face challenges in finding work due to high competition, lack of networking connections, or necessary skills for employment in new environments. Social integration of IDPs and building relationships with the local population are important for the harmonious development of communities [8].

Sometimes, tensions may arise in communities due to a lack of mutual understanding between immigrants and the local population, as well as the challenges of adapting to a new society. Responding to these challenges requires a comprehensive approach that includes the development of adaptive policies, resource mobilization, awareness-raising, and the involvement of all stakeholders [16–18]. The displacement of the population in conditions of war has created unprecedented challenges for Ukrainian communities, but at the same time has opened up significant opportunities for strengthening their resilience, adaptability, and development. The active inclusion of IDPs in the social, economic, and cultural life of communities can guarantee not only their recovery but also further development. An analysis of the challenges associated with the internal displacement of the population in Ukraine during the war reveals a deep and multifaceted impact on the communities hosting internally displaced persons [8].

It is worth noting that some of the people who resorted to forced migration have returned to their place of permanent residence. As of the beginning of 2024, there are 4,455,000 returnees in Ukraine. 47% of them returned from another region of Ukraine, 26% returned from abroad, another 26% stayed within their region and then returned to their permanent place of residence [5]. It should be noted that persons who returned to their place of permanent residence were displaced for an average of 142 days [3, 4].

Most of the returned persons are living in the city of Kyiv (23%) and the Kyiv region (17%) [5]. In our opinion, the high percentage of return of residents to the city of Kyiv and the Kyiv region is related to the absence of active hostilities in the region, its distance from the front line, and the presence of the best air defense system in the country. The presence or absence of an effective air defense system becomes one of the key factors influencing the decision of Ukrainians to return to their place of permanent residence, taking into account regular missile attacks from the Russian Federation.

Overall, among the main reasons for returning to their permanent place of residence, 69% of people indicate an improvement in the security situation, 68% cite the desire to be closer to family and friends, and 51% state that a guaranteed job opportunity in their place of permanent residence is one of the main reasons for returning [5].

Also, one of the leaders in the number of displaced persons returning to their permanent place of residence is the Kharkiv region (564,000 returnees) [5]. In our opinion, this is due to the successful actions of the Armed Forces of Ukraine, which liberated most of the region from the Russian occupying army during counteroffensive actions in 2022. Additionally, a significant number of returnees were recorded in Mykolaiv (205,000) and Odesa (211,000) regions, mainly due to the successful offensive actions of the Armed Forces of Ukraine, which forced the Russian army to retreat from the right bank of the Dnieper, minimizing the risk of occupation of Mykolaiv and Odessa by the Russian Federation. Also, a significant number of displaced persons returned to one of the most populated regions of Ukraine—Dnipropetrovsk (359,000) [5]. This is due to the stabilization of the front and the absence of manifestations of the approach of the combat contact line to the borders of this region, which provides a sufficient level of security that encourages local residents to return home.

To comprehensively evaluate the strengths and weaknesses of the research topic, as well as identify potential opportunities and threats, the authors conducted a SWOT analysis (Fig. 27.1). This analytical tool enabled the systematization of the

Fig. 27.1 SWOT analysis of forced migrations for the sustainability of host communities [5, 8, 10]

main groups influencing population migrations on the stability of communities, identification of positive and negative factors of influence, opportunities, and threats for the development of further plans and programs, and the substantiation of complex development models, particularly of host communities and many other issues that require urgent solutions at all management levels from local to national.

27.4 Conclusions

In the near future, internal population displacement will remain an urgent problem, given the scale and duration of the war in Ukraine. This is one of the main challenges of wartime, requiring a comprehensive approach. Understanding the scale, dynamics, and consequences of this process is critical for developing effective response strategies, ensuring community resilience, and supporting IDPs on their path to integration and recovery. Despite significant challenges, many Ukrainian communities have shown extraordinary resilience and adaptability. They were able to mobilize resources to help IDPs and ensure their basic needs and integration into life in a new place. An important aspect of community resilience is not only the provision of immediate needs but also the development of long-term development and recovery strategies. Successful integration of IDPs involves access to work, education, health, and social services, as well as fostering mutual understanding between IDPs and the local population.

The situation with mass migration is extremely complex. However, with the right approach, it can be transformed into benefits for host communities. This can ensure the sustainable development of these settlements, as well as establish a proper standard of living for displaced persons. The ways of successful integration of IDPs into host communities and the search for points of local socio-economic growth are promising areas for further research.

References

1. Analytical report: Internal displacement in Ukraine: mapping flows and challenges [Analitychnyi zvit: Vnutrishnie peremishchennia v Ukraini: kartohrafuvannia potokiv ta vyklykiv] [Electronic resource]. Access mode: https://www.pragueprocess.eu/ru/migration-observatory/publications/33-reports/249-2020-09-07-10-47-09
2. Elizbarashvili N, Niemets LM, Meladze G, Sehida KY, Lohvynova MO, Meladze M, Sidamonidze D (2020) Forced migrations of Georgia and Ukraine: reasons for the emergence, process features, implications for countries. Series Geol Geogr Ecol 52:139–155. https://doi.org/10.26565/2410-7360-2020-52-11
3. International Organization for Migration (IOM) (2024, April 19) DTM Ukraine REGIONAL PROFILES. [Ukraina PROFILI OBLASTEI]. https://dtm.iom.int/reports/ukraina-profili-oblastey-opituvannya-zagalnogo-naselennya-raund-15-grudnya-2023?close=true
4. International Organization for Migration (IOM) (2024, April 19) DTM Ukraine—basic territorial assessment of registered IDPs (at the district level) [Ukraina Bazova terytorialna otsinka

zareiestrovanykh VPO (na rivni raionu)] round 34 (March 2024). IOM, Dnipro. https://dtm.iom.int/reports/ukraina-bazova-teritorialna-ocinka-zareestrovanikh-vpo-na-rivni-rayonu-raund-34-berezen?close=true

5. International Organization for Migration (IOM) (2024, March 18) DTM Ukraine report on population distribution in Ukraine survey of the general population [Zvit propovernennia naselennia v Ukraini Opytuvannia zahalnoho naselennia] round 15 (November 27–December 27, 2023). IOM, Dnipro. https://dtm.iom.int/reports/ukraina-zvit-propovernennya-naselennya-v-ukraini-opituvannya-zagalnogo-naselennya-raund-15?close=true
6. Niemets L, Husieva N, Sehida K, Lohvynova M, Suptelo O (2019) Forced internal displacement in Ukraine: realities, socio-economic problems, perspectives/proceedings of the 33rd international business information management association conference, IBIMA 2019: education excellence and innovation management through vision 2020, 2019, P.3064 3081 https://doi.org/10.26565/2410-7360-2019-51-10
7. Novikova O, Amosha O, Antoniuk V (2016) Internally displaced persons: from overcoming obstacles to a strategy for success, monograph [Vnutrishno peremishcheni osoby: vid podolannia pereshkod do stratehii uspikhu: monohrafiia]; NAN Ukraine. – Kyiv, 2016, p 448. https://iie.org.ua/wp-content/uploads/monografiyi/2016/2016_mono_Novikova_Amosha_Antonyuk_2016.pdf
8. Potapenko V (2023) The impact of migration on the socio-economic situation in Ukraine: analyst. Add. [Vplyv mihratsii na sotsialno-ekonomichnu sytuatsiiu v Ukraini]/ https://doi.org/10.53679/NISS-analytrep.2023.01
9. Report International Organization for Migration (IOM) Ukraine National Monitoring System: The Situation of IDPs (March 2016)/ https://dtm.iom.int/reports/ukraine-%E2%80%94-national-monitoring-system-situation-idps-march-2016?close=true
10. Report on the results of monitoring the functioning and use of the Unified Information Database on IDPs by the social protection administrations [Zvit za rezultatamy monitorynhu funktsionuvannia ta vykorystannia upravlinniamy sotsialnoho zakhystu naselennia Yedynoi informatsiinoi bazy danykh pro VPO] [Electronic resource]. Access mode: http://vpl.com.ua/uk/materials/zvit-za-rezultatamy-monitorynhu-funktsionuvannya-ta-vykorystannyauprav
11. Sehida K, Husieva N, Niemets L, Kobylin P, Kliuchko L (2018) Migration challenges of the 21st century: the experience of Ukraine/proceedings of the 32nd International Business Information Management Association Conference, IBIMA 2018 – vision 2020: sustainable economic development and application of innovation management from regional expansion to global growth, 2018, p 328–337 https://ibima.org/accepted-paper/migration-challenges-of-the-21st-century-the-experience-of-ukraine/
12. Tesliuk R (2024) Demographic sustainability of Ukraine and its changes caused by the Russian-Ukrainian war/19 January 2024 volume 32, 2024. pp 431–447. https://doi.org/10.25133/JPSSv322024.026
13. Wątróbski J, Bączkiewicz A, Ziemba E, Sałabun W (2022) Sustainable cities and communities assessment using the DARIA-TOPSIS method sustainable cities and society volume 83, August 2022, https://doi.org/10.1016/j.scs.2022.103926
14. Website of the State Statistics Service of Ukraine [Electronic resource]. Access mode: https://www.ukrstat.gov.ua
15. Website of Ukrainian State Employment Service [Electronic resource]. Access mode: https://dcz.gov.ua/
16. Morar C, Nagy G, Dulca M, Boros L, Sehida K (2019) Aspects regarding the military cultural-historical heritage in the city of Oradea, Romania, ANNALES – Annals for Istrian and Mediterranean studies – series. Historia et Sociologia 19(2):303–322. https://doi.org/10.19233/ASHS.2019.21
17. Morar C, Nagy G, Boros L, Gozner M, Niemets L, Sehida K (2021) Heritage, culture and regeneration of the former military areas in the city of Oradea (Romania). Architektúra & urbanizmus 55(1–2):76–87. https://doi.org/10.31577/archandurb.2021.55.1-2.6
18. Morar C, Tiba A, Jovanovic T, Valjarevic A, Ripp M, Vujicic MD, Stankov U, Basarin B, Ratković R, Popović M, Nagy G, Boros L, Lukic T (2022) Supporting tourism by assessing

the predictors of COVID-19 vaccination for travel reasons. Int J Environ Res Public Health 19(2):918. https://doi.org/10.3390/ijerph19020918
19. Tiba AI (2010) Grounded cognition perspective on irrational beliefs in rational emotive behaviour therapy. J Cogn Behav Psychother 10(2):87–99
20. Tiba AI, Johnson C, Vadineanu A (2012) Cognitive vulnerability and adaptation of parents of children with autistic spectrum disorder. J Cogn Behav Psychother 12(2):209–218
21. Tiba AI, Manea L (2018) An embodied simulation account of cognition in rational emotive behavior therapy. New Ideas Psychol 48:12–20. https://doi.org/10.1016/j.newideapsych.2017.08.003
22. Palinchak M, Zayats O, Tsalan M, Toca CV (2023) Determinants of UMSCA competitive force formation. Balt J Econ Stud 9:200–205. https://doi.org/10.30525/2256-0742/2023-9-5-200-205
23. Mihai M, Darabaneanu D, Chirodea F, Constantin-Vasile Ț (2022) Analysis of social support as an argument for the sustainable construction of the European community space. Sustainability 14(12):7448. https://doi.org/10.3390/su14127448
24. Tiba AI, Manea L (2018) The vividness of imagining emotional feelings is attenuated in dysphoria and predicts the experience of positive emotional feelings. J Clin Psychol 74:2238–2263. https://doi.org/10.1002/jclp.22676
25. Tiba AI, Trip S, Bora CH, Drugas M, Borz F, Miclăuş DC, Voss L, Iova SC, Pop S (2023) Positive irrational beliefs are associated with hypomanic personality. Front Psychol 14:1053486. https://doi.org/10.3389/fpsyg.2023.1053486FI3.8
26. Al I, Wendt AJ (2014) The evolution of historical regions and borderlands (1916–2011) at the Northern Romanian Ukrainian border. In: Milan B, Julian M, Anssi P (eds) The new European frontiers: social and spatial (re)integration issues in multicultural and border regions, 2014. Cambridge Scholars Publishing, Newcastle upon Tyne. ISBN 978-1-4438-5650-8

Chapter 28
Change Management of the Extracurricular Activities in Higher Education Institutions for Sustainable Development in Times of Crisis

Tetiana Cherednychenko

Abstract The article discusses the change management in the context of the crisis in extracurricular activities in Ukrainian higher education institutions. Based on the changes in the global social and economic environment, the article focuses on the need for adaptability and flexibility of management strategies to ensure the effective functioning and further development of higher education institutions. The article examines the 17 Sustainable Development Goals and the Ukrainian response to this initiative through the national report "Sustainable Development Goals: Ukraine". In particular, the article examines the impact of these goals on the educational process in Ukraine, especially on the quality of higher education and extracurricular activities. The basic concepts of change management are explored, starting with the work of Kurt Lewin, who in the mid-twentieth century first identified approaches to change in groups. Further analysis is devoted to the methods of change management, in particular, the Kotter and ADKAR models. The article emphasises the importance of understanding and using different change management models to achieve success in change management. The article analyses the challenges faced by higher education institutions in Ukraine today in the context of military operations. After a detailed analysis, recommendations for change managing in times of crisis are offered, including adaptation, staff support, development of new forms of activity, partnership, innovation, strategic planning and social responsibility.

Keywords Change management · Extracurricular activities · Institution of higher education

T. Cherednychenko (✉)
V.N. Karazin Kharkiv National University, Kharkiv, Ukraine
e-mail: cherednichenko@karazin.ua; https://karazin.ua/

C. Morar et al. (eds.), *Achieving Sustainability in Ukraine through Military Brownfields Redevelopment*, NATO Science for Peace and Security Series C: Environmental Security, https://doi.org/10.1007/978-94-024-2278-8_28

28.1 Introduction

The quality of education is one of the most important tasks for the sustainable development of the education sector and its reform on the way to achieving the Sustainable Development Goals proposed by the United Nations (UN) in 2015. In recent years, higher education institutions in Ukraine have often faced unpredictable challenges and the need to constantly adapt to changes in the economic and social environment. As a result of unforeseen events, in particular those related to the fight against the spread of COVID-19 (Severe Acute Respiratory Syndrome Coronavirus—2019) and the full-scale invasion of Ukraine by the Russian Federation in early 2022, significant transformations have taken place in both the education sector in general and higher education in particular. And while in the former case, teachers, lecturers, pupils and students had to spontaneously switch to distance learning, in the latter case, the issue of education became unimportant for a while, as it was necessary to first address the deficit needs of Abraham Maslow's pyramid related to basic survival, which include physiological needs (such as food and sleep) and safety needs (such as freedom from danger and security). The high speed of events, which result in fundamental changes, increase the level of risks and uncertainty, necessitates an adequate response and management decisions to ensure the functioning and further development of training of quality specialists in the face of new challenges generated by global factors and national peculiarities.

Therefore, change management has become a critical component of the successful functioning and development of higher education institutions in Ukraine.

This research article is aimed at analysing the problems and developing strategies for managing changes in the management of extracurricular activities in Ukrainian higher education institutions with a focus on sustainable development in times of crisis. This topic is of particular relevance due to the spread of global crises that have caused serious challenges for educational systems around the world.

In the context of extracurricular activities, which cover the administrative and organisational aspects of management in universities and other educational institutions, it is important to understand not only the change management process itself, but also its impact on the sustainable development of these institutions and their ability to perform their core functions [11–16]. The article will provide an overview of theoretical approaches to change management, highlight the specifics of managing extracurricular activities in higher education institutions, and consider the challenges and opportunities of change management in times of crisis. The results of the analysis will allow identifying key aspects of the problem and developing recommendations for practical application in higher education institutions in order to ensure sustainable development in times of crisis.

With the overall goal of improving the quality of education and ensuring sustainable development of society, research in the field of change management of the extracurricular activities in higher education institutions is an important step towards addressing the complex challenges facing the modern educational system.

28.2 Theoretical Overview

In September 2015, the United Nations convened the 70th session of the UN General Assembly in New York, where it adopted the resolution "Transforming our world: the 2030 Agenda for Sustainable Development", which approved 17 Sustainable Development Goals that stimulate social, economic and environmental progress and 169 targets [1]. Achieving the Sustainable Development Goals should significantly improve the situation in the world.

Ukraine also joined the resolution, which recommended that all UN member states develop national programmes with ambitious goals covering all aspects of the implementation of the Agenda as soon as possible. Taking into account this recommendation, the Ministry of Economic Development and Trade of Ukraine, with the support of UNDP, developed the national report "Sustainable Development Goals: Ukraine" [2]. The report adapted the 17 Sustainable Development Goals for Ukraine in the form of 86 targets and, most importantly, defined indicators for each target to assess the implementation of the targets. In particular, Goal 4 "Quality Education" is recognised as one of the main goals.

According to paragraph 23 of Article 1 of Section I of the Law of Ukraine "On Higher Education" "the quality of higher education is the compliance of the conditions of educational activities and learning outcomes with the requirements of legislation and standards of higher education, professional and/or international standards (if any), as well as the needs of stakeholders and society, which is ensured through the implementation of internal and external quality assurance procedures" [3]. And according to paragraph 1 of Article 47 of Section IX of the Law of Ukraine "On Higher Education", "the educational process is an intellectual, creative activity in the field of higher education and science carried out in a higher education institution (research institution) through a system of scientific, methodological and pedagogical measures and aimed at transferring, assimilating, multiplying and using knowledge, skills and other competencies of students, as well as at forming a harmoniously developed personality" [3]. Therefore, the training of a quality specialist by a higher education institution capable of competing at the international level must necessarily include the harmonious development of the individual. The harmonious development of the individual should be understood as a process in which educational activities contribute not only to the acquisition of professional knowledge and skills, but also to the comprehensive development of various aspects of the student's personality. This means that, in addition to academic success, learning also contributes to the development of a student's cultural, social, physical, moral and spiritual potential.

It is worth noting that the term "*extracurricular activities*" refers to a set of national-patriotic, spiritual and mental, cultural, artistic and sports activities in an educational institution aimed at the harmonious development of the personality of students. The implementation of extracurricular activities can also be aimed at participating in public initiatives.

Change management in the context of extracurricular activities management is a strategic and systematic approach to managing the change process in higher education institutions in order to achieve the set goals and improve performance and includes the development of strategies, tools and methods for effective implementation of changes, maintaining stability during the transition and ensuring the active participation of all stakeholders. Some researchers define change management as the process of planning, implementing and controlling changes in order to achieve the organisation's strategic goals and ensure its competitiveness in a constantly changing environment [4]. Other scholars consider change management as a systemic process that includes goal setting, situation analysis, strategy development, change implementation and results evaluation [5].

However, change management in higher education institutions also has its own unique features due to certain specifics of this sector. In particular, the higher education sector faces a high level of complexity due to the diversity of stakeholders (students, faculty, administration, authorities, potential employers, etc.), the importance of ensuring quality of education and adherence to academic standards, and the constant need to adapt to changes in the economic, social and technological environment. Accordingly, management in higher education institutions should be flexible, innovative and focused on achieving the strategic goals of the institution.

Sustainability in change management is becoming increasingly important in today's context and involves taking into account environmental, social and economic aspects when implementing change. It means ensuring a balanced development that meets the needs of the current generation without depriving future generations of opportunities [6]. In the context of higher education, this may mean developing programmes, teaching and management methods aimed at creating a sustainable, healthy and socially responsible environment.

These aspects reflect the importance of integrating the concepts of change management in higher education institutions and sustainable development to achieve successful results in managing change in the university environment.

Kurt Lewin, one of the founders of social psychology, is considered to be the founder of the scientific field of change management, who in the mid-twentieth century proposed the concept of change management, which was later widely used in practice and helped to understand the essence of resistance to change and the stages of the process of behaviour change in groups [7]. His idea is that the basis of any change in an organisation is the change of people associated with that organisation. Therefore, in order to implement changes, it is necessary to make sure that these changes are understood by all its employees for their further implementation. At the same time, Levin's concept states that any changes in most cases cause psychological resistance in many people, which must be taken into account, because it is this resistance that hinders the implementation of changes to the greatest extent. Levin divides the change process into three stages for successful implementation of changes (see Fig. 28.1), which the company must go through [8], namely

1. Unfreezing (destruction of outdated values and approaches);
2. Working on changes (introducing new values and approaches);
3. Refreezing (fixing new values and approaches).

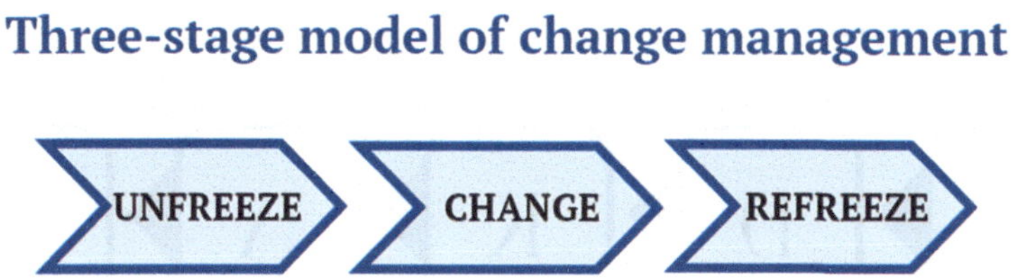

Fig. 28.1 Three-stage model of change management. (Created by the author using the data [7, 8])

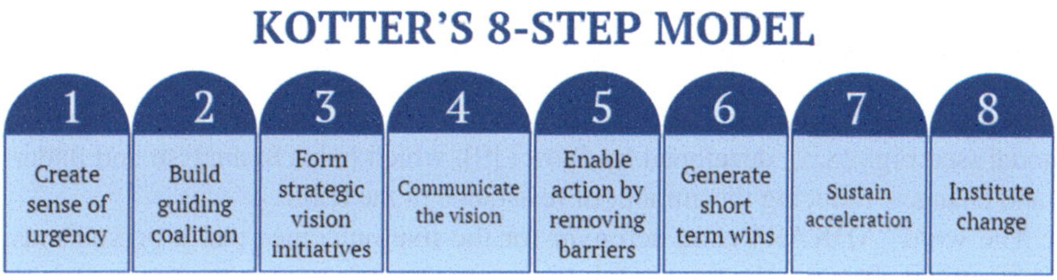

Fig. 28.2 Kotter's 8-step model. (Created by the author using the data [5])

Today, Lewin's concept is considered to be the basis for all subsequent change management schemes and methods.

One of the main methods of change management is also considered to be the **Kotter model,** developed by Harvard Business School professor John P. Kotter [5], which defines eight stages of change management (see Fig. 28.2), namely:

1. Creating a sense of urgency for change to prepare employees for future changes and encourage them to participate in them.
2. Building a team of people who will implement the changes. The team should have a common understanding of the changes, clear roles and encourage other employees to change.
3. Forming a vision and strategy to motivate the team to act.
4. Promoting a new concept of the future to communicate the vision and strategy in a way that encourages the rest of the company to accept and support the change initiative.
5. Creating conditions for broad employee participation in changes to minimise employee resistance to change, management impotence and other risks.
6. Setting short-term goals and achieving them to keep the team motivated while achieving them.
7. Sustaining the gains and deepening the changes to ensure that the work to achieve change is sustained and progress is monitored.
8. Enshrining changes in the corporate culture, including adjustments to organisational norms and values, procedures, reward systems and other infrastructure components to ensure alignment with the new strategy.

Kotter believed that breaking this sequence or abandoning some steps only leads to the illusion of rapid change, but does not lead to the desired result.

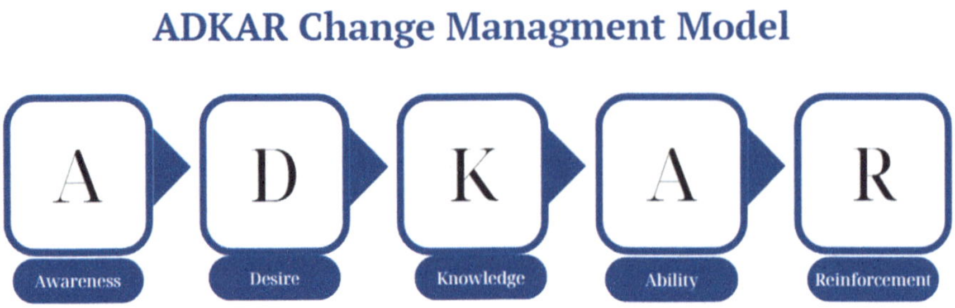

Fig. 28.3 ADKAR change management model. (Created by the author using the data [9])

Also, one of the most effective change management models is the ADKAR model (see Fig. 28.3), developed by Prosci [9], which helps to analyse and understand changes, reducing the amount of resistance in the team.

The word "ADKAR" is an acronym for the five outcomes that a person must achieve for a change to be successful:

1. Awareness—creating awareness of the need for change among employees
2. Desire—focus on the desire and motivation of employees to change
3. Knowledge—At this stage, employees acquire the necessary knowledge and skills to successfully implement change
4. Ability—development of skills and abilities that will allow employees to work successfully in a new context
5. Reinforcement—stabilisation of changes and their consolidation in the organisational culture.

The ADKAR model is used by thousands of change leaders around the world [10]. This powerful model allows organisations to take a systematic approach to change management, ensuring that each stage of the process is effectively executed and that every employee feels supported and motivated to implement change.

In addition to these models of change management, there are many others that are considered in the change management literature, such as the models by Barnes, McConnell, Jeannie Daniel Duck and others. Each of these models has its own specific features and can be applied in a particular change management context.

28.3 Analysis of the Current State of Change Management in the Field of Extracurricular Activities of Higher Education Institutions in Ukraine

Today, higher education institutions in Ukraine face complex challenges that threaten their effectiveness and, in some cases, even their existence. This is due not only to the ongoing military conflict in the country, but also to a number of other factors. First of all, the current situation in the country is characterised by instability and lack of sustainable security, which negatively affects educational institutions and their development opportunities. Ongoing hostilities lead to the destruction of infrastructure, restricted access to educational institutions and danger to participants in the educational process.

Another significant factor is the massive brain drain—outflow of staff and students abroad in search of security and opportunities for quality education. This leads to a decrease in the number of students, as well as the loss of qualified teachers and researchers, which complicates the educational process and research. It is also worth noting the distance learning format in educational institutions in eastern Ukraine, which has become necessary due to the restrictions and health risks associated with their proximity to the contact line. This requires higher education institutions to quickly adapt to new technologies and teaching methods, which is not always possible due to limited technical and financial resources. The unstable financial situation is also a serious challenge for higher education institutions. Limited financial resources make it difficult to implement certain extracurricular initiatives and develop necessary programmes and projects. Another important aspect is the lack of qualified staff with change management and strategic planning skills. This makes it difficult to implement the necessary changes and reforms in higher education.

All these factors point to the need to develop and implement effective change management strategies in the field of extracurricular activities in higher education institutions of Ukraine.

28.4 Practical Aspects of Change Management in the Field of Extracurricular Activities of Higher Education Institutions in Ukraine

To ensure sustainable development in times of crisis, it is important to consider opportunities and strategies for managing change in the field of extracurricular activities of Ukrainian higher education institutions. Change management in times of crisis requires special attention to some key aspects, among which the development of flexible and adaptive strategies that would help to respond quickly to changes is important. At the same time, effective communication with stakeholders and their active involvement in the process is important. The administration and staff of higher education institutions should be provided with the necessary knowledge, skills and competencies to effectively manage change and adapt to new conditions. Effective risk management requires identification and assessment of risks, as well as a strategy to minimise the negative consequences of crisis situations. And for the effective implementation of changes and timely identification of problems to adjust the strategy, continuous evaluation and monitoring of intermediate results is also required. These aspects play an important role in successful change management in a crisis and contribute to achieving positive results in the change process.

Based on the above materials, it is possible to draw some recommendations for change management in the field of extracurricular activities of higher education institutions in Ukraine in times of crisis:

28.5 Adaptation to New Conditions

Representatives of the extracurricular activities of higher education institutions should be prepared to respond quickly to changes in the environment and current trends in society. In order to adapt more effectively, it is important to continuously monitor changes in the economic, social and political environment in order to respond in a timely manner. This may include studying new legislation, economic trends and socio-cultural changes, etc.

28.6 Staff Support and Retraining

This will create a team that is ready to face the challenges of the crisis and ensure sustainable growth for both individuals and the organisation. This can be done by organising trainings and seminars on change management, strategic planning and other important skills to help staff prepare for change. An important aspect is also the organisation of refresher courses, which will allow staff to acquire new knowledge and skills necessary for successful functioning in the new environment. These courses can be specifically aimed at preparing for specific changes or include general development. Providing opportunities for mentoring and coaching will help employees receive individual support and advice on managing change. This will help develop leadership skills and increase motivation for self-development. It is important that management provides support and encouragement for staff to learn and develop. This may include creating a supportive environment for self-improvement and providing resources for learning.

28.7 Development of New Forms of Extracurricular Activities

It is important to create and implement new initiatives that respond to the current challenges and needs of society. Today, these can be online events and/or events with a hybrid form of participation. This will help to develop communication skills between students who want to participate in person and those who can only do so online. It is important to use modern communication technologies that are convenient for all participants in the process.

28.8 Partnership and Cooperation

It is important to engage stakeholders, including government, business, NGOs and the local community, to jointly address issues and implement initiatives. This collaboration can take many forms and provide a variety of benefits. Partnerships with government agencies allow higher education institutions to influence the development and implementation of educational policies and strategies. Working together with local governments can facilitate the involvement of students in various city, regional, national and international events. Cooperation with enterprises and business structures will help to implement measures that will help students develop skills that meet the needs of the labour market. Partnership with NGOs allows higher education institutions to attract additional resources and expertise to implement social and civic initiatives. Joint activities may include the development of volunteer programmes, environmental projects and initiatives to support vulnerable groups. Interaction with the local community allows higher education institutions to influence the development of the region and solve its problems. Working together with local authorities, NGOs and residents helps to create a favourable environment for learning and development.

28.9 Innovative Approaches

The introduction of new and innovative methods of extracurricular activities will contribute to the harmonious development of the individual, and as a result, improve the quality of education and management efficiency. For example, the use of modern information and communication technologies will expand learning opportunities and promote student involvement in extracurricular initiatives. Creating interactive events will also help to increase student engagement. Ensuring a favourable environment for the development and implementation of innovative ideas and projects will help stimulate the creative potential of staff and students.

28.10 Strategic Planning

Developing long-term strategies and action plans should be aimed at achieving sustainable development goals in the face of the crisis.

Strategic planning in the change management of extracurricular activities of higher education institutions is a key element of successful adaptation and sustainable development in times of crisis. For effective strategic planning, it is important to pay attention to the analysis of the situation, including the external environment (economic, social, political and technological factors), internal potential and resources of the institution, as well as current problems and opportunities. Based on the analysis of the situation, strategic goals and objectives should be identified to achieve sustainable development in the face of the crisis. These goals should be specific, measurable, achievable, relevant and time-bound (SMART). Based on the defined goals, strategies are developed, which include areas of activity, methods and approaches to achieve these goals. Strategies should be adaptive and flexible to take into account changes in the environment. Action planning should include the development of specific projects and initiatives aimed at achieving the identified strategic goals. These plans should be detailed, with clearly defined tasks, timelines, and responsible parties. It is also important to constantly monitor the implementation of strategies and action plans, as well as evaluate their effectiveness. Regular monitoring will help to identify problems in time and adjust strategies, if necessary. Implementation of strategic plans requires effective communication with all stakeholders. It is important to ensure the involvement and support of staff, stakeholders and other interested groups for the successful implementation of strategies.

28.11 Social Responsibility

Carrying out activities should take place with due regard to social, environmental and ethical aspects, contributing to the sustainable development of society. This will allow higher education institutions to develop programmes and initiatives aimed at supporting and developing local communities, as well as paying attention to reducing the environmental impact of their activities through the introduction of green initiatives, energy-efficient technologies and renewable energy sources. Promote the involvement of students in projects aimed at ensuring equal opportunities for all, regardless of their social status, race, gender, disability or other characteristics. The importance of adherence to ethical standards should be emphasised. In this way, higher education institutions can create favourable conditions for the development and implementation of social innovations and support initiatives aimed at solving social problems and improving the quality of life.

28.12 Conclusions

In this scientific article, a study was conducted on the issues of change management in the management of extracurricular activities in higher education institutions of Ukraine in order to ensure sustainable development in times of crisis. The study has shown that change management in this area is critical, as it requires adaptability, flexibility and efficiency to successfully overcome crisis situations. The results of the study indicate that a successful change management strategy requires a comprehensive approach that takes into account not only the internal factors of the higher education institution, but also the external environment and the impact of crisis phenomena.

Successful change management in the field of extracurricular activities of higher education institutions in times of crisis requires the development of flexible and adaptive strategies. These strategies should be aimed at solving specific problems arising from crisis situations and take into account the requirements of society and the needs of the higher education institution itself. Active participation of stakeholders and partners is an important component of a successful change management strategy. Joint problem-solving and cooperation contribute not only to the effective implementation of initiatives but also to the creation of a favourable environment for the sustainable development of higher education institutions. In addition, equipping staff with the necessary knowledge and skills in change management is an integral part of the successful implementation of strategies. Supporting staff and providing them with ongoing development and professional development opportunities will help to strengthen the internal culture of change and increase the effectiveness of management in times of crisis.

In view of the above, recommendations for change management in the extracurricular activities of higher education institutions include the active development of flexible strategies, the involvement of stakeholders and partners, and support for staff in relevant training and development. This will facilitate effective change management and ensure the sustainable development of Ukrainian higher education institutions in the context of the crisis.

References

1. UNDP Homepage, https://www.undp.org
2. Ministry of Economy of Ukraine. National Report "Sustainable Development Goals: Ukraine", https://bit.ly/41VEmkl
3. Verkhovna Rada of Ukraine. Law of Ukraine "On Higher Education". https://zakon.rada.gov.ua/laws/show/1556-18#Text
4. Cummings TG, Worley CG (2014) Organisational development and change. Cengage Learning, Boston, MA
5. Kotter JP (1996) Leading change. Harvard Business Review Press, Boston
6. Senge, P. The fifth discipline: the art & practice of the learning organisation (1990)
7. Lewin K (1952) Field theory in social science: selected theoretical papers. Tavistock, London
8. Lewin K (1947) Resolving social conflicts. Harper & Row, New York
9. PROSCI. ADKAR Resources, https://empower.prosci.com/the-prosci-adkar-model-ebook-bundle
10. Hiatt J, Creasey T (2012) Change management. Ingram, Hardback
11. Caciora T, Jubran A, Ilies DC, Hodor N, Blaga L, Ilies A, Grama V, Sebesan B, Safarov B, Ilies G et al (2023) Digitization of the built cultural heritage: an integrated methodology for preservation and Accessibilization of an art nouveau museum. Remote Sens 15:5763. https://doi.org/10.3390/rs15245763
12. Herman GV, Grama V, Ilieş A, Safarov B, Ilieş DC, Josan I, Buzrukova M, Janzakov B, Privitera D, Dehoorne O et al (2023) The relationship between motivation and the role of the night of the museums event: case study in Oradea municipality. Romania Sustain 15:1738. https://doi.org/10.3390/su15021738

13. Grama V, Ilies G, Safarov B, Ilies A, Caciora T, Hodor N, Ilies DC, Kieti D, Berdenov Z, Josan I et al (2022) Digital technologies role in the preservation of Jewish cultural heritage: case study Heyman house, Oradea. Romania Build 12:1617. https://doi.org/10.3390/buildings12101617
14. Marcu F, Ilieş DC, Jan WA, Indrie L, Ilies A, Burta L, Caciora T, Herman GV, Todoran A, Baias S, Albu A, Gozner M (2020) Investigations regarding the biodegradation of cultural heritage. Case study of traditional embroidered peasant shirt (Maramures, Romania). Rom Biotechnol Lett 25(2):1362–1368. https://doi.org/10.25083/rbl/25.2/1362.1368
15. Ilieş DC, Safarov B, Caciora T, Ilieş A, Grama V, Ilies G, Huniadi A, Zharas B, Hodor N, Sandor M et al (2022) Museal indoor air quality and public health: an integrated approach for exhibits preservation and ensuring human health. Sustain For 14:2462. https://doi.org/10.3390/su14042462
16. Eva B-B, Liliana I, Camelia ID, Nicolaie H, Zharas B, Grigore H, Tudor C (2020) Chitosan – a non-invasive approach for the preservation of historical textiles. Ind Textila J 71(6):576–579. https://doi.org/10.35530/IT.071.06.1756

Chapter 29
Trends in the Degradation of Water Resources of Ukraine (on the Example of the Kharkiv Region) During the War

Sviatoslav Dmitriiev

Abstract The article examines the consequences of active hostilities during 2022–2024 on the territory of Ukraine using the Kharkiv region as an example. Taking into account the importance of all types of natural waters (inland surface, coastal, underground, transitional) with the aim of their sustainable use on the basis of strengthening protection and improvement, supporting the vital activity of living organisms, electric power, agriculture, communal housing, transport and their key role in the formation, functioning of regional and global ecosystems, in the presented study they are the object of scientific interest. For the pre-war period, the deterioration of the natural waters of the territory was noted against the background of climate changes and socio-economic activity. The war significantly accelerated the pollution of water resources by heavy metals and oil products and caused difficulties in monitoring and analyzing the characteristics of the Siversky Dinets river basin within the Kharkiv region. As a result of hostilities, many water bodies disappeared. A modern assessment of changes in natural waters for the purpose of their sustainable development and protection will help to develop an action plan for their restoration and preservation. The study highlights the impact of military actions on the surrounding air layer, the occurrence and spread of forest fires, which is a significant factor in the change in the state of water resources. An overview of the pollution of natural waters as drinking water resources, which affects the health of the population that consumes them, is made. Prospects and problems related to the post-war restoration of the state of natural waters in the territory of the Kharkiv region of Ukraine are considered.

Keywords Water resources · Ecological condition · Climate changes

S. Dmitriiev (✉)
V.N. Karazin Kharkiv National University, Kharkiv, Ukraine
e-mail: s.dmitriiev@student.karazin.ua; https://karazin.ua/

C. Morar et al. (eds.), *Achieving Sustainability in Ukraine through Military Brownfields Redevelopment*, NATO Science for Peace and Security Series C: Environmental Security, https://doi.org/10.1007/978-94-024-2278-8_29

29.1 Introduction

Water resources are represented by fresh surface and underground waters that are in water bodies and are used or can be used by humans. Average annual data on Ukraine's water resources are characterized by local river runoff of 52.4 km^3, groundwater available for use—7.0 km^3, total water resources—94.1 km^3 (excluding the Danube). Almost 80% of Ukraine's drinking water supply is provided by surface water. The transit flow is partially used. The rivers Dnipro, Dniester, Siverskyi Donetsk, and Southern Bug are of special importance for water supply. Today, one resident of Ukraine accounts for about 1.2 thousand m^3 of local runoff and 0.16 thousand m^3 of groundwater. This is less than in most European countries. Water consumption per unit of production in Ukraine significantly exceeds indicators in developed European countries: compared to France—by 2.5 times, with Germany—by 4.3 times, with Great Britain and Sweden—by 4.2 times. In the country, the problem of rational use of water resources is urgent: the water load on the Siverskyi Dinets river basin is 41.8%, the Dnieper—20.7%, the Dniester—7.1%, the Southern Bug—13.8%.

The main consumers of water resources are industry—48% of total consumption (power industry, metallurgy, chemical industry), agriculture—40%, where 12% is for housing and communal services. Ukraine's supply of fresh water differs significantly by region: the northern and western regions have high indicators, the southern regions and Crimea have low indicators. Consequently, Ukraine's provision of water resources in comparison with other European countries is characterized as insufficient.

Ukraine's water resources are significantly differentiated by region: they are unevenly distributed across the territory, which is determined by climatic factors and the nature of the surface relief. Climate change and military aggression have put Ukraine's water security at risk, which may lead to water shortages in the future.

The Kharkiv region belongs to the regions with an average degree of water resources security due to a small amount of precipitation and a large population. They play an important role in the regional economic complex: industry, energy, agriculture, and housing and communal services. In addition, they are of absolute importance for the drinking needs of the population and are the most important strategic resource, where any changes in the state of water resources carry catastrophic consequences for the population of the region and its ecosystem.

Russian military aggression in Ukraine, which turned into a full-scale war phase after February 2022, caused an extremely rapid degradation of the ecological state of the natural environment. Along with this, there is constant destruction of infrastructure and industry facilities, as well as intensification of migration processes. A characteristic feature of all the listed processes is their suddenness and huge scale.

The consequences of active military actions are man-made pollution, clogging of water resources with heavy metals, harmful substances, their arbitrary use, which creates a threat to the normal supply of the population's drinking, household, and medical and health needs. The impact of military actions on the country's water resources covers all sectors of the economy: the destruction of bridges, dams, reservoirs, coastlines, where water intake has decreased, discharges of polluted return water have increased, and there is a disparity in the development of water supply and sewage networks in the most vulnerable regions.

An urgent task is to conduct monitoring studies of water resources in order to assess them and create short-term and long-term strategies for their recovery and preservation.

This study is devoted to the analysis of the impact of military actions on the natural waters of Ukraine using the example of the territory of the Kharkiv region, which was the most affected by Russian aggression and is on the line of contact: 4% of its territory at the time of the study (March 2024) is under occupation.

Thus, the purpose of the study is to carry out a comprehensive analysis of the state of the quality of water bodies of Ukraine (on the example of the Kharkiv region) as components of the regional ecosystem and to substantiate possible measures for their restoration, preservation and sustainable use in the post-war period.

The object of this study is the water resources of Ukraine on the example of the Kharkiv region.

To fulfill the set goal, several tasks must be completed:

- to study the change of climatic conditions in the pre-war period;
- to establish changes in the ecological state of surface water resources that occurred in the pre-war period;
- assess the direct consequences of military actions for the condition of surface water bodies;
- assess the role of the consequences of war as an accelerator of changes in the state of the natural environment, highlight the main problems related to the restoration of the state of natural waters.

29.2 Analysis of Literary Sources

Studies of the impact of military actions on the state of natural waters are presented in numerous scientific publications. However, it is characteristic that they mostly relate to the events of 2014–2015 in Ukraine. The military events of 2022–2024 are covered mainly through journalistic articles or interviews of scientists. Attention is drawn to the work of the Ukraine War Environmental Consequences Work Group, which highlights the environmental consequences of the war in Ukraine. The works of N. Lisova, A. Strapak, A. Skok, V. Khrudba and others are also devoted to this topic [6, 11, 12].

A considerable amount of research by domestic and foreign scientists is devoted to research on the reuse of abandoned military facilities (brownfields). Among them, we can highlight the works of Niemets et al. [18] (a study of urban brownfields and grayfields), Berman et al. [19] (visual characteristics of brownfields in the USA and Romania), Kochanov and Tovstiy [17] (in the study of the optimization of the territories of military facilities for the purposes of social inquiry (on the example of Chuguyiv district of Kharkiv region), Morar et al. [20] (regeneration urban brownfields in the aspect of sustainable urban development and cultural heritage, Morar et al. [21] (aspects regarding military cultural and historical heritage in the city of Oradea (Romania), Peric and Miljus [22] (regeneration of military brownfields in Serbia), Klusáček et al. [23] (the nature of the reuse of the former military training ground from the point of view of local players), etc. [17–23].

Among foreign scientists, the works of J. Hupy, Quan Li, A.S. Mihalache-O'Keef, K. Biswas and others [13, 15–33], however, they mostly refer to those events that took place relatively long ago, but mainly popular science or mass media publications are still appearing regarding Ukraine.

29.3 Database and Methods

Like most studies devoted to the study of the natural environment, this work is based on long-term statistical data of stationary observations of the state of the adjacent layer of the atmosphere, surface water bodies and ecological monitoring points. The study of the pre-war climatic, hydrological and ecological conditions of the studied territory includes the analysis of data for the period 1961–2020 for hydrological and climatological monitoring and for the period 1995–2020 for ecological monitoring.

The sample of hydrological and climatological information is representative, since the period covered is more than 30–40 years and corresponds to the recommendations of the World Meteorological Organization. For environmental information, no similar requirements are made, as these studies are relatively new and are presented in this study as fully as possible, based on the available data. In order to highlight the complete list of observation points that were involved in this study, accompanying cartographic material was compiled (Fig. 29.1).

The research methodology includes system analysis, as well as system, complex and cartographic approaches. The research also used general scientific, cartographic, statistical, comparative-geographical, comparative-historical methods. Among general scientific methods, the methods of analysis and synthesis, induction and deduction, systematic, comparison, systematization, classification and cartographic visualization prevail.

29.4 Research Results

29.4.1 General Characteristics of the Studied Territory

Kharkiv region is one of the largest regions of Ukraine, before the war it was the fourth largest in terms of population and area [5]. In terms of its size, its territory can be compared with such European countries as Belgium, Montenegro, and

Fig. 29.1 Kharkiv region. Relief and monitoring points (created by the authors using the data [14])

Albania. The region is located in the eastern part of Ukraine, on the Eastern European Plain in the temperate continental region of the temperate zone of the Northern Hemisphere. Accordingly, the climate is characterized by relatively wet winters with unstable frosts and snow cover, hot and moderately dry summers.

The river system of the region is not very dense compared to the western regions of Ukraine due to the lower amount of precipitation and consists of two large basins—the Dnipro and the Don. They are represented by the rivers Oril and Siverskyi Donets, respectively, with many tributaries. The water regime of local rivers can be characterized as Eastern European (according to Zaikov's classification. link) with a clearly defined spring flood, as well as summer and winter demarcation. Significant regulation of the Siverskyi Donets and Oskil rivers by reservoirs is indicated. For example, according to the post in Pecheneg, the spring flood is less pronounced than in others, since it is located directly near the dam of the Pecheneg Reservoir. The water is released gradually over several months, which reduces the contrast between the summer low and the flood.

Forest resources of Kharkiv region, which act as a powerful temperature regulator and stabilizer of the ecosystem, are very limited, the forest cover of the territory is only 13.3% and is characterized by a long decline during the 1990s and 2000s. areas due to a decrease in the volume of their felling in the territory of the Kharkiv region.

29.4.2 Climate Changes in the Pre-war Period

For a more complete overview of the pre-war dynamics of the state of the environment, it is worth dividing it into two parts—climatological and hydrological. Therefore, the dynamics of air temperature, rainfall, and then water temperature, water level and flow rate are considered first.

To determine the specified changes, the total sample period was divided into two periods of equal duration: 1961–1990 and 1991–2020. Thus, it made it possible to more carefully assess the nature and speed of changes in temperature indicators and precipitation indicators. In addition, the period 1961–1990 is considered the climate norm according to the World Meteorological Organization, and the period 1991–2020 from 2019 is considered the "new climate norm".

Average air temperatures in January increased the fastest among average monthly temperatures, accompanied by a sharp decrease in the number of days with snow cover and frost. The cases when average daily temperatures of +4°C and above were kept for several weeks became not unique. According to the calculated trends (Fig. 29.2), January temperatures increase at a rate of 0.7–0.9 °C every 10 years.

Average monthly temperatures in July are also rapidly increasing at all weather stations in the region at a rate of 0.3–0.4 °C every 10 years (Fig. 29.2). The distribution of temperature increase by season is characterized by the largest values in summer and winter and smaller values in spring and autumn. So, in autumn, their growth

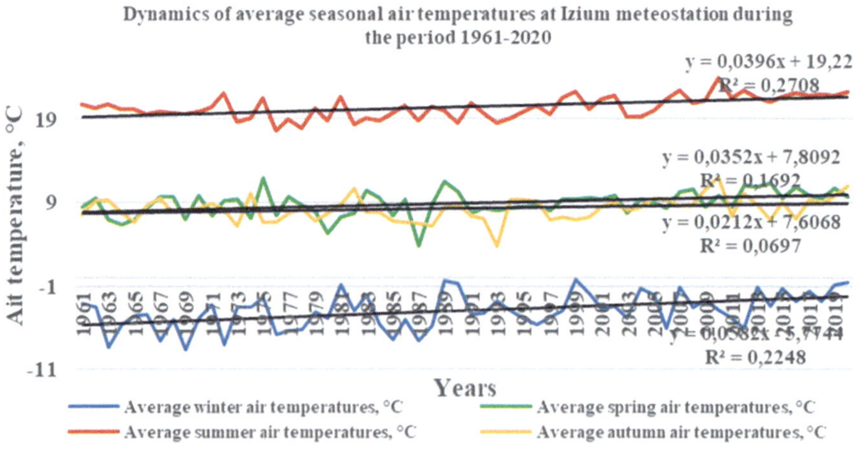

Fig. 29.2 Dynamics of average seasonal air temperatures at Izium meteostation during the period 1961–2020 (created by the authors using the data [7–10])

by 0.5–1.0°C is characteristic. In the spring, they grow faster - from 0.9 to 1.4°C. This is due to the variability of weather conditions in these seasons, due to which they are characterized by very large intraseasonal temperature fluctuations, and therefore the overall growth is not as fast as in more stable seasons (Fig. 29.2).

The dynamics of the amount of precipitation in the studied territory in the pre-war period was characterized by a decrease in their amount, except for the districts of Slobozhansk and Kupyansk. The heterogeneity of this dynamics is explained by the fact that it completely depends on long-term baric fields and features of the atmospheric circulation, which are very unstable. Thus, the climatic dynamics of the studied area corresponds to the general trends of global climate changes, increases the pollution of natural waters, since the warming of the climate and the decrease in the amount of precipitation contribute to the reduction of the flow volumes in the rivers.

29.4.3 Changes in the Condition of Surface Water Bodies

Due to the lack of measurements in the winter period, since measurements are not carried out if the water bodies are covered with ice. July water temperatures are a significant indicator that demonstrates changes in the summer regime and indicate an increase in water temperature indicators, except for Kozachoa Lopan. The rate of this growth is insignificant compared to air temperature indicators, because water warms up and cools down more slowly due to its high heat capacity, and therefore, the multi-year dynamics will also be less noticeable. Seasonal indicators of water temperature are additional markers of changes in the thermal regime of the surface waters of the Dinets basin. They grow the fastest in the summer period (by 0.37°C every 10 years), autumn temperatures also have changes in the direction of growth.

The nature of river water dynamics is described by two indicators—flows and water level. Since they are interdependent and have common trends, it is appropriate to consider them together. The most noticeable decrease in water consumption and water level during the study period is characteristic of the Ohirtseve, Zmiyiv, Protopivka and Izyum hydroposts, on the tributaries of Siverskiy Dinets (Bezlyudivka and Vovchansk). The least—for Kozachoya Lopan and Pechenegs. It is worth noting that the nature of river water dynamics, in addition to the climatic factor, also depends on management, because all large rivers in the Kharkiv region are significantly regulated by numerous dams and reservoirs, which strongly influence the water regime.

The dynamics of the river's water level is best illustrated by the difference integral curve constructed for all the main watercourses of the region (Fig. 29.3). Like the curve for the Siversky Dinets and the rivers of the Dnipro basin in the Kharkiv region, it shows the presence of two complete cycles and one incomplete one. Since 2007, there has been a low-water phase, which continues even now. It is characterized by rather rapid shallowing of rivers, which is accompanied by their

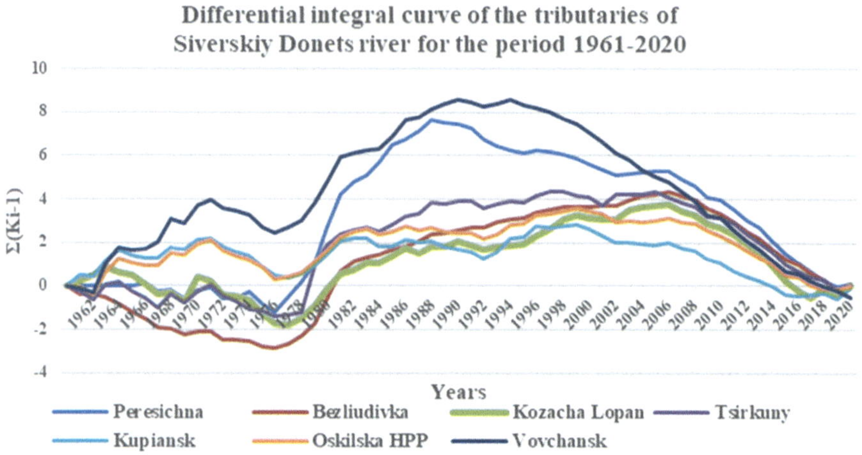

Fig. 29.3 Differential integral curve of the tributaries of Siverskiy Donets river for the period 1961–2020 (created by the authors using the data [1–3])

waterlogging, division into branches, more rapid warming, and, as a result, intensification of bioactivity. Because of this, small rivers (Fig. 29.3) suffer especially, because they have less total water content, less depth, and even minor temperature fluctuations and water regime fluctuations are fatal for them. This has become a significant problem for the local population, because in addition to the disappearance or pollution of surface watercourses, deterioration is also observed in the condition of underground water resources, which are the main source of drinking water, and in rural areas, in the absence of other sources, the disappearance of water in wells and columns has put many villages on the edge of survival. This problem is especially evident in the southern part of the region, which is located on the watershed and is characterized by less rainfall.

Thus, the pre-war dynamics of the state of surface water bodies in the Kharkiv region is characterized by an increase in water temperature, which is not as fast as in the atmosphere, due to the physical properties of water, as well as a rapid decrease in the water flow of rivers, which is accompanied by their shallowing, more intense warming, and, as a result, more intense shoaling. That is, this process is cyclical, and its elements reinforce each other. The situation is especially dangerous with small rivers, which, together with underground waters, served as the only source of water in rural areas, where there is no organized water transport.

29.4.4 Assessment of the Pre-war Dynamics of the Ecological State

Before the war, the Kharkiv region was not one of the leading regions in Ukraine in terms of atmospheric air and surface water pollution, because the large industrial complex that was located here was almost completely dismantled during 1991–2020, and therefore the industrial load decreased. Along with this, due to population depopulation, the demographic burden has also decreased. However, the traffic load has increased due to the increase in the number of private vehicles, as well as the level of surface water pollution with municipal waste and garbage has increased due to a decrease in the efficiency of their cleaning, lack of control over the condition of rivers, lakes and reservoirs. Having a forest cover lower than the average for Ukraine, the Kharkiv region was nevertheless one of the leading regions in the number of illegal logging. However, during the last decade, the rate of planting trees began to outpace the rate of felling for the first time, but this still did not solve the problem of forest resources, because it takes many years to restore the forest resources that have been cut down, moreover, it is impossible to be sure that all the planted trees will definitely grow.

The question of the state of water resources arises separately. According to the results of the research, it was possible to establish that the rivers around the city of Kharkiv (Table 29.1) and the Oskil River are the most polluted. In addition to pollution within the borders of Ukraine, transboundary pollution also occurs, because almost all rivers in the region begin on the territory of Russia and continue to flow through the territory of Ukraine. For example, in the village of Ohirtseve on the border with Russia, the water pollution index is twice as high as the average for the Siverskyi Donets River (Table 29.1), which indicates the introduction of harmful substances by the flow of the river from another country.

Table 29.1 Water pollution (on the example of the Siverskiy Donets river basin) (created by the authors using the data [4, 5])

River	Monitoring point	Average water pollution index (WPI, 1995–2020)	Water quality class
Siverskiy Donets	Ogirtseve	4,00	V. Dirty
Siverskiy Donets	Chuguiv (h.t.)	0,99	II. Clean
Siverskiy Donets	Chuguiv	2,24	III. Moderately polluted
Siverskiy Donets	Zmiiv	1,75	III. Moderately polluted
Siverskiy Donets	Izium	3,72	IV. Polluted
Kharkiv	Tsirkuny	4,18	V. Dirty
Lopan	Kharkiv	5,53	V. Dirty
Lopan	Kozacha Lopan	3,36	IV. Polluted
Udy	1 km	7,00	VI. Very dirty
Udy	10 km	3,70	IV. Polluted
Oskil	Kupiansk	3,61	IV. Polluted
Vovcha	Vovchansk	1,17	III. Moderately polluted

Therefore, the chemical composition of air pollutants in the Kharkiv region in the pre-war period is characterized by the predominance of formaldehyde and sulfur dioxide. They gravitate to the southern basin, to the area of the city of Izyum and the Dnipro-Donbas channel, respectively. The concentration of sulfur dioxide is observed in places where there are forests in the Kharkiv region, which are potentially dangerous areas for the occurrence of fires that annually cover the Kharkiv region in the summer.

The surface waters of the studied area are most polluted with nitrogen dioxide, copper and nitrates, which indicates pollution mainly by agricultural products, namely, mineral fertilizer residues. In addition, there is a characteristic phenomenon when substances introduced into the soil as early as the 60 s of the twentieth century, gradually seeping through the soil, enter the surface water 50–60 years after their introduction, since they are not well infiltrated by the soil cover.

29.5 Military Operations in the Kharkiv Region and Their Consequences

29.5.1 The Nature of Military Operations in the Studied Territory

Kharkiv region is one of the first regions where the war began on February 24, 2022. Almost immediately, the main clashes began in the northern suburbs of Kharkiv, as well as near Kupyansk. By the end of March 2022, almost a third of the region was under occupation.

A feature of this military conflict is the active use of ground artillery, as well as missile weapons. In addition, the mining of territories and the destruction of infrastructure facilities became almost universal. All this endangered the ability of the population to live in this area due to the lack of housing, rapid pollution of water resources and problems in the cultivation of agricultural fields due to mining. In September 2022, a quick Ukrainian counteroffensive took place, and by October 2022, only 4% of the region, located in the northeast of the region, remained under occupation. After that, the nature of the war changed. Massive attacks by missiles and drones, mainly on infrastructure objects and housing stock, have become commonplace. However, despite this, gradual demining of the territory has begun, and currently the region is gradually returning to normal living conditions. However, along the northern and eastern borders of the region, constant artillery shelling and air raids are observed, which cause significant destruction of infrastructure and high atmospheric air pollution.

29.5.2 Modern Dimension of Environmental Consequences of War

As mentioned above, the conduct of the war is intensive, all types of weapons are used, except for nuclear weapons, which caused significant pollution of the air, water bodies and soil. In this study, water bodies and their condition are considered, however, taking into account the absolute influence of atmospheric air on the condition of water bodies, it is appropriate to consider its ecological condition under the influence of military actions.

Atmospheric Air Pollution There are two ways of military-provoked influence on the state of the air: direct (explosions of various types of weapons, use of air bombs, artillery and rockets) and indirect (hitting shells at oil depots, fires in forest areas and settlements). Given the impossibility of conducting full-fledged environmental monitoring according to the usual algorithm in the conditions of active military operations, there are almost no data on the state of the atmospheric air for 2022–2024, and those that are available are incomplete due to the fact that monitoring was resumed at many points only at the end of autumn 2022 r. Therefore, satellite data covering the whole of Ukraine was more widely used, but it is quite possible to interpolate it to the territory of the Kharkiv region as well. According to these data, the air is most polluted by nitrogen oxides, sulfur, carbon dioxide, lead and soot, which are released during the reaction of explosive compounds from devices (artillery shells, rockets, aerial bombs). Almost 4000 drones and 5000 missiles launched over Ukraine (Fig. 29.4) leave a significant carbon footprint, which, in addition to simply polluting the air, also accelerates climate change. The fact is that most of the gases released during explosions are greenhouse gases and contribute to an increase in the average air temperature.

Pollution and Destruction of Water Bodies Emissions of harmful substances into the atmospheric air also contribute to surface water pollution, since water bodies are in direct interaction with the surface layer of the atmosphere. In addition, water pollution by heavy metals and mainly oil products is a serious problem. Motor vehicles left near the coast, the destruction of bridges with fuel tankers on them, attacks on oil depots—all this contributes to the spillage of fuel and lubricants into the water. A striking example is the attack on the oil depot in Kharkiv in February 2024, when as a result the Nemyshlia and Kharkiv rivers, and then Lopan and Udy rivers were covered with an oil film, which caused the mass death of aquatic life and significantly polluted the water. A separate problem remains the issue of the Oskil reservoir, which was destroyed in April 2022 (Fig. 29.4), and currently only a tenth of its volume remains. This became the biggest ecological disaster during the war in the Kharkiv region, as it caused extensive flooding downstream, and contributed to the radical change of local landscapes, and also caused the impoverishment of the local organic world and contributed to the lowering of the groundwater level. Pollution also affected groundwater. Thus, in the city of Kharkiv, among 11 officially regis-

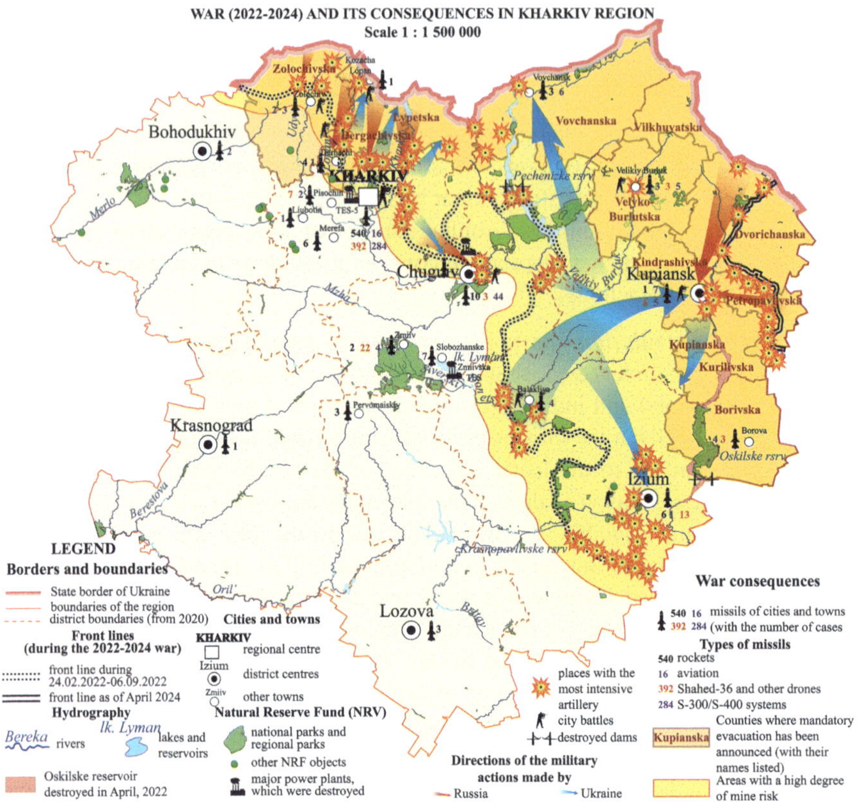

Fig. 29.4 War (2022–2024) and its consequences in Kharkiv region (created by the authors using the data [4, 15])

tered sources of drinking water, it is suitable for drinking only in 1, which is a huge problem for a city with a population of more than 1 million people.

Forest Fire The hostilities caused numerous forest fires, for example, the famous Izyum forest, which before the war was supposed to become another national natural park, burned almost completely. Another well-known example is the forests near the city of Kupyansk, where hostilities have been taking place for more than a year. The PZF objects that suffered from the war should be considered separately (Fig. 29.4). Thus, during the occupation, the Dvorichansky national park, the Izyumska luka regional park, many reserves and natural monuments, including Mount Kremyanets, the highest point of the Kharkiv region, were in the zone of direct hostilities.

The Impact of the War on Drinking Water Resources and the State of Health of the Population As already mentioned above, the pollution of water bodies as a result of military operations became the reason for the decrease in the number of sources with potable water suitable for drinking. This problem was particularly acute in Kharkiv, where the population density is significant, and which, even before the war, had the problem of a lack of drinking water. In addition, in the northern and eastern parts of the region, where active hostilities are still taking place, it is impossible to establish a stationary water supply, and water has to be brought in from other regions of the region, which increases its cost and worsens the living conditions of the population. However, sometimes in conditions of a shortage of drinking water, local residents were forced to consume dirty untreated water, as was the case in the northern regions of the region during their occupation in 2022. In addition to water, the health of local residents is negatively affected by constant shelling, during which heptyl compounds are released, which damages the central nervous system, mucous membranes, and respiratory organs. And a person can lose consciousness from a large amount of the component.

Thus, military actions caused significant pollution of natural waters with oil products and heavy metals, air with nitrogen oxides, sulfur, carbon dioxide, lead and soot and made it impossible to consume drinking water in most sources in the region, and especially in the city of Kharkiv.Харкові.

29.5.3 Prospects for Restoring the Pre-war State of Water Resources

Today, it is necessary to take into account both the impact of military actions on water resources and the risks of climate change. In recent decades, the country has seen a rapid decline in the number of water bodies: more than 10,000 small rivers have completely disappeared, and the water level in the more full-water ones has dropped by a meter and a half below the seasonal minimum. Not only drinking water supplies are under threat, with the disappearance of rivers, unique ecosystems in the lower river valleys are dying. Agriculture accounts for about 70% of water consumption, where water losses due to inefficient irrigation technologies account for up to 80%. Excessive abstraction of surface water for irrigation of agricultural land, illegal construction of dams has catastrophic consequences: the flow of an already silted river is disturbed, the bottom is silted up—the reservoir turns into a swamp.

According to FAO, about 40% of the world's population suffers from a shortage of water resources. In the near future, Ukraine risks adding to the list of countries with particularly difficult access to water. If the current trend continues, the majority of the country's territory risks turning into a semi-desert, the southern regions—into a desert. The question arises of restoring the water potential of the territory of Ukraine, polluted soils and air, burned forests, as well as management of the transboundary basins of the Dniester, Danube, and Prut rivers. After all, the consequences of climate change, floods and droughts are felt in river basins and pose a threat to many countries. 50% of the regions of Ukraine have a natural shortage of fresh water, where for 80% of the population of Ukraine, the state of drinking water is unsatisfactory.

Constant attacks on water facilities and infrastructure are a threat to the sustainable use and protection of water resources. The destruction of the dam of the Kakhovka hydroelectric power station caused an ecological disaster in the south of the country, the Black Sea, which requires further monitoring studies and management decisions. There is also the issue of restoring the Oskil reservoir, which is the most destroyed facility in the Kharkiv region. The existing management system has changed river ecosystems, which today are mostly streams or dehydrated deserts. Therefore, the implementation of Ukraine's water policy, sustainable management of water resources, international experience of environmental action programs for each river basin will become important stages on the way to the post-war restoration of Ukrainian water bodies.

29.6 Conclusions

Ukraine expects a further decrease in the reserves of surface and underground water available for use, which will also be accompanied by a deterioration in their quality. The deterioration of water quality is influenced by climate changes, anthropogenic stress and military actions, which brought a negative impact and significantly worsened the situation. Also, in recent years, the role of the climatic factor is constantly growing.

The pre-war state of natural waters in the Kharkiv region is characterized by a warming trend, which causes changes in the ecological state due to increased pollution of water resources in the area of the city of Kharkiv and in the north of the region. The military actions caused irreparable damage to the ecosystem of the territory: the Oskil reservoir was destroyed, the quality of drinking water deteriorated, large areas of forests were burned, and many objects of the nature reserve fund were damaged.

The decrease in water content of rivers due to climate fluctuations, the impact of military actions on water resources should be taken into account in the further long-term planning of water supply and water consumption of the territory, in their post-war restoration.

References

1. State Water Cadastre (1981) Long-term data on the regime and resources of land surface waters. Part 1. Rivers and canals, vol 3. Basin of Siversky Dinets, Azov and Crimean rivers (1961–1980). Kyiv, 2-406
2. State Water Cadastre (2001) Long-term data on the regime and resources of land surface waters. Part 1. Rivers and canals, vol 3. Basins of the Siversky Dinets, the Azov and Crimean rivers (1981–2000). Kyiv, 2-409

3. State Water Cadastre (2021) Long-term data on the regime and resources of land surface waters. Part 1. Rivers and canals, vol 3. Basins of the Siversky Dinets, the rivers of the Azov region (2001–2020). Kyiv, 2-401
4. Ecological bulletin of the state of surface waters. Kharkiv, 1995–2020, 15 p
5. Ecological passport of the Kharkiv region. Kharkiv, 2015–2020, 208 p
6. Lisova N (2017) Impact of military actions in Ukraine on the ecological state of the territory. In: Volodymyr Hnatyuk Volodymyr Hnatyuk Ternopil National Pedagogical University notes. Series: geography, vol. 43 No. 2, pp 165–173. Ternopil
7. Meteorological monthly (2021) Kyiv, 1992–2021. Vol. No. 1 January–March, 490 p
8. Meteorological monthly (2021) Kyiv, 1992–2021. Vol. No. 2 April–June, 501 p
9. Meteorological monthly (2021) Kyiv, 1992–2021. Vol. No. 3 July–September, 491 p
10. Meteorological monthly (2021). Kyiv, 1992–2021. Vol. No. 4 October–December, 500 s
11. Skok A, Hrutba V, Rak O (2023) Atmospheric air quality in Ukraine before and during a full-scale invasion. International Revival Fund, Kyiv
12. Strapak A (2023) Ecological dimension of the Russian-Ukrainian war. In: Bulletin of the DonNU student scientific society named after Vasyl Stus, vol 1, No. 15, pp 77–83. Vinnytsia. Biswas, Asit K: Scientific assessment of the long-term environmental consequences of war. In: Austin JE, Bruch CE (eds) The environmental consequences of war: legal, economic, and scientific perspectives. Cambridge University Press, Cambridge, pp 303–315
13. Earthdata Search. URL: https://search.earthdata.nasa.gov/search. Last accessed 2016/04/01
14. Hryhorczuk D, Levy BS, Prodanchuk M et al (2024) The environmental health impacts of Russia's war on Ukraine. J Occup Med Toxicol 19:1. https://doi.org/10.1186/s12995-023-00398-y. Last acessed 2016/04/06
15. Hupy JP (2008) The environmental footprint of war. Environ History 14(3):405–421. http://www.jstor.org/stable/20723680. Last accessed 2016/04/05
16. Reuveny R, Mihalache-O'Keef AS, Li Q (2010) The effect of warfare on the environment. J Peace Res 47(6):749–761. https://doi.org/10.1177/0022343310382069. Last accessed 2016/04/02
17. Kochanov EO, Tovstiy YM (2012) Optimization of the territories of military facilities for the purposes of social demand (on the example of Chuguyiv district of Kharkiv region) // Human and environment. Problems of neoecology: coll. of science pr. – Kharkiv: V.N. Karazin KhN. – No. 3–4, pp 145–152. http://luddovk.univer.kharkov.ua/sites/default/files/Papers/19-12-18.pdf. Last accessed 2016/04/02
18. Niemets L, Husieva N, Suptelo O, Sehida K, Kravchenko K (2018) Research of brownfields and greyfields of the city: theory and practice. Sustainable economic development and application of innovation management from regional expansion to global growth. Proceedings of the 32nd International Business Information Management Association Conference (IBIMA) (Seville, Spain, 15–16 November 2018). Editor Khalid S. Soliman. Seville, International Business Information Management Association (IBIMA), pp 409–435. https://ibima.org/accepted-paper/research-of-brownfields-and-greyfields-of-the-city-theory-and-practice/. Last accessed 2016/04/02
19. Berman L, Morar C, DeGrane L, Unkart S, Erdal S (2022) Brownfields in Romania and the United States: a visual tour. J Environ Health 85(5):28–38. PMID: 37204986; PMCID: PMC10190036
20. Morar C, Dulca M, Nagy G (2016) Brownfields regeneration, between sustainable urban development and cultural heritage. The former military sites in Oradea, Romania. J Urban Reg Anal VIII(1):75–84. https://www.jurareview.ro/resources/pdf/volume_17_brownfields_regeneration_between_sustainable_urban_development_and_cultural_heritage_the_former_military_sites_in_oradea_romania_abstract.pdf. Last accessed 2016/04/02
21. Morar C, Nagy G, Dulca M, Boros L, Sehida K (2019) Aspects regarding the military cultural-historical heritage in the City of Oradea (Romania). Annales-Anali Za Istrske in Mediteranske Studije-Series Historia et Sociologia (Annals for Istrian and Mediterranean Studies Series

Historia et Sociologia) 29(2):303–322. https://doi.org/10.19233/ASHS.2019.21. Last accessed 2016/04/02
22. Peric A, Miljus M (2021) The regeneration of military brownfields in Serbia: moving towards deliberative planning practice? Land Use Policy 102:105222. https://doi.org/10.1016/j.landusepol.2020.105222. Last accessed 2016/04/02
23. Klusáček P, Martinát S, Krejčí T, Bartke S (2019) Re-development of a former military training area—the case of Brdy told from a local actors' perspective. Land Use Policy 82:147–157. https://doi.org/10.1016/j.landusepol.2018.11.045. Last accessed 2016/04/02
24. Ponjiger MT, Tin L, Basarin B, Maja J, Wilby RL, Pavic D, Mesaroš M, Valjarevic A, Milanovic MM, Morar C (2021) Detailed analysis of spatial–temporal variability of rainfall erosivity and erosivity density in the Central and Southern Pannonian Basin. Sustain For 13:13355. https://doi.org/10.3390/su132313355
25. Lukić T, Ponjiger TM, Basarin B, Sakulski D, Gavrilov M, Marković S, Zorn M, Komac B, Milanović M, Pavić D, Mesaroš M, Marković N, Durlević U, Morar C, Petrović A (2021) Application of Angot precipitation index in the assessment of rainfall Erosivity: Vojvodina region case study (North Serbia). Acta Geogr Slov 61(2):123–153. https://doi.org/10.3986/AGS.8754
26. Morar C, Tiba A, Jovanovic T, Valjarevic A, Ripp M, Vujicic MD, Stankov U, Basarin B, Ratković R, Popović M, Nagy G, Boros L, Lukic T (2022) Supporting tourism by assessing the predictors of COVID-19 vaccination for travel reasons. Int J Environ Res Public Health 19(2):918. https://doi.org/10.3390/ijerph19020918
27. Valjarević A, Popovici C, Djekić T, Morar C, Filipović D, Lukić T (2022) Long-term monitoring of high optical imagery of the stratospheric clouds and their properties new approaches and conclusions. Egypt J Remote Sens Space Sci 25(4):1037–1043. https://doi.org/10.1016/j.ejrs.2022.11.006
28. Blešić I, Ivkov M, Tepavčević J, Raljić JP, Petrović MD, Gajić T, Tretiakova TN, Syromiatnikova JA, Bajrami DD, Aleksić M, Vujačić D, Kričković E, Radojković M, Morar C, Lukić T (2022) Risky travel? Subjective vs. objective perceived risks in travel behaviour—influence of hydro-meteorological hazards in South-Eastern Europe on Serbian tourists. Atmosphere 13(10):1671. https://doi.org/10.3390/atmos13101671
29. Valjarević A, Algarni S, Morar C, Grama V, Stupariu M, Tiba A, Lukić T (2023) The coastal fog and ecological balance for plants in the Jizan region, Saudi Arabia. Saudi J Biol Sci 30(1):103494. https://doi.org/10.1016/j.sjbs.2022.103494
30. Valjarević A, Popovici C, Štilić A, Radojković M (2022) Cloudiness and water from cloud seeding in connection with plants distribution in the Republic of Moldova. Appl Water Sci 12:262. https://doi.org/10.1007/s13201-022-01784-3
31. Valjarević A, Milanović M, Gultepe I, Filipović D, Lukić T (2022) Updated Trewartha climate classification with four climate change scenarios. Geogr J 188:506–517. https://doi.org/10.1111/geoj.12458
32. Valjarević A, Filipović D, Milanović M, Valjarević D (2020) New updated world maps of sea-surface salinity. Pure Appl Geophys 177:2977–2992. https://doi.org/10.1007/s00024-019-02404-z.5
33. Valjarević A (2024) Long-term remote sensing-based methods for monitoring air pollution and cloud cover in the Balkan countries. Environ Sci Pollut Res 31:27155–27171. https://doi.org/10.1007/s11356-024-32982-y

Chapter 30
Urbicide in Eastern Ukraine (Case Study of the City of Kharkiv, Ukraine)

Liudmyla Kliuchko, Elizaveta Bedriy, Hanna Kucheriava, and Nataliia Husieva

Abstract Contemporary cities play pivotal roles in economic, geopolitical, scientific, and technological realms, driving innovation, fostering productivity, and shaping global dynamics. In addition, cities are centers of education, culture, art, and technological progress, influence public opinion, and shape worldviews. They attract talent, investment, and business activity, facilitating the exchange of goods, services, and ideas on a global scale. Geopolitically, cities often serve as nodes of power and influence, hosting key institutions, government agencies, and diplomatic missions. They can shape regional and global dynamics due to their strategic location, economic potential, and cultural significance. However, they are also vulnerable targets for military aggression and terrorism. The shelling of the civilian population, the systematic destruction of residential buildings and historical and cultural monuments, and the destruction of social and critical infrastructure—this is how russia's urbicide and genocide against the civilian population and cities of Ukraine are manifested. Kharkiv, as a major economic, cultural, and educational hub, has endured relentless shelling due to its proximity to the russian border. This article examines the directed urbicide of the Russian army against Ukrainian cities, focusing on Kharkiv as a case study. The authors delineated the stages and forms of urbicide in Kharkiv, analyzed factors influencing the intensity of shelling, and investigated the temporal and spatial features of urbicide. The most affected areas of the city have been identified, and the direct and lasting impacts of urbicide on Kharkiv have been explored.

Keywords Urbicide · Destruction of cities · Kharkiv

L. Kliuchko (✉) · E. Bedriy · N. Husieva
V.N. Karazin Kharkiv National University, Kharkiv, Ukraine
e-mail: ludmila.klychko@karazin.ua; bedriy2020.8950449@student.karazin.ua; nataliya.guseva@karazin.ua

H. Kucheriava
Kyiv National Linguistic University, Kyiv, Ukraine
e-mail: hanna.kucheriava@knlu.edu.ua

C. Morar et al. (eds.), *Achieving Sustainability in Ukraine through Military Brownfields Redevelopment*, NATO Science for Peace and Security Series C: Environmental Security, https://doi.org/10.1007/978-94-024-2278-8_30

30.1 Familiarization and Review of Previous Scientific Developments in This Field

30.1.1 Introduction

The twenty-first century is marked by dynamic processes and phenomena such as urbanization, globalization, technological progress, and innovation, impacting various aspects of social life from demography to geopolitics [1–5]. However, despite human achievements, armed conflicts, terrorist attacks, and wars persist and even intensify, posing significant threats to global stability, security, and development [6–8].

In an interconnected world, the repercussions of conflicts extend beyond immediate areas, affecting neighboring regions and global networks. Addressing these challenges necessitates collective action, informed decisions, international cooperation, and a commitment to human rights, peace, and security.

Contemporary cities play a crucial role in shaping global trends and influencing public discourse. The dynamics of global geopolitical confrontations are reflected in urban environments, creating unique challenges [9]. Conflicts and military operations increasingly occur in cities, presenting complexities and posing risks to civilian populations. Paradoxically, the pursuit of national security can sometimes exacerbate insecurities within cities [10].

One form of violence and military aggression taking place in cities is *urbicide*. Urbicide entails the deliberate and systematic destruction of cities, including their architecture and cultural identity. In urbicide, the city itself becomes the primary target, with perpetrators aiming to eradicate not only physical structures but also the collective memory, heritage, and identity tied to urban spaces [11]. This deliberate assault on cities distinguishes urbicide from other forms of violence as it aims to undermine the essence of urban life and civilization. The term "urbicide" is closely associated with "genocide" and often co-occurs in discussions regarding systematic destruction and violence against populations.

Genocide is a crime under international law. Article 2 of the Convention on the Prevention and Punishment of the Crime of Genocide [12] defines genocide as specific acts committed with the intent to destroy, in whole or in part, a national, ethnic, racial, or religious group [12]. The term was first coined by Polish-Jewish lawyer Raphael Lemkin in 1944 in his book "The Axis Rule in Occupied Europe" [13] and has since been enshrined in international law, especially the UN Convention on the Prevention and Punishment of the Crime of Genocide, adopted in 1948.

The contemporary era is marred by excessive brutality both on the battlefield and in peaceful areas. Regrettably, the cities of Ukraine have become a "testing ground" for the unchecked actions of the russian army.

The war in Ukraine has persisted for over two years, marked by egregious violations of international humanitarian and human rights law. Deliberate attacks on civilians, indiscriminate shelling of settlements and cities, and targeting of both critical infrastructure and hospitals and schools have caused significant civilian casualties, forced displacement, and immense suffering.

Urbicide in Kharkiv, akin to similar conflicts in other urban centers, yields far-reaching consequences that extend beyond its immediate vicinity. This situation highlights the interconnectedness of global security and underscores the imperative of swift and coordinated crisis response.

The ramifications of urbicide in Kharkiv pose a direct threat to regional stability, the security of the entire country, and have implications for the international community at large.

30.1.2 Theoretical and Methodological Foundations of the Study of Urbicide

The purpose of the research is to study and analyze the spatial aspects and consequences of urbicide, with a focus on the city of Kharkiv. In this regard, the authors set out several objectives:

1. To identify the factors influencing the spatial features of the urbicide in Kharkiv.
2. To delineate the stages and forms of urbicide in Kharkiv. This involves outlining the various stages and forms of urbicide in Kharkiv, with the aim of comprehensively understanding this process.
3. To examine the spatial characteristics of the consequences of military shelling of the city.

Through addressing these objectives, the study aims to contribute to a better understanding of the phenomenon of urbicide and its implications for the urban environment.

The concept of urbicide has evolved and been applied in various contexts, resulting in a broadened and at times ambiguous interpretation [14]. Initially understood as the deliberate destruction of cities, urbicide has since been adapted and expanded to include additional attributes and nuances.

Beyond warfare, urbicide now encompasses other forms of violence, such as urban development projects displacing marginalized communities or policies undermining the social fabric of cities [14].

As urbicide continues to find application across different contexts and disciplines, its meaning and implications require careful consideration to effectively address the challenges faced by urban environments and communities.

Researchers distinguish between strategic and ideological forms of violence, such as ethno-confessional genocide or the deliberate destruction of cultural heritage [15, 16]. For instance, studies by Zorana Knezevic and Luis Felipe Mantilla highlight how ISIS destroyed cultural sites in Syria influenced by ideology and territorial dominance [17].

In "Urbicide: The Politics of Urban Destruction," Martin Coward delves into the phenomenon of urbicide, with a particular focus on the intentional destruction of the urban environment in conflict situations [18].

Particular attention should be given to Ukrainian researchers [19], who employ a functional approach to elucidate various facets of urbicide, concentrating on its causes and magnitude. The authors delineate several forms of urbicide, including: direct urbicide, which is characterized by the physical destruction of both symbolic landmarks and household buildings; indirect urbicide, which typically arises from the occupation of cities and the erosion of urban identity among residents; delayed urbicide, which is manifested by the reluctance or inability of displaced residents to return to occupied cities, resulting in a gradual loss of the collective human experience associated with these places [19].

Malchykova & Pylypenko examine the transformations in the functionality of Kherson and the daily survival strategies of its inhabitants during the occupation [20]. The authors identify the employment of a "mild" form of indirect urbicide during the occupation, marked by selective terror and strategies leading to economic decline. Following the occupation, there is a shift to a stricter form of direct urbicide, characterized by relentless shelling, extensive damage to infrastructure, demographic shifts, and a decline in citizens' well-being. Significant challenges for residents living under occupation include the militarization of urban areas, a prevalent sense of lawlessness, human rights violations, restricted access to humanitarian aid, economic disturbances, changes in currency, control over information, and transformations in identity.

30.2 Manifestations of Urbicide in the City of Kharkiv

30.2.1 Phases and Forms of Urbicide in Kharkiv

The city of Kharkiv has been under terrorist attacks from russia since the beginning of the full-scale invasion, which began on February 24, 2022.

The urbicide in the city of Kharkiv during the more than two years of full-scale invasion can be roughly divided into several stages:

- *Initial Stage* (February—April 2022): Characterized by attempts to surround and capture the city, urban battles, fighting on the ring road, shelling with rocket artillery, and bombing from airplanes.
- *Systematic Terror Stage* (May—early September 2022): Marked by the systematic terrorization of the urban population through missile strikes and artillery fire, as well as the movement of Russian troops from the outskirts of Kharkiv.
- *Kharkiv Operation* (September 2022): Involving specific military operations focused on the city of Kharkiv.
- *Continuation of Urban Terror Stage* (October 2022—present): Continuing the terrorization of the urban population primarily through rocket attacks and UAV strikes.

The primary targets in the city included airfields, military units, and air defense complexes. Concurrently, there were attacks on civilian infrastructure such as schools, residential buildings, and hospitals. On February 26–27, attempts were made to penetrate the city, including the first attempt at an amphibious operation to advance deep into the city, notably the seizure of School No. 134 (Fig. 30.1). Subsequently, russian special forces involved in this operation were either eliminated or captured at the same school.

During the battles for the city, Kharkiv's civilian infrastructure was badly damaged. Various districts of the city were bombed, including Northern Saltivka, Kholodnohirskyi district, and the nearest suburbs.

On March 1, two sea-based Caliber missiles struck the Kharkiv Regional State Administration. The following day, the occupiers carried out an air strike targeting the National University of the Air Force named after I. Kozhedub, the Faculty of Economics of V. Karazin University, and the adjacent building housing the regional police department and the Security Service of Ukraine (SSU).

At that stage, the combined resistance of the local population and the defensive actions of the troops played a crucial role in preventing the complete capture of the city. During the subsequent stage of urbicide, counteroffensive actions by the Ukrainian defense forces succeeded in pushing the enemy back from the nearest suburbs.

The systematic shelling of Kharkiv resulted in the destruction of numerous residential buildings, schools, medical institutions, and other civilian facilities.

Fig. 30.1 The ruins of school No.134 in the city of Kharkiv [21]

Between May and September 2022, shelling was carried out using multiple launch rocket systems such as the BM-30 “Smerch” and BM-27 “Uragan,” artillery systems like the 2S7 “Pion,” and anti-aircraft missile systems including the S-300. These attacks, orchestrated by the russian federation, targeted civilian infrastructure institutions such as the “Vostorg” supermarket, the Kharkiv Metro depot, school No.17 (Fig. 30.2), the KhPI sports complex, gymnasium No.47, the building of the Kharkiv National Pedagogical University named after G. Skovoroda, the Lokomotiv Sports Palace, and numerous multi-story and private houses.

This stage of urbicide reflects the potential threat it poses to both civilians and civilian infrastructure.

During the next stage of the urbicide, which encompassed the Kharkiv operation, Ukrainian defense forces successfully liberated almost the entire occupied part of the Kharkiv region from russian invaders.

The most severe consequences for the city included the shelling of Kharkiv’s Combined Heat and Power Plant-5 and the Zmiivska Thermal Power Plant, along with as high-voltage stations. This resulted in significant disruptions to the power systems across both the city and the region, leading to blackouts in several regions of Ukraine.

The urbicide phase that began in October 2022, is characterized by the terrorization of the urban population through rocket attacks and unmanned aerial vehicle (UAV) attacks. This stage of the conflict is marked by intense remote violence in the city, instilling fear and intimidation among the civilian population.

Rocket attacks became the predominant means of attack on Kharkiv, which led to the destruction of residential buildings and all types of infrastructure in the city.

Fig. 30.2 Destroyed school No.17 in Kharkiv as a result of the air strike by an Iskander ballistic missile. (Photo taken by the author on September 22, 2022)

Fig. 30.3 Destroyed Kharkiv Palace in the city of Kharkiv as a result of a hit by a North Korean KN-23 missile [22]

The S-300 anti-aircraft missile system for ballistic missiles, operational-tactical missile systems that fire Iskander-M ballistic missiles, and North Korean Hwasong-11Ga (KN-23) missiles are actively used to shell the city (Fig. 30.3).

UAVs are also actively used to carry out pinpoint attacks on civilian objects and infrastructure. Their high mobility and capacity for accurate targeting render them effective weapons of military aggression within the city. The "Shahed-131/136" is a type of UAV that russians constantly use to attack the territory of Ukraine, including Kharkiv specifically.

Undoubtedly, the use of rocket attacks and UAVs for urbicide violates the principles of international humanitarian law and exacerbates the suffering of the civilian population. It threatens the peaceful life and security of the local population and poses severe challenges to the global community in efforts to reduce hostilities and restore peace in the region.

During the full-scale invasion, many infrastructure facilities were destroyed.

30.2.2 Factors Influencing Shelling Intensity and Spatial Aspects

Urbicide has been a defining feature of russian aggression since the first days of the full-scale invasion. The spatial distribution of shelling consequences is uneven, and various political, economic, and strategic considerations influence the motivations and dynamics of destruction. In Kharkiv, the spatial and structural manifestations of urbicide are shaped by several significant factors. These include the city's geographical location, the broader geopolitical context, the historical aspects of urban development, the presence of industrial enterprises and critical infrastructure facilities, the essential functions the city performs, and indirectly, the city's morphology.

Kharkiv has a rich historical past, establishing itself as one of the leading centers of trade, culture, and education not only in Ukraine but also across the European landscape.

The geopolitical context, encompassing territorial disputes, historical perceptions, and power dynamics, significantly influenced the spatial aspects of urbicide. In the case of Kharkiv, its proximity to the russian border and its strategic importance make it a prime target for military aggression and urbicide. The military strategy employed by russian troops plays a crucial role in shaping the spatial aspects of urbicide in Kharkiv.

Urban planning and city structure, including its urban morphology, transport networks, and overall development, significantly influence the spatial aspects of urbicide. Certain areas are more vulnerable to attack because of location, accessibility, or symbolic meaning.

In Kharkiv, the distribution of the population, encompassing residential areas, commercial areas, and industrial zones, also affects the geography of shelling. Targeting densely populated areas, such as North Saltivka and downtown, has led to higher casualties and greater destruction. Moreover, attacks on critical infrastructure have disrupted essential services and exacerbated the humanitarian crisis.

The cultural and historical significance of Kharkiv, including its architectural heritage, cultural monuments, and identity as a Ukrainian city, are also targeted by constant attacks. Strikes on symbolic places or areas of cultural importance carry profound psychological and symbolic impacts on the population.

Understanding these factors is crucial for assessing the impact of the urbicide, mitigating its effects, and promoting peace and stability in the region. The interplay of these factors, in a specific manner, shapes spatial aspects of urbicide in Kharkiv as a consequence of the full-scale russian invasion of Ukraine.

30.2.3 Spatial Features of Urbicide in Kharkiv

The most heavily affected areas of Kharkiv are those through which the russians launched their attacks in the initial days of a full-scale invasion. During the first stage of urbicide, intense battles occurred directly in the city. Specifically, the Saltivskyi district, particularly the residential area of Nothern Saltivka, bears the brunt of the urbicide's consequences. The Industrial district also suffered significantly as the occupiers attempted to breach it from the city of Chuhuiv. Additionally, Kyivskyi district, a focal point of the Russian invasion, witnessed heavy fighting as they sought to seize key administrative buildings in Kharkiv through this area. Subsequent stages of urbicide saw all areas of the city affected due to ongoing

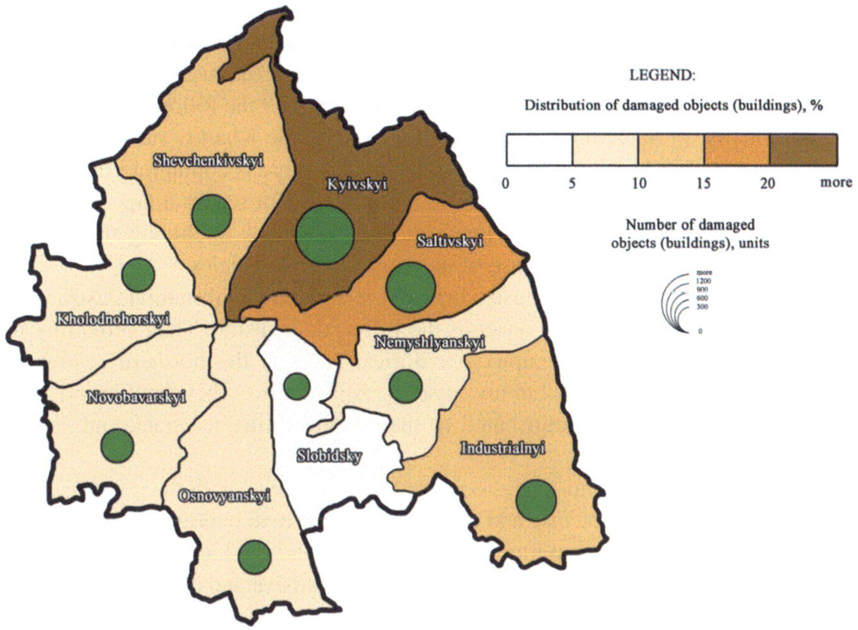

Fig. 30.4 Distribution of the number of damaged apartment buildings of all types and forms of ownership as well as private houses, by districts as of November 10, 2023 (compiled by the authors according to [23]

missile attacks and reprisals against the civilian population in response to military setbacks.

Considering the spatial features of the effects of urbicide in Kharkiv (Fig. 30.4), it should be noted that the relatively smaller number of damaged buildings in areas such as Slobodsky, Osnovyansky, Nemyshlyansky is attributed to their physical remoteness from the frontline of combat. However, these areas still experienced significant destruction due to missile and unmanned attacks. During the full-scale invasion, social infrastructure facilities were among those damaged, including educational institutions, Kharkiv airport, Nemyshlyanska oil depot, hotels, dormitories, and administrative buildings. Additionally, the Nemyshlyansky electric depot was also destroyed as a result of rocket fire by russian terrorists in the summer of 2022.

The impact of the Shahed 131/136 UAV striking the oil depot in the Nemyshlyanskyi district in March 2024 resulted in a fire, the release of oil products into the Nemyshlya River, and casualties among the civilian population.

Over 500 buildings were damaged in the Kholodnohirskyi and Novobavarskyi districts. Terrorists targeted the military fuel and energy infrastructure of these areas, severely affecting the city's operations and the lives of local residents. Despite the enemy's focus on military-strategic targets, the use of outdated and inaccurate weapons led to the destruction of numerous civilian structures, including multistory buildings, schools, and hospitals.

The military school in the Kholodnohirskyi district has been subjected to constant shelling since the initial days of the invasion, starting on March 1, 2022, when the occupiers launched rockets. An airstrike on March 2 followed this attack, and the shelling continued with various types of weapons over the following years.

Additionally, in September 2022, a strike targeted the Kharkiv railway station. The purpose of this attack was to target refrigerated wagons containing the bodies of Russian military personnel, which were being stored for repatriation.

However, even in remote areas, there are risks of attack, including from long-range artillery, rocket fire, or the use of unmanned aerial vehicles.

The Shevchenkivskyi and Industrialnyi districts endured substantial destruction as a result of the fighting at the onset of the full-scale invasion. Their proximity to the border and previously occupied territories, such as the northern region of Kharkiv and parts of the Chuhuiv district adjacent to Shevchenkivskyi and Industrialnyi, respectively, contributed to their vulnerability to attack and subsequent devastation.

From the outset, active hostilities occurred in these areas, with the invaders aiming to advance deep into the city and capture administrative buildings. The shelling involved various weapons, ranging from small arms and field artillery to multiple-launch rocket systems, missile systems, and high-explosive and cluster bombs.

In March 2022, an attack targeted the Kharkiv TV tower in the Shevchenkivskyi district. A terrorist pilot flying an SU-34 fighter dropped 8 FAB-500 air bombs, damaging the metal structures of the TV tower and the nearby administrative building. The TV tower was struck again directly in April 2024.

Missiles and UAVs continue to inflict the greatest destruction to this day. As of March 2024, launches and hits of modified winged bombs on residential buildings in the district have also been recorded.

Additionally, several other key locations suffered damage, including the Epicenter shopping center, school No.47, UNIFEHT, boarding school for children with hearing impairment No.5, Sarzhyn Yar (heavily shelled with the assistance of large-caliber Pion SPG), Kharkiv Palace hotel, Planeta Mall shopping center, Central Park, the main and northern buildings of V. N. Karazin National University, along with numerous residential buildings, among others.

In the Industrialnyi district, the Horizon residential area, the Elektrovazhmash SE plant, the Kharkiv Tractor Plant, the Vostorg supermarket, the Loseve station, and numerous other buildings have been severely affected.

Presently, the area continues to endure rocket attacks and UAV strikes. The concentration of impacts in this area can be attributed to the presence of large industrial and energy enterprises. The ongoing missile attacks pose significant challenges to the restoration of urban infrastructure.

The most significant effects of urbicide are evident in the Saltivskyi and Kyivskyi districts. These areas, situated closest to active hostilities, bore the brunt of the initial impact. Almost every building near the city's ring road in Kharkiv has been destroyed by enemy shelling. These attacks have resulted in extensive destruction, compromising infrastructure and disrupting the normal functioning of the local population.

Numerous key locations in the Saltovsky district, including the shopping mall "French Boulevard," the Saltovsky tram depot, Urban Park, the Barabashov market, the Continent shopping mall, a hostel for people with hearing impairments, a trolleybus depot, along with many residential buildings, hospitals, and schools, have been destroyed. Additionally, the supermarket "Rost Heroes of Labor" and branch No.6 of the Nova Poshta were destroyed during the issuance of humanitarian aid. The Kyivskyi district has witnessed the highest number of destroyed buildings due to its coverage of the residential areas of Northern Saltivka, Pyatykhatky, the central part of the city, and Shevchenko Street, which were all subjected to battles and attacks.

At present, it is nearly impossible to find an undamaged building in the residential area of Northern Saltivka due to the full-scale invasion and shelling. The russians utilized artillery and mortars to support their offensive from the direction of the village of Tsyrkuny towards the city, resulting in severe damage to Northern Saltivka.

One particularly notorious incident occurred at a 16-storey building on Natalia Uzhviy Street, where terrorists dropped an FAB-500 aerial bomb, causing the entrance to collapse completely.

The impact of missiles and shells has left almost every house in the residential area damaged. Several apartments, and even entire entrances, have been consumed by fire. In order to save their lives, many residents were forced to flee Northern Saltivka and seek refuge in the subway (Heroiv Pratsi station) and basements of buildings.

As of April 2024, many citizens have returned home, and shops and some shopping centers have reopened. Bus routes have also been restored. Currently, houses destroyed by aerial bombs are being dismantled, with plans to restore those deemed suitable for future habitation.

In the Kyiv district, the following objects were also destroyed: Kharkiv Regional State Administration, Shopping mall "Dafi," Kharkiv Aviation University, supermarket "Class" on Lesya Serdyuk Street, KhNPU named after Skovoroda, Saltivske Electric Depot, Nikolsky Shopping Mall, School No. 134 (where russian troops initially hid at the beginning of the invasion), Karavan Shopping Mall, along with many schools, hospitals, high-rise, and private buildings.

The ongoing restoration efforts in the affected areas require substantial effort, time, and financial resources. While some institutions and buildings have already been restored, many others are awaiting repairs. This is a complex and lengthy procedure, but efforts and resources are dedicated to restoring urban areas and returning them to their usual rhythm of life. However, the restoration process is hindered by the persistent shelling of the city and the prevailing economic crisis.

The ongoing restoration efforts in the affected areas require substantial effort, time, and financial resources. While some institutions and buildings have already undergone restoration, many others are awaiting repairs. Despite being a complex and lengthy procedure, endeavors and resources are dedicated to restoring urban areas and reinstating their normal pace of life. However, these restoration efforts are impeded by the persistent shelling of the city and the prevailing economic crisis.

The socio-economic consequences of urbicide in Kharkiv are profound and palpable across various aspects of the local population's lives and the city's economy. Urbicide has resulted in extensive damage to infrastructure, population evacuation, loss of life and injuries, economic setbacks, psychological distress among civilians, and disruption of the city's social functioning [24–28].

A register of damaged objects, including buildings affected by hostilities, is maintained in the city of Kharkiv and surrounding settlements within the Kharkiv region. This register will be instrumental in future restoration and rebuilding efforts [23].

From the onset of hostilities, terrorists targeted critical infrastructure, notably the fuel and energy systems. In Kharkiv, this led to the complete destruction or severe damage of all combined heat and power plants and their distribution networks. Consequently, these attacks left every consumer in the city without electricity, heating, and water supply at various times. The lack of electricity also crippled electric transport and other civilian infrastructure, inflicting significant economic damage on the city.

Enemy attacks in the city of Kharkiv led to the destruction of many enterprises, prompting the relocation of several key businesses to other regions within Ukraine. For example, the bakery company Kulinichi moved to the Lviv region, while the engineering firm Turboatom was partially relocated to both Lviv and Khmelnytskyi. The Malyshev Plant, known for its heavy machinery production, was moved to Dnipro and Kryvyi Rih. Additionally, the Slobozhansky Meat Processing Plant found a new home in the Poltava region.

By the end of 2023, nearly 100,000 residents had fled the city of Kharkiv as a result of intensified enemy shelling. The socio-economic impact of this urbicide on Kharkiv has been both dramatic and extensive. The destruction of enterprises and critical energy infrastructure, coupled with the mass evacuation of the population, has had profound effects. Additionally, the loss and injury of lives, significant economic damage, and psychological stress have deeply affected both the city and its inhabitants, exacerbating the challenges they face in recovery and rebuilding.

In the context of post-war reconstruction in Ukraine, it is crucial to actively secure diverse financial resources to facilitate the successful rebuilding of the country after the conflict. The primary sources of funding for this reconstruction should include reparations from the russian federation, as it is deemed the aggressor, alongside substantial assistance from international allies. A coalition of global communities, including the European Union (EU), the International Monetary Fund (IMF), the World Bank, and others, should be pivotal in the efforts to rebuild destroyed cities. On top of that, the implementation of comprehensive sanctions against the aggressor country is essential, serving both as a deterrent against further aggressive actions and as a means of applying pressure.

Ukraine's Euro-Atlantic trajectory plays a crucial role in securing the necessary funding and support for its post-war recovery. Accession to the European Union is particularly significant, as it would unlock numerous financial mechanisms vital for the country's reconstruction and economic support. Simultaneously, joining the North Atlantic Treaty Organization (NATO) would provide security guarantees that help prevent the possibility of new conflicts against Ukraine. Moreover, membership in the Euro-Atlantic Alliance is likely to enhance the investment climate and improve the economic environment, fostering greater stability and attracting additional investment opportunities.

30.3 Conclusions

The phenomenon of urbicide in our era is primarily a consequence of military conflicts, where urban areas are targeted for destruction. This destruction manifests through shelling, bombing, and various forms of aggression, leading to significant damage to urban infrastructure. Such hostilities not only physically dismantle cities but also deeply impact the socio-economic fabric and the psychological well-being of the affected populations.

Additionally, it is important to acknowledge that the full-scale Russian invasion of Ukraine in 2022 involved extensive use of urbicide as a tactic to systematically destroy cities and infrastructure. This strategy not only caused profound human suffering and loss of life but also led to significant destruction of economic and cultural assets.

Urbicide exerts a profound global impact on the world community, resulting in significant consequences across various spheres of life. This impact is evident in humanitarian crises, environmental challenges, economic shifts, tensions in international relations, and other facets. As such, the support and cooperation of the global community are increasingly vital for conflict prevention, war resolution, and the reconstruction of affected regions. These efforts can contribute to attaining new objectives for the sustainable development of the global community.

The intensity and nature of urbicidal actions in Kharkiv exhibit significant variation depending on the geographical location of the city's districts. Saltivsky and Kyivsky districts, situated closer to the frontline, endured the most destruction due to direct combat and artillery shelling. The Industrial and Shevchenkivskyi districts suffered as a result of their strategic significance and the occupiers' efforts to breach the city's defenses.

Constant shelling and the ongoing instability exacerbate the recovery process, rendering it time-consuming and demanding substantial efforts and resources. Rebuilding the city necessitates not only considerable financial investments but also meticulous planning to adjust to new circumstances and mitigate future risks of urbicide.

The use of urbicide tactics by the occupiers in Kharkiv not only resulted in the destruction of the city's rich infrastructure but also set the stage for enduring demographic challenges. The impact on the population—through displacement and loss—compounds the difficulties in urban and social recovery.

Full post-war reconstruction of Kharkiv, and similarly affected areas, hinges on the cessation of hostilities and will require extensive support from European countries and the global community. The path to recovery is complex and demands a coordinated, compassionate international response to effectively address the multifaceted consequences of urbicide.

References

1. Morar C, Nagy G, Dulca M, Boros L, Sehida K (2019) Aspects regarding the military cultural-historical heritage in the City of Oradea, Romania. ANNALES Ann Istrian Mediterranean Stud Ser Historia et Sociol 19(2):303–322. https://doi.org/10.19233/ASHS.2019.21
2. Morar C, Nagy G, Boros L, Gozner M, Niemets L, Sehida K (2021) Heritage, culture and regeneration of the former military areas in the city of Oradea (Romania). Architektúra Urbanizmus 55(1–2):76–87. https://doi.org/10.31577/archandurb.2021.55.1-2.6
3. Niemets L, Mezentsev K, Morar C, Sehida K, Husieva N, Peresadko V (2018) Innovation and investment potential of region as a factor of its "smart transformation": Kharkiv region case study (Ukraine), Visnyk of V N Karazin Kharkiv National University, series. Geol Geograp Ecol 49:137–159. https://doi.org/10.26565/2410-7360-2018-49-11
4. Morar C, Berman L, Unkart S, Erdal S (2021) Sustainable brownfields redevelopment in the European Union: an overview of policy and funding frameworks. J Environ Health 18(21):11169. https://www.neha.org/node/62305
5. Morar C, Lukic T, Valjarević A, Niemets L, Kostrikov S, Sehida K, Telebienieva I, Kliuchko L, Kobylin P, Kravchenko K (2022) Change detection based spatiotemporal analysis of urban green areas: a case study of Kharkiv-City. Front Environ Sci 10:1–27. https://doi.org/10.3389/fenvs.2022.823129
6. Meszaros EL, Toca CV (2020) The EU's multifaceted approach to resilience building in the Eastern Neighbourhood. Security sector reform in Ukraine. Eastern J Europ Stud 11:120–145. Special Issue SI, https://www.webofscience.com/wos/woscc/full-record/WOS:000593531000007?SID=EUW1ED0BEEp2kP2NxOAqUPsOYoFDc
7. Mészáros EL, Ţoca CV (2023) The EU's resilience and the management of hybrid threats coming from the Eastern neighbourhood: Belarus and the deliberate facilitation of irregular immigration. In Eastern. J Eur Stud 14(1):5–30. https://ejes.uaic.ro/articles/EJES2023_1401_MES.pdf
8. Toca C-V (2015) The regional dimension of the lifelong learning policies program in Romania. Transylvanian Rev 24(1):145–161. https://www.webofscience.com/wos/woscc/full-record/WOS:000364727800012?SID=EUW1ED0BEEp2kP2NxOAqUPsOYoFDc
9. Graham S (2004) Introduction: cities, warfare, and states of emergency. In: Graham S (ed) Cities, war, and terrorism: towards an urban geopolitics. Blackwell Publishing, Oxford, pp 1–25
10. Sassen S (2010) When the city itself becomes a technology of war. Theory Cult Soc 27(6):33–50. https://doi.org/10.1177/0263276410380938
11. Wright JL (2015) Urbicide: the ritualized killing of cities in the Ancient Near East. In: Olyan SM (ed) Ritual violence in the Hebrew Bible: new perspectives. New York, 2016; online edn, Oxford Academic, 17 Dec 2015. https://doi.org/10.1093/acprof:oso/9780190249588.003.0009
12. Convention on the Prevention and Punishment of the Crime of Genocide. United Nations. United Nations Human Rights Office of the High Commissioner. https://www.ohchr.org/en/instruments-mechanisms/instruments/convention-prevention-and-punishment-crime-genocide. Last accessed 2024/02/21
13. Genocide. United Nations. Office on Genocide Prevention and the Responsibility to Protect. https://www.un.org/en/genocideprevention/genocide.shtml. Last accessed 2024/03/10
14. Alvarado A (2023) Urbicide, violence, and destruction against cities by criminal organizations. In: Carrión Mena F, Cepeda Pico P (eds) Urbicide, The Urban Book Series. Springer, Cham. https://doi.org/10.1007/978-3-031-25304-1_30
15. Boyle M (2009) Explaining strategic violence after wars. Stud Conflict Terrorism 32:209–236
16. Kalyvas S (2006) The logic of violence in civil war. Cambridge University Press, Cambridge
17. Mantilla LF, Knezevic Z (2022) Explaining intentional cultural destruction in the Syrian civil war. J Peace Res 59(4):562–576
18. Coward M (2010) Urbicide. The politics of urban destruction. Global Discourse 1(2):186–189. https://doi.org/10.1080/23269995.2010.10707864
19. Mezentsev K, Mezentsev O (2022) War and the city: lessons from urbicide in Ukraine. Czasopismo Geograficzne 93(3):495–521. https://doi.org/10.12657/czageo-93-20

20. Malchykova D, Pylypenko I (2022) Occupation urbicide: urban experience and everyday practices of the population (a case of Kherson, Ukraine). Ekonomichna ta Sotsialna Geografiya / Економічна та соціальна географія 88:6–15. https://doi.org/10.17721/2413-7154/2022.88.6-15
21. During the fighting for Kharkiv, the 134th school was destroyed. Maidan Website 18.08.2023, https://maidan.org.ua/en/2023/08/pid-chas-boiv-za-kharkiv-bula-znyshchena-134-ta-shkola/. Last accessed 2024/04/15
22. Kharkiv Palace Hotel Destroyed in Russian Rocket Attack. Ukraine Travel News January 2, 2024, https://www.ukrainetravelnews.com/kharkiv-palace-hotel-destroyed-in-russian-rocket-attack/. Last accessed 2024/03/28
23. In Kharkiv and the region, the districts most affected by the war were named. Comments.ua. November 10, 2023, https://kharkov.comments.ua/ua/news/war/2023/23083-u-harkovi-ta-oblasti-nazvali-rayoni-yaki-naybilshe-postrazhdali-vid-viyni.html. Last accessed 2024/05/02
24. Tiba AI (2010) Grounded cognition perspective on irrational beliefs in rational emotive behaviour therapy. J Cogn Behav Psychother 10(87–99):2
25. Tiba AI, Johnson C, Vadineanu A (2012) Cognitive vulnerability and adaptation of parents of children with autistic Spectrum disorder. J Cogn Behav Psychother 12(2):209–218
26. Tiba AI, Manea L (2018) An embodied simulation account of cognition in rational emotive behavior therapy. New Ideas Psychol 48 C:12–20. https://doi.org/10.1016/j.newideapsych.2017.08.0032
27. Tiba AI, Manea L (2018) The vividness of imagining emotional feelings is attenuated in dysphoria and predicts the experience of positive emotional feelings. J Clin Psychol. https://doi.org/10.1002/jclp.226765
28. Tiba AI, Trip S, Bora CH, Drugas M, Borz F, Miclăuş DC, Voss L, Iova SC, Pop S (2023) Positive irrational beliefs are associated with hypomanic personality. Front Psychol 14:1053486. https://doi.org/10.3389/fpsyg.2023.1053486FI3.8

Chapter 31
Study of the Agricultural Development Based on the Method of Extrapolation (Case Study of Ukraine)

Pavlo Kobylin and Ievgeniia Telebienieva

Abstract Forecasting the development of agriculture is critical for effective management, especially in the face of climate change and technological innovation. Forecasting the development of agriculture involves the use of contemporary complex research methods based on the interdisciplinary approach of various scientific fields. This question lies, in particular, in the subject-object field of study of human geography as a complex, interdisciplinary science about nature, society, and economics.

The authors conducted a thorough analysis of previous studies on human-geographical studies of agricultural development worldwide and found that there is a certain lack of research on the use of different interdisciplinary methods, in particular, the extrapolation method, in forecasting various statistical indicators of agricultural development.

The purpose of this article is to apply the extrapolation method to the specific case of Ukraine.

The future of agriculture in Ukraine depends on many factors and requires research based on contemporary interdisciplinary approaches and methods. In this study, we applied the extrapolation method, a widely used technique in forecasting, to the agricultural sector of Ukraine. We projected a specific series of indicators, namely the gross harvest of individual crops (corn, wheat, sunflower), into the future. The base period for our analysis was 2000–2021, and the forecast period included 2022–2026. The results of our forecasting indicate a general trend towards an increase in gross harvest in Ukraine during the study period and a rise in indicators in the forecast period.

Keywords Agriculture · Human geography · Extrapolation method · Ukraine · Forecasting

P. Kobylin (✉) · I. Telebienieva
V.N. Karazin Kharkiv National University, Kharkiv, Ukraine
e-mail: pavlo.kobylin@karazin.ua; eugene.telebeneva@karazin.ua

C. Morar et al. (eds.), *Achieving Sustainability in Ukraine through Military Brownfields Redevelopment*, NATO Science for Peace and Security Series C: Environmental Security, https://doi.org/10.1007/978-94-024-2278-8_31

31.1 Introduction

Agriculture is a crucial sector of development in many countries and regions of the world. The well-being of many population segments, the state of health, and the ability to overcome a significant problem of civilization, namely hunger, depend on its development. This is one of the essential issues indicated among the Sustainable Development Goals. Agriculture plays a complex function in developing the world economy since it provides the population with food products and jobs, contributes to the financial replenishment of local budgets, etc. The development and sustainable functioning of agriculture depend on many natural-geographical, climatic-agronomic factors and the implementation of contemporary, powerful technologies at all stages of its functioning. It is critically unprofitable in most regions of the world and requires significant state support. It requires comprehensive scientific research to identify specific problems in certain natural, climatic, environmental, and economic circumstances with further forecasting at the stage of the requirements and goals of sustainable development [1–8].

Ukraine has favorable natural, geographical, and climatic conditions, which provides a strong level of agricultural development. Agricultural products are an essential component of Ukraine's exports. However, the current conditions of military aggression lead to the decline of many branches of agriculture; the territories of many regions are not suitable for the production of agricultural products due to mining of fields, land pollution, etc [9–11].

Agriculture, a cornerstone of the global economy, ensures food security and resources for the population. Forecasting its development cannot be overstated, as it is crucial for effective planning and management, particularly in the face of evolving climatic conditions, demographic trends, and technological innovations. Moreover, forecasting agricultural development is instrumental in making investment decisions, developing new agricultural technologies, optimizing resource utilization, and creating a robust economic foundation for rural communities. The use of the extrapolation method allows analyzing trends in the agriculture development on the basis of historical data and predicting the future state of the industry. This approach becomes especially important in the context of the rapid development of technologies and the use of data, allowing for more accurate modeling of complex interactions in agriculture. That is why the use of the extrapolation method in agricultural forecasting will be the focus of this article.

31.2 Analysis of Previous Research

Agriculture in human-geographical and economic science is considered extensively, and various aspects of its functioning are disclosed. Thus, Szőke and Kovács [12] created and analyzed networks of settlements representing the relations of purchase and sale of products in Hungary. Akdim et al. [13] grouped the administrative-territorial units of Morocco by types of land-use change, Vijulie et al. [14] assessed the fragmentation of agricultural land in Izvoarele commune, Olt county in Romania. Istudor et al. [15] investigated the state of food security in Romania and its place in the world, and Wang et al. [16] analyzed the relationship between the eco-efficiency of agriculture and the food security system in China. Konečný [17] characterized the dynamics of spatial polarization of agriculture in the Czech Republic during integration into the European Union. Zhenyu and Minghong [18] characterized the spatial differences in the development of specialized agriculture in the mountainous regions of China.

Ukrainian human geographers have significantly contributed to studying agriculture in Ukraine and its regions. The general methodological problems of agricultural research were considered in the study of the integrative role of human geography in solving global problems of modernity [19], the analysis of the information and synergetic components of the human-geographical research [20], contemporary approaches to the methodology of the human-geographical research [21]. Zayachuk [22] considered the theory and practice of human-geographical research on the formation and geospatial organization of farming in Ukraine, and Sukhyi [23] investigated the development of the agro-food complex of the Western Ukrainian region. Lohvinova [24] characterized the profitability of agriculture as a whole in Kharkiv region, and in the context of its districts. Lykhvan et al. [25] grouped the districts of the Kharkiv region according to the peculiarities of agricultural specialization, Niemets et al. [26] outlined the problems of food security in Ukraine through the analysis of indicators of volumes, norms of consumption, and production of food products.

Forecasting is essential in studying various social spheres, including agriculture. Thus, Mezentsev [27] formulated the theoretical and methodological foundations of the human-geographical forecasting of regional development. Forecasts of socio-economic phenomena using the extrapolation method are actively used in scientific research. In particular, this method was used to carry out a short-term forecast of the population of Kharkiv region as a whole and its individual districts [28], tourist flows in Ukraine [29], enterprise profit [30], market conditions in marketing research [31], consumption volumes of domestic food complex products [32], potential sales volumes [33], etc. At the same time, there is a lack of research on the development and substantiation of human-geographical forecasts of the agricultural development on the basis of various statistical indicators (in particular, the volume of gross harvest of agricultural crops, etc.) using the extrapolation method. Therefore, this article aims to develop forecasts of agricultural development in Ukraine using the extrapolation method.

31.3 Materials and Methods

The statistical basis of the study was the Food and Agriculture Organization of the United Nations (FAO) database, whose activities aim to ensure food security, regular access to high-quality food, and promote an active, healthy lifestyle (FAO). The statistics division of this organization has created the FAOSTAT portal, which provides free access to food and agriculture data for countries and territories since 1961. The article was used for analysis of the indicators of gross harvest of individual crops (corn, wheat, sunflower) in Ukraine for 2000–2021. The choice of these crops is because they are characterized by the most significant volumes of gross harvest for 2021 and, accordingly, are very important in the development of agriculture and are an essential component of Ukraine's food exports.

The forecast of the gross harvest of crops is important for managing agricultural development in the state and developing its future strategies. Forecasting methods are a set of ways and techniques of thinking that allow the conclusion of the development of the forecasting object in the future under specific conditions based on the retrospective analysis of trends and patterns in the development of endogenous (internal) and exogenous (external) data of the forecasting object [34]. Statistical information is used for forecasting and describing the processes over the past years, and an expert assessment of trends in changes in macroeconomic indicators is carried out.

Currently, there are over 200 different forecasting methods, but only a fraction of them, around 15–20, are actually used in practice. The choice of a forecasting method depends on several factors, including the purpose and task of the forecast, the forecast period, the nature of the forecasting object, and the types, completeness, and reliability of the available information [35].

The application of extrapolation methods to estimate economic indicators in the future is based on the assumption that the identified trends in economic phenomena will remain unchanged over the forecast period [27]. This method's advantages are low costs, quick data acquisition, and ease of use. The disadvantages are the following: possible inaccuracies in changing the trend; changes in factors affecting the studied indicators are not considered; the forecast is fulfilled for the short term [30, 33].

The application of the extrapolation method requires a relatively significant amount of data. According to the empirical rule, the forecasting period should not exceed a third of the basic period. For example, it is desirable to have statistical data for at least three years for a 1-year forecast [31]. If the dynamics of the forecasted indicator has a steady tendency to increase or decrease, and fluctuations around this trend are insignificant, then the extrapolation of the indicator is carried out according to the average rate of change (formulas 31.1 and 31.2) [31]:

$$Y_{t1} = Y_t{}^{*}T \tag{31.1}$$

where, Y_{t+1} is a predicted level of the dynamic series;
Y_t is the last level of the series to be studied;

T is the average rate of levels change of the series:

$$T = \sqrt[n-1]{Y_t / Y_n} \tag{31.2}$$

where, Y_n is the initial level of the series under study; n is the number of levels in the series.

In this study, we used the least squares method to extrapolate the analytical alignment of the time series. Each year (both within the basic and forecast periods) was assigned a sequence number, which will be used as a change of x in the future. The calculations were carried out in the Microsoft Excel software environment. After plotting the studied indicator, we established the type of development (we use the linear kind of development), added a trend line to the graph, and displayed its formula. The development trend is described by the linear function Y = a + bT. Constants a and b are calculated using formulas 31.3 and 31.4 [31]:

$$\mathrm{b} = \frac{\Sigma TY - n\underline{T}\underline{Y}}{\Sigma T^2 - n\left(\underline{T}\right)^2} \tag{31.3}$$

$$\mathrm{a} = \underline{Y} - \mathrm{b}\underline{T} \tag{31.4}$$

Using the equation's constants, we substitute the serial number of the period for which the forecast is developed and get the indicator's forecast value.

In addition, we made pessimistic and optimistic forecasts. For this purpose, the standard deviation between the real studied indicators and the trend indicators produced by the linear function was calculated according to the formula 31.5 [36]:

$$\mathrm{S} = \sqrt{\frac{\sum_{i=1}^{n}\left(x_i - x_{\text{mean}}\right)^2}{n-1}} \tag{31.5}$$

S is standard deviation.
N is a volume of sample.
X_i is a variable (value of the indicator).
X_{mean} is the mean of the sample.

After that, the forecast pessimistic and optimistic values of the indicators are calculated by subtracting/adding, respectively, the obtained standard deviation and trend values.

31.4 Research Results

The extrapolation method is used in many interdisciplinary studies and has its advantages. Therefore we have taken it as a basis for forecasting the further development of agriculture in Ukraine.

Accordingly, if the general trend during the base period is characterized by a contraction/growth, then the forecast will tend to decrease/increase. The dynamics of the gross harvest of corn in Ukraine are optimistic for 2000–2021 (Fig. 31.1).

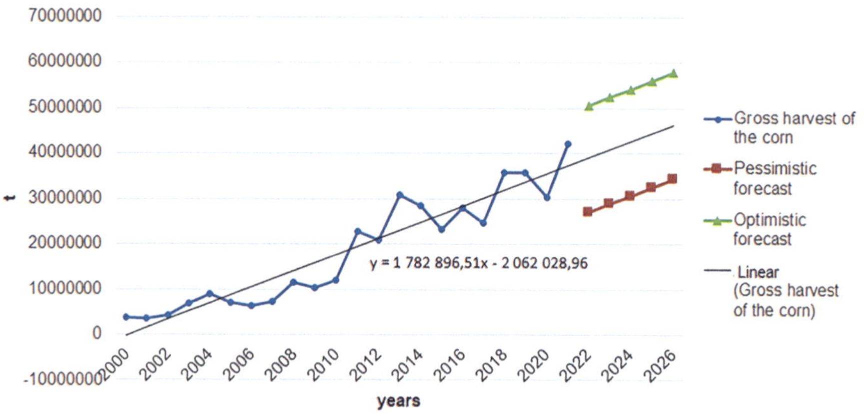

Fig. 31.1 Forecast of gross harvest of the corn until 2026 (using the extrapolation method) (constructed by the author according to [37])

Table 31.1 Summary table of data on forecasts of gross harvest of corn until 2026 (using the extrapolation method)

Years	Pessimistic values	Optimistic values
2024	27164017	50725164
2025	28946914	52508061
2026	30729810	54290957
2027	32512707	56073854
Type of dependence	Linear	
Algebraic formula	y = 1782896,51x – 2062028,96	
Standard deviation	S_y = 11780573,47	

Such dynamics is explained by the high profitability of this crop at insignificant costs (the profitability of corn averaged 25%, while for wheat and barley, this indicator is estimated at 10–15%) [38, 39], the growth of demand on the world market for corn grain and the need to increase exports of this crop [40]. Considering the forecasts of the gross harvest of corn for 2022–2026, we can see that this indicator will grow for the specified period (Fig. 31.1). Thus, according to the pessimistic forecast, the gross harvest of corn will increase from 27.2 million tons to 34.3 mln. tons, i.e., it will increase by 7.1 million tons (Table 31.1). According to the optimistic forecast, the gross harvest of corn will increase from 50.7 million tons to 57.9 mln. tons, i.e., it will also increase by 7.2 million tons. The difference between the optimistic and pessimistic values of the forecasts is 23.5 million tons, or optimistic values exceed pessimistic ones by 1.86 times.

For 2000–2021, the dynamics of the gross harvest of wheat in Ukraine varied but with a general upward trend (Fig. 31.2). That is, there are periods of growth and decrease in the gross harvest of this crop. Until 2000, there was a rapid decrease in

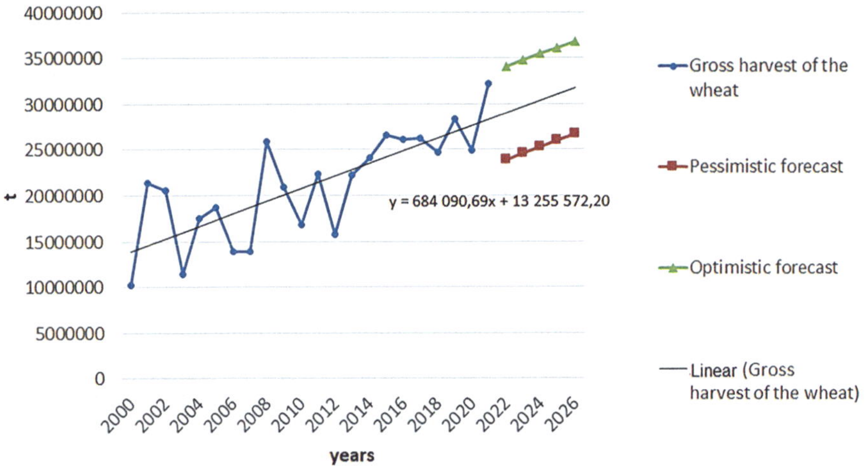

Fig. 31.2 Forecast of the gross harvest of the wheat until 2026 (using the extrapolation method) (constructed by the author according to [37])

Table 31.2 Summary table of data on forecasts of gross harvest of the wheat until 2026 (using the extrapolation method)

Years	Pessimistic values	Optimistic values
2024	23901925	34077391
2025	24586016	34761481
2026	25270107	35445572
2027	25954197	36129663
Type of dependence	Linear	
Algebraic formula	y = 684090,69x +13255572,20	
Standard deviation	S_y = 5087732,722	

production, which was the result of crisis processes in the agricultural sector as a whole, and then a slow growth [38], which is due to stable demand in the interregional market and the highest profitability among other crops [41]. Considering the forecasts of the gross harvest of the wheat for 2022–2026, we can see that this indicator will grow for the specified period (Fig. 31.2). Therefore, according to the pessimistic forecast, the gross harvest of the wheat will increase from 23.9 mln. tons to 26 mln. tons, i.e., it will increase by 7.1 mln. tons (Table 31.2). According to the optimistic forecast, the gross harvest of the wheat will also increase from 34.1 mln. tons to 36.1 mln. tons, i.e, it will also increase by 2 million tons. The difference between the optimistic and pessimistic values of the forecasts is 10.2 mln. tons, or optimistic values exceed pessimistic ones by 1.4 times (Table 31.2).

The dynamics of the gross harvest of sunflowers in Ukraine for 2000–2021 has a general upward trend (Fig. 31.3). According to Marushinets [38], this trend is associated with market conditions and profitability of the production of this crop, the use of advanced agricultural technologies, high-quality seed. In addition, Ukraine is the world's leading exporter of sunflower [42]. Considering the forecasts of the gross

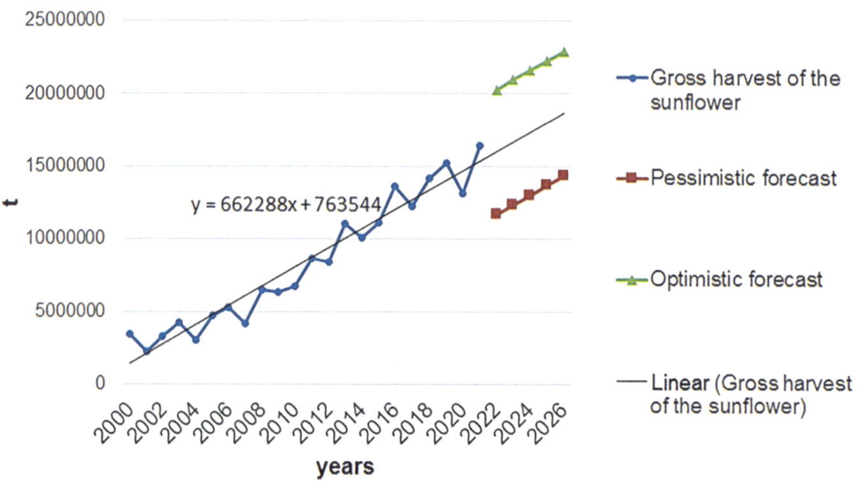

Fig. 31.3 Forecast of the gross harvest of the sunflower until 2026 (using the extrapolation method) (constructed by the author according to [37])

Table 31.3 Summary table of data on forecasts of gross harvest of the sunflower until 2026 (using the extrapolation method)

Years	Pessimistic values	Optimistic values
2024	11682370	20309966
2025	12344658	20972254
2026	13006946	and21634542
2027	13669234	22296830
Type of dependence	Linear	
Algebraic formula	y = 662288x + 763544	
Standard deviation	S_y = 4313797,728	

harvest of the sunflower for 2022–2026, we can see that this indicator will grow for the specified period (Fig. 31.3). Thus, according to the pessimistic forecast, the gross harvest of the sunflower will increase from 11.7 mln. tons to 13.7 mln. tons, i.e., will increase by 2 mln. tons (Table 31.3). According to the optimistic forecast, the gross harvest of the sunflower will also increase from 20.3 mln. tons to 22.3 mln. tons, i.e, it will also increase by 2 mln. tons. The difference between the optimistic and pessimistic values of the forecasts is 8.6 mln. tons, or optimistic values exceed pessimistic ones by 1.7 times.

31.5 Conclusions

The study conducted gave an opportunity to justify the use of the extrapolation method in forecasting the development of agriculture in Ukraine. The analysis of previous studies by scientists from different countries has shown that the topic of agricultural research is very broad and multifaceted in terms of the object-subject area. Such studies include the analysis of land use, land relations, food security, and spatio-temporal features of agricultural development. An important place in the study of agriculture is occupied by forecasting of its development. The author found out that there are forecasting models of population, tourist flows, enterprise profits, and volumes of product consumption in the human-geographical studies. However, there is a lack of research on the use of the extrapolation method in the forecasting agricultural development based on statistical indicators. This is exactly what was focused on in this paper.

The extrapolation method is based on projecting a specific series of indicators into the future. It assumes that the identified trends in economic phenomena will remain unchanged over the forecasting period. The methodology includes determining the type of development (using the linear function of the trend line), calculating the projection and forecast values according to the established linear function of the trend line, the standard deviation between the actual and projection values of the indicators, and the development of pessimistic and optimistic forecasts by subtracting and adding the value of the standard deviation from the projection values. This method's advantages are low costs, quick data acquisition, and ease of use. The disadvantages are the following: possible inaccuracies in changing the trend; changes in factors affecting the studied indicators are not considered; the forecast is fulfilled for the short term. Given the complexity of socio-economic, environmental, and military processes in Ukraine, short-term forecasts of development, especially of agriculture, are needed.

The study's statistical basis was the Food and Agriculture Organization of the United Nations (FAO) database. The indicators of gross harvest of individual crops were used for the analysis for 2000–2021 (basic period). We considered corn, wheat, and sunflower. These crops are leading in terms of gross harvest in Ukraine for 2021). The forecast period is set for 2022–2026.

The forecasting results of three crops characterize the general trend towards an increase in gross harvest in Ukraine during the study period and a rise in indicators in the forecast period.

Naturally, it is impossible to predict the forecast values of the gross harvest of crops since this can be influenced by many factors that are relatively unstable and difficult to predict. For a country such as Ukraine, which is in a state of war, this factor determines the high degree of uncertainty in the future. However, forecasting the development of agriculture makes it possible to analyze the future development of agriculture and make appropriate management decisions, which are very important in contemporary conditions in Ukraine. Further research is required on implementing the latest innovative and information technologies in the agriculture of

Ukraine, forecasting the industry's export potential for the near and distant future, and the place and role of individual regions in the implementation of sustainable development goals.

References

1. Lukić T, Ponjiger TM, Basarin B, Sakulski D, Gavrilov M, Marković S, Zorn M, Komac B, Milanović M, Pavić D, Mesaroš M, Marković N, Durlević U, Morar C, Petrović A (2021) Application of Angot precipitation index in the assessment of rainfall Erosivity: Vojvodina region case study (North Serbia). Acta Geogr Slov 61(2):123–153. https://doi.org/10.3986/AGS.8754
2. Valjarević A, Popovici C, Djekić T, Morar C, Filipović D, Lukić T (2022a) Long-term monitoring of high optical imagery of the stratospheric clouds and their properties new approaches and conclusions. Egypt J Remote Sens Space Sci 25(4):1037–1043. https://doi.org/10.1016/j.ejrs.2022.11.006
3. Blešić I, Ivkov M, Tepavčević J, Raljić JP, Petrović MD, Gajić T, Tretiakova TN, Syromiatnikova JA, Bajrami DD, Aleksić M, Vujačić D, Kričković E, Radojković M, Morar C, Lukić T (2022) Risky travel? Subjective vs. objective perceived risks in travel behaviour—influence of hydro-meteorological hazards in South-Eastern Europe on Serbian tourists. Atmosphere 13(10):1671. https://doi.org/10.3390/atmos13101671
4. Valjarević A, Algarni S, Morar C, Grama V, Stupariu M, Tiba A, Lukić T (2023) The coastal fog and ecological balance for plants in the Jizan region, Saudi Arabia. Saudi J Biol Sci 30(1):103494. https://doi.org/10.1016/j.sjbs.2022.103494
5. Valjarević A, Popovici C, Štilić A, Radojković M (2022b) Cloudiness and water from cloud seeding in connection with plants distribution in the Republic of Moldova. Appl Water Sci 12:262. https://doi.org/10.1007/s13201-022-01784-3
6. Valjarević A, Milanović M, Gultepe I, Filipović D, Lukić T (2022c) Updated Trewartha climate classification with four climate change scenarios. Geogr J 188:506–517. https://doi.org/10.1111/geoj.12458IF
7. Valjarević A, Filipović D, Milanović M, Valjarević D (2020) New updated world maps of sea-surface salinity. Pure Appl Geophys 177:2977–2992. https://doi.org/10.1007/s00024-019-02404-z.IF
8. Valjarević A (2024) Long-term remote sensing-based methods for monitoring air pollution and cloud cover in the Balkan countries. Environ Sci Pollut Res 31:27155–27171. https://doi.org/10.1007/s11356-024-32982-y
9. Niemets L, Mezentsev K, Morar C, Sehida K, Husieva N, Peresadko V (2018) Innovation and investment potential of region as a factor of its "smart transformation": Kharkiv region case study (Ukraine), Visnyk of V N Karazin Kharkiv National University, series. Geol Geograp Ecol 49:137–159. https://doi.org/10.26565/2410-7360-2018-49-11
10. Morar C, Berman L, Unkart S, Erdal S (2021) Sustainable brownfields redevelopment in the European Union: an overview of policy and funding frameworks. J Environ Health 18(21):11169. https://www.neha.org/node/62305
11. Morar C, Lukic T, Valjarević A, Niemets L, Kostrikov S, Sehida K, Telebienieva I, Kliuchko L, Kobylin P, Kravchenko K (2022) Change detection based spatiotemporal analysis of urban green areas: a case study of Kharkiv-City. Front Environ Sci 10:1–27. https://doi.org/10.3389/fenvs.2022.823129
12. Szőke V, Kovács L (2023) Networks, agriculture and geography: how business connections of agricultural enterprises shape the connection of settlements in Western Hungary. Geographica Pannonica 27(1):10–24. https://doi.org/10.5937/gp27-39849

13. Akdim M, Akdim H, Sabiri A, Alami A (2023) Land-use change and sustainability in the south-eastern oases of Morocco. Bull Geograp Socioecon Ser 59:95–109. https://doi.org/10.12775/bgss-2023-0007
14. Vijulie I, Matei E, Manea G, Cocoş O, Cuculici R (2012) Assessment of agricultural land fragmentation in Romania, A case study: Izvoarele commune. Olt County Acta Geographica Slovenica 52(2):403–430. https://doi.org/10.3986/AGS52206
15. Istudor N, Ion RA, Sponte M, Petrescu IE (2014) Food security in Romania—A modern approach for developing sustainable agriculture. Sustain For 6:8796–8807. https://doi.org/10.3390/su6128796
16. Wang R, Chen J, Li M (2023) Coupling and coordinating relationship between agricultural eco-efficiency and food security system in China. Int J Environ Res Public Health 20(1):431. https://doi.org/10.3390/ijerph20010431
17. Konečný O (2017) Spatial polarization of agriculture of Czechia during the integration into the European Union. Aust Geogr 122:257–280. https://doi.org/10.37040/geografie2017122030257
18. Zhenyu H, Minghong T (2023) Spatial differences of specialty agriculture development in the mountainous areas of China – "one village, one product" as an example. Heliyon 9(8):E18391. https://doi.org/10.1016/j.heliyon.2023.e18391
19. Niemets L, Sehida K, Kobylin P, Şahin TM, Bartosh O, Telebienieva I, Kravchenko K, Kliuchko L (2023) Integrative role of human geography in solving global problems of modernity. Visnyk of V N Karazin Kharkiv National University, Series. Geol Geograp Ecol 58:171–187. https://periodicals.karazin.ua/geoeco/article/view/22199, https://doi.org/10.26565/2410-7360-2023-58-14
20. Niemets L, Kravchenko K, Sehida K, Telebienieva I, Kliuchko L (2022a) Informational and synergetic component of human-geographical research in the scientific heritage of Kostiantyn Niemets. Visnyk of V. N. Karazin Kharkiv National University, Series. Geol Geograp Ecol 57:153–172. https://doi.org/10.26565/2410-7360-2022-57-12
21. Niemets K, Sehida K, Niemets L, Kravchenko K, Kobylin P, Telebienieva I, Kliuchko L (2022b) Methodology of human-geographical researches: contemporary approaches and methods. Visnyk of V. N. Karazin Kharkiv National University, Series. Geol Geograp Ecol 56:143–158. https://doi.org/10.26565/2410-7360-2022-56-10
22. Zayachuk MD (2015) Formation and geospatial organization of farming in Ukraine (theory and practice of the human-geographical research): thesis for obtaining the degree of D.Sc. in Geography: 11.00.02. Chernivtsi. http://scc.univ.kiev.ua/upload/iblock/f27/zayachuk%20_dis.pdf
23. Sukhyi PO (2008) Agri-food complex of the Western Ukrainian region: monograph. Chernivtsi, Ruta. (in Ukrainian)
24. Lohvinova MO (2018) Profitability of agriculture in Kharkiv region as an indicator of its economic efficiency. Sci Bull Kherson State Univ Ser Geograph Sci 8:84–91. https://gj.journal.kspu.edu/index.php/gj/article/view/86 (in Ukrainian)
25. Lykhvan V, Dobrovolska N, Kandyba Y (2013) Determination of the specialization of agriculture in Kharkiv region. Human. Geogr J 14(1):97–103. https://periodicals.karazin.ua/socecongeo/article/view/6582 (in Ukrainian)
26. Niemets L, Kandyba Y, Kobylin P, Kucheriava H, Krainiukov O (2020) Food security problems in ukraine. In: Education excellence and innovation management: a 2025 vision to sustain economic development during global challenges. Proceedings of the 35th International Business Information Management Association Conference (IBIMA) (Seville, Spain 1–2 April, 2020). Editor Khalid S. Soliman. Seville, International Business Information Management Association (IBIMA), pp 4955–4963. https://ibima.org/accepted-paper/food-security-problems-in-ukraine/
27. Mezentsev KV (2005) Human-geographical forecasting of regional development: monograph. Kyiv, Publishing and Printing Center "Kyiv University". (in Ukrainian)

28. Sehida KY (2015) Geodemographic forecast of Kharkiv region (using the extrapolation method). Visnyk of V. N. Karazin Kharkiv National University, series. Geol Geograp Ecol 43:164–173. https://doi.org/10.26565/2410-7360-2015-43-26 (in Ukrainian)
29. Munkachii I, Hrytsyshyn A, Bilous S (2023) Methodology of trends extrapolation as a tool for forecasting tourist flows in Ukraine. Int Sci J Manag Econom Financ 2(4):65–76. https://doi.org/10.46299/j.isjmef.20230204.08 (in Ukrainian)
30. Revak IO, Vlasyuk NI, Mytsak OV (2017) Forecasting the company's profit based on the extrapolation method. Sci Bull Lviv State Univ Inter Affair Econom Seri 2:238–247. https://dspace.lvduvs.edu.ua/bitstream/1234567890/1114/1/%2brevak.pdf (in Ukrainian)
31. Yarenko AV (2015) Systematization of quantitative methods for forecasting market conditions in marketing research // Bulletin of KNUTD, Series. Econom Sci 3(87):11–18. https://knutd.edu.ua/publications/pdf/Visnyk/2015-3/11_18.pdf (in Ukrainian)
32. Senyshin OS (2014) Extrapolation forecasting methods as a tool for predicting the optimal volumes of products consumption of the domestic food complex. Youth Econom Digest 1:26–32. https://ir.kneu.edu.ua:443/handle/2010/3940 (in Ukrainian)
33. Zvonar YP, Fetsynets VV (2018) Features of the use of forecasting methods to assess potential sales volumes. Econ Soc 14:366–370. https://economyandsociety.in.ua/journals/14_ukr/50.pdf (in Ukrainian)
34. Shvaika LA (2006) State regulation of the economy: a textbook. Kyiv, Knowledge. (in Ukrainian)
35. Stelmashchuk AM (2000) State regulation of the economy: a textbook. Ternopil, TANG. (in Ukrainian)
36. STDEV Function. Microsoft support https://support.microsoft.com/uk-ua/office/%D1%84%D1%83%D0%BD%D0%BA%D1%86%D1%96%D1%8F-stdev-51fecaaa-231e-4bbb-9230-33650a72c9b0. Last accessed 2024/04/28
37. Official website of the Food and Agriculture Organization of the United Nations (FAO). Retrieved from: https://www.fao.org/. Last accessed 2024/04/28
38. Marushinets AV (2016) Transformation of the agrarian sector and socio-economic development of the Kyiv Dnieper region: thesis for obtaining the degree of PhD in Geography: 11.00.02. Kyiv. https://shron1.chtyvo.org.ua/Maruschynets_Andrii/Transformatsiia_ahrarnoi_sfery_ta_sotsialno-ekonomichnyi_rozvytok_Kyivskoho_Prydniprovia.pdf
39. Slastion R. Corn becomes the main bread of Ukrainian farmers. Ukrainian Agribusiness Club UCAB Retrieved from: https://ucab.ua/ua/pres_sluzhba/novosti/kukurudza_stae_osnovnim_khlibom_ukrainskikh_agrariiv?category=29778, last accessed 2024/04/28
40. Shymkova M (2020) The global corn market and Ukraine's place on it. Price Rev 10(288) https://pricereview.com.ua/articles/svitovij-rinok-kukurudzi-ta-miscze-ukra%D1%97ni-nanomu (in Ukrainian)
41. Kravets TM (2012) Social and geographical aspects of the functional transformation of the agrarian sector of Lviv region. Geograp Tourism 23:209–216. http://www.irbis-nbuv.gov.ua/cgi-bin/irbis_nbuv/cgiirbis_64.exe?I21DBN=LINK&P21DBN=UJRN&Z21ID=&S21REF=10&S21CNR=20&S21STN=1&S21FMT=ASP_meta&C21COM=S&2_S21P03=Fila=&2_S21STR=gt_2012_23_35 (in Ukrainian)
42. Ukrainian sunflower: everything starts with quality seeds. AIC-Inform. Retrieved from: https://www.apk-inform.com/uk/exclusive/topic/1500844. Last accessed 2024/04/28

Chapter 32
Comparative Analysis of the Transport Infrastructure of the Cities of Warsaw and Kharkiv as a Component of the Innovative Development and Post-war Reconstruction of the Cities of Ukraine

Kateryna Kravchenko and Yevhen Shpak

Abstract The article provides a comparative analysis of the transport infrastructure of the cities of Warsaw and Kharkiv as a component of post-war reconstruction. The transport systems of the cities of Kharkiv and Warsaw are characterized, the spatial features of the organization of transport routes are revealed, and the transport accessibility of the districts of Kharkiv and Warsaw is given. The passenger traffic of the studied cities was analyzed according to the ratio according to the types of public transport. It was established that the metro dominates in the city of Kharkiv, and buses dominate in Warsaw regarding the structure of the volume of passenger transport and the types of transport. The dynamics of indicators of passenger transportation volume and passenger traffic by mode of transport in Kharkiv have a downward trend. The volume of passenger transportation is decreasing, and passenger traffic is more or less stable in Warsaw. In the spatial aspect, the transport system of the cities of Kharkiv and Warsaw is best developed in the central areas of the city. A study of the bicycle infrastructure of the cities of Kharkiv and Warsaw was carried out. It was found that the bicycle infrastructure is much better developed in the city of Warsaw than in the city of Kharkiv. It can be used as a development model for the city of Kharkiv.

The authors conducted a SWOT analysis of the transport systems development of Kharkiv and Warsaw, proving the presence of significant development potential in both cities. The presence of innovative solutions that work effectively in the transport system of Warsaw can be impleme.

Keywords Transport infrastructure · Innovative development · Urban renewal

---

K. Kravchenko (✉) · Y. Shpak
V.N. Karazin Kharkiv National University, Kharkiv, Ukraine
e mail: kateryna.kravchenko@karazin.ua; yevhen.shpak@student.karazin.ua

C. Morar et al. (eds.), *Achieving Sustainability in Ukraine through Military Brownfields Redevelopment*, NATO Science for Peace and Security Series C: Environmental Security, https://doi.org/10.1007/978-94-024-2278-8_32

32.1 Introduction and Review of Previous Scientific Developments in This Area

In the modern world, the number of large cities is growing extremely rapidly, exerting great pressure on the environment, causing many different problems for the population and the economy. This requires the development of optimal complex models, programs for the functioning of urban systems with powerful infrastructure to ensure the normal life of the population. Comprehensive studies of the city as a large, open, dynamic system with related subsystems based on the analysis of direct and related, internal and external relationships between the infrastructure of the subsystems and the external environment. Among the individual infrastructure subsystems of a large city, an important place and role belongs to the transport infrastructure, which to a large extent ensures the optimality of its vital activities. The issue of optimal development of the world's major cities is in the center of attention of scientists of various fields, international forums, and powerful international organizations. In particular, in many documents of the Sustainable Development Strategy, these issues are noted, and goals are set for implementation by all countries of the world [9]. The following goals are directly related to the development of cities: 8. Decent work and economic growth, 9. Innovation and infrastructure, 11. Sustainable development of cities and communities [9]. There are programs for the formation and development of smart cities, creative cities, and green cities. Ukraine is a large European country, located practically in the center of Europe and is building its future in terms of European integration. Therefore, the experience of developed large European cities is very important for Ukrainian development, especially in the post-war period.

In this study, a comparative analysis of the European city of Warsaw and the city of Kharkiv, Ukraine was performed. In particular, the issues of the role, location and specifics of the development of the transport infrastructure of these two cities are comprehensively considered.

As previously mentioned, transport infrastructure is an extremely important subsystem of the socio-economic system of the city as a whole.

The development of infrastructure, in particular—transport, is one of the key factors of economic growth and development of the city, region, and country. Transport infrastructure is the defining basis of urban infrastructure. This is a complex element of the city's economic system of strategic importance, which ensures the transportation of passengers and goods, the connection of territories of different hierarchical levels: the districts of the city, the city and surrounding settlements. Transport infrastructure acts as a construction and planning frame of the city, its internal and external relationships, which is important for the optimal functioning of the city as a whole. In modern conditions, among all settlements, large cities and megacities—world cities—are the most attractive for the population. High-quality public transport, efficient logistics and smart development of communication technologies are important conditions for their successful functioning and further development. The stability of the city's transport system is the key to a successful comprehensive transformation of the city space. For the city of Kharkiv, a city of a million people, the restoration and development of transport infrastructure on the basis of efficiency, equality and accessibility, innovation, and environmental friendliness is an important component of the further evolution of the city system on the way to sustainable development. To transform the transport infrastructure in accordance with the principles of sustainable development, we suggest using elements of the positive experience of the city of Warsaw. Today, Warsaw is recognized as the alpha city of the world, a major international tourist center and a significant cultural, political and economic center. Warsaw is also currently experiencing a surge of new commercial developments, and the European perspective of spatial development in relation to the new member states of the European Union, including the accession candidate Ukraine, demonstrates the need for the creation of large urban centers that will play a strategic role in the development of the peripheral regions of the European Union. The Warsaw metropolis, being the most eastern, has the opportunity to achieve the position of "locomotive of European development" for the eastern part of the territory of the European Union, to obtain the status of a metropolis of European importance, and its experience of transformation on the way to sustainable development can serve as an example for the cities of Ukraine, in particular—the city of Kharkiv. Both cities have a rich history, are cities of millionaires, powerful industrial, service, educational, cultural and innovation centers. Warsaw is the capital of Poland, for a certain time the city of Kharkiv also performed the function of the capital. Both cities have a historical center, buildings of different historical stages of development. Also,both Kharkiv and Warsaw are centers of powerful agglomerations. Also, Warsaw is already officially recognized as a world alpha city,one of the most sustainable cities in the world, a smart city, therefore its experience of development and, especially, post-war recovery is extremely relevant and useful in the modern realities of Ukraine, because during the Second World War the Polish capital suffered the greatest destruction among European cities.

32.2 Contemporary State of Public Transport in Kharkiv and Warsaw

Transport infrastructure is an integral part of the transport system of any city, region and country and plays an important role in the development of the economy and the formation of the social well-being of the population. The study of the effectiveness of the transport infrastructure, its functioning and development is the foundation for establishing the prospects for the development of the transport network, the possibilities of providing logistics services and determining the contribution of the transport sector to the economy of the region. A wide range of works of domestic and foreign scientists are devoted to the study of the transport infrastructure of cities and the peculiarities of the organization of logistics activities/ In particular, a significant number of works are devoted to the study of transport infrastructure and logistics at the level of regions or cities as poles of growth.

The purpose of the work is to determine the modern characteristics of the development of the transport infrastructure of the city of Kharkiv, to develop measures for its transformation on the way to achieving the Sustainable Development Goals, in particular—goal 9 "Innovation and infrastructure" based on a comparative analysis of the transport system of the cities of Kharkiv and Warsaw for the further implementation of European experience the development and functioning of the transport infrastructure of the city of Warsaw, especially in the conditions of the post-war reconstruction of Ukraine. As part of the study, it is planned to perform an analysis of the public transport system of the cities of Kharkiv and Warsaw, to identify the peculiarities of the development of transport infrastructure, to justify measures to improve the transport infrastructure of the city of Kharkiv in the aspect of achieving the goals of sustainable development and its post-war recovery [11–17].

A cartographic research method was chosen to assess the transport accessibility of bicycle paths and density of routes in the cities of Kharkiv and Warsaw. In particular, ArcGis 10.7 and QGIS 3.16 software were used. The database for the study was an interactive map of OpenStreetMap. Data on the network of bicycle paths of Kharkiv and Warsaw were loaded from the OpenStreetMap with the help of the Overpass Turbo service and layers (shapefiles) of they were created with the help of QGIS 3.16 software [5]. Kharkiv has all types of public transport: tram (since 1906), trolleybus, bus and metro (since 1975). Three intersecting metro lines connect the main residential, industrial and transport hubs. With the development of the metro network from the city, especially its central part, the tram ceased to be the main public transport [2].

The basis of the historically formed street and road network of the city of Kharkiv is a system of radial streets converging to the center and leading to intercity directions (to Kyiv, Sumy, Dergachi, Izyum, Zmiiv, Dnipro). As you approach the central part of the city, the width of the streets decreases. Connection of the peripheral areas of the city with each other is, as a rule, possible mostly only through the center. Bicycle tracks as transport tracks are practically absent. There are a small number of walking bike paths. The city is crossed by main railway lines in directions to Kyiv, Simferopol, Izyum, Kupyansk, from which there are numerous branches to industrial enterprises. Some of them have now stopped working [3].

The structure of passenger traffic is significantly dominated by the metro (more than half of the passengers transported in the city) (see Fig. 32.1a). Other types of transport (buses, trolleybuses, trams) have approximately the same total weight (11–16%). This can be explained by the fact that the subway allows to transport passengers from different parts of the city relatively quickly. The subway is also significantly dominated in the structure of passenger turnover (more than half of the passenger turnover of the city, see Fig. 32.1b).

We will analyze the features of the concentration of transport routes in the public transport system in Kharkiv (Fig. 32.3). Regarding the spatial distribution of the subway in Kharkiv, the subway covers mostly the central, northern and eastern parts of the city [4]. Metro stations provide connections between most districts of the city, but Novobavarskyi and Osnovyanskyi districts are not well served. Trolleybuses and buses run from metro stations to sleeping areas.

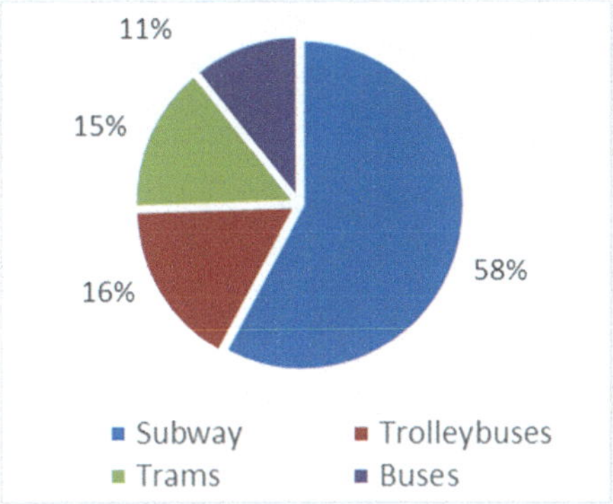

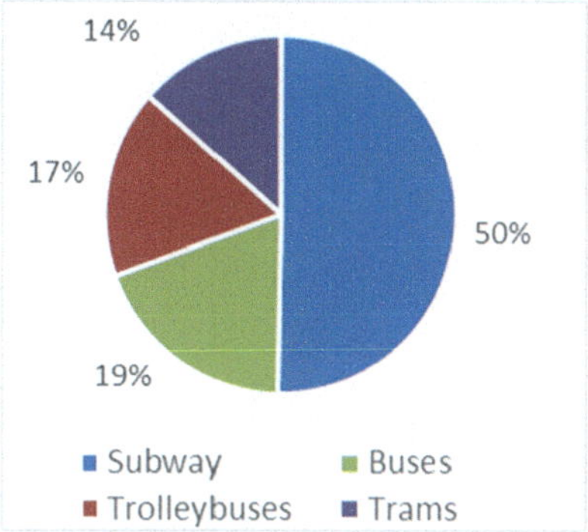

Fig. 32.1 Number of passengers transported by modes of public transport in Kharkiv, 2022 (**a**) and Passenger turnover by modes of transport in Kharkiv, 2022 (**b**)

As of 2024, it is functioning in the city of Kharkiv28 routes were operating in Kharkiv, the length of the trolleybus network is 263 km, there are 298 passenger trolleybuses in the city [3]. Regarding the spatial distribution of bus trolleybuses in the city of Kharkiv, their greatest density is characterized for Saltivskyi, Slobidskyi, Shevchenkivskyi, Kyivskyi, and Industrial districts [4]. We are also talking about large sleeping areas (Pavlovo Pole, Alekseevka, Saltivka, New Houses, Horizon, Rohan, etc.). In addition, some routes of the city involve trolleybuses with autonomous running, which increases their maneuverability and allows running trolleybus routes to those areas where there is no stationary trolleybus line,or when there is a lack of electricity. The length of tram tracks is 217 km, there are 273 passengers and 40 service cars in the city. Regarding the spatial distribution of tram routes in Kharkiv, their greatest density is characterized for Saltivskyi, Slobidskyi, Shevchenkivskyi, Kyivskyi, and Industrial districts,they combine housing estates and workplaces, factories, enterprises, etc. [3].

Currently, the city of Kharkiv has 60 bus routes. Regarding the spatial distribution of bus routes in the city of Kharkiv, their greatest density is also characterized for the central, northern and eastern parts of the city. During the power outage, 34 temporary bus routes will be introduced in Kharkiv. Buses will replace the subway and ground electric transport [2]. In Kharkiv has an international airport, which is currently not functioning as a result of hostilities.

The city of Kharkiv has a powerful and extensive transport infrastructure, however, as a result of military operations, a significant part of the transport fleet was destroyed, logistics hubs and depots were destroyed. Emergency power outages, which occur as a result of shelling of the energy infrastructure, also create a significant problem, because it is difficult for a city of a million people to ensure the complete overlap of electric transport routes (metro, trolleybuses, trams) with buses.

Public transport in Warsaw is currently developing dynamically. The capital of Poland, as the country's largest city, is constantly increasing transport costs and implementing major investment projects, such as the extension of the metro, which

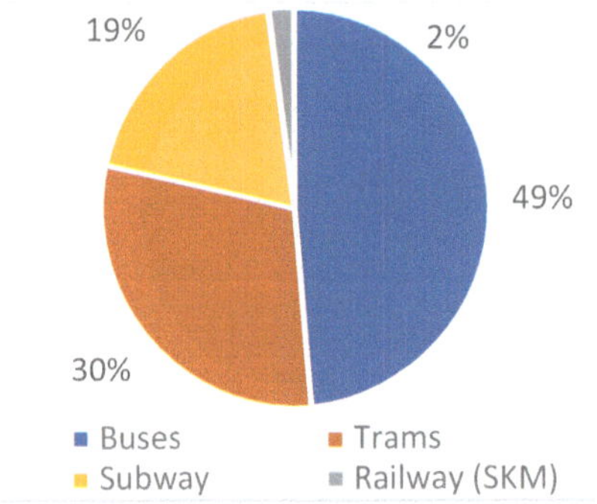

a

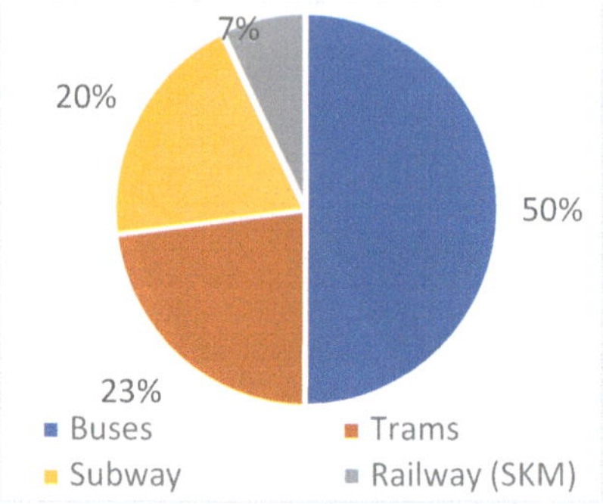

b

Fig. 32.2 Number of passengers transported by modes of public transport in Warsaw (**a**) and Passenger turnover by modes of transport in Warsaw (**b**)

gives residents more opportunities to move efficiently within the agglomeration. Warsaw is served by two international airports, the busiest being Warsaw's Chopin Airport and the smaller Warsaw Modlin. The main public transport services operating in the city include the Warsaw Metro, buses, suburban railways and an extensive network of trams [6]. The total volume of passenger transportation by all modes of transportation in the city of Warsaw in 2022 amounted to 863.4 million people. The structure of passenger transportation is significantly dominated by buses (almost half of the passengers transported in the city) (see Fig. 32.2). The second rank is occupied by tram transport (30%), then the subway (19%), and the smallest share is occupied by trains (2%) [8]. The predominance of buses in the transportation of passengers is explained by the maneuverability of buses.

Regarding the spatial distribution of bus routes in the city of Warsaw, then their greatest density is characterized by the central most densely populated areas of the city (Ochota, Śródmieście, Praga Północ, Praga Południe) [3]. Subway of the city of Warsaw consists of 2 lines and 39 stations. Regarding the spatial distribution of the subway in Warsaw, its stations are mainly located in the central most densely populated areas of the city (their density is characterized by the central areas of the city (Bemowo, Wola, Śródmieście, Targówek,Praga Północ, Żoliborz, Bielany, Ursynów).In areas such asOchota,Praga Południe, Ursus, Wilanów, Wawer, Wesoła, Rembertów, there is no subway.Regarding the spatial distribution of tram routes in the city of Warsaw, their greatest density is characterized by the central most densely populated areas of the city (Ochota, Wola, Śródmieście, Praga Północ, Praga Południe, Żoliborz). There are no tram lines in areas such as Ursus, Ursynów, Wilanów, Wawer, Wesoła, Rembertów [3]. Concerning the spatial distribution of fast urban railways in the city of Warsaw, then they pass along the city from north to south and from west to east. Its stations are located in such areas of the city as Ursus, Ochota, Wola, Włochy, Śródmieście, Targówek, Praga Północ, Praga Południe, Żoliborz, Rembertów, Wesoła, Wawer [10].

So, the transport infrastructure of the cities of Kharkiv and Warsaw has a lot in common, but there are also differences. In Kharkiv, the largest passenger traffic is characteristic of the metro, for Warsaw, the priority mode of public transport is the bus. There are no trolleybuses in Warsaw, but there is a city high-speed train. In recent years, we have seen that Warsaw's transport infrastructure, in particular, the transport routes, have become even longer, allowing to increase the transport accessibility of the peripheral areas of the city. For the city of Kharkiv, this experience should be useful and relevant in the aspect of post-war reconstruction and restoration of transport infrastructure.

32.3 Peculiarities of the Development of Bicycle Infrastructure in the Cities of Kharkiv and Warsaw as an Indicator of Sustainable Development

A bicycle, as a public transport, is more efficient than a car in terms of using street space, occupying 5–10 times less space during movement and 10–20 times less during parking. The bicycle does not pollute the air and does not create excessive noise, which makes the city streets more comfortable for being in the open space [1]. In the city of Kharkiv, the Concept of the development of bicycle traffic and the arrangement of bicycle infrastructure has been developed, which has been implemented since 2004. During this time years, in some districts of Kharkiv, multifunctional paths have appeared, which combine the movement of both cyclists and pedestrians. These paths have provided more comfortable and safer conditions for cyclists, especially in recreation areas. The city has introduced bike-sharing programs, which allow residents and guests of the city to rent bicycles for short trips. These programs usually offer convenient pick-up and drop-off locations at various points in the city. As of 2024, there is only one lane in Kharkiv dedicated specifically to the movement of bicycles, therefore, in the case of bicycle paths in the city of Kharkiv, they are mostly located in recreational areas and near them, and do not have a strategic importance in the transport system and logistics of the city (see Fig. 32.3).

Comparing the transportation network of the city of Kharkiv without bicycle paths and with them, it can be seen that there are no public transport routes in the middle of Lisopark (Shevchenkivskyi district), since it is a continuous forest massif. The priority direction for the development of cycling in Kharkiv is the combination of the city's areas with the highest population density and the largest centers of gravity. Due to the remoteness of the sleeping areas from the city center, it is necessary to strive for the routes to take the shortest and most direct routes with the most continuous movement of bicycle traffic [7].

Fig. 32.3 (**a**) The transportation system of the city of Kharkiv without bicycle lanes. (**b**) Transportation system of the city of Kharkiv with bicycle lanes

Warsaw is investing in improving its cycling infrastructure, including dedicated bike lanes, bike sharing programs and bike racks in key locations. The city aims to popularize the bicycle as an environmentally friendly mode of transport. The analysis of the length of bicycle lanes and their density within the districts of the city of Warsaw shows that they are relatively evenly distributed throughout the city (Fig. 32.4a, b). At the same time, their longest length is characteristic of such districts as Mokotów, Ursynów, Bilyany, Byalolenko, Sredmistya, and the shortest—in the districts of Vlokha, Okhota, Ursus, Rembertów, Vesola. At the same time, according to the indicator of the density of cycle paths, the highest density is characteristic of such areas as Žoliborz, Sredmistja, Praha-Povnoc, Okhota, Mokotów, and the lowest—in the areas of Wilanów, Bilolenka, Rembertów, Wawer, Wesoła [8].

Comparing the transport network of the city of Warsaw without cycle paths and with them (Fig. 32.4a, b), it can be seen that they run along highways, and their buffer zones fill white spots only in some places where the buffer zones were not

Fig. 32.3 (continued)

covered by public transport routes. Warsaw, like many other European cities, actively implemented measures aimed at improving the city's cycling infrastructure in order to popularize cycling as an environmentally friendly and healthy form of transport [7].

So, in the city of Warsaw, the development of bicycle infrastructure, the branching of the bicycle network is much greater than that of the city of Kharkiv. Warsaw's cycling infrastructure is more comprehensive, safer and better integrated with public transport. Some efforts have been made in Kharkiv to improve the bicycle infrastructure, but the city is still in the process of forming a comprehensive network. For the city of Kharkiv, the experience of the city of Warsaw in this matter is a good example for imitation and adaptation to the realities of Ukraine. This experience include expanding the network of bicycle paths, improving safety and integrating cycling into the general transport network to achieve the goals of sustainable development.

Fig. 32.4 (**a**) The transportation system of the city of Warsaw without bicycle lanes. (**b**) in Transportation system of the city of Warsaw with bicycle lanes

32.4 Problems and Prospects of the Transition of Transport Infrastructure of the City of Kharkiv to Sustainable Development

In terms of the conducted research, we have done SWOT analysis of the development of the transport system of the city of Kharkiv (see Table 32.1).

Thus, the transportation system of Kharkiv has many strengths, but it also faces significant challenges and potential threats. Strategic investments in infrastructure modernization, sustainability and technology, together with efforts to eliminate congestion and promote intermodal transport, can help the city further develop its role as a key transport hub while mitigating weaknesses and vulnerabilities. The biggest problem that threatens both the city and its transport system is Russian aggression, war and shelling of the city of Kharkiv. In terms of the conducted research, we have done SWOT analysis of the development of the transport system of the city of Warsaw (see Table 32.2).

Fig. 32.4 (continued)

Thus, Warsaw's transport system has a wide list of strengths and development opportunities, but at the same time, the city faces significant problems and potential threats. Strategic investments in infrastructure modernization, sustainability, technology and intermodal transport, together with efforts to eliminate congestion and promote economic growth and increase investment attractiveness, can help the city to further develop its role as a key transport hub in Europe, while addressing the challenges. Undoubtedly, the war in Ukraine created an additional burden on the transport system of Poland and the city of Warsaw itself. However, the increase in passenger traffic is also an incentive for development.

For the city of Kharkiv, it is possible to implement the innovative practices of the city of Warsaw in terms of the development of transport infrastructure on the basis of sustainability, in particular:

1. *Increasing the level of integration of public transport.* Warsaw has a well-integrated public transport system that includes buses, trams and the metro. Kharkiv can take advantage of Warsaw's experience through the further integration of different types of public transport.

Table 32.1 SWOT analysis of the development of the transport system the city of Kharkiv

Strengths	**Weaknesses**
Kharkiv is a key transport and logistics hub connecting the East with the regions of Ukraine, key transport highways E40 and E105 pass through the city. Stability of the main elements of the supporting framework of the transportation system of the city of Kharkiv: The high compactness and convenient configuration of the city territory allow connecting all areas of the city with the center and with each other. High level of urbanization, coverage of all areas of the city by the transport network. Integration of SMART technologies in the use of the transport system. Significant investment potential of the city of Kharkiv.	The peripheral position of the city of Kharkiv in relation to the capital of Ukraine and its remoteness from European countries, the city currently provides interregional communication. Destruction of vehicles, destruction of infrastructure (roads, depots) as a result of military operations. As a result of the destruction of the energy infrastructure, there are problems with urban electric transport. At the moment, in the city of Kharkiv there are practically no or poorly developed ecologically clean modes of transport that could ensure the evolution of the city's transport system in the concept of "sustainable development". The low attractiveness of the city's transport infrastructure due to its proximity to the aggressor country.
Opportunities	**Threats**
Modernization and restoration of infrastructure at the expense of investments from partner countries. Wide implementation of SMART technologies in the management of the city's transport system. Development of ecological types of transport, in particular—bicycle. Development of intermodal transport to increase the efficiency of the urban transport system. Adaptation to the realities of the time: the development and modernization of the metro to ensure better transport accessibility for residents of all areas of the city, the opening of new stations that will also serve as shelters for the population, the revitalization of the metro from alternative and autonomous energy sources, the widespread introduction of autonomous trolleybuses, the construction of underground parking lots and depot for public transport.	Escalation of hostilities, deterioration of the security situation, deterioration of the investment climate. Further destruction of transport infrastructure and destruction of public transport units. Decommissioning of problematic sections of roads (in particular, tram tracks). There is a possibility that the city will remain without a tram connection. Due to the fact that travel in transport is free, there are no conductors to monitor order or surveillance cameras, there are cases of vandalism and damage to the interiors of vehicles that are only cosmetically repaired. With a possible further increase in the number of private vehicles due to the low capacity of roads, the functioning of public transport will be complicated.

Table 32.2 SWOT analysis of the development of the transport system the city of Warsaw

Strengths	**Weaknesses**
Warsaw's location in Central Europe makes it a key transport hub connecting Eastern and Western Europe, making it a center for regional and international trade. A comprehensive system of public transport, including trams, buses and metro, city railway. Availability of integrated intermodal means of transport. Investments in infrastructure. The presence of pedestrian tourist zones in the city center, which allows the transport system to function better. Wide implementation of SMART technologies: traffic management, synchronized traffic lights.	The city has a large area, and traffic jams often occur during peak hours. Some elements of Warsaw's transport infrastructure are physically and morally outdated. Due to the presence of a large number of vehicles, there are many problems with parking. Limited investment in bicycle infrastructure. Overcrowding of public transport, in particular, as a result of migrations from Ukraine as a result of the war. Problems in communication between industrial, peripheral and central areas of the city.
Opportunities	**Threats**
Further expansion of the public transport network, reduction of "automobilization" of urban space. Improvement of bicycle infrastructure. Further implementation of intelligent management systems to optimize traffic flow and reduce traffic jams. Management of parking in the city, improving the availability of parking spaces. Expansion of pedestrian zones in the central areas of the city, restriction of access for cars. Implementation of new green logistics initiatives, use of environmentally friendly means of transport, favoring the sustainability and development of the city's transport system.	Constant traffic jams, especially during peak hours, hinder the efficiency of the transport system. Depreciation of transport infrastructure (roads, bridges and tunnels). Exacerbation of environmental problems, air pollution by emissions. Economic downturns, crises or fluctuations that may affect transportation financing and investment in transportation infrastructure. Competition from other transport hubs. Changing mobility patterns: population, excessive popularization of private vehicles.

2. *Development of bicycle infrastructure*. The City of Warsaw is actively investing in cycling infrastructure, including dedicated bike lanes and bike sharing programs. Kharkiv could follow suit by developing a more extensive network of bike lanes, creating dedicated bike lanes, and promoting cycling as an environmentally friendly mode of transportation in the city. Bicycle transport can also be considered as an element of a continuous intermodal network, in particular—combining different types of transport, such as buses, trams and bicycles, metro.
3. *Further use of "smart technologies"*. Intelligent traffic management systems and technologies have been implemented in Warsaw to improve traffic flow and safety. Kharkiv may consider investing in similar smart transportation solutions, such as traffic signal synchronization and real-time traffic monitoring, to improve overall transportation efficiency.

4. *Implementation of the "green initiative.* Warsaw is making strong efforts to promote green transport, including the introduction of electric buses and the promotion of green vehicles. Kharkiv can explore sustainable development initiatives such as introducing hybrid public transport and encouraging the use of clean, green energy sources for urban transport. Warsaw actively promotes the culture of sustainable use of public transport. Kharkiv can take measures to educate and engage its residents in sustainable transport practices through information campaigns and incentives.
5. *Modernization of road infrastructure.* Warsaw has implemented road infrastructure modernization projects, including the construction of new highways and the reconstruction of existing roads and bridges. For the city of Kharkiv, it is currently important to restore the transport infrastructure using new, innovative technologies.
6. *Public-private partnership.*Warsaw has successfully attracted public-private partnerships to finance and develop transport infrastructure projects. Such experience can be useful for Kharkiv, in particular, attracting European and international funds for the restoration and development of the territory.

32.5 Conclusions

Conducting this research is connected with several important aspects. In particular, these are: the extreme importance of the transport system in the development of large cities, the need for a radical transformation in the city of Kharkiv, as one of the most powerful socio-economically in Ukraine in terms of its European integration in the nearest future, the need to justify the ways of its development in accordance with the goals of sustainable development and post-war recovery. Considering that the level of development of transport infrastructure and transport accessibility are indicators of the level of development of all subsystems of the urban system, transport systems of large cities and agglomerations are interesting objects of scientific research that has practical significance and planning perspective orientation. On the basis of the conducted thorough analysis, the territorial and sectoral features of the transport system of the city of Kharkiv were established. In terms of the active political and economic integration of Poland and Ukraine, Ukraine's desire to join the European Union, it is important to study and implement the positive development of Polish cities and regions, in particular, transport as an element of territorial connectivity, as well as the presentation of positive practices in the development of Ukraine's transport system. The issues of transport accessibility and its improvement and restoration are relevant for Ukraine, in particular for the city of Kharkiv. A comparative analysis of the transport infrastructure of the cities of Kharkiv and Warsaw, in particular, a SWOT analysis, proved the presence of common features and some important differences between them. Characteristic for Warsaw, as well as for many large and medium-sized cities in Europe, is the significant development and use of bicycles as a means not only of sports, but also of transport. Warsaw's cycling infrastructure is more comprehensive, safer and better integrated with public transport. For the city of Kharkiv, the experience of the city of Warsaw in this matter is a good example for imitation and adaptation to the realities of Ukraine.

The identified differences in the development of the transport infrastructure of the cities of Kharkiv and Warsaw are related to many factors, in particular, the features of the city planning system. This question requires further research. and analysis of transport infrastructure in the aspect of implementation of the Sustainable Development Strategy in other European cities, in particular the cities—partners of Kharkiv, in particular the cities: Porto, Cincinnati, Nuremberg, Lille, between which there is intensive cooperation.

References

1. Apasnore P, Ismail K, Kassim A (2017) Bicycle-vehicle interactions at mid-sections of mixed traffic streets: examining passing distance and bicycle comfort perception. Accid Anal Prev 106:141–148
2. Official website of Kharkiv City Council, Mayor, Executive Committee, https://www.city.kharkov.ua/. Last accessed 2024/04/05
3. Official website of the Main Department of Statistics in the Kharkiv region, http://kh.ukrstat.gov.ua/. Last accessed 2024/04/05
4. OpenStreetMap, https://www.openstreetmap.org/. Last accessed 2024/04/05
5. OverpassTurbo, https://overpass-turbo.eu/. Last accessed 2024/04/05
6. Public transport step by step. Warsaw Public Transport, https://www.wtp.waw.pl/en/public-transport-step-by-step/. Last accessed 2024/04/05
7. The concept of the development of bicycle traffic and the arrangement of bicycle infrastructure in the city of Kharkiv, https://www.city.kharkiv.ua/uk/document/kontseptsiya-rozvitku-velosipednogo-rukhu-i-oblashtuvannya-velosipednoi-infrastrukturi-um-kharkovi-50008.html. Last accessed 2024/04/05
8. ZTM annual reports. Public Transport Authority in Warsaw Available at: https://www.ztm.waw.pl/raporty-roczne-ztm/. Last accessed 2024/04/05,
9. 17 Sustainable Development Goals, https://globalcompact.org.ua/tsili-stijkogo-rozvytku/. Last accessed 2024/04/05
10. Morar C, Nagy G, Dulca M, Boros L, Sehida K (2019) Aspects regarding the military cultural-historical heritage in the city of Oradea, Romania. ANNALES Ann Istrian Mediterranean Stud Ser Historia et Sociol 19(2):303–322. https://doi.org/10.19233/ASHS.2019.21
11. Morar C, Nagy G, Boros L, Gozner M, Niemets L, Sehida K (2021) Heritage, culture and regeneration of the former military areas in the city of Oradea (Romania). Architektúra Urbanizmus 55(1–2):76–87. https://doi.org/10.31577/archandurb.2021.55.1-2.6
12. Morar C, Tiba A, Jovanovic T, Valjarevic A, Ripp M, Vujicic MD, Stankov U, Basarin B, Ratković R, Popović M, Nagy G, Boros L, Lukic T (2022) Supporting tourism by assessing the predictors of COVID-19 vaccination for travel reasons. Int J Environ Res Public Health 19(2):918. https://doi.org/10.3390/ijerph19020918
13. Niemets L, Mezentsev K, Morar C, Sehida K, Husieva N, Peresadko V (2018) Innovation and investment potential of region as a factor of its "smart transformation": Kharkiv region case study (Ukraine). Visnyk of V N Karazin Kharkiv National University, series. Geol Geograp Ecol 49:137–159. https://doi.org/10.26565/2410-7360-2018-49-11

14. Morar C, Berman L, Unkart S, Erdal S (2021) Sustainable brownfields redevelopment in the European Union: an overview of policy and funding frameworks. J Environ Health 18(21):11169. https://www.neha.org/node/62305
15. Morar C, Lukic T, Valjarević A, Niemets L, Kostrikov S, Sehida K, Telebienieva I, Kliuchko L, Kobylin P, Kravchenko K (2022) Change detection based spatiotemporal analysis of urban green areas: a case study of Kharkiv-City. Front Environ Sci 10:1–27. https://doi.org/10.3389/fenvs.2022.823129
16. Palinchak M, Zayats O, Tsalan M, Ţoca CV (2023) Determinants of UMSCA competitive force formation. Baltic J Econom Stud 9:200–205. https://doi.org/10.30525/2256-0742/2023-9-5-200-205
17. Mihai M, Darabaneanu D, Chirodea F, Ţoca CV (2022) Analysis of social support as an argument for the sustainable construction of the European Community space, in sustainability. Special Issue Sustainable Well-Being 14(12):7448. https://doi.org/10.3390/su14127448

Chapter 33
Brownfields in Ukraine: Depressed Territories or Spaces of Multifaceted Opportunities

Viktoriia Patsiuk

Abstract Derelict territories, including industrial ones, despite their negative image, hold great potential for transformation into innovative physical structures with high intellectual, cultural, ecological, and visual value. This process requires comprehensive theoretical research for their further implementation in practical terms. Military events in Ukraine, while having a significantly negative impact on various aspects of life, have necessitated a reevaluation of various areas, including achieving a balance between societal, environmental, and economic development.

The present study explores the concept of "brownfields" and highlights theoretical aspects of addressing this issue in scientific sources. It examines the specific considerations for using brownfields in Ukraine. Using the case study of Kryvyi Rih, the country's most powerful industrial center currently located in a front-line territory, pre-war trends in researching the possibilities of brownfield revitalization are revealed. Directions for the further use of different types of brownfields depending on the development of the Russian-Ukrainian conflict are highlighted. Field study of the city's brownfields is conducted, and PEST analysis reveals that Kryvyi Rih has significant potential for their redevelopment. However, under the current unfavorable security situation, achieving sustainable development in these abandoned territories remains challenging.

Keywords Brownfields · Sustainability · Military use of landscapes · Industrial spaces

V. Patsiuk (✉)
Taras Shevchenko National University of Kyiv, Kyiv, Ukraine

Kryvyi Rih State Pedagogical University, Kryvyi Rih, Ukraine
e-mail: viktoriia.patsiuk@kdpu.edu.ua

C. Morar et al. (eds.), *Achieving Sustainability in Ukraine through Military Brownfields Redevelopment*, NATO Science for Peace and Security Series C: Environmental Security, https://doi.org/10.1007/978-94-024-2278-8_33

33.1 Introduction

The problem of reusing previously developed territories is actively discussed in both scientific research and practical applications across Europe. Brownfield renovation is a crucial mechanism for achieving sustainable development, and military-origin brownfields are no exception. However, while developed countries have seen a reduction in military facilities over the past decades, with their transformation towards cultural, educational, or sports purposes, the full-scale war initiated by Russia has slowed down sustainable trends of military facilities redevelopment.

Military events in Ukraine have created three distinct development lines regarding brownfield sustainability. Firstly, the area of disturbed lands has significantly increased, and their sustainable transformation will be a necessary component of post-war reconstruction. Secondly, the relocation of enterprises from the eastern and occupied southern territories has stimulated redevelopment of non-functioning industrial spaces in the western part of the country. Thirdly, which significantly differs from the previous trends, the constant threat from the aggressor country does not contribute to demilitarization of Ukraine's territory. On the contrary, it necessitates a reevaluation of using existing production spaces, both non-functioning enterprises and various anthropogenic landscapes, for military purposes [1, 2].

While many foreign publications emphasize that the need for brownfield regeneration is driven by urbanization processes and the growth of urban populations, military events in Ukraine may lead to the total abandonment of territories, including individual cities and more complex formations. This situation highlights the critical need for a more detailed study of the sustainable use of brownfields in Ukraine, taking into account both national specifics and the military situation.

33.2 Publication Overview

The term "brownfield" first came into use on June 28, 1992, at a U.S. congressional field hearing hosted by the Northeast Midwest Congressional Coalition. Also in 1992, the first detailed policy analysis of the issue was convened by the Cuyahoga County, Ohio Planning Commission. EPA selected Cuyahoga County as its first brownfield pilot project in September 1993. The term applies more generally to previously used land or to sections of industrial or commercial facilities that are to be upgraded [3].

The US Environmental Protection Agency [EPA] defines brownfields as "abandoned, idle, or under-used industrial and commercial facilities where expansion or redevelopment is complicated by real or perceived environmental contamination" [4].

In England, a brownfield is a previously used land—other than for agriculture—which is currently underused or unused (DCLG 2006), due to the presence of one or more factors, which may include contamination, affecting its reuse [5].

In the European scientific discourse, this concept emerged after the report prepared by Working Group "Brownfields and Redevelopment of Urban Areas" of the Concerted Action "Contaminated Land Rehabilitation Network for Environmental Technologies" (CLARINET) [6]. CLARINET is a coordinated activity of the European Commission's Environment and Climate Program. CLARINET Working Group "Brownfield Redevelopment" has agreed on the following definition, which is intended to describe the full context of the environmental, economic and land use issues that are involved:

Brownfields are sites that: • have been affected by the former uses of the site and surrounding land; • are derelict or underused; • have real or perceived contamination problems; • are mainly in developed urban areas; require intervention to bring them back to beneficial use.

This interpretation is shared by CABERNET (Concerted Action on Brownfield and Economic Regeneration Network). A key aspect of the CABERNET definition is the emphasis it places on the need for intervention as a common characteristic of brownfields [7].

At the same time, the National Brownfield Strategy UK, seeking to differentiate itself from the definition proposed in the USA, proposes a more appropriate definition that outlines the essence of a brownfield: "previously developed land" (or PDL) [8]. However, the author does not consider "previously developed land" to be synonymous with abandoned or degraded lands.

The subject matter of brownfields has been explored through various aspects.

J. Litt and T.A. Burke [9] investigate the ecological and physical characteristics of brownfields in the southeast of Baltimore, Maryland, to identify the potential hazards of such sites and potential risks to the health of residents in adjacent communities.

A. Longo and D. Campbell [10], using England as an example, research the determinants and barriers to the reuse of derelict territories, attempting to identify which characteristics make a site more conducive to regeneration.

The impact of brownfields on regeneration rights to land ownership is highlighted by D. Adams et al. [11] The authors hypothesize that property rights restrictions stem from the peculiarities of land as a commodity, imperfections of the land market, behavioral characteristics of landowners, and the institutional context of property ownership, exchange, and land development. During the research, property rights restrictions are divided into those related to deficiencies or limitations of ownership rights to potential development land and those directly related to the strategies, interests, and actions of those who hold such rights [11].

Brownfields as a component of green infrastructure for implementing ecosystem services in cities are investigated by J. Mathey et al. [12]. As a result, they conclude that since abandoned areas are often the only spaces in densely built urban areas that can contribute to the creation of new green zones, it makes sense to integrate abandoned territories into the existing city green infrastructure system. A similar approach is demonstrated by L. Loures [13], who notes that as land in the city is a scarce resource, the once used but currently abandoned parcels of land scattered across cities, known as brownfields, form a latent resource that holds the potential of being used for urban greenspaces (UGS). Largely due to the previous use, brownfields are potentially contaminated but are usually supported by existing infrastructure. They are often the only available option for redevelopment in densely populated cities. In their study, S. Chowdhury et al. [14] justify the feasibility of transforming brownfields into urban green spaces, citing the results of stakeholder surveys.

However, in the context of the present research, the aspects of sustainable use of brownfields of military origin are more intriguing. Specifically, S. G. Bagaeen [15] justifies the redevelopment of military bases as an example of regenerating abandoned lands and conducted a comparative analysis using examples from the UK, Germany, and Jordan to highlight experiences, lessons learned, and outline potential for further research. C. Hill [16] dedicates her research to analyzing the success of redeveloping former military bases in Florida. Fundamental research into the peculiarities of redeveloping former military facilities in US cities is carried out by M. Touchton and A. J. Ashley [17]. B. McKee [18] focuses on considering architectural approaches and territorial planning specifics when analyzing the possibilities of transforming closed military bases into new civilian facilities.

In Ukraine, the research base on this issue is just beginning to form. M. Soldak [19] thoroughly examines this question, emphasizing the urgent need for Ukraine to address the formation of modern policies in the revitalization of abandoned areas of former industrial use (brownfields), particularly in the area of industrial waste processing. The scholar notes that the absence of clear national goals and standards for state and private actors, uncertainty in funding and stimulating these processes, and the lack of a unified registry of derelict industrial areas hinder the resolution of many environmental issues.

33.3 Results

Brownfields are recognized as a problem in many European cities that requires increased attention from the EU. Abandoned industrial spaces arise due to structural changes, such as the decline of traditional mining industries. Brownfields deteriorate the environment in visual, aesthetic, sociological, economic, psychological, and other aspects, and their dereliction invariably leads to risks to people's lives.

According to estimates by J. Litt and T.A. Burke, there are approximately 400,000 to 500,000 abandoned and underused commercial and industrial facilities with real or perceived contamination across the United States, which they refer to as brownfields [9]. There are more than 66,000 hectares of previously developed land, or brownfields in England, mostly located in the high-growth areas of greater London, the South East and East, that would be suitable to accommodate more than 200,000 new homes (DCLG 2015) [20]. Most European countries cannot provide estimates on the size of problem in their areas. Even where countries can provide figures—for example Germany (about 128,000 hectares), The Netherlands (between 9000 and 11,000 hectares), Belgium/Wallonia (about 9000 hectares)—it is clear that the data are not directly comparable, and include different kinds of site [21].

At the same time, Ukraine has a large number of brownfields; however, a territorial count of their quantity was not conducted even in peacetime. The current situation with full-scale invasion makes it impossible to estimate the number of these territories, and what's worse is that the share of brownfields, especially in eastern regions of Ukraine, continues to grow steadily. Therefore, it is not without reason that on the map "Global number of brownfield sites", presented in the studies by D. Hou et al. [22], Ukraine is represented as a blank spot alongside most countries in Africa and Asia.

Frantál et al. [23] show that brownfields located in municipalities with a higher local development potential are more likely to be redeveloped. Confirmation of this can be found in Ukraine, as the most successful brownfield redevelopment projects are presented in the capital city of Kyiv and the city of Lviv. Iconic revitalized objects in Kyiv include the Art Factory "Platforma", located on the territory of a former Darnytsya Silk Factory; "UNIT.City" which was created on the territory of a former Kyiv Motorcycle Plant; the National Art and Culture Museum Complex "Mystetsky Arsenal", located in the building of the Old Arsenal; "Pivzavod", thanks to which the historic building of a former brewery was preserved.

In Lviv the most successful examples of revitalization are: "Lem Station" on the basis of a former tram depot, "! FESTrepublic"—on the site of a former plant "Halychsko", "ReZavod"—formely a plant of electronic medical equipment. Another exemplary case of industrial space revitalization in Ukraine is the innovation center "Promprylad. Renovation" (formerly a Soviet instrument plant), located in the city of Ivano-Frankivsk. This city began to demonstrate rapid development in the 2010s, surprisingly accelerating its pace even since the onset of the war, driven by a large number of internally displaced persons who chose this city for living and a significant number of relocated industrial enterprises. The "Promprylad. Renovation" is one of the first examples of transforming an abandoned enterprise into a modern business cluster with the involvement of public investment in Ukraine [24].

At the same time, in powerful industrial centers of Ukraine like Kryvyi Rih, Zaporizhzhia, Mykolaiv, Kamyanske, despite availability of a large number of brownfields, successful examples of their redevelopment are virtually absent. However, the awareness of the necessity and sometimes the economic viability of revitalizing industrial spaces has been growing. In particular, in the city of Kryvyi Rih, through the initiative of the community, specifically the public organizations "Kryvyi Rih Foundation of the Future" and "Urban City", with financial support of the Metinvest group of companies, the "Urban Hackathon" was organized on November 15–16, 2019. This was a thematic event in the city advocating for a reconsideration of revitalization possibilities for brownfields. During this event, the leaders of the "Urban CAD" project from Kherson presented their experience in giving a new life to industrial heritage. In addition, five objects of the city promising for revitalization were put forward for discussion. They were the mine "Artem-2", the premises and shaft of the mine "Ventylyatsiyna-3", the premises of the former management office of the Lenin mine, the abandoned greenhouse and Burshchytsky dump.

Teams of experts developed concepts for the transformation of these objects. The project proposed transforming the non-functioning mine into the Industrial Culture Park (ICP) "SHAKHTA". This project of revitalizing the mine "Artem-2" is aimed to preserve identity and authenticity of the industrial object (the mine) by creating conditions for modern rethink of industrialization processes. The implementation of the project is possible subject to a flexible planning structure. To achieve the purpose, the following was is proposed:

1. Creation of locations on mining (historical) themes: the Museum of Mining Engineering and a skansen of rocks and minerals located outdoors. an exhibition hall for changing expositions (photos on industrial themes, art objects made of metal, etc.), a training geological laboratory placed in the premises of the on-surface facility on the first level; the attraction-excursion "Descending into the mine";

   The second level can host a free recreation area, divided into: a souvenir shop of the ICP "SHAKHTA", a hall "Free microphone", a cafe "Tormozok".
2. Creation of post-industrial tourist locations: a tourist information center (TIC) and the branch of a children tourism center—on the first level; a rope adventure park; a climbing wall, a sports ground; a free art-zone and information stands on industrial tourism in Kryvyi Rih—in the yard.
3. Creation of locations for small-scale and private functions (administration offices, tenants' offices in a separate room.
4. Creation of an open space. The project provides for "soft" planning of the courtyard—we suggest preserving its function of the central space from where all buildings can be entered. It will be adapted for lectures, movie screening, performances, exhibitions, community meetings, hackathons, etc. [25].

There were hopes that among the powerful enterprises in Kryvyi Rih, there would be those willing to invest in the practical implementation of the project and create a revitalized object of European level in our industrial city. Specifically, during the mayoral election campaign in 2020, one of the candidates, who is the director of the PJSC "Central Mining and Processing Plant", proposed a comprehensive development program for Kryvyi Rih. One of the points of this program was the creation of a tourist and educational complex based on the non-functioning mine Artem-2, which would combine a modern industrial museum and an interactive science center called the Experimentarium (see Fig. 33.1) [26].

This project envisioned sustainable development of the area with multi-vector use of creative spaces. However, present realities have influenced the situation, and now in the city of Kryvyi Rih, which is in a pre-frontline position, significant investments in the development of creative spaces is not even a matter of consideration. On the contrary, the question of the feasibility of using brownfields for military purposes is being raised. In particular, the mine Artem-2 has a unique feature as the ore was transported to the surface not through the traditional vertical skip mechanism but via two belt conveyors installed in inclined (not vertical) shafts. These mechanisms reached a depth of 700 meters. Therefore, this mine could easily be adapted for military needs, providing, under the necessary humidity regime,

Fig. 33.1 Project of creating an interactive industrial museum and scientific and educational Experimentarium [26]

underground storage for equipment and weapons. However, this complex is privately owned by the Metinvest group of companies, and naturally, in implementing its projects, the company considers economic benefits, which are unlikely to be feasible in the militarization of this object.

Moreover, the onset of the full-scale invasion has sharpened the question of the need for military personnel training and practical skills development. The search for territories for comprehensive training has begun, and in this context, the various mining and industrial landscapes of the city offer significant resources for organizing various activities.

According to V. Kazakov's estimates [27], the total area of mining and industrial landscapes in Kryvyi Rih area is 18.1 thousand hectares. This includes: over 4.2 thousand hectares of open pit areas, 7.0 thousand hectares of dumps, 6.5 thousand hectares of extractive landscapes (tailings storage facilities), and 3.4 thousand hectares of mine subsidence and disturbance zones. These figures are constantly changing due to ongoing mining activities and formation of new dumps. This allows considering Kryvyi Rih region the best area for studying mining and industrial landscapes, restructuring sustainable environmental policies, and developing ecological networks based on newly created anthropogenic landscapes.

Before the war, various forms of extreme tourism were actively developing in Kryvyi Rih: urban exploration and infiltration of abandoned sites, rooftop climbing (roofing), including industrial structures, spelunking into artificial mine shafts, quad biking in dump and subsidence areas, mountain biking on dump slopes, paintballing in abandoned factories, and diving into flooded open pits. Given the trends indicating Russia's unchanging desire to annex Ukrainian territory, the issue of enhanced civilian preparedness for military action is becoming more relevant.

An excellent example of sustainable use of reclaimed industrial spaces for military purposes is the Männiku training ground, created on the site of an exhausted sand quarry. Information about this area is not classified and is even available on Wikipedia [28].

The boundary of the Männiku training field is marked by boundary markers or trees with white and red stripes. There are barriers on three driveways to the training field from the Tallinn-Saku road, which close in case of hazardous activities. Also, barriers are installed near them to prevent people from entering the hazardous zone. Flagpoles are set up on the training field, and one hour before training, red warning flags are raised. The area is used for shooting exercises, shooting at distances up to 600 meters, anti-tank grenade launcher firing, hand grenade throwing, explosive and demining work. Tactical exercises can be conducted on the range, both on foot and with armored vehicles, using simulated equipment. Shooters and security companies can also train on the training field. Motorsport enthusiasts can use the track for mobile technology training.

The infrastructure of the training field includes three shooting ranges, a patrol range, a grenade launcher range, a detonation range, a tactical range with a mobile equipment training course and a pioneer range, as well as a service center.

Information boards with field marking and closure instructions, along with a schedule of use for up to two weeks, are installed around the training field. Contact information of the person responsible for the practice is provided on the information board. In the northwest part of the Männiku training field, when there are no training sessions, people often go for walks. The author of the study visited this area, albeit not under very favorable weather conditions (see Fig. 33.2).

Fig. 33.2 Information signs on the territory of the landfill Männiku. (Photographed by author)

At the same time, efforts are being made to combine military use of the Männiku training field with the maximum possible preservation of the natural environment. The Männiku practice field is home to several protected species. These species are very rare and are at risk of disappearing in Estonia. Permanent habitats are created for protected species to ensure favorable conditions for their survival by preserving their habitats. It should be noted that individual animals of protected species should not be captured or intentionally disturbed during breeding, raising offspring, hibernation, and migration [29].

Therefore, the brownfields of Kryvyi Rih can be used for various purposes, Table 33.1. However, specific objects that can currently be used for military purposes have not been explicitly identified.

To understand how external factors may influence the process and mechanism of brownfield redevelopment in Kryvyi Rih and the frontline regions of Ukraine as a whole, a PEST analysis is conducted.

In terms of politics, a key factor affecting the further existence and emergence of brownfields is military actions and a continuing threat from Russia. If the threat persists, the development of brownfields will occur in a destructive manner: their number will increase and the possibility will be reconsidered to attribute to former industrial objects military functions. Political will at the local level will also play a significant role since a large number of abandoned territories are communal or state-owned, and local authorities will determine the direction of their further development. Effective use of brownfields considering multi-vector aspects is possible only through the collaboration of authorities and communities. Another lever is availability of a clear and favorable legal environment. Political decisions aimed at adopting and implementing relevant legislative acts can stimulate investment and development of brownfield revitalization projects. The formation of a land market in Ukraine will also be significant, although it is currently difficult to predict whether this will positively or negatively impact availability and use of brownfields.

In the economic terms, brownfield redevelopment is possible only with financial support and the creation of favorable conditions for investment. Analyzing Estonia's experience, J. Tintěra et al. [30] note that transitioning to brownfield regeneration in this country has led to economic growth, EU accession, and access to EU structural funds. It is clear that Ukraine, especially its frontline regions, cannot overcome the negative economic consequences of war without international support and the ability to participate in various grant donor programs. However, such support should not just address situational problems but focus on creating a favorable economic environment for investors and prioritizing sustainable development projects.

Analyzing the social aspect, it is worth noting that brownfields worsen the surrounding environment in aesthetic, visual, sociological, economic, psychological, and other contexts, and their abandonment can lead to risks for people's lives. Therefore, for the cities in the front-line regions of Ukraine to reduce depressive tendencies, it is necessary to focus on revitalizing abandoned territories, which usually have industrial origins. During redevelopment, it is necessary to consider the socio-cultural characteristics of the local community. This includes attention to the historical significance of the territory, preservation of cultural identity, as well as the

Table 33.1 Directions for using brownfields in Kryvyi Rih under various security conditions (developed be the author)

Anthropogenic landscape	Use in wartime	Use in post-war time	
		Under Russia's threat	Without Russia's threat
Dumps	Installation of air defense systems; disactivation of unexploded missile parts; use as defensive lines and firing points	Military personnel training; practicing movement tactics and combat within rough terrain; tactical shooting practice	Creation of landscape parks and recreational areas with panoramic views on dump territory
Abandoned industrial structures	Storage of weapons and equipment; military personnel deployment	Creation of training grounds for practicing combat tactics in urban conditions; reprofiling industrial structures into military plants	Redevelopment with formation of creative spaces, cultural and artistic complexes, commercial and entertainment centers; establishment of military-themed museums in frontline areas
Flooded open pits	Creation of test ranges for marine drones.	Using the water surface for training in controlling water drones; teaching underwater swimming and diving techniques; training military divers.	Development of recreational complexes for various activities: swimming, fishing, kayaking, and stand-up paddleboarding, etc.; areas for free nature regeneration.
Tailings storage facilities	–	Practicing combat shooting, shooting at different distances, firing from an anti-tank grenade launcher, throwing hand grenades; practicing sapper work	Technical and biological reclamation; in the distant future, transformation into landscape parks
Non-functioning mines	–	Creating conditions for underground storage of weapons and equipment; equipping underground workshops for weapon production	Museumification of non-functioning mines applying European practices; creation multifunctional cluster objects for cultural and tourist purposes; conducting tours, including underground ones

needs and views of residents. In particular, the residents of Kryvyi Rih already require the emergence of new objects both of cultural and recreational formats, which will not only improve the city's amenities but also provide the residents with new jobs and provide representatives of small and medium-sized businesses with a platform to realize their own products.

Regarding the technological aspect, redevelopment of outdated structures is much more complex and difficult than constructing new buildings. However, besides financial gain, it is necessary to focus on considering the principles of sustainable development of territories, preserving local identity, promoting the formation of industrial culture, and taking into account the spatial specifics of the development of various territories. Moreover, available construction technologies in the front-line regions of the country allow for the implementation of various projects, the realization of which depends solely on the available budget.

Thus, analyzing the PEST analysis methodology, availability of a whole range of positive factors regarding the redevelopment of brownfields in the front-line territories of Ukraine in general and in the city of Kryvyi Rih in particular is evident. However, in today's conditions, all these factors have a predominantly negative coloration, and it is unlikely that anything will change drastically until the security situation in Ukraine improves.

Overall, as noted by S. Perovic and N.K. Folić [31], brownfield regeneration includes defining sustainable development goals, their indicators for pollution control and land reclamation, managing post-construction projects and infrastructure, urban design, as well as directly managing brownfield reclamation projects. The scientists emphasize that promoting public-private partnerships is of great importance for urban efficiency and brownfield regeneration programs, as well as for communication, building trust, and cooperation among relevant stakeholders.

K. Williams and C.A. Dair [32] emphasise that important role in the regeneration of industrial brownfields is awareness of interested parties (stakeholders, agents). These parties may be involved in drawing up specific legislation (parliament and ministries), making plans for and using the sites (local authorities, interest groups), investment and development (companies, businessmen), project planning and construction (architects, construction engineers, builders, urban planners), etc. [33–35].

Therefore, consolidation of efforts from various stakeholders is a necessary condition for the successful implementation of brownfield sustainable development projects. In contrast, urban planners should strive for a flexible use of abandoned territories to provide space for maneuvering and responding to future development challenges. Ideally, these challenges should stimulate constructive development rather than satisfying destructive military motives [36–38].

33.4 Conclusions

In Ukraine, the past decade has seen a growing focus on the sustainable transformation of brownfields. The decline of industrial production has led to the closure of businesses, deterioration of industrial buildings, and adoption of new models for organizing industrial activity. A review of leading practices from around the world demonstrates the effectiveness of brownfield redevelopment. The adaptation of this experience, with a strong emphasis on sustainable development principles, is necessary and achievable in Ukraine. A balanced spatial analysis of non-functioning industrial spaces and the consolidation of opportunities among authorities, scientists, and communities to develop comprehensive solutions for integrating brownfields into urban fabric are crucial for a sustainable regeneration process for brownfield redevelopment becomes increasingly acute in front-line regions of Ukraine, where the aggressive destruction of the territory has led to a rise in abandoned areas. While the practical implementation of sustainable regeneration opportunities for abandoned lands is questionable, studying this issue in detail, with an eye toward post-war reconstruction, is essential. Furthermore, due to Ukraine's clear European course, the transformation of urban policy should be based on the implementation of key principles of EU regional and environmental policy into Ukrainian legislation.

This study represents an initial step towards creating a robust foundation for assessing the suitability of various brownfields for redevelopment. It will also lay the groundwork for developing a comprehensive methodology tailored to the unique challenges faced by a territory subjected to heavy war-torn conditions. This approach will ensure well-informed strategic decisions regarding the optimization of land use in these territories. Ultimately, brownfield regeneration holds immense potential to foster new opportunities, economic growth, environmental restoration, cultural vibrancy, and the adoption of innovative public space models. It will undoubtedly be a key driver in rebuilding Ukraine and laying the foundation for a prosperous future.

References

1. Meszaros EL, Toca CV (2020) The EU's multifaceted approach to resilience building in the Eastern Neighbourhood. Security sector reform in Ukraine. Eastern J Eur Stud 11:120–145, Special Issue SI, https://www.webofscience.com/wos/woscc/full-record/WOS:000593531000007?SID=EUW1ED0BEEp2kP2NxOAqUPsOYoFDc
2. Mészáros EL, Ţoca CV (2023) The EU's resilience and the management of hybrid threats coming from the Eastern neighbourhood: Belarus and the deliberate facilitation of irregular immigration. Eastern J Eur Stud 14(1):5–30. https://ejes.uaic.ro/articles/EJES2023_1401_MES.pdf
3. Maliene V, Wignall L, Malys N (2012) Brownfield Regeneration: Waterfront Site Developments in Liverpool and Cologne. J Environ Eng Landsc Manag 20(1):5–16. https://doi.org/10.3846/16486897.2012.659030
4. United States Environmental Protection Agency (1997) Brownfields Glossary of Terms. Homepage, http://www.epa.gov/swerosps/bf/glossary-htm. Date accessed: September 15
5. English Partnerships (2008) Brownfield land and deprivation. http://www.englishpartnerships.co.uk/nlud.htm
6. EU CLARINET (2002) Brownfields and redevelopment of urban areas, a report from the contaminated land rehabilitation network for environmental technologies – working group 2001, Umweltbundesamt, Vienna. https://haemers-technologies.com/wp-content/uploads/2018/02/Brownfields_and_Redevelopment-Urban_Areas.pdf
7. Cabernet Coordination Team (2006) Sustainable Brownfield Regeneration: CABERNET Network Report. University of Nottingham [online]. [cit: 17.9.2014]. http://www.cabernet.org.uk/resourcefs/427.pdf
8. Towards a National Brawnfield Strategy (2003) Research Findings for the Deputy Prime Minister. https://image.guardian.co.uk/sys-files/Society/documents/2003/11/06/brownfield.pdf

9. Litt JS, Burke TA (2002) Uncovering the historic environmental hazards of urban brownfields. J Urban Health 79:464–481. https://doi.org/10.1093/jurban/79.4.464
10. Longo A, Campbell D (2017) The determinants of brownfields redevelopment in England. Environ Resour Econ 67:261–283. https://doi.org/10.1007/s10640-015-9985-y
11. Adams D, Disberry A, Hutchison N, Munjoma T (2001) Ownership constraints to brownfield redevelopment. Environ Plann A Econ Space 33(3):453–477. https://doi.org/10.1068/a33200
12. Mathey J, Rößler S, Banse J, Lehmann I, Bräuer A (2015) Brownfields as an element of green infrastructure for implementing ecosystem services into urban areas. J Urban Plan Develop 141(3):A4015001. https://doi.org/10.1061/(ASCE)UP.1943-5444.0000275
13. Loures L (2015) Post-industrial landscapes as drivers for urban redevelopment: public versus expert perspectives towards the benefits and barriers of the reuse of post-industrial sites in urban areas. Habitat Int 45(P2):72–81. https://doi.org/10.1016/j.habitatint.2014.06.028
14. Chowdhury S, Kain JH, Adelfio M, Volchko Y, Norrman J (2023) Transforming brownfields into urban greenspaces: a working process for stakeholder analysis. PLoS One 18(1):e0278747. https://doi.org/10.1371/journal.pone.0278747
15. Bagaeen SG (2006) Brownfield sites as building blocks for sustainable urban environments: a view on international experience in redeveloping former military sites. Urban Des Int 11:117–128
16. Hill C (2000) Measuring success in the redevelopment of former military Ba-ses: evidence from a case study of the Truman annex in Key West. Florida Econom Develop Quarter 14(3):267–277. https://doi.org/10.1177/089124240001400306
17. Touchton M, Ashley AJ (2019) Salvaging community: how American cities rebuild closed military bases. Cornell University Press
18. McKee B (1994) Military Base conversions. Architecture (American Institute of Architects) 83(8):99–106
19. Soldak MO (2021) Promyslova ekosystema i revitalizatsiia braunfildiv. [Industrial ecosystem and brownfields revitalisation]. Econ Indust 3(95):70–101. https://doi.org/10.15407/econindustry2021.03.070
20. DCLG (2015) Building more homes on brownfield land. Departmentfor Communities and Local Government, London
21. Grimski D, Ferber U (2001) Urban brownfields in Europe. Land Contaminat Reclamat 9(1):143–148
22. Hou D, Al-Tabbaa A, O'Connor D, Hu Q, Zhu YG, Wang L, Rinklebe J (2023) Sustainable remediation and redevelopment of brownfield sites. Nat Rev Earth Environ 4(4):271–286
23. Frantál B, Kunc J, Nováková E, Klusáček P, Martinát S, Osman R (2013) Location matters! Exploring brownfields regeneration in a spatial context (case study of the South Moravian Region, Czech Republic). Moravian Geograph Rep 21(2):5–19
24. Promprylad. Renovatsiia. URL https://promprylad.ua/ua
25. Patsiuk V, Kazakov V, Skorupskas R, Ostapchuk I, Petrova A (2022) Revitalization of the industrial heritage: guidelines for Kryvyi Rih. OP Conference Series: Earth and Environmental Science (EES). 1049. https://doi.org/10.1088/1755-1315/1049/1/012082
26. Horka O. U Kryvomu Rozi bude stvoreno interaktyvnyi industrialnyi muzei i naukovo-osvitnii Eksperymentarium [In Krivyi Rih will be created an interactive industrial museum and scientificeducational Experimentalium]. https://krivbass.city/news/view/u-krivomu-rozi-bude-stvoreno-interaktivnij-industrialnij-muzej-i-naukovo-osvitnij-eksperimentariumd mitro-shevchik
27. Kazakov VL (2011) Na shliakhu do povnoho vyvchennia hirnychopromyslovykh landshaftiv Kryvbasu. Teoretychni, rehionalni, prykladni napriamy rozvytku antropohennoi heohrafii ta heolohii: mater. III mizhnar. nauk. Konfer [Towards a complete study of the mining landscapes of Kryvbas. Theoretical, regional, and applied directions of development of anthropogenic geography and geology: materials. III international scientific conference.]. Kryvyi Rih, pp 35–47

28. Männiku harjutusväli. https://et.wikipedia.org/wiki/M%C3%A4nniku_harjutusv%C3%A4li
29. Kaitseväe harjutusväljad: Männiku harjutusväli. https://mil.ee/kaitsevagi/harjutusvaljad/#t-manniku
30. Tintěra J, Ruus A, Tohvri E, Kotval Z (2014) Urban brownfields in Estonia: scope, consequences and redevelopment barriers as perceived by local governments/Urbánní brownfieldy v Estonsku: Rozsah, dopady a revitalizační bariéry z pohledu městských samospráv. Moravian Geograph Rep 22(4):25–38. https://doi.org/10.1515/mgr-2014-0021
31. Perovic S, Folić NK (2012) Braunfild regeneracija-imperativ za održivi urbani razvoj. Građevinar 64(05):373–383
32. Williams K, Dair C (2007) A framework for assessing the sustainability of brownfield developments. J Environ Plan Manag 50(1):23–40. https://doi.org/10.1080/09640560601048275
33. Morar C, Dulca M, Nagy G (2016) Brownfields regeneration, between sustainable urban development and cultural heritage, the former military sites in Oradea, Romania. J Urban Reg Anal 8(1):75–84
34. Kostrikov S, Niemets L, Sehida K, Niemets K, Morar C (2018) Geoinformation approach to urban geographic system research (case studies of Kharkiv region, Ukraine). Visnyk of V. N. Karazin Kharkiv National University, series. Geol Geograp Ecol 49:107–124. https://doi.org/10.26565/2410-7360-2018-49-09
35. Niemets L, Mezentsev K, Morar C, Sehida K, Husieva N, Peresadko V (2018) Innovation and investment potential of region as a factor of its "smart transformation": Kharkiv region case study (Ukraine). Visnyk of V N Karazin Kharkiv National University, series. Geol Geograp Ecol 49:137–159. https://doi.org/10.26565/2410-7360-2018-49-11
36. Morar C, Berman L, Unkart S, Erdal S (2021) Sustainable brownfields redevelopment in the European Union: an overview of policy and funding frameworks. J Environ Health 18(21):11169. https://www.neha.org/node/62305
37. Berman L, Morar C, Unkart S, Erdal S (2022) An overview of brownfields redevelopment in the United States through regulatory, public health, and sustainability lenses. J Environ Health 84(9) https://www.neha.org/Images/resources/JEH5.22-Feature-Overview-Brownfields-Redevelopment.pdf
38. Morar C, Lukic T, Valjarević A, Niemets L, Kostrikov S, Sehida K, Telebienieva I, Kliuchko L, Kobylin P, Kravchenko K (2022) Change detection based spatiotemporal analysis of urban green areas: a case study of Kharkiv-City. Front Environ Sci 10:1–27. https://doi.org/10.3389/fenvs.2022.823129

Chapter 34
IT, Geography and Education: Prospects for High-Tech Industry Sectors in the Context of Sustainable Development

Olha Suptelo and Oleksii Niemets

Abstract The rapid evolution of technology and globalization necessitates urgent reforms in the education sector to adapt to the demands of the modern world. This study emphasizes the critical role of geographic education in addressing these challenges and fostering sustainable development. It highlights the interdisciplinary nature of geography, which encompasses natural, social, and exact sciences, and emphasizes the need for integrating geographic knowledge into various educational domains. The integration of geographic education with information technology (IT) and high-tech industries emerges as a key strategy for promoting intellectual human potential and advancing the goals of sustainable development. Through hands-on learning experiences and interactive simulations, students gain practical skills in using modern IT tools like Geographic Information Systems (GIS) and remote sensing for spatial analysis and planning. This integration fosters critical thinking and problem-solving abilities necessary for addressing complex socio-economic and environmental challenges. The study underscores the importance of collaboration between academia, government agencies, industry stakeholders, and local communities in implementing geographic education initiatives. By leveraging geospatial technologies and adopting interdisciplinary approaches, educators can equip students with the knowledge and skills needed to navigate the complexities of the modern world. Furthermore, it calls for further research to explore the potential of geographic education in shaping the knowledge economy and fostering socio-economic development. In conclusion, geographic education serves as a cornerstone for sustainable development, offering insights into the relationships between humans, the environment, and socio-economic systems. By integrating geographic knowledge into educational curricula and leveraging modern technologies, educators can empower students to become informed global citizens capable of driving positive change in their communities and beyond.

O. Suptelo (✉) · O. Niemets
V.N. Karazin Kharkiv National University, Kharkiv, Ukraine
e-mail: olha.suptelo@karazin.ua; oleksii.niemets@student.karazin.ua

C. Morar et al. (eds.), *Achieving Sustainability in Ukraine through Military Brownfields Redevelopment*, NATO Science for Peace and Security Series C: Environmental Security, https://doi.org/10.1007/978-94-024-2278-8_34

Keywords IT · Geography · Education · High-tech industry sectors · Innovations · Geographical education · Innovations in education creative economy · Sustainability

34.1 Relevance of Research

The modern globalized world is evolving rapidly. It is changing the established vision of the Universe, the planet, human interaction with the natural environment, and socio-economic development in general. The progressive global community is compelled to address the growing threats of the global socio-geoecological crisis by seeking scientific and organizational solutions to correct the situation. As the main axiom of the Sustainable Development Strategy, adopted by most countries of the world in 1992 in Rio de Janeiro, is the need to preserve the state of the biosphere and provide future generations with the opportunity to live in an ecologically safe world and have the ability to meet their vital needs, albeit in the current volume.

In the rapidly evolving landscape of the twenty-first century, technology is not only influencing our daily lives, but also fundamentally changing major sectors of the economy. Thus, the sustainability of the modern world is the main task not only of governments and global organizations, but also of every member of society. This problem requires a radical restructuring of the education system at all levels, starting with preschool. Education should become continuous throughout a person's life, ensuring ongoing learning and adaptation to changing circumstances.

Education becomes one of the primary links in the implementation of the principles and principles of sustainable development. Maintaining the stability and innovation of the educational sphere creates the foundations for the further development and modernization of economic sectors. This transformation involves reshaping industries and fostering the development of new sectors that leverage IT technologies and artificial intelligence, while adhering to the principles of the creative economy.

Completely new and complex questions about the justification of the model of the further development of global civilization in the near and distant evolutionary development appear before scientists. Human geography can assist in solving these problems. It is a complex, interdisciplinary science equipped with methods and approaches for building models, forecasts, and monitoring the development of complex, open, synergistic systems, which form the basis of human and social life.

The importance of the relationship between IT, geography, education, and high-tech industrial sectors is determined by the fact that this relationship creates a multifaceted environment for innovation, efficient use of resources and sustainable development [12]. The general trends that shape these relationships include the digitalization of all aspects of society, the growing importance of geo-information technologies in decision-making, the need for highly qualified personnel to work in high-tech sectors, as well as the desire for sustainable development, which is becoming a priority in all areas of human life [16]. The use of IT and geoinformation technologies helps to improve resource management and ensure sustainable development. Also, these technologies contribute to the formation of an environmentally responsible approach to industry and urban planning. Accordingly, the introduction of digital and information tools leads to the transformation of both the educational sphere and the introduction of innovations in industrial production. Innovations of this kind are based on the effective use of data, ensuring better product quality and increased competitiveness.

The purpose of this study is to to investigate the relationship between information technologies (IT), geographic education, and high-tech industrial sectors, and their impact on fostering intellectual human potential for global economic success and meeting the requirements of the Sustainable Development Strategy [12, 19–23, 25, 26].

34.2 Analysis of Previous Research

It is relevant, in the context of sustainable development to study the field of interaction between technology and education and their influence on the formation of economic transformations. Therefore, various aspects of this issue were investigated by Ukrainian and foreign scientists. Traditionally, the pioneers of studying the impact of technological education on industry and their role in the environment have been experts from the polytechnic and economic spheres. In particular, Melnikova and Koshevoy [10] study the impact of science-intensive and the high-tech industries on the structure of the economy of Ukraine, determine the problems and prospects of investing in these sectors of the economy. Bykova and Shchabelska [4] investigate the peculiarities of the development of scientific and innovative activities and high-tech industrial enterprises in the conditions of the transformation of the industrial sphere of the urban environment of Bilotserkivskyi [3] studies the peculiarities of forecasting the development of the high-tech sector of Ukraine's economy based on available statistical data, using mathematical and statistical models.

Over the last decade, the development of the IT sector in Ukraine has shown significant growth rates, both in terms of economic performance indicators and in the number of employed specialists. The analysis of the development of the IT sphere in Ukraine is disclosed in the article by Niemets et al. [12].

The connection between education and the introduction of technological innovations in production is a relevant topic of research today which is revealed in many domestic works. The most widely covered topic of innovation is directly in education. In particular, A. Lytvyn writes about the problems and prospects of introducing innovations in the field of vocational and technical education in the monograph "Informatization of vocational and technical educational institutions of the construction profile" [8]. Features of modern IT education in Ukraine are investigated by M.-O. Yershov, defines the structure, problems and prospects of formal and informal IT education [18].

At the same time the international experience of this topic focuses on specific methods and cases of integration of education, technology and industry to achieve sustainable development. The book "Teaching for the Future: Effective Classroom Practices To Transform Education", published by the authors and experts of the Organization for Economic Cooperation and Development [14], contains methodological and practical recommendations for the implementation of innovative pedagogical practices, including the use of IT in education. Susan Caldis and Grant Kleeman explore the use of the principles of STEM methods in teaching geography [5]. The publication and approval of the International Declaration on Research in the Field of Geographical Education is a strong argument for the introduction of modern technologies of geographical education. In particular, through the development of applied cases and their use, the sustainability of the environment [7]. However, the study of regional patterns and integrated IT-geography-industry-sustainable development cases is quite limited.

34.3 Presentation of the Main Material

The future of humanity hinges on the widespread implementation of such industries and the transition to a creative economy. Signs of the creative economy can already be seen today in the predominance of the production of intellectual products over material ones, as well as in the priorities of intellectual and creative resources, assets, and the rapidly growing free scientific and technological exchange on the world market. Today, the need for training creative management specialists is urgent. This necessitates restructuring traditional educational programs in economics, social geography, and related fields. The method of cultivating students' knowledge and competencies requires restructuring and improvement. This entails integrating elements of individual and collective creativity into pedagogy. We are talking about the establishment of a system of creative education to ensure the orientation of the individual in order to develop a person's creative abilities, the formation of professional consciousness and the generation of innovation and their practical implementation. This type of education can cultivate robust intellectual potential grounded in universal human values.

It is fair to say that geography, in particular, its socio-economic component and geo-information technologies play an important role in the modern world, especially in the context of the development of industry and IT. Since this science studies and analyzes such aspects as regional development, geographic distribution of resources, trade, transport and other factors affecting economic development, as well as spatial aspects of the economy and society, in particular, population distribution, development of cities and regions, interrelationships connections between people and their environment. Simultaneously, the education sector is undergoing a wave of digital transformation, marked by cutting-edge and practical innovations [17].

In the context of modern changes in global society, geographical education, due to its complexity and its interdisciplinary nature, is able to give students the ability to holistically assess the environment and implement effective tools for its sustainability. Let's consider the connection of high-tech industries with advanced and innovative methods of geographical education of schoolchildren. After all, it is at the lower levels of education that basic soft and hard skills are formed.

In fact, among dozens of leading methods and approaches, in our opinion, the most effective, specifically in geographic education, are the following [13, 17]:

- direct use of information and computer technologies: IT, GIS, 3D, augmented and virtual reality, in general, the transformation of geographic education in the context of EdTech;
- integration: the introduction of differentiated geographic disciplines into school education (not limited to physical and human geography in general components), and the integration of geographic knowledge into other subjects and courses, including non-science majors;
- implementation of the results of school research: immediate practicality and visibility of the knowledge gained.
- Next, an analysis of each of the approaches is presented. Practical cases are presented in more detail, which can become the basis for further implementation in higher education, and be promising for the training of specialists in high-tech industries.

34.3.1 Direct Use of Information and Computer Technologies

In today's globally computerized world, the use of computer technologies in education is obvious. In particular, the challenges of distance learning have expanded the instrumental arsenal of teachers both in the context of communication and cooperation with students, and for the purpose of evaluating and correcting knowledge. However, concrete scientific research in many fields of education still remains unaffected by computer technology. The driver of changes and an example of the practical use of geographical knowledge is the implementation of GIS not only in higher education, but also in school education.

The direct use of information and computer technologies has revolutionized the process of studying geography in many ways, offering unparalleled opportunities for research, analysis and understanding of the world around us. GIS software allows geographers to acquire, store, analyze, and effectively present geographic data. This makes it possible to integrate different layers of geographic information, such as satellite images, terrain elevation, land use, population density, etc. The use of GIS in the educational environment facilitates the process of visualization of geographic data, which contributes to a better understanding of geographic processes and phenomena. Therefore, students and pupils can use GIS to solve real-world geographical problems, such as population distribution, assessment of natural resources or features of infrastructure development.

The introduction of GPS was one of the breakthroughs of the last century. Which has improved our daily lives and contributed to the development of many fields, from navigation to overcoming global challenges. GPS technology allows precise location tracking and navigation using signals from satellites orbiting the Earth. It allows geographers to accurately map geographic features, track movement, and conduct field research. In geographic research, GPS improves data collection and fieldwork by providing georeferenced information. Researchers can collect data points with spatial coordinates, facilitating the creation of detailed maps and spatial analysis. Thus, geographic education is able to prepare leading specialists who will have the skills to use GPS and GIS to solve today's global industrial, social, and environmental challenges. Of course, the greatest impact should be made in high-tech industries.

High-tech industries such as urban planning, environmental management, and logistics rely heavily on GIS for spatial solutions. For example, urban planners use GIS to optimize urban planning, while logistics companies use it to plan routes and allocate resources. GPS for various purposes including vehicle tracking, asset management, geofencing and precision agriculture is also used by a number of high-tech industries. For example, logistics companies use GPS to track shipments in real time and optimize delivery routes to improve efficiency.

The integration of virtual (VR) and augmented reality (AR) tools into the educational process and further implementation into professional activities, in particular, in the context of geographic education, is promising. VR and AR technologies offer immersive experiences that enhance the visualization and understanding of geographic concepts. Geographers can use VR to simulate the geographic environment, while AR overlays digital information on the real world. When studying geography, VR can transport students to remote locations, allowing them to virtually explore landscapes, ecosystems and cultural sites. AR, on the other hand, can provide contextual information about the environment in real time [6, 13].

High-tech industries use VR and AR for training, simulation and visualization. For example, architects are using VR to visualize building designs in 3D space, while tourism companies are using AR to enhance visitors' experiences at historical sites.

After all, the ability to work with large data bases, as the main hard skills of the twenty-first century, needs to be implemented and consolidated in the educational process. Professional use of Big Data analysis skills is well established and widely implemented. At the same time, there is a significant shortage of specialists in the field. In our opinion, geographic education, as a multidisciplinary field of knowledge, is able to fill this gap and lay the foundation for mastering Big Data analysis skills in schoolchildren [6]. The proliferation of big data combined with advances in machine learning has opened up new opportunities in the study of geography. Geographers can now analyze vast arrays of spatial data to derive information for developing forecasts, modeling development plans, and more. Machine learning algorithms can be trained to classify land cover, predict environmental changes, and detect anomalies in geographic datasets. These techniques allow geographers to automate repetitive tasks, and focus on high-level analysis. High-tech industries use big data and machine learning for geospatial analysis, predictive modeling and risk assessment. For example, insurance companies use machine learning algorithms to assess natural disaster risk and price policies accordingly [6].

34.3.2 Integration

The integration of geographic education into school programs involves a comprehensive approach that encompasses differentiated geographic disciplines and the inclusion of geographic knowledge in various subjects of natural, social, and exact sciences. In addition to the traditional study of physical-geographic and socio-economic aspects of the surrounding space, school geography requires the involvement of more special geographic disciplines, in particular, GIS, geoecology, geopolitics, urban studies, climatology, geomorphology, etc. [11]. The implementation of a differentiated, and at the same time, integrative and complex approach will make it possible to develop in pupils and students the ability to analyze and evaluate the set of processes of natural and social geosystems, as well as:

- study the impact of human activity on ecosystems and learn about environmental protection strategies;
- to develop skills of spatial thinking, data visualization and geographic analysis;
- explore how geography shapes political unions, economic systems and international relations;
- evaluate the influence of natural and social factors on global and local economic processes.

The implementation of geographical knowledge in other areas of education and integration in natural, social and exact disciplines is also very important. This, in turn, allows for a deeper understanding of each of the disciplines, as well as develops the skills of complex and critical thinking [2]. Here are some practical examples of integrating geography with school courses:

- *Natural sciences*: in biology, students study how geographical factors affect biodiversity, the dynamics of geosystems and the distribution of species; in chemistry, they learn about the chemical composition of the Earth's atmosphere, hydrosphere, and lithosphere; in physics, they study the principles of climatology, hydrology, and geomorphology.
- *Social sciences*: in history, students analyze the role of geography in the formation of historical events, migrations and civilizations; in economics examine how geographic factors affect trade patterns, resource allocation, and economic development; in sociology and anthropology they examine the cultural, social, and spatial dimensions of human societies.

- *Exact sciences*: in mathematics, students use spatial reasoning and geometry to solve problems related to map projections, coordinate systems, and geometric analysis; in informatics, GIS methods are used to analyze spatial data, develop cartographic applications, and solve problems based on location determination; in engineering and architecture consider geographic factors such as topography [24], climate, and natural hazards when designing infrastructure and buildings.

Educators can provide students and learners with a comprehensive understanding of the world around them and equip them with global problem-solving skills to make informed decisions by integrating differentiated geographic disciplines and introducing geographic knowledge into various subjects and courses in the natural, social and exact sciences. Ultimately, this will help lay the foundation for education and development of leading specialists in high-tech industries capable of decisive and conscious innovation.

34.3.3 Implementation of the Results of School Research

Providing hands-on learning experiences, interactive simulations, and case studies that illustrate the interdisciplinary nature of geographic concepts, engage social majors, and reinforce the value of geographic education beyond traditional academic disciplines is not just a timely challenge for the modern education system today, it is also an opportunity for widespread implementation geographical knowledge in other spheres of activity, in particular, high-tech industry. Today, this presents not only a timely challenge for the modern education system but also an opportunity for the widespread implementation of geographical knowledge in other spheres of activity, particularly in the high-tech industry [1].

Therefore, geographical research carried out by students and pupils can provide valuable information and data that can be used in various fields, ranging from spatial planning to the development of high-tech industries, territorial development and sustainable development. Involvement of students in real projects in cooperation with the industrial enterprises allows them to gain practical experience and skills in working with modern IT technologies. Industrial companies can use geographical analytical systems and other technologies for planning the location of production sites, marketing research, market analysis and logistics. Let's consider specific examples of the implementation of geographical research in various social spheres of activity.

- Spatial planning:
  - Urban Planning: Geographical studies can offer insights into urban dynamics, population distribution, land use patterns and infrastructure needs. This information can guide urban planners in designing more efficient and sustainable

cities. For example, studies of traffic patterns or spatial analysis of urban sprawl can inform transportation planning and zoning regulations.
  - Regional Development: understanding regional differences and resource allocation is critical to balanced development. Social geographic research can identify areas of economic potential, ecological importance, or social vulnerability, helping policymakers allocate resources more effectively. This may include mapping natural resources, analyzing economic activity, and assessing demographic trends.

- High-tech industry:
  - Location selection: geographical research can help in choosing the optimal location for high-tech industries based on factors such as proximity to skilled labor, transportation networks, access to markets, and availability of resources. Geographic Information Systems (GIS) can be used to analyze spatial data and identify suitable locations for technology parks or research centers.
  - Cluster analysis. The identification of clusters of high-tech industries can promote innovation and cooperation. Geographical research can map existing clusters, analyze their characteristics, and identify factors that contribute to their success. This information can guide policies aimed at promoting the growth of technology clusters and attracting investment.

- Territorial development:
  - Rural development: Geographical research can support rural development initiatives by identifying opportunities for diversification, increasing agricultural productivity or promoting ecotourism. Understanding the spatial distribution of resources and amenities can help policymakers design targeted interventions to improve rural livelihoods and infrastructure.
  - Border Regions: research on border regions can provide a basis for cross-border cooperation and integration efforts. Geographical analysis can identify areas with potential for joint development projects, facilitate infrastructure planning across borders, and address cross-border issues such as environmental conservation or migration.

- Sustainability [9]:
  - Environmental management: geographical research plays a critical role in environmental management by mapping ecosystems, monitoring land-use change, and assessing environmental risks. This information can inform policies aimed at conserving biodiversity, mitigating natural hazards and promoting sustainable land use practices.
  - Adaptation to climate change: understanding the spatial impacts of climate change is essential for adaptation planning. Geographical research can identify areas vulnerable to sea-level rise, extreme weather events or habitat loss, helping communities develop strategies to increase resilience and reduce risk.

Implementation of these research findings involves collaboration between academia, government agencies, industry stakeholders, and local communities. The use of geospatial technologies, such as GIS, remote sensing, and spatial modeling, increases the accuracy and efficiency of geographic research and its application in various fields. In addition, the interdisciplinary approaches that combine geographic analysis with socio-economic and environmental strategies are essential to solving complex challenges and achieving sustainable development goals and implementing innovative approaches to spatial management.

34.4 Conclusions

There is an urgent need to reform the education system, particularly given the rapid changes in global society. This includes restructuring technologies that underpinned the industrial economy and transitioning to the foundations of the post-industrial information society. Education becomes a powerful engine and the basis of these changes. It is not only about the need to change human qualities, but also the formation of the necessary professional qualities, creative views on the processes taking place on the planet,according to A. Pechchei, the founder of the Club of Rome [15]. This is also required by the decisions of the Sustainable Development Strategy.

This study explores the relationship between various levels and areas of education, particularly focusing on geography. The relationship between information technologies (IT), geographic education and high-tech industry sectors is revealed in view of their impact on the formation of intellectual human potential for the success of the global creative economy and the requirements of the Sustainable Development Strategy.

By implementing these methods and approaches, and adhering to the principles of STEM and EdTech education, laying the foundations for education integration becomes feasible. This will contribute both to the implementation of the principles of sustainable development and the education of specialists for high-tech industries. In turn, the integration of geographic education into school programs and the use of the results of geographic research in various fields of activity requires a comprehensive approach to the study of geography. It covers various geographical disciplines and the incorporation of geographical knowledge into the subjects of natural, social and exact sciences. This approach not only provides pupils and students with the latest knowledge but also forms very important competencies and skills for professional activity, increasing their ability to analyze and evaluate the processes occurring in natural and social geosystems, as well as in the globalized informational and creative society.

Geographical science plays a critical role in the implementation of sustainable development and innovation. Its importance lies in the fact that it helps to understand the relationships between man and the environment, economic processes and social structures. Geographical science provides the necessary data and analysis to make informed decisions about development and planning.

There is a close connection between geographic education, IT and high-tech industries. The integration of geographic knowledge into various fields of education and the use of modern technologies, such as geographic information systems (GIS), remote sensing, and spatial modeling, allows for effective analysis of space, forecasting changes, and developing strategies for sustainable development.

Thus, geographic education lays the groundwork for developing critical thinking and understanding complex problems in the modern world, including those related to ecology, economy, politics, and society. The integration of geographical knowledge into a wide range of social spheres of activity contributes to the adoption of reasoned decisions and the development of innovative approaches to solving problems.

Further research is needed to explore the formation of intellectual potential at various educational levels, including geographic education. Special attention is required for the transition to a knowledge economy, which forms the basis for developing a creative society grounded in innovative technologies. Education for the knowledge economy becomes an intellectual product, and knowledge is an important resource for the socio-economic development of the global society. Further research by the authors will focus on examining this aspect. In line with this, research is planned to reform the education system in Ukraine, particularly focusing on post-war times. This endeavor aims to restore the country's intellectual potential for the reconstruction of the national economy.

References

1. Abed N (2014) Effective inquiry-based learning strategies in teaching science to a group of high school learners. Retrieved from https://www.academia.edu/29694618/Effective_Inquiry_Based_Learning_Strategies_in_Teaching_Science_to_a_Group_of_High_School_Learners
2. Åström M (2007) Integrated and subject-specific. An empirical exploration of science education in Swedish compulsory schools. Retrieved from https://www.diva-portal.org/smash/get/diva2:422902/FULLTEXT01.pdf
3. Bilotserkivskyi O (2017) Forecasting the development of high-tech sector of Ukrainian economy. https://economyandsociety.in.ua/journals/11_ukr/92.pdf [in Ukrainian]
4. Bykova M, Shchabelska V (2021) Scientific and innovative activities and high-tech industrial industries of Kyiv: current state, problems and prospects. Grail Sci 8:485–496. Retrieved from https://ojs.ukrlogos.in.ua/index.php/grail-of-science/article/view/14867 [in Ukrainian]
5. Caldis S, Kleeman G (2019) Geography and STEM. Geogr Educ 32. Retrieved from https://files.eric.ed.gov/fulltext/EJ1238489.pdf
6. Demkovych T (2023) Shaping the future of education: the role of technology. Retrieved from https://forbytes.com/blog/future-of-technology-in-education/
7. IGU CGE (2015) International declaration on research in geography education. Retrieved from https://www.igu-cge.org/2015-declaration/
8. Lytvyn A (2011) Informatization of vocational and technical educational institutions of the construction profile [Monograph]. Lviv: Kompaniè "Manuskrypt". Retrieved from https://sci.ldubgd.edu.ua/bitstream/123456789/6034/1/%D0%9C%D0%BE%D0%BD%D0%BE%D0%B3%D1%80_%D0%9B%D0%B8%D1%82%D0%B2%D0%B8%D0%BD_%D0%9

0_%D1%84%D1%80%D0%B0%D0%B3%D0%BC%D0%B5%D0%BD%D1%82.pdf [in Ukrainian]

9. Meadows ME (2020) Geography education for sustainable development. Geograp Sustainab 1:88–92. https://doi.org/10.1016/j.geosus.2020.02.001
10. Melnikova M, Koshevoy M (2013) Aspects of sector high-tech and knowledge-intensive production in Ukraine. Efektyvna ekonomika, 11. Retrieved from http://www.economy.nayka.com.ua/?op=1&z=2720
11. Mohr K, Welker R (2017) The role of integrated curriculum in the 21st century school. Retrieved from https://www.researchgate.net/publication/325795457_The_Role_of_Integrated_Curriculum_in_the_21st_Century_School
12. Niemets L, Sehida K, Kravchenko K, Nosyriev O, Lohvynova M, Suptelo O (2021) IT-industry in Ukraine: trends and development centers. In Proceedings of the 38th International Business Information Management Association Conference (IBIMA) (November 2021, Seville, Spain). Retrieved from https://ibima.org/accepted-paper/it-industry-in-ukraine-trends-and-development-centers/
13. Notermans M (2023) 7 Top EdTech Industry trends shaping the future. Retrieved from https://www.thinkorion.com/blog/edtech-industry-trends
14. OECD (2018) Teaching for the future: effective classroom practices to transform education. OECD Publishing. https://doi.org/10.1787/9789264293243-en
15. Peccei A (1977) The human quality. Pergamon Press, Oxford/New York. Retrieved from https://archive.org/details/humanquality0000pecc/page/n5/mode/2up
16. Stankov U, Vasiljević AĐ, Jovanović V, Kranjac M, Vujičić MD, Morar C, Bucur L (2019) Shared aerial drone videos — prospects and problems for volunteered geographic information research. Open Geosci 11(1):462–470. https://doi.org/10.1515/geo-2019-0037
17. VASS (2023) Trends in tech shaping the future of education industry. Retrieved from https://vasscompany.com/nordics/en/insights/blogs-articles/trends-in-education-industry/
18. Yershov M-O (2023) Development trends of IT education in independent Ukraine [Monograph]. Kyiv: "Vydavnytstvo Lyudmyla". Retrieved from http://eprints.zu.edu.ua/38844/1/Ershov.pdf [in Ukrainian]
19. Lukić T, Ponjiger TM, Basarin B, Sakulski D, Gavrilov M, Marković S, Zorn M, Komac B, Milanović M, Pavić D, Mesaroš M, Marković N, Durlević U, Morar C, Petrović A (2021) Application of Angot precipitation index in the assessment of rainfall Erosivity: Vojvodina region case study (North Serbia). Acta Geogr Slov 61(2):123–153. https://doi.org/10.3986/AGS.8754
20. Valjarević A, Popovici C, Djekić T, Morar C, Filipović D, Lukić T (2022) Long-term monitoring of high optical imagery of the stratospheric clouds and their properties new approaches and conclusions. Egypt J Remote Sens Space Sci 25(4):1037–1043. https://doi.org/10.1016/j.ejrs.2022.11.006
21. Blešić I, Ivkov M, Tepavčević J, Raljić JP, Petrović MD, Gajić T, Tretiakova TN, Syromiatnikova JA, Bajrami DD, Aleksić M, Vujačić D, Kričković E, Radojković M, Morar C, Lukić T (2022) Risky travel? Subjective vs. objective perceived risks in travel behaviour—influence of hydro-meteorological hazards in South-Eastern Europe on Serbian tourists. Atmosphere 13(10):1671. https://doi.org/10.3390/atmos13101671
22. Valjarević A, Algarni S, Morar C, Grama V, Stupariu M, Tiba A, Lukić T (2023) The coastal fog and ecological balance for plants in the Jizan region, Saudi Arabia. Saudi J Biol Sci 30(1):103494. https://doi.org/10.1016/j.sjbs.2022.103494
23. Morar C, Lukic T, Valjarević A, Niemets L, Kostrikov S, Sehida K, Telebienieva I, Kliuchko L, Kobylin P, Kravchenko K (2022a) Change detection based spatiotemporal analysis of urban green areas: a case study of Kharkiv-City. Front Environ Sci 10:1–27. https://doi.org/10.3389/fenvs.2022.823129

24. Morar C, Nagy G, Dulca M, Boros L, Sehida K (2019) Aspects regarding the military cultural-historical heritage in the city of Oradea, Romania. ANNALES Ann Istrian Mediterranean Stud Ser Historia et Sociol 19(2):303–322. https://doi.org/10.19233/ASHS.2019.21
25. Morar C, Nagy G, Boros L, Gozner M, Niemets L, Sehida K (2021) Heritage, culture and regeneration of the former military areas in the city of Oradea (Romania). Architektúra Urbanizmus 55(1–2):76–87. https://doi.org/10.31577/archandurb.2021.55.1-2.6
26. Morar C, Tiba A, Jovanovic T, Valjarevic A, Ripp M, Vujicic MD, Stankov U, Basarin B, Ratković R, Popović M, Nagy G, Boros L, Lukic T (2022b) Supporting tourism by assessing the predictors of COVID-19 vaccination for travel reasons. Int J Environ Res Public Health 19(2):918. https://doi.org/10.3390/ijerph19020918

Chapter 35
City Sustainability: Driving Ukraine's Post-war Reconstruction

Daria Venhryn

Abstract The article is dedicated to developing practical approaches to effective socio-economic development of Ukraine post-war based on the concept of sustainable development. The research substantiates that sustainable development is the foundation of post-war reconstruction in Ukraine. Building upon previous studies and legislative acts on Reconstruction of Ukraine post-war, a set of indicators has been developed, which can serve as a useful tool for determining indicators reflecting not only the state of cities under normal conditions but also their condition post-war and the need for recovery. The obtained results serve as a basis for the development and implementation of specific managerial decisions, substantiation of strategic directions, and mechanisms for ensuring sustainable development of the city.

Keywords Sustainable developmen · Indicators · City · Ukraine

35.1 Introduction and Review of Previous Scientific Developments in This Area

Russia's military aggression in Ukraine shattered the country's plans for progressive development, pushing it back decades. The catastrophic losses inflicted on the industrial, residential, and infrastructural sectors amount to billions of US dollars, and the war continues, causing further devastation. History offers numerous examples of post-war reconstruction and development in countries. The Sustainable Development Goals play a crucial role in Ukraine's post-war reconstruction. The success of such reconstruction depends on several factors, but without international assistance and support, Ukraine cannot handle it alone. Sustainable urban

D. Venhryn (✉)
V.N. Karazin Kharkiv National University, Kharkiv, Ukraine
e-mail: daria.venhryn@student.karazin.ua

C. Morar et al. (eds.), *Achieving Sustainability in Ukraine through Military Brownfields Redevelopment*, NATO Science for Peace and Security Series C: Environmental Security, https://doi.org/10.1007/978-94-024-2278-8_35

development creates conditions for the reconstruction and improvement of people's lives, while also preventing future destruction.

President of Ukraine V. Zelensky established the National Council for Recovery of Ukraine from the Consequences of War in April 2022. Later, the government presented the vision and basic principles of recovery during the Recovery Conference in Lugano. The vision for Ukraine's «Recovery is to become a strong European country—a magnet for foreign investment». The program plan aims to accelerate sustainable economic growth. Within the plan, a list of National Programs identified to achieve key results. It is based on 5 core principles [1]:

- Immediate commencement and gradual development;
- Increase of fair prosperity;
- Integration into the EU;
- Reconstruction better than before on a national and regional scale;
- Encouragement of private investments.

All these ideas should be incorporated into future city development strategies, based on economic, socio-demographic, and ecological considerations. The social dimensions of sustainable urban development include creating jobs, ensuring access to education and healthcare, supporting vulnerable groups and promoting socio-cultural development [28–30]. This strengthens social ties in cities, improves the population's quality of life and creates conditions for long-term sustainable development [24–26, 31, 32]. The economic dimension of sustainable urban development involves revitalizing and developing the local economy [21–23, 27]. This includes creating new businesses, providing access to financial resources and supporting entrepreneurship within the community. Economic development contributes to increased employment, improved income levels and poverty reduction. The environment is also a key aspect of sustainable urban development. Military actions have led to significant destruction of ecosystems and pollution of water and air resources on the territory of Ukraine. Therefore, it is necessary to focus on preserving natural resources using effective methods and technologies. This is an important task in the post-war recovery of the country's territory.

Indeed, the issue of post-war economic recovery in Ukraine is complex and has been the focus of attention for many scholars.

N. Klymenko and others examine the impact of risks associated with combat actions and conditions of macroeconomic uncertainty on Ukraine's agro-industrial complex. The researchers also assess the investment attractiveness of agriculture and explore the potential for attracting investments during the post-war recovery period [4]. K. Nikolaets and others examine the issue of migration of Ukrainians to Western European countries. The authors emphasize that human capital will play a significant role in the post-war economic recovery of Ukraine. However, addressing this issue will require the government to promote re-emigration and tackle the problem of uneven population distribution across the territory [6]. M. Chepeliyev and others examine the transition to green energy as a driver of economic growth in Ukraine during the post-war period [2]. A number of scientific papers discuss the post-war reconstruction of Ukraine in a global context. For instance, V. Zalizko and others focus on the issue of restarting Ukraine's economy in the post-war period. Researchers see investment attraction as the main driver of this recovery, which could be implemented in the form of a new "Marshall Plan" for Ukraine [14]. V. Shvedun and others believe that one of the promising drivers for post-war recovery could be the development of a waste processing program in cooperation with the European Union (EU). This would not only accelerate the integration process but also create an additional anchor point for stabilizing the country's energy system [8]. O. Cherniaieva and others believe that the creation of a so-called innovative economy could be promising for rapid post-war reconstruction. In this context, the role of the government in creating favorable conditions for companies specializing in renewable energy and the IT sector is increasing [3]. The importance of maintaining stability in state finances during wartime and in the post-war period is discussed in the article by N. Petrukha and others [7]. The authors emphasize the creation of a social model for post-war recovery, which will be aligned with the paradigm of EU policy principles and standards. The significance of the role of the banking system in maintaining macroeconomic stability and state finances is discussed by N. Sokrovolska and others [11].

Sustainable urban development should be tailored to local conditions and needs. Each city has its own peculiarities and requires an individualized approach [5]. It is considered that using indicators to assess the performance of sustainable urban development is an important and effective method employed in many countries worldwide. Despite the fact that each city has its own set of indicators, which significantly differ from one another, the common goal—achieving sustainable urban development—remains the same. The selection of indicators should be done with a clear understanding of the city's needs and their application, considering the differences in practices [9].

The aim of this research is to develop social, economic, and environmental indicators for the effective recovery of Ukraine after the war, focusing on both the territories in need of reconstruction and those in sustainable development. The object of the study is the post-war development of Ukraine based on the concept of sustainable development. The subject of the research includes theoretical, methodological, and practical approaches to the effective development of Ukraine after the war, guided by the concept of sustainable development.

35.2 Set of Indicators as a Means to Assess the Sustainable Development of Post-war Reconstruction in Ukraine

The presented material is based on the preliminary results of a study that analyzed six development strategies of Ukraine's largest cities [13]. The main goal of development strategies for regions is the achievement of sustainable development. The employment of indicators to assess the efficacy of sustainable urban development is widely recognized as a paramount and impactful methodology, adopted by numerous nations worldwide [12]. Given the inherent diversity of implementation practices, the selection of indicators should be guided by a comprehensive understanding of the city's unique requirements and their intended application [10]. This work presents a comprehensive inventory of indicators that are consistently employed across six Ukrainian city development strategies. These indicators have been meticulously categorized into three distinct groups: economic, social, and environmental, providing a structured framework for analysis. Within each of these established categories, a standardized unit of measurement has been meticulously assigned to each indicator. Additionally, indicators that were not previously employed but are featured in internationally recognized lists of urban sustainability indicators have been thoughtfully incorporated. This carefully curated assemblage of indicators is not intended to be an exhaustive compilation. It can and should be dynamically expanded or conversely have indicators removed, contingent upon the city's evolving short-term and long-term sustainable development plans.

Amidst the prevailing challenges, sustainable development stands as the bedrock upon which Ukraine's post-war recovery will be built. However, the current selection of indicators falls short in adequately reflecting the transformed landscape and prioritized objectives of post-war reconstruction, necessitating an expansion of the indicator list to encompass these evolving priorities. According to the law of Ukraine On amendments to certain legislative acts of Ukraine concerning the principles of state regional policy and the policy of restoration of regions and territories, territories to be restored (hereinafter referred to as TR) are micro-regions, territorial communities on the territory of which military operations took place and/or which were temporarily occupied, and/or territories, the critical infrastructure, social infrastructure, housing stock facilities of which were destroyed asa result of military operations, as well as which are character-ized by a sharp deterioration in the level of socio-economic development and significant movement of the popu-lation to other regions and/or other states [1].

As evident in the aforementioned legislation, particular emphasis is placed upon the deterioration of the socio-economic landscape, stemming from the destruction of infrastructure and the mass displacement of populations due to hostilities. Consequently, pivotal indicators for post-war recovery encompass the pace of infrastructure reconstruction, alongside migration patterns, with a specific focus on the rate of population repatriation to Ukraine.

The author proposes a comparative table of sustainable development indicators. The first column will include indicators selected in our previous study, which are effective under normal conditions in regions unaffected by the war. The second column will include indicators for regions requiring recovery (Table 35.1).

According to joint estimates by the Ukrainian government and the World Bank, Ukraine's economic losses from the two years of Russia's full-scale invasion amount to $499 billion. The same amount is needed for post-war reconstruction. This information was shared by Ukrainian Prime Minister Denys Shmyhal [3]. The war in Ukraine has inflicted severe damage upon the nation's trade, industry, agriculture, energy, and transportation sectors. Revitalizing industrial facilities and small businesses is paramount to revitalizing Ukraine's post-war economy and ensuring its

Table 35.1 A comprehensive set of indicators for evaluating sustainable urban development

For non-war-affected regions	For regions requiring recovery
Economic indicators	
Gross regional product	Restoration of industrial facilities
Municipal Budget	Repatriation of capital assets to the country
Foreign Direct Investment (FDI)	Number of small businesses
Volume of realized innovative products	Private remittances from abroad, million UAH
Industrial Facility Count	
Social indicators	
Population of administrative centers of Ukraine	Proportion of population migration flow to the territory requiring reconstruction
Average life expectancy	Restoration of healthcare facilities
Unemployment rate	Restoration of sports facilities
Assessment of the adequacy of educational institutions	Restoration of administrative buildings
External independent evaluation level	Restoration of shopping centers
Number of enrolled students	Restoration of educational institutions
The share of the homeowners' association in the multifamily residential building among residential buildings	Restoration of roads of various categories, including railways
City's energy consumption	Restoration of airports
Number of tourist arrivals	Bridge restoration
Number of healthcare facilities	Combined Heat and Power plant restoration
Number of healthcare institutions	Power line restoration
	Gas supply network restoration
	Government social support for citizens
	Number of war veterans
Environmental indicators	
Emissions of pollutants into the atmosphere per year	Restoration of the natural reserve fund after deoccupation.
Total water consumption by the city	Restoration of soil cover after explosions
The share of passenger transportation by environmentally friendly modes	Water pollution indicator
Provision of public green spaces	Number of cleared areas
The share of recycled waste compared to total waste generation	The area of military garbage dumps (burnt equipment, remnants of aerial bombs, etc.)

Developed by the Author Utilizing Data from Venhryn [13]

sustainable long-term growth trajectory. To foster Ukraine's post-war economic revitalization, a comprehensive strategy encompassing domestic enterprise revitalization and foreign investment attraction is paramount. In this regard, the unwavering support from international communities and Ukrainians abroad, particularly in the form of financial contributions, has played a pivotal role in bolstering the country's economy during these challenging times.

The war has presented a formidable challenge to Ukraine's social sphere, primarily inducing significant demographic shifts. The precise extent of the increased mortality rate remains unclear due to the following factors: limited data availability from occupied territories; inconsistent data from frontline areas; relatively reliable data for non-occupied areas. Similarly, researchers do not have data on the dead participants of the hostilities. It is crucial to recognize that the additional losses (war-related deaths) extend beyond those directly resulting from combat, encompassing casualties from a broader range of factors like chronic stress, hypothermia, malnutrition, lack of healthcare, vaccine-preventable diseases. Mass migration, both internal and external, poses a significant challenge in the context of the war in Ukraine. The process of Ukrainians returning home is a long-term and complex undertaking that necessitates addressing numerous issues, including those at the legislative level. Despite the significant number of Ukrainians who have returned home during the two years of full-scale war, a far larger portion remains abroad. Therefore, the development of a range of policy measures is crucial to encourage their return, even after an extended period of time spent overseas. The introduction of an index such as the proportion of the population migration flow into a territory requiring reconstruction is indeed a valuable proposal. Additionally, monitoring all demographic indicators (birth rate, death rate, emigration, migration) is crucial under normal circumstances as well. Mass attacks by the Russian army lead to the destruction of social infrastructure, without which it is difficult to imagine modern cities. Therefore, it is advisable to monitor its recovery with the help of the following indicators: restoration of health care facilities, sports facilities, administrative buildings, shopping centers, restoration of educational institutions. For the convenient functioning of the city and its residents, critical infrastructure that undergoes incredible destruction is of utmost importance. It is necessary to track the restoration of roads of various categories, including railway tracks, airports, bridges, as well as electric and gas supply lines. Special attention should be paid to the restoration of the energy infrastructure.

The full-scale invasion has contributed to the emergence of a large number of environmental problems. One of the most important is pollution of air, water, and soil. Continuous bombing and shelling of cities and populated areas further lead to the release of a large amount of toxic chemicals into the environment. A striking example is the incident of the sabotage explosion of the Kakhovka hydroelectric power plant. It was a true ecocide, with 620 square kilometers of land submerged in water across four regions—Kherson, Mykolaiv, Dnipropetrovsk, and Zaporizhzhia. Therefore, it is necessary to continue ecological monitoring, including measuring emissions of pollutants into the atmosphere, as well as adding indicators of water pollution and soil recovery after explosions. The destruction of forests and habitats of wildlife another serious environmental problem in Ukraine during the war. Continuous bombing and shelling of forests have caused significant damage to ecosystems, and many plant and animal species have been forced to leave their homes. This has greatly affected the biodiversity of the region and the livelihoods of local communities, which depend on the forest as a source of food, fuel, and other resources. Therefore, it is advisable to introduce such an indicator as the restoration of natural reserves after de-occupation. A large part of Ukraine's territory is mined after mass attacks and occupation. The indicator of demined territories is important. In addition, war leaves behind a large amount of scrap metal (in the form of destroyed equipment, burnt ammunition, etc.) and concrete blocks from demolished multi-story buildings, increasing the area of landfills in cities.

This set of indicators should be adaptable to changing needs, location, and conditions. Additionally, the indicators do not evaluate actions taken but rather indicate the effectiveness of implementing specific practices. The data obtained serves as a foundation for developing and implementing concrete management decisions, justifying strategic directions, and establishing mechanisms to ensure sustainable urban development [15–20].

35.3 Conclusions

Sustainable urban developent serves as a crucial foundation for Ukraine's post-war recovery. Urban development strategies are employed to achieve sustainable development. These strategies enable the restoration and growth of the local economy, support the social sphere, and preserve the environment. Utilizing indicators to assess the effectiveness of sustainable urban development is a crucial and effective method employed in many countries around the world.

The full-scale invasion of Ukraine has resulted in significant economic losses, with the trade, industry, agriculture, energy, and transportation sectors being particularly hard hit. The social sphere has undergone significant changes, particularly in demographic aspects and migration processes. Additionally, the war has increased the environmental burden, including excessive air, water, and soil pollution, deforestation, and so on. All of these factors pose a threat to achieving sustainable development goals.

According to the Law on the Reconstruction of Ukraine, special attention is paid to regions with destroyed infrastructure and mass outflow of population. A comparative table of sustainable development indicators can be a useful tool for identifying metrics that reflect not only the state of cities under normal conditions but also their state after the war and the need for recovery. Since previous urban development strategies do not take into account the changes and priorities of post-war reconstruction, this list of indicators can be useful for finding optimal solutions. Of course, it is not universal and can be changed (either expanded or, on the contrary, indicators can be removed) depending on the city's sustainable development plans. The results obtained serve as the basis for the development and implementation of specific management decisions, substantiating strategic directions and mechanisms for ensuring the sustainable development of the city.

References

1. CABINET of Ministers of Ukraine (2022). https://zakon.rada.gov.ua/laws/show/856-2015-%D0%BF (12.09.2022)
2. Chepeliev M, Diachuk O, Podolets R, Semeniuk A (2023) Can Ukraine go "green"on the post-war recovery path? Joule 7(4):606–611. https://doi.org/10.1016/j.joule.2023.02.0071
3. Cherniaieva O, Kulinich T, Gutsul T, Korolkov V, Gubar O (2023) Innovative transformations of Ukraine's economy in the context of globalization: theoretical and analytical aspects. Finan Credit Activ Problems Theory Pract 1(48):249–257. https://doi.org/10.55643/fcaptp.1.48.2023.3977
4. Klymenko N, Voronenko I, Nehrey M, Rogoza K, Rogoza N (2023) Risk assessment of shock periods and investment attractiveness of agroholdings of Ukraine. Agric Resour Econom Int Sci E-J 9(2):163–182. https://doi.org/10.51599/are.2023.09.02.07
5. Niemets K, Kravchenko K, Kandyba Y, Kobylin P, Morar C (2021) World cities in terms of the sustainable development concept. Geograp Sustanab 2(4):304–311. https://doi.org/10.1016/j.geosus.2021.12.003
6. Nikolaiets K, Shkuropadska D, Ozhelevskaya T, Shtunder I, Khrustalova V (2023) External migration in the conditions of the Russian – Ukrainian war. Finan Credit Activ Problems Theory Pract 2(49):445–456. https://doi.org/10.55643/fcaptp.2.49.2023.4005
7. Petrukha N, Petrukha S, Alekseienko N, Kushneruk O, Mazur A (2023) Social imperatives of public finance: war adaptation and principles o fpost-war recovery. Finan Credit Activ Problems Theory Pract 3(50):358–371. https://doi.org/10.55643/fcaptp.3.50.2023.40311
8. Shvedun V, Bulba V, Bozhko L, Kucher L, Kholodok V, Ihnatiev O (2023) Circular economy in Ukraine on the way to European integration: directions for sustainable management during the ear and post-war recovery. J Environm Manag Tour 14(1):194–206. https://doi.org/10.14505/jemt.v14.1(65).19
9. Shen LY, Jorge Ochoa J, Shah MN, Zhang X (2011) The application of urban sustainability indicators-a comparison between various practice. Habitat Int 35(1):17–29
10. Shen L, Peng Y, Zhang X, Wu Y (2012) An alternative model for evaluating sustainable urbanization. Cities 29:32–39
11. Sokrovolska N, Korbutiak A, Oleksyn A, Boichenko O, Danik N (2023) State regulator's role in the country's banking system during wartime. Finan Credit Activ Problems Theory Pract 2(49):43–55. https://doi.org/10.55643/fcaptp.2.49.2023.3985
12. Sustainable Development Goals. UNDP. https://www.undp.org/ukraine/sustainable-development-goals
13. Venhryn DV (2023) Assessing the sustainability of urban spaces based on city strategy analysis. Human Geograp J 35:28–36. https://doi.org/10.26565/2076-1333-2023-35-03
14. Zalizko VD, Dobrowolski RH, Myskin YI, Sokolova OM, Diachenko SA (2023) Definlandization and the new economic policy of Ukraine in the post-warperiod. Naukovyi Visnyk Natsionalnoho Hirnychoho Universytetu 2:184–189. https://doi.org/10.33271/nvngu/2023-2/184
15. Morar C, Dulca M, Gy N (2016) Brownfields regeneration, between sustainable urban development and cultural heritage, the former military sites in Oradea, Romania. J Urban Reg Anal 8(1):75–84
16. Kostrikov S, Niemets L, Sehida K, Niemets K, Morar C (2018) Geoinformation approach to urban geographic system research (case studies of Kharkiv region, Ukraine). Visnyk of V. N. Karazin Kharkiv National University, series. Geol Geograp Ecol 49:107–124. https://doi.org/10.26565/2410-7360-2018-49-09
17. Niemets L, Mezentsev K, Morar C, Sehida K, Husieva N, Peresadko V (2018) Innovation and investment potential of region as a factor of its "smart transformation": Kharkiv region case study (Ukraine). Visnyk of V N Karazin Kharkiv National University, series. Geol Geograp Ecol 49:137–159. https://doi.org/10.26565/2410-7360-2018-49-11

18. Morar C, Berman L, Unkart S, Erdal S (2021) Sustainable brownfields redevelopment in the European Union: an overview of policy and funding frameworks. J Environ Health 18(21):11169. https://www.neha.org/node/62305
19. Berman L, Morar C, Unkart S, Erdal S (2022) An overview of brownfields redevelopment in the United States through regulatory, public health, and sustainability lenses. J Environ Health 84(9) https://www.neha.org/Images/resources/JEH5.22-Feature-Overview-Brownfields-Redevelopment.pdf
20. Morar C, Lukic T, Valjarević A, Niemets L, Kostrikov S, Sehida K, Telebienieva I, Kliuchko L, Kobylin P, Kravchenko K (2022) Change detection based spatiotemporal analysis of urban green areas: a case study of Kharkiv-City. Front Environm Sci 10:1–27. https://doi.org/10.3389/fenvs.2022.823129
21. Caciora T, Jubran A, Ilies DC, Hodor N, Blaga L, Ilies A, Grama V, Sebesan B, Safarov B, Ilies G et al (2023) Digitization of the built cultural heritage: an integrated methodology for preservation and accessibilization of an art nouveau museum. Remote Sens 15:5763. https://doi.org/10.3390/rs15245763
22. Herman GV, Grama V, Ilieş A, Safarov B, Ilieş DC, Josan I, Buzrukova M, Janzakov B, Privitera D, Dehoorne O et al (2023) The relationship between motivation and the role of the night of the museums event: case study in Oradea municipality. Romania Sustainab 15:1738. https://doi.org/10.3390/su15021738
23. Grama V, Ilies G, Safarov B, Ilies A, Caciora T, Hodor N, Ilies DC, Kieti D, Berdenov Z, Josan I et al (2022) Digital technologies role in the preservation of Jewish cultural heritage: case study Heyman house, Oradea, Romania. Buildings 12:1617. https://doi.org/10.3390/buildings12101617
24. Marcu F, Ilieş DC, Jan WA, Indrie L, Ilies A, Burta L, Caciora T, Herman GV, Todoran A, Baias S, Albu A, Gozner M (2020) Investigations regarding the biodegradation of cultural heritage. Case study of traditional embroidered peasant shirt (Maramures, Romania). Romanian Biotechnol Lett 25(2):1362–1368. https://doi.org/10.25083/rbl/25.2/1362.1368
25. Ilieş DC, Safarov B, Caciora T, Ilieş A, Grama V, Ilies G, Huniadi A, Zharas B, Hodor N, Sandor M et al (2022) Museal indoor air quality and public health: an integrated approach for exhibits preservation and ensuring human health. Sustain For 14:2462. https://doi.org/10.3390/su14042462
26. Bou-Belda E, Indrie L, Ilieş DC, Hodor N, Berdenov Z, Herman G, Caciora T (2020) Chitosan – a non-invasive approach for the preservation of historical textiles. Industria Textila J 71(6):576–579. https://doi.org/10.35530/IT.071.06.1756
27. Elhosiny SM, Hassan TH, Josan I, Salem AE, Abdelmoaty MA, Herman GV, Wendt JA, Janzakov B, Mahmoud HME, Abuelnasr MS (2023) Oradea's cultural event management: the impact of the 'night of the museums' on tourist perception and destination brand identity. Sustain For 15:15330. https://www.mdpi.com/2071-1050/15/21/15330
28. Meszaros EL, Toca CV (2020) The EU's multifaceted approach to resilience building in the Eastern Neighbourhood. Security sector reform in Ukraine. Eastern J Eur Stud 11:120–145. Special Issue SI, https://www.webofscience.com/wos/woscc/full-record/WOS:000593531000007?SID=EUW1ED0BEEp2kP2NxOAqUPsOYoFDc
29. Meszaros EL, Toca CV (2023) The EU's resilience and the management of hybrid threats coming from the Eastern neighbourhood: Belarus and the deliberate facilitation of irregular immigration. Eastern J Eur Stud 14(1):5–30. https://ejes.uaic.ro/articles/EJES2023_1401_MES.pdf
30. Toca C-V (2015) The regional dimension of the lifelong learning policies program in Romania. Transylvanian Rev 24(1):145–161. https://www.webofscience.com/wos/woscc/full-record/WOS:000364727800012?SID=EUW1ED0BEEp2kP2NxOAqUPsOYoFDc
31. Ilies DC, Blaga L, Hassan TH, Ilies A, Caciora T, Grama V, Herman GV, Dejeu P, Zdringa M, Marshall T et al (2023a) Indoor microclimate and microbiological risks in heritage buildings:

a case study of the Neologic Sinagogue, Oradea, Romania. Buildings 13:2277. https://doi.org/10.3390/buildings13092277
32. Ilies DC, Herman GV, Safarov B, Ilies A, Blaga L, Caciora T, Peres AC, Grama V, Bambang SW, Brou T et al (2023b) Indoor air quality perception in built cultural heritage in times of climate change. Sustain For 15:8284. https://doi.org/10.3390/su15108284

Correction to: GIS Based Decision Support System for Brownfield Redevelopment: The Case Study Municipality of Prokuplje, Serbia

Aleksandar Valjarević, Kh. Md Nahiduzzaman, and Filip Vujović

Correction to:

Chapter 14 in: C. Morar et al. (eds.), *Achieving Sustainability in Ukraine through Military Brownfields Redevelopment*, NATO Science for Peace and Security Series C: Environmental Security, https://doi.org/10.1007/978-94-024-2278-8_14

The original version of Chapter 14 was inadvertently published with an incorrect author name. It appeared as "Kh. Md Nahiduzzamanand," but has now been corrected to "Kh. Md Nahiduzzaman."

The updated version of this chapter can be found at
https://doi.org/10.1007/978-94-024-2278-8_14

C. Morar et al. (eds.), *Achieving Sustainability in Ukraine through Military Brownfields Redevelopment*, NATO Science for Peace and Security Series C: Environmental Security, https://doi.org/10.1007/978-94-024-2278-8_36

Index

C. Morar et al. (eds.), *Achieving Sustainability in Ukraine through Military Brownfields Redevelopment*, NATO Science for Peace and Security Series C: Environmental Security, https://doi.org/10.1007/978-94-024-2278-8

The manufacturer's authorised representative in the EU is Springer Nature Customer Service Centre GmbH, Europaplatz 3, 69115 Heidelberg, Germany. If you have any concerns regarding our products, please contact ProductSafety@springernature.com

Printed and bound by CPI Group (UK) Ltd, Croydon, CR0 4YY
01/10/2026
02225934-0016